Biology *of the* Three-Spined Stickleback

MARINE BIOLOGY

SERIES

Peter L. Lutz, Editor

PUBLISHED TITLES

Biology of Marine Birds
E. A. Schreiber and Joanna Burger

Biology of the Spotted Seatrout
Stephen A. Bortone

The Biology of Sea Turtles, Volume II
Peter L. Lutz, John A. Musick, and Jeanette Wyneken

Biology of Sharks and Their Relatives
Jeffrey C. Carrier, John A. Musick, and Michael R. Heithaus

Early Stages of Atlantic Fishes: An Identification Guide for the Western Central North Atlantic
William Richards

The Physiology of Fishes, Third Edition
David H. Evans

Biology of the Southern Ocean, Second Edition
George A. Knox

Biology of the Three-Spined Stickleback

Edited by
Sara Östlund-Nilsson
Ian Mayer
Felicity Anne Huntingford

CRC Press
Taylor & Francis Group
Boca Raton London New York

CRC Press is an imprint of the
Taylor & Francis Group, an informa business

CRC Press
Taylor & Francis Group
6000 Broken Sound Parkway NW, Suite 300
Boca Raton, FL 33487-2742

Printed in the United States of America on acid-free paper
10 9 8 7 6 5 4 3 2 1

International Standard Book Number-10: 0-8493-3219-2 (Hardcover)
International Standard Book Number-13: 978-0-8493-3219-7 (Hardcover)

Library of Congress Cataloging-in-Publication Data

Biology of the three-spined stickleback / editors, Sara Ostlund-Nilsson, Ian Mayer, Felicity Anne Huntingford.
p. cm. -- (Marine biology)
Includes bibliographical references and index.
ISBN-13: 978-0-8493-3219-7 (0-8493-3219-2 : alk. paper)
1. Threespine stickleback. I. Ostlund-Nilsson, Sara. II. Mayer, Ian. III. Huntingford, Felicity.

QL638.G27B56 2007
597'.672--dc22 2006019944

Visit the Taylor & Francis Web site at
http://www.taylorandfrancis.com

and the CRC Press Web site at
http://www.crcpress.com

Preface

The three-spined stickleback, *Gasterosteus aculeatus,* is one of nature's most recognizable species. In part owing to its abundance, wide distribution, and ease of collection but, more so, to its unique array of behavioural and morphological characteristics, this small teleost fish has attracted the attention of researchers for many years. Although the reproductive behaviour of the stickleback has been extensively studied for over a hundred years, culminating in Tinbergen's now-classical work leading to the 1973 Nobel prize, this species is now attracting growing interest in emerging fields of biology, most notably comparative and functional genomics.

To date, the increasingly diverse stickleback research community has produced one of the largest research literatures for any nonmammalian vertebrate model organism, including more than 2000 research papers and several textbooks on the life history, behaviour, morphology, distribution, and ecology of different stickleback populations. Further testament of the growing interest in the unique characteristics of the stickleback is the now regular hosting of the International Conference on Stickleback Behaviour and Evolution, which was last held at the University of Alaska, Anchorage, in 2006. The increasing popularity of these meetings accurately reflects the growing interest in stickleback biology beyond the more traditional research fields of ecology, behaviour, and evolution, where interest still remains strong.

The stickleback have undergone dramatic adaptive radiation since the recent glacial period, and with its diverse array of behavioural and morphological traits offers a unique opportunity to study the genetic architecture, gene expression, and developmental mechanisms that underlie evolutionary change in vertebrates. In the relatively short time period of 10,000–15,000 years since the end of the last glacial period, isolated populations of sticklebacks have experienced dramatic changes, with different populations showing pronounced alterations in body size, number of dorsal spines, pattern and number of lateral plates, and pelvic fin development, as well as differences in behaviour. Several studies involving genetic crosses between sticklebacks from different localities have now indicated that many of the major morphological transformations in the vertebrate skeleton are probably controlled by only a small number of genes. The stickleback genome is now close to being successfully sequenced, which offers a new and powerful tool in the field of genetics and functional genomics. This will allow comparative genomics using the four teleost genomes already available, and it will likely be possible to adapt molecular and transgenic tools developed for the model teleosts medaka and zebra fish. More importantly, for the first time, we will have the molecular substrate to investigate the genetics of natural populations, of reproductive isolation, and of ecological adaptation in terms of physiology, behaviour, and the evolution of body shape in a vertebrate model organism.

In addition to its growing importance in the field of functional genomics, the three-spined stickleback is rapidly becoming an important sentinel and surrogate model in the field of environmental toxicology. In having a quantifiable androgenic endpoint, the stickleback is now being adopted as a sensitive indicator species for the identification of both antiandrogenic and androgenic contaminants. With the validation of an oestrogenic endpoint, the stickleback is now being promoted as a universal indicator species of both environmental androgens and oestrogens. Further, as opposed to the other small teleosts, such as zebra fish and medaka, presently being promoted as model species for chemical screening and testing, the stickleback has the advantage in that it is endemic to both Europe and North America, allowing for its use in both laboratory testing and *in situ* biomonitoring programs.

The contents of this book accurately reflect the growing importance of the stickleback as a model species in the new emerging fields of biology, including genetics and genomics, and environmental toxicology. At the same time, the stickleback continues to attract investigators from the more traditional fields of biology, including behaviour, sexual selection, and evolutionary biology. It is hoped that this book will offer the reader a greater insight into the fascinating biology of the sticklebacks.

We would like to thank all our authors for writing the chapters and our colleagues who put much effort into reading and commenting on the chapters. We are very thankful for the help of Alison Bell, Anders Berglund, Neils Dingemanse, Frank von Hippel, Erik Höglund, Lorna Kennedy, Craig Miller, Göran Nilsson, Finn-Arne Weltzien, Kjartan Østbye, Tom Reimchen, Joe Ross, Ben Rushbrook, Sandy Scott, Mike Shapiro, Claus Wedekind, and a number of anonymous referees.

Sara Östlund-Nilsson
Ian Mayer
Felicity Anne Huntingford
Oslo, Norway

The Editors

Sara Östlund-Nilsson received her Ph.D. degree in animal ecology and behavioural ecology from Uppsala University in Sweden in 2000 with a thesis focusing on female choice and paternal care in the 15-spined stickleback (*Spinachia spinachia*) and on whether the nest may also serve as an ornament. In parallel, she investigated similar questions on the three-spined stickleback (*Gasterosteus aculeatus*). During 2001–2002 she worked at the University of Queensland, studying mate choice, ultraviolet (UV) coloration, fish ecology and hypoxia, and the effects of isopods parasitising on coral reef fish. She continued working on parental questions in sticklebacks and on UV coloration in fish living in shallow temperate waters at the University of Oslo in Norway. Currently, she is serving in a research position as an associated professor at the National Library in Oslo.

Ian Mayer received his Ph.D. degree in fish reproductive physiology from the University of Wales in 1987. For the last 15 years Dr. Mayer has been studying the reproductive endocrinology and physiology in fishes at the Department of Zoology, Stockholm University. During this period much of his research was centred on studying the neuroendocrine control of reproduction including reproductive behaviour in the three-spined stickleback. For the last 3 years Dr. Mayer has headed a group promoting the stickleback as a universal biomarker of environmental contamination. He was the organizer of the Fourth International Conference on Stickleback Behaviour and Evolution, which was held in Sweden in 2003. The same year, he took up a position at the University of Bergen where he continues to pursue his work on the stickleback.

Felicity Anne Huntingford has been engaged in research into behaviour and ecology of fishes for more than 30 years. In particular she has a long-standing interest in the origin and function of individual variability in morphology and behaviour in three-spined sticklebacks. She was one of the first to show correlations between risk taking in different functional context (behavioural syndromes) and has shown how between-population variation in risk-taking reflects local predation regimes, demonstrated inherited differences in risk taking (amplified by experience) and examined factors promoting reproductive success in male sticklebacks.

Contributors

Iain Barber
Department of Biology
University of Leicester
Leicester, United Kingdom

Bertil Borg
Department of Zoology
Stockholm University
Stockholm, Sweden

Janette Wenrick Boughman
Department of Zoology
University of Wisconsin
Madison, Wisconsin, USA

Susan Coyle
Fish Biology Group
Division of Environmental and Evolutionary Biology
Institute of Biomedical and Life Sciences
University of Glasgow
Glasgow, United Kingdom

Felicity Huntingford
Fish Biology Group
Division of Environmental and Evolutionary Biology
Institute of Biomedical and Life Sciences
University of Glasgow
Glasgow, United Kingdom

Ioanna Katsiadaki
Cefas Weymouth Laboratory
Weymouth
Dorset, United Kingdom

David M. Kingsley
Department of Developmental Biology and Howard Hughes Medical Institute
Stanford University
Stanford, California, USA

Michelle Y. Mattern
Department of Ecology and Evolutionary Biology
University of Toronto
Toronto, Ontario, Canada

Ian Mayer
Department of Biology
University of Bergen
Bergen, Norway

Deborah A. McLennan
Department of Zoology
University of Toronto
Toronto, Ontario, Canada

Sara Östlund-Nilsson
Department of Biology
University of Oslo
and
National Library
Oslo, Norway

Miklós Páll
Institute of Marine Research
Research Station Austevoll
Storebø, Norway

Catherine L. Peichel
Division of Human Biology
Fred Hutchinson Cancer Research Center
Seattle, Washington, USA

Table of Contents

1 Phylogeny, Systematics, and Taxonomy of Sticklebacks

Michelle Y. Mattern

CONTENTS

This chapter begins with a presentation of what is currently known about the geographic and genetic variation within each genus and species of stickleback fish and explores the possibility of any subdivision within currently recognised species. The second part of this chapter focuses on the phylogeny of gasterosteids, providing a historical review of the relationships between and within the different genera and discusses the most recent and complete phylogeny for this family. The chapter concludes with a detailed synonymy of all previously assigned species and genus names to provide researchers with a "road map" with which to navigate the historical literature on this fascinating group of fish.

1.1 GENERAL INTRODUCTION

With the possible exceptions of salmonids, no group of fish has been as well studied as the sticklebacks. Over 2000 papers and books have been published to date. Although the greatest concentration of these papers centres on the behaviour of these fascinating fishes, a substantial literature has accumulated on their evolutionary history as well. With such a large scientific database at our disposal, sticklebacks have become a model system for studying many evolutionary processes, including speciation and adaptive radiation,[1–3] sexual selection,[4–7] alloparental care,[8] and egg cannibalism.[9–11] Given the importance of these fishes, it is imperative that the ecological, evolutionary, and physiological information collected over the years be examined within a phylogenetic framework to construct rigorous hypotheses of character evolution.[12]

All authors agree that the Gasterosteidae comprise five genera residing in northern temperate habitats (but see Keivany and Nelson[13]). *Spinachia*, *Apeltes*, and *Culaea* are currently recognised as monotypic and geographically restricted. *Spinachia spinachia* inhabits the shallow coastal waters of Western Europe, *Apeltes quadracus* is a brackish water species distributed across the eastern coast of North America from the Gaspé Basin of Quebec to Virginia, and *Culaea inconstans* is restricted to freshwaters of North America from the eastern coast to the Rocky Mountains. *Gasterosteus* and *Pungitius*, on the other hand, are both geographically widespread and morphologically variable. There is still much debate in the literature about the actual number of species of *Gasterosteus* and *Pungitius*.

It will become evident from this review that our knowledge about sticklebacks is unevenly distributed and that much more is known about *Gasterosteus* and *Pungitius* than the three other genera. Similarly, the different species have been the focus of different types of investigation. For example, most of the research on *Culaea* has concentrated on behaviour, whereas investigations involving *Pungitius* are much more likely to involve biogeography and either genetics or morphology.

1.2 *SPINACHIA* CUVIER 1817

1.2.1 *SPINACHIA SPINACHIA* (LINNAEUS 1758)

Besides a recent flurry of behavioural and foraging papers,[14–28] very little is known about this monotypic genus. Confined to the marine coastal waters of Europe, it is the only species restricted in its distribution to Europe.

1.2.2 GEOGRAPHIC VARIATION

Only one study has been published on the geographic variability of *Spinachia*. Gross[29] surveyed 501 fish from 28 populations and found a weak correlation between mean breeding season temperatures and total number of vertebrae, dorsal fin rays, and gill rakers, with all three variables showing a V-shaped relationship with temperature (minimum values for all three variables at 16°C). There was no indication of species subdivision based on any of the ten morphological characters studied.

1.3 *APELTES* DEKAY 1842

1.3.1 *APELTES QUADRACUS* (MITCHELL 1815)

Apeltes is the first of two entirely North American genera. Until recently, it was thought that *Apeltes* was confined strictly to brackish and marine waters off the coast of eastern North America, ranging from Virginia in the south to Newfoundland in the north.[30] However, a series of freshwater finds have expanded the known range of this species to include freshwater lakes in New Brunswick,[31] Nova Scotia,[32] Quebec,[33] Pennsylvania,[34,35] Newfoundland,[36–38] and Ontario.[39] *Apeltes* has also been accidentally introduced into Lake Superior, Ontario,[40] and the Avalon Peninsula, Newfoundland.[41]

1.3.2 GEOGRAPHIC VARIATION

The species was originally described by Mitchell[42] as *Gasterosteus quadracus*, named for its four spines. Krueger[43] pointed out that, ironically, Mitchell's original specimen had five spines, not four. Recognition of this spine variation has led to a series of papers investigating the geographic variability of *Apeltes quadracus*.

Cox[44] studied regional variation in dorsal spine number in *Apeltes* and found a correlation between the percentage of fish with five free spines (four free spines and one attached to the soft dorsal fin) and salinity and summer temperature. Generally, populations inhabiting areas of high salinity and lower summer temperatures had a higher percentage of individuals with five spines. Krueger[43] found that four-spined sticklebacks varied in the number of vertebrae (30–33), dorsal fin rays (9–14), anal fin rays (7–11), and dorsal spines (2–6). He noted that vertebral number decreased from north to south (or with increased temperature) and, like Cox, found that the number of five-spined individuals increased with increased salinity.

Blouw and Hagen,[45] studying fish from Maritime Canada (Nova Scotia, New Brunswick, and Prince Edward Island), reported extensive geographic variation with (sometimes dramatic) frequency shifts over short distances. They found that four-spined sticklebacks had between one and seven dorsal spines, with four-, five-, and three-spined morphs making up 98% of fish surveyed (76.4, 20.6, and 2.8%, respectively). Also, spine number was independent of the sex or age of the fish. In New Brunswick and Prince Edward Island, there was a high frequency of five-spined individuals, with the three-spined morph being very rare or absent. Nova Scotia had intermediate numbers of three- and five-spined fish (3 and 11%) except in Bras d'Or Lake, Cape Breton Island, where there are remarkably high percentages of three-spined *Apeltes* (average 15% of individuals; as high as 51% at one site) coupled with a sharp decrease in five-spined fishes (only 5%). There are a number of sites in nine geographic areas where either the three- or five-spined morph predominates (1 and 46 sites, respectively). Overall, comparison of 117 freshwater sites with 453 saltwater sites, in which environmental variables ranged from clear, cold Atlantic saltwater with no current to warm and tea-stained freshwater with a fast current, revealed no significant differences in spine number. This supported Hagen and Blouw's previous hypothesis[46] that dorsal spine number was more strongly influenced by heritable (polygenic) than environmental factors.

1.4 *CULAEA* WHITLEY 1950

1.4.1 *CULAEA INCONSTANS* (KIRTLAND 1840)

Culaea is the only stickleback genus confined entirely to freshwater. It is the second of the stickleback genera to be found only in North America. Its range extends east to the coast of the Atlantic Ocean, west to the Rocky Mountains, south to the Ohio and Missouri Rivers, and north to Great Slave Lake. Brook stickleback are capable of living in lakes, rivers, streams, creeks, bogs, ditches, and underground pools.

1.4.2 GEOGRAPHIC VARIATION

Inspired, perhaps, by Kirtland's species name *inconstans* (Latin meaning "variable"),[47] there have been many studies conducted on the geographic variability of morphological traits in the brook stickleback. The modal dorsal spine count for *Culaea* is five, but counts of four and six are fairly common.[48–50] Counts as low as one and as high as seven have been observed but represent less than 1% of fish.[48,50] Lawler[51] found that *Culaea* from the Hudson's Bay drainage had a higher percentage of individuals with six spines (24.1%, n = 13 populations) vs. Great Lakes draining fishes (7.5%, n = 5 populations); the average spine count was 5.4 in Hudson's Bay fishes and 5.0 in Great Lakes fishes. This is in line with Hansen's report that 84% of fish from nine populations in Wisconsin and Illinois had five spines. Similarly, Edge and Coad[50] reported reduced spine numbers in two populations from Great Lakes drainages' fishes in Ontario and Quebec, with counts of only one and two in the Quebec population. Hansen[48] also recorded that 95% of these fish had 9–11 pectoral, dorsal, and anal fin rays (the average being 10 for all the three fins), with occasional counts of 4–13 (pectoral), 6 or 7 (dorsal), and 5 or 7 (anal) fin rays.

Although there seems to be a genetic basis for reduction of the pelvic girdle in *Culaea*,[52] the distribution of girdleless populations is not correlated with geographic proximity. A single drainage may have some populations with predominantly fully girdled individuals and other populations in which most individuals lack the pelvic girdle either partially or completely. It does appear, however, that the loss of the pelvic girdle is more common in populations on the extreme western end of the species' range (Alberta and Saskatchewan[53]) and less common to the east and south. Interestingly, Edge and Coad[50] found that all the fish from the Quebec population with only one or two dorsal spines had normal pelvic girdles, indicating that changes to dorsal spines and the pelvic girdle are decoupled. This seems odd, given that Nelson[54] (see also Andraso and Barron[55]) reported a strong north-west to south-east cline in pelvic and dorsal spine length, with longer spines in the south-eastern populations (exceptions include populations in Nebraska and Cayuga Lake, New York, with unusually short dorsal spine lengths for their southerly locations). These data seem to indicate that spine length is coupled at a mechanistic level, but that reduction or loss of spines is not correlated.

There has also been one study on the geographic variability of male agnostic responses. Burks et al.[56] sampled four populations of brook stickleback, one from Ohio (Urbana), two from Wisconsin (Oshkosh and Ft. Atkinson), and one from Saskatchewan (Saskatoon). The Urbana fish had only 1.38 aggressive displays per encounter between two territorial males (25 displays for 18 encounters in 5 h of observation) as opposed to an average of 3.01 displays per encounter for the other populations (3184 displays for 1058 encounters for 15 h). The Urbana population failed to display 9 of the 12 threat behaviours presented in all other populations (charging, biting, circle fighting, frontal approach, head down, follow, tail beating, sigmoid atkinson, and dragging). They retained only the broadside, lateral, and sigmoid displays, which they used almost equally (bs = 40%, la = 36%, and s = 24%). In the Ft. Atkinson, Oshkosh, and Saskatoon fish, charging was the predominant threat display (49%, 25%, and 48%, respectively), followed by biting, sigmoid display, and broadside (although the order and frequencies differed between the three populations). In all, the Urbana population showed the fewest displays, fewest number of different displays, fewest attack postures, and shortest encounters.

Burks et al.[56] also noted that the spines were significantly longer in the Urbana fish and that they were significantly lighter in colouration than other territorial fish. The observation that Mad River drainage fish (which includes the Urbana population) have longer spines is supported by Andraso and Barron[55] and Nelson,[54] who found these fish to have unusually long spines, even for the south-east extreme of this species. Brook sticklebacks are generally more flexible when they lack a pelvic girdle and pelvic spines, because this facilitates a quicker startle response and escape from a predator.[57] It may be that the unusually long spines of Mad River drainage fish reduce the flexibility required to perform threat postures or that the elongated spines have replaced more elaborate behavioural displays. It would be interesting to study whether the elongation of spines or the absence of a pelvic girdle are somehow correlated with nuptial displays.

1.4.3 Relationships within the Genus

Gach[58] investigated the interspecific relationships among 32 different populations of *Culaea inconstans*. Using mtDNA RFLP analysis (of 96 restriction enzymes), she found that the 11 resultant haplotypes were subdivided into two lineages, A (Alberta, Minnesota, Wisconsin, the Upper Michigan peninsula, and Quebec) and B (the Lower Michigan peninsula, Ohio, and Pennsylvania). The lineages differed by 6.4% sequence divergence and were diagnosable by several restriction enzymes. She approximated the age of split between the two lineages as older than the Pleistocene, but noted that the present species distribution was also influenced by postglacial dispersal from separate refugia from the Mississippi and Ohio River basins. Gach concluded that the overlap between the two lineages in the Detroit, MI, area was the result of secondary introductions. She also solved the riddle of the mysterious, extremely disjunctive New Mexico population,[59] showing that it shared a haplotype with fish from the Lower Michigan peninsula. Evidently, these fish were introduced via a general bait-release programme, something that is a widespread problem for freshwater fishes in North America, and are not a Pleistocene relict.

I undertook a more detailed investigation of the geographic and phylogenetic variability in *C. inconstans*, using three mitochondrial genes, cytochrome b, ATPase 6, and the mitochondrial control region (totalling 1888 aligned base pairs).[60] The results of this study indicate that the complex genetic diversity patterns found within *Culaea* are the product of both pre-Pleistocene and Pleistocene forces. The phylogenetic analysis identified the existence of three significant clades with *Culaea*: a Mississippi group and two Atlantic groups. The deepest split within the genus (ca. 4.2 Mya), the Mississippi–Atlantic split, predated the Pleistocene and its associated glaciations, whereas a more recent split between the two Atlantic groups occurred during the glaciations (ca. 1.4 Mya). The Mississippi–Atlantic split is consistent with the divergence found by Gach.[58] However, Gach did not detect the second geographic split, because she did not sample populations as extensively in the eastern portion of the species' range. Current distributions within each clade are the direct result of recolonisation following the Wisconsin glaciation.

1.5 *Pungitius* Coste 1848

Although no fewer than 36 species have been described in the genus *Pungitius*, only 8 are considered serious contenders for specieshood: *P. pungitius* (Linneaus 1758), *P. platygaster* (Kessler 1859), *P. occidentalis* (Cuvier 1829), *P. sinensis* (Guichenot 1869), *P. tymensis* (Nikolskii 1889), *P. hellenicus* (Stephanidis 1971), *P. laevis* (Cuvier 1829), and *P. kaibarae* (Tanaka 1915). Despite the best efforts of researchers, there is still no consensus as to how many species of *Pungitius* there actually are, or how they are related to each other (see Table 1.1).

1.5.1 Geographic Variation

The first systematic investigation into the geographic variation of *P. pungitius* came with Münzing.[61] He found that the distribution of *Pungitius* differs markedly from

TABLE 1.1
Number of Species of *Pungitius* Recognised by Various Authors. See the Appendix for a Complete List of Names

Author	Number of Species Recognised	Comments
Linnaeus 1758[a]	1	
Cuvier and Valenciennes 1828[b]	2	Differentiated New World fish from European fish
Sauvage 1874[c]	12	Named 2 of them
Jordan and Gilbert 1882[d]	1 (with 2 subspecies)	North America only
Jordan and Evermann 1896[e]	1 (with 2 subspecies)	North America only
Müning 1969[f]	2 (*P. pungitius* with 4 subspecies and *P. platygaster* with 2 subspecies)	
Monod 1973[g]	3 (*P. pungitius* with 4 subspecies)	
Paepke 1996[h]	4 (*P. pungitius* and *P. platygaster* each with 2 subspecies)	
Keivany and Nelson 2000[i]	3 (*P. pungitius* with 5 subspecies)	
Banarescu and Paepke 2002[j]	3 (no subspecies)	Europe only

[a] Linnaeus, C., *Systema Naturae*. X ed. Vol. Pisces in Volume 1, 1758.

[b] Cuvier, G. and Valenciennes, A., *Histoire Naturelle des Poissons*. Vol. 4, 1829.

[c] Sauvage, H.E., Révision des espèces du groupe des Épinoches. Nouvelles Archives Museum Histiore Naturelle Paris, 10, 5, 1974.

[d] Jordan, D.S. and Gilbert, C.H., Synopsis of the fishes of North America. Bulletin of the U.S. National Museum, 16, 1, 1882.

[e] Jordan, D.S. and Evermann, B.W., *The Fishes of North and Middle America*. T.F.H. Publ., New Jersey, 1896.

[f] Münzing, J., Variabilität, Verbreitung and Systematik der Arten une Unterarten in der Gattung *Pungitius* Coste, 1848 (Pisces, Gasterosteidae), *Zeitschrift für Zoologische Systematik and Evolutionsforschung*, 7, 208, 1969.

[g] Hureau, J.-C. and Monod, T., *Checklist of the Fishes of the Northeastern Atlantic and of the Mediterranean*. Vol. 2. Unesco, Paris, 1973.

[h] Paepke, H.-J., *Die Stichlinge*. Vol. 10. Westarp Wissenschaften, Magdeburg, 1996.

[i] Keivany, Y. and Nelson, J.S., Taxonomic review of the genus *Pungitius*, ninespine sticklebacks (Gasterosteidae), *Cybium*, 24, 107, 2000.

[j] Banarescu, P.M. and Paepke, H.-J., eds. *The Freshwater Fishes of Europe. Cyprinidae 2, Part III: Carassius to Cyprinus. Gasterosteidae.* 2002.

Gasterosteus, with the latter being primarily a marine fish that has penetrated freshwater, whereas *Pungitius* is primarily a freshwater fish. Münzing recorded substantial morphological diversity in European populations of *Pungitius*. In northern Germany, *Pungitius* specimens have a naked body with a weakly developed caudal peduncle keel, whereas in England and France they are completely naked. *Pungitius* is more heavily plated in Eastern Europe. Münzing hypothesized that there are corresponding genotypes controlling these varying phenotypes.

McPhail[62] studied the geographic variability of *P. pungitius* in North America and discovered that the number of lateral plates was correlated with tidal waters (high plate counts) and inland waters (low plate counts). He hypothesised that the variation in the number of dorsal spines and gill rakers indicated that two forms of the species exist, which were isolated in different refugia (Bering and Mississippi) during the Pleistocene glaciations. McPhail felt, however, that differences between the forms and the unreliability of placing specimens in the correct group did not warrant the division of the North American ninespines in two distinct subspecies.

Gross[63] found a different pattern in Europe, suggesting that the number of vertebrae, dorsal and anal fin pterygiophores, and dorsal spines vary latitudinally whereas gill raker number, body depth, and pelvic spine length are site specific, implying ecological control of these characters (i.e., diet or predation pressure). Although his study did include specimens of *P. laevis* and *P. platygaster*, Gross did not deal directly with their taxonomic status, except to say that some previous authors doubted the taxonomic status of *P. laevis*.

Studying variation in *P. pungitius* and *P. sinensis* on the island of Honshu, Japan, Tanaka[64] found that *P. pungitius* has between 5 and 13, and *P. sinensis* between 28 and 35 lateral plates, with *P. pungitius* having considerably more variation in plate number and arrangement. He found similar clinal variations in length of the last dorsal, pelvic, and anal spines for both species, with average spine lengths longer by the coast and shorter inland. Tanaka also studied one sympatric site, where he found hybrids with an intermediate phenotype occurring at a frequency of 9.1%. He suggested that although there is some hybridization, the low level indicates some form of mating isolation.

Takata et al.[65] performed a similar study on the variation of *P. pungitius, P. sinensis,* and *P. tymensis* on the island of Hokkaido, Japan. They found that *P. tymensis* had significantly shorter pelvic spines, more dorsal spines, and fewer gill rakers than either *P. pungitius* or *P. sinensis*. There was only one significant difference between *P. pungitius* and *P. sinensis*; *P. sinensis* was always fully plated and *P. pungitius* was never so. They suggested that these morphological characteristics distinguished *P. tymensis* as a distinct species but questioned whether *P. pungitius* and *P. sinensis* were, in fact, distinct and proposed only to recognise them as subspecies.

Takahashi et al.[66] researched the evolution of lateral plate dimorphism in 47 populations of the *P. pungitius-P. sinensis* complex (*P. tymensis* as the outgroup), using a phylogeographic tree based on mtDNA RFLPs of 7 restriction enzymes. They concluded that the partially plated morph was plesiomorphic, with complete plating evolving several times within the complex. There was no significant geographic correlation with plate morph or haplotype.

1.5.2 Relationships within the Genus

Yang and Min[67] studied the variability of 25 allozyme loci for 12 Korean localities of *P. sinensis*, *P. kaibarae*, and their reputed hybrids (from a single locality) and found that specimens of *P. kaibarae* were more similar to each other than to specimens of *P. sinensis* and *vice versa*. The hybrids fell between the two groups. One must note here that *P. kaibarae* was originally described from Kichisho-in, southwest of Kyoto, Japan, not from Korea, and is a fully plated, freshwater form.[68] *Pungitius kaibarae* is now considered locally extirpated from Japan and, therefore, probably extinct.[69]

Haglund et al.[70] conducted a study on the allozyme variation and phylogenetic relationships of Asian, North American, and European populations of *P. pungitius* and *P. sinensis*. From the outset of the study, they considered *P. platygaster* and *P. tymensis* valid species and excluded them from the analysis. Although they used *G. aculeatus* as the outgroup, they did not use phylogenetic systematic methodology, so the results they reported are a hypothesis of overall similarity. Allelic variation was displayed by 19 of 21 loci in the 14 allozyme sites that they sampled. The authors found that *P. sinensis* as currently recognised was paraphyletic, with *P. sinensis* from Japan falling out as the sister group to *P. pungitius* from Japan and *P. sinensis* from Korea as the sister group to the *P. pungitius-P. sinensis* clade. However, because *P. pungitius* from Europe and Asia were not monophyletic, the authors recognised all Asian ninespines as *P. sinensis* and all European ninespines as *P. pungitius*. The European ninespines did not form a clade with the North American representative, and so Haglund et al. concluded that the latter must be considered a separate taxonomic unit, *P. occidentalis*.

Takahashi and Goto[71] studied the relationships of East Asian ninespines by sequencing approximately 900 bp of the mtDNA control region. They included members of *P. tymensis, P. sinensis,* and *P. pungitius* (including one European specimen) that they identified based mostly on defensive armour. They found that *P. tymensis* comprised a strongly supported monophyletic clade (Lineage A) that was the sister group to mainland *P. sinensis* (Lineage B). The remaining (mostly Japanese island dwelling) *P. sinensis* were indistinguishable from the *P. pungitius* specimens (Lineage C, which included the European representative). The *P. pungitius-P. sinensis* clade was characterised by two large insertions into the mitochondrial genome. These results are in basic agreement with Haglund et al.,[70] and supported Takata et al.,[65] who found significant morphological differences between *P. tymensis* and *P. pungitius-P. sinensis* on the island of Hokkaido (*P. pungitius* and *P. sinensis* fish only differed on the basis of lateral plate morphology). There are two interpretations of the results of Takahashi and Goto: (1) *P. sinensis* as currently described is invalid, because it is paraphyletic. To resolve this, Lineage B, mainland *P. sinensis*, should continue to be identified accordingly, whereas Kamchatkan and Japanese *P. sinensis* should be reclassified as *P. pungitius* regardless of defensive armour; or (2) there has been massive genetic introgression of *P. pungitius* haplotypes into *P. sinensis* populations through hybridization. Interestingly, Takahashi and Takata[72] reported mtDNA introgression between *P. pungitius* and *P. tymensis*. On phylogenetic analysis, all *P. pungitius* haplotypes formed a single clade, whereas *P. tymensis* haplotypes were paraphyletic (one monophyletic group and the rest interspersed with *P. pungitius*). The authors interpreted this pattern to mean that hybridization had allowed nine haplotypes to move from *P. pungitius* to *P. tymensis*. Because specimens were identified based only on dorsal spine number and not on diagnostic allozyme loci,[73,74] an error could have been made in the identification process.

The Korean localities of Yang and Min are contained within Takahashi and Goto's study. It may be that the populations identified as *P. kaibarae* and *P. sinensis* are in fact representatives of the mainland *P. sinensis* and *P. pungitius-P. sinensis* clades (Takahashi and Goto's Lineages B and C). Further study is required to clarify this and their sympatric presence in Korea.

1.5.3 Reproductive Isolation

After discovering intermediate phenotypes in several of Hokkaido's rivers, Kobayashi[75,76] demonstrated that hybrids of *P. pungitius* and *P. tymensis* are fertile and are, indeed, capable of back-crossing. Ziuganov and Gomeluk[77] demonstrated that no premating isolating mechanisms exist between *P. platygaster* and *P. pungitius*, and that hybridization between the two produced completely fertile F1s, F2s, and backcrosses.

Niwa[74] found significant differences in gene frequencies at four allozymic loci of *P. pungitius* and *P. tymensis* and suggested that this indicated reproductive isolation. However, allozyme phenotypes indicated the presence of hybrids, and there was also evidence of nuclear gene flow.

1.5.4 How Many Species of *Pungitius* Are There?

Keivany and Nelson[78] performed a taxonomic review of the genus *Pungitius*. They concluded that *Pungitius* comprised three species: *P. hellenicus, P. platygaster*, and *P. pungitius. P. hellenicus* (restricted to three sites in the Sperchios drainage, Greece) was diagnosed by a combination of five characters, lack of a keeled caudal peduncle and pelvic girdle, reduced ectocoracoid, fewer than seven dorsal spines, and large lateral plates; *P. platygaster* was diagnosed by a combination of two characters, lack of a keeled caudal peduncle and large lateral plates; and *P. pungitius* was diagnosed by the presence of a keeled caudal peduncle. They did not recognise the two subspecies of *P. platygaster* (*platygaster* and *aralensis*), because the distinguishing character for *P. platygaster aralensis,* weak serration of the pelvic spines, was also discovered in a *P. platygaster platygaster* specimen from the Sea of Azov. The authors did, however, recognise five subspecies of *P. pungitius*: *P. p. laevis, P. p. occidentalis, P. p. sinensis, P. p. pungitius,* and *P. p. tymensis*. They argued that the significant overlap of four major meristic characters (number of dorsal spines, pelvic spines, pelvic rays, and lateral plates) among the subspecies indicated that the taxa did not warrant full species status. The subspecies were diagnosed by the following traits: *P. p. pungitius* — lack of large lateral plates, long and oblique haemal, and neural spines on preural 4; *P. p. laevis* — caudal peduncle keel is not evident in unstained specimens, and caudal peduncle is relatively deep; *P. p. occidentalis* — short and horizontal haemal and neural spines on preural 4 and a (usually) truncated caudal fin; *P. p. sinensis* — large lateral plates and (usually) two pelvic soft rays on each side; and *P. p. tymensis* — usually 11 or more dorsal spines.

The great amount of variation and the lack of clarity in subspecific diagnoses in *P. pungitius* may indicate that it represents a cryptic species flock rather than a species with clearly defined subspecies or geographic variants.

1.6 *GASTEROSTEUS* LINNAEUS 1758

There are currently only two recognised species in the genus *Gasterosteus*: *G. wheatlandi* and *G. aculeatus.*

1.6.1 *Gasterosteus wheatlandi* (Putnam 1867)

Gasterosteus wheatlandi is found in the coastal waters off eastern North America from Newfoundland to New York. Unlike its sister species, there is no taxonomic confusion surrounding this species (but see Hubbs[79]). Originally described by Putnam in 1867 from Massachusetts,[80] it was only once subsequently described from Maine.[81] Goode and Bean[82] questioned whether *G. wheatlandi* was truly a species unto itself and not just a geographic variant of *G. aculeatus*. However, Kendall's 1896 redescription provided several diagnosable characteristics removing doubt as to its specific status.[81]

It is easily distinguished from other threespines by its nuptial colouration; males are gold with distinctive black spots, a low number of lateral plates (5–11) on the anterior portion of the body, and a lack of a caudal peduncle keel and posttemporal and supracliethra bones.

1.6.1.1 Geographic Variation in *Gasterosteus wheatlandi*

Coad and Power[83] examined a sympatric population of *G. aculeatus* and *G. wheatlandi* in Amory Cove, Quebec, and found that *G. wheatlandi* was smaller and had fewer lateral plates, gill rakers, soft dorsal and anal fin rays, and vertebrae than *G. aculeatus*. Females also produced fewer and smaller eggs, which was not surprising given that body size is correlated with fecundity in sticklebacks.[84]

Sargent et al.[85] reported that low, partial, and fully plated morphs existed in populations from Maine to southern New York. The low-plate phenotype predominated north of Cape Cod, MA, and was rare or absent south of that point, with a small area of overlap around the Cape. This pattern is opposite to the cline found in *G. aculeatus,* in which the completely plated morph increases in frequency with latitude. Vertebral count increases north of Cape Cod, as in *G. aculeatus*. Interestingly, they also found that *G. wheatlandi* is sexually dimorphic for several characters including lateral plates, vertebrae number, overall length, second dorsal spine length, gill rakers, and anal fin ray number.

1.6.1.2 Reproductive Isolation of *Gasterosteus wheatlandi*

Perlmutter[86] reported that *G. aculeatus* tends to breed earlier in the season and is almost finished by the time *G. wheatlandi* begins its breeding season when the two species are sympatric. Reisman[87] theorised that the two species are potentially reproductively isolated, based on differences in courtship, nuptial colouration, and nest size and were, therefore, "good" species. McInerney[88] investigated the reproductive behaviour of blackspotted sticklebacks and found that although *G. aculeatus* and *G. wheatlandi* males do not discriminate between con- and heterospecific females, females showed a strong preference for conspecific males.

Ayvazian[89] discovered asymmetric reproduction between two populations of *G. wheatlandi* from Massachusetts and Connecticut. Crosses between Massachusetts fish were the most successful, with 100% of nests resulting in spawning and 23 of 24 spawnings producing offspring. Although only 54% of Connecticut crosses spawned, all of those spawnings (7) produced offspring. Nine of ten male Connecticut–female Massachusetts crosses spawned and seven of those produced offspring, whereas none of the male Massachusetts–female Connecticut crosses produced successful spawnings.

1.6.2 *Gasterosteus aculeatus* Linnaeus 1758

Although today we recognise *Gasterosteus aculeatus* as the only scientific name for all three-spined sticklebacks, between 1792 and 1910 no fewer than 40 scientific names in addition to Linnaeus' original *G. aculeatus* designation were proposed: 9 on the Pacific coast of North America, 2 in Asia, 11 on the Atlantic coast of North America, and 18 in Europe (see Table 1.2 and this chapter's appendix). At the height of taxonomic splitting in 1874, Sauvage recognised 30 species of three-spined stickleback.[90] Six years later, Dr. Francis Day [p. 747 in Jordan and Evermann[91]] observed that it is "remarkable how many species of sticklebacks have been named, outnumbering even those of the Salmonidae of the fresh waters, and it becomes a first consideration whether any general principles are perceptible in the distribution of these species or varieties ... Heckel and Kner [1858], in their account of the fishes of Austria, did not admit the foregoing to be more than varieties differentiated by the development of the lateral scutes or plates, which they found varied in number between 3 and 28." Following Day's advice, Jordan and Evermann[91] recognised only four species of three-spined stickleback in North America: *G. aculeatus, G. bispinosus* (with two subspecies), *G. cataphractus,* and *G. williamsoni* (with two subspecies).

1.6.2.1 Lateral Plate Variation

Most of the preceding taxa were classified based, in part, on the extent of their lateral plating. For example, Cuvier and Valenciennes[92] used plate development to identify three different species of *Gasterosteus*: *G. trachurus*, *G. leiurus*, and *G. semiarmatus*. The trachurus form was characterised by a complete row of 30 to 35 lateral plates, beginning just in front of the pectoral fin and running continuously to the tail, resulting in a keeled caudal peduncle. Leiurus (also known as the gymnurus form in older literature) is on the other end of the spectrum, characterised by 1 to 9 lateral plates in only the anterior region of the body. A few of these fish have no plates at all, and all lack a keeled caudal peduncle. Semiarmatus is intermediate between the trachurus and leiurus forms, with from 8 to 30 plates, a gap between the anterior and posterior plate groups, and a less well-developed caudal peduncle keel compared to trachurus. There are two theories concerning the origin of the semiarmatus morphs. The first one states the semiarmatus fish are hybrids of low and fully plated fish. This theory is partially correct in that crossing experiments between low-plated and completely plated morphs do produce a wide range of intermediates. However, that is not what is maintaining the large numbers of semiarmatus fish on the east coast of North America, because low-plated morphs are all but absent from this region. The second theory of partial plating is that it is a character in and of itself.[93]

Bertin[94] reduced these species designations to intraspecific forms, recognising four forms: trachura (fully plated), semiarmata (partially plated), gymnura (low-plated), and hologymnura (nonplated). Hubbs[79] criticised Bertin's classification as "pre-Darwinian pigeonholing" and equated all three-spined sticklebacks from the

TABLE 1.2
Number of Species of *Gasterosteus* Recognised by Various Authors, Excluding *G. Wheatlandi* and Its Synonyms

Author	Number of Species Recognised	Comments
Linnaeus 1758[a]	1	
Shaw 1803[b]	3	
Cuvier and Valenciennes 1829[c]	12	Named ten of them
Yarrell 1836[d]	5	
Sauvage 1874[e]	30	Named four of them
Jordan and Gilbert 1882[f]	4 (*G. williamsoni* with two subspecies)	North America only
Jordan and Evermann 1896[g]	4 (*G. bispinosus* with two subspecies and *G. williamsoni* with two subspecies)	North America only
Monod 1973[h]	1	Northeastern Atlantic and Mediterranean only
Paepke 1996[i]	2	One unnamed, the white stickleback of Nova Scotia
Banarescu and Paepke 2002[j]	1	Europe only

Sources:

[a] Linnaeus, C., *Systema Naturae*, 10th ed., Vol. Pisces in Vol. 1, 1758.

[b] Shaw, G., *General Zoology or Systematic Natural History*, Vol. 4, G. Kearsley, London, 1803.

[c] Cuvier, G. et al., *Hist. Nat. Poissons*, 4, 1829.

[d] Yarrell, W., *A History of British Fishes*, London, 1836.

[e] Sauvage, H.E., *Nouv. Arch. Mus. Hist. Nat. Paris*, 10, 5, 1874.

[f] Jordan, D.S. et al., *Synopsis of the Fishes of North America*, Bulletin of the U.S. National Museum, 16, 1, 1882.

[g] Jordan, D.S. et al., *The Fishes of North and Middle America*, T.F.H. Publ., NJ, 1896.

[h] Hureau, J.-C. and Monod, T., *Checklist of the Fishes of the Northeastern Atlantic and of the Mediterranean*, Vol. 2, UNESCO, Paris, 1973. With permission.

[i] Paepke, H.-J., *Die Stichlinge*, Vol. 10, Westarp Wissenschaften, Magdeburg, 1996. With permission.

[j] Banarescu, P.M. et al., Eds., *The Freshwater Fishes of Europe*. 2002, p. 1. With permission.

North American Atlantic coast with the *G. aculeatus* complex. Today, these forms are known as complete-, partial-, and low-plated morphs.[95] Unfortunately, plate-morph terminology has never been used consistently in the literature (for a review, see Bakker and Sevenster[96]). For example, trachurus became synonymous with marine and saltwater forms whereas leiurus was linked to freshwater populations, even though many freshwater populations are fully plated, especially on the east coasts of Asia, North America, and Europe.[93]

Determining what exactly controls plate variation in sticklebacks has been the subject of study for nearly 50 years. Developmentally, the anterior plates form first, followed by the posterior (including the keel on the caudal peduncle), then the middle, plates, so the low-plated and partially plated morphs may be paedomorphic.[97–100]

Correlations between plate morphology and the presence or absence of predators,[101–104] calcium availability,[105] and stream gradients[106] have been noted. Hagen and Moodie[93] discovered a loose correlation between climate (see also Heuts[107]) and the distribution of plate morphs in freshwater, with more heavily plated forms at higher frequencies on the east coasts of Asia and North America, Alaska, and in eastern Europe. Low-plated morphs are rare or totally absent from these waters. These regions all have a similar climate, corresponding to the 6.6 to 10°C winter or summer isoclines. Conversely, on the west coasts of North America and Europe, in the warmer 4.5 to 15.5°C winter and 10 to 21°C summer isoclines, the lower-plated morphs predominate.

Genetic explanations began when Münzing[108] theorised that plate morph was controlled by a single locus with incomplete dominance (fully plated [AA], partially plated [Aa], and low-plated [aa]). From this point, numerous mechanisms have been proposed, including plate morph being controlled by the number of dominant alleles inherited in a two-locus system (three or four [complete], two [partial], one or none [low and nonplated]),[109] complete dominance at a single locus (Friant, CA, population),[110] epistatic interactions between a single major locus and one modifier locus,[111] and two or more alternative alleles at the major locus.[112] Using genetic mapping and allelic complementation experiments, Colosimo et al.[113] confirmed that a single major locus (Gac4174 in linkage group 4) contributes to most of the variation in lateral plate pattern and number in both the Friant population and in a marine-benthic cross.

1.6.2.2 Species vs. Subspecies

Attempts to differentiate between species and subspecies within the *G. aculeatus* complex have generally focused on the evolution of pre- and postmating isolating mechanisms. For example, Ziuganov[114] reported complete and nearly complete (93%) positive assortative mating among complete and low-plated morphs from both Lake Azabachije and the White Sea in Russia. He failed, however, to find any positive assortative mating between any of the other crosses, indicating that reproductive isolation may be site specific. On the other side of the Atlantic, Ross[115] recognised three subspecies in Southern California: *G. aculeatus aculeatus*, *G. a. microcephalus*, and *G. a. williamsoni*. Standard meristic characters placed 96% of *williamsoni* and *microcephalus* specimens in nonoverlapping groups, with reduced fertilization success between them (intrasubspecific crosses, 90%; intersubspecific crosses, 69%). Hagen[116] studied isolating mechanisms in Pacific three-spined sticklebacks in the Little Campbell River, where a freshwater leiurus and an anadromous trachurus form coexist. He found significant ecological segregation between the two forms, with the freshwater residents confined primarily to the upper reaches of the river and the anadromous form occupying the lower reaches during the breeding season. Hybrids only occurred in a narrow zone, which Hagen attributed to selection against hybrids outside of the ecologically intermediate contact zone. Differences between the two forms in the timing of breeding and habitat or nesting site preferences (see also Kynard[117]) contributed to the maintenance of population identity (significant differences in the electrophoretic patterns of muscle proteins), even though they courted each other and produced hybrids, which could successfully backcross in the laboratory. Given these findings, Hagen concluded that the trachurus and leiurus forms satisfied the biological species concept (sensu Mayr[118]) and proposed that the freshwater residents be known as *G. aculeatus* and the marine form as *G. trachurus*.

Miller and Hubbs[119] disagreed with these designations. They argued that the morphs could not be designated as full species according to the biological species concept, because introgression was present (but see Hagen and McPhail[120]), and thus advocated a more conservative approach to nomenclature: retention of the specific name *G. aculeatus* for all three-spined sticklebacks, with the distinct forms identified as subspecies. They proposed *G. aculeatus aculeatus* Linnaeus for the Holarctic, fully plated marine form (trachurus), *G. a microcephalus* Girard for the partially plated (leiurus) form found along the Pacific coast of North America, and *G. a. williamsoni* Girard for the plateless form found in southern California. Miller and Hubbs noted that similar trachurus-leiurus distributions existed in Europe and expanded subspecific status to include freshwater residents *G. a. algeriensis* Sauvage in the southern Mediterranean and Algeria, *G. a. hologymnus* Regan in Italy, and *G. a. islandicus* Sauvage in Iceland. Bell[121] pointed out that if a character is subject to local adaptation, it is not very informative taxonomically (after Mayr[122]). As these characters include some of the traits that are used most commonly to identify three-spined species or subspecies, lateral plates, pelvic and dorsal spines, and gill rakers, he proposed that *G. aculeatus* be considered a phenotypically diverse superspecies. At the heart of this superspecies is the marine form[123] that appears to have changed very little over the past 10 million years, as evidenced by the fossil record (reviewed in Bell[124]). Bell argued that the marine ancestor had given rise to freshwater descendants multiple times following cycles of dispersal and isolation in freshwater, so Miller and Hubbs' subspecies were polyphyletic and, therefore, invalid.

Bell's hypothesis was echoed by researchers investigating the different three-spined ecomorphs on the west coast of North America. These morphs can be divided into roughly two general types: the streamlined, blue-green dorsally or silver ventrally, marine or anadromous form and the smaller, deeper bodied, mottled grey-green, freshwater forms. There are, however, two additional types living in some lakes. "Benthic" ecomorphs are very deep-bodied, robust fishes with few, short gill rakers, wide mouths, and short, broad snouts, living relatively close to shore, and feeding on small bottom-dwelling invertebrates. "Limnetic" ecomorphs are smaller, slender fishes with numerous, long gill rakers, narrow mouths, and long, slim snouts. They are found in more open waters of the lake, where they pursue a variety of planktonic prey. Pairs of benthics and limnetics are found in six lakes on three islands off the coast of southern British Columbia. Three decades of research have established that:

1. The two ecomorphs differ significantly in morphological traits associated with efficiency in capturing their preferred prey.[1,3,95,125–131]
2. The freshwater populations have reduced genetic variation compared with marine populations.[132,133]
3. There are consistent allozyme and microsatellite differences between the two morphs, even though some pairs are indistinguishable in terms of mtDNA restriction sites.[1,127,134]
4. Premating isolating mechanisms exist between the two morphs (the morphs mate assortatively).[135–137]

5. The morphological differences between the two morphs have a heritable basis.[127,128,138]
6. F_1 and F_2 hybrids between the two morphs show no reduction in viability, fertility,[127,128] or the ability to acquire mates.[139]
7. Such hybrids are at a disadvantage compared to their parental ecomorphs, demonstrating poorer foraging abilities and a reduced growth rate.[2,3,126–128,140,141]

Bell's "superspecies"[121] could be considered synonymous with the macrospecies of Brooks and McLennan,[12,142] in which macrospecies comprise microspecies that may one day become macrospecies in their own right, collapse back into the ancestral macrospecies, or, being often ephemeral, go extinct (see, e.g., Hadley Lake, British Columbia, benthics). Microspecies are identified phylogenetically and can act as species but have not undergone a permanent split (see also Hagen and McPhail[120]). Certainly, the increased number of hybrids between benthics and limnetics in Enos Lake, British Columbia (17% in 1999 contrasted with only 1% in 1984 and 1992),[143] would appear to indicate that speciation may have been initiated, but is not yet complete. Whether or not these forms are in the process of, or have actually completed, speciation is thus still open to debate (for an extensive discussion see Brooks and McLennan[12]).

1.6.2.3 Colour Variation

McPhail[144] first reported three-spined sticklebacks with black nuptial colouration (similar to *Culaea* or *Pungitius*) from the Olympic peninsula of Washington. Completely black and polymorphic red or black populations were subsequently reported from Lake Wapato, Washington,[145] the Queen Charlotte Islands, British Columbia,[146] and Holcomb Creek, California.[147] The evolution of black nuptial colour in three-spined sticklebacks has been attributed to:

1. Predation pressure (decreased conspicuousness[144]).
2. Competition with the Olympic mudminnow, *Novumbra hubbsi*, for territories.[148,149]
3. Transmission properties of the habitat in which the fish live.[150]

Scott and Foster[151] demonstrated that black sticklebacks did not have a selective advantage over red sticklebacks in the competition for nesting locations with *Novumbra*, eliminating hypothesis 2 for at least one of the black populations. Indeed, interactions between sticklebacks and mudminnows were rare in both the laboratory and the field. They also noted that the black colouration in one population does not develop until the parental phase. This result supports McPhail's hypothesis that predation pressure on the fry during the parental guarding stage may have favoured a less conspicuous male signal. This presupposes, of course, that black is less conspicuous than red, something that may not be true for populations living in heavily tea-stained waters.[150] Whatever the ultimate explanation, it is still unclear whether the black populations are all descended from one common ancestor or have evolved independently a number of times.[149]

The development of black nuptial colour is controlled by two autosomal alleles at one locus with no dominance, which are sex limited in their expression and affected by polygenic modifier genes in heterozygotes.[149] Hagen and Moodie showed that when handled or frightened, black males flush in a manner similar to *Culaea* or *Pungitius*, whereas red males fail to lose their bright costume. Ten of the black males showed red ventral colouration, which was covered by the black signal but was visible when the black faded from disturbance. They postulated that these fish represented hybrids between black morphs and red marine morphs but did not investigate the genetics of the unusual fish.

1.6.2.4 Colour Morph as an Isolating Mechanism

Males with black nuptial colouration tend to inhabit tannin-stained waters, which do not transmit red wavelengths very well.[150] Scott[152] hypothesised this change in transmission properties of the habitat is the basis for sensory drive evolution in this system, because black is more conspicuous in such habitats. Interestingly, colour morphs from the Chehalis River drainage do show some reproductive isolation in the laboratory; melanic females prefer melanic males, and mosaic females prefer mosaic males.[153]

In 1984, Blouw and Hagen identified a "white" form of *Gasterosteus* that co-occurs with "typical" *G. aculeatus* in Nova Scotia.[154] Males of this form display iridescent white nuptial colouration both dorsally and laterally, making them extremely conspicuous. White sticklebacks appeared to be reproductively isolated from other sticklebacks in the laboratory and in nature. They nest above the substrate in filamentous algae, show increased levels of activity and a prolonged courtship phase, frequently leave their territories during the day and disperse their eggs among the filamentous algae immediately after spawning, then show no more parental care.[155–157] Paepke[158] recognised the white sticklebacks as an unnamed distinct species based on their unusual behavioural characteristics, despite the fact that there are no significant differences in either allozymes[159] or mitochondrial DNA sequences[160] between whites and regional *G. aculeatus*. Based on those data, it seems more likely that the white stickleback is merely a colour variant in a panmictic three-spined stickleback population, and that the significant reproductive changes displayed by white sticklebacks are extremely recent.

1.6.3 Relationships within the Genus

Haglund et al.[161] studied allozyme variation among 16 populations of *G. aculeatus* from North America, Europe, and Asia. They found that 13 of the 18 loci they examined were polyallelic, and that there was significant interpopulational divergence. Populations could be divided into two primary clades: (1) European, North American, and some Japanese populations, which were further subdivided into an Atlantic basin clade and a "basal Pacific basin assemblage" of western North American and Japanese populations (Hokkaido island and one inland Gifu population); and (2) a divergent Japanese group (from both the Pacific Ocean and Sea of Japan drainages). They felt that the larger Holarctic clade represented *G. aculeatus*, and that the divergent Japanese clade warranted further study to determine its specific, and therefore taxonomic, status.

Ortí et al.[160] took the genetic analysis of *G. aculeatus* one step further and sequenced the mtDNA gene cytochrome *b* from 25 populations around the world. Their findings, although similar to those of Haglund et al., painted a slightly different picture, identifying: (1) a Japanese clade with a few representatives in British Columbia and Alaska and (2) a widespread clade, comprising an Atlantic basin group (with the exceptional inclusion of southern California populations) and a basal Pacific assemblage restricted to populations from Alaska and British Columbia. The major differences between these two studies related to the proposed structure of the Pacific basin clade and whether populations of Japanese sticklebacks are mono- or paraphyletic. The differences between the two studies may be, in part, due to the varying methods of analysis they employed. Haglund et al. used a distance Wagner procedure to build their tree, a phenetic algorithm identifying degree of similarity, which may or may not reflect phylogenetic relationships, whereas Ortí et al. used phylogenetic systematics to reconstruct their phylogeny.

Yamada et al.[162] performed an RFLP analysis of the mtDNA ND5/6 gene comparing 15 Japanese populations of *G. aculeatus*. Their results were more in line with those of Ortí et al., a monophyletic Japanese clade with Russian and Alaskan populations as sister groups. Although previous studies suggested restricted gene flow between Japan Sea and Pacific Ocean forms, the study by Yamada et al. revealed evidence of substantial gene flow between the two.

1.6.4 How Many Species of *Gasterosteus* Are There?

Bell[121] argued that stickleback classification should reflect phylogeny and that it is therefore impossible to answer this question until a thorough investigation of the relationships within *Gasterosteus* is undertaken. The current answer to this question is "two species," the widespread, polymorphic *Gasterosteus aculeatus* and the blackspotted stickleback *Gasterosteus wheatlandi*. Given our current state of knowledge about *Gasterosteus*, it would be unwise to apply species labels to any population, lineage, or "species pair."

1.7 FAMILY-LEVEL RELATIONSHIPS

1.7.1 Historical Review of Gasterosteid Systematics

Gasterosteid systematics has been on the minds of scientists since Linnaeus included them in his *Systema Naturae* in 1758.[163] Whereas early biologists[90–92,164,165] concentrated on the identification and enumeration of species, later researchers shifted their focus to elucidating the relationships among the genera and species. Bertin[94] hypothesized that *Pungitius* represented the ancestor to both *Culaea* and *Gasterosteus* and that *Spinachia* and *Apeltes* were sister groups (Figure 1.1a). Leiner,[166] on the other hand, believed *Spinachia* was the basal member of the family and *Apeltes* the ancestor of *Gasterosteus* and *Pungitius* (which itself represented the ancestor of *Culaea*) (Figure 1.1b).

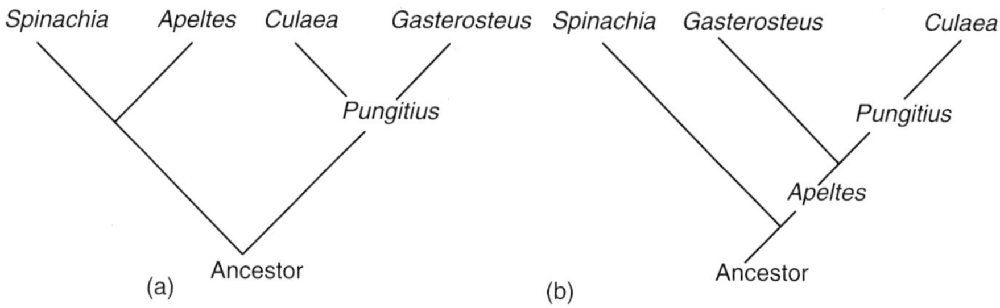

FIGURE 1.1 (a) Relationships proposed by Bertin (from Bertin, L., *Ann. Inst. Oceanogr. Monaco*, 2, 1, 1925. With permission). (b) Relationships proposed by Leiner (from Leiner, M., *Z. Morphol. Okol. Tiere*, 28, 107, 1934. With permission).

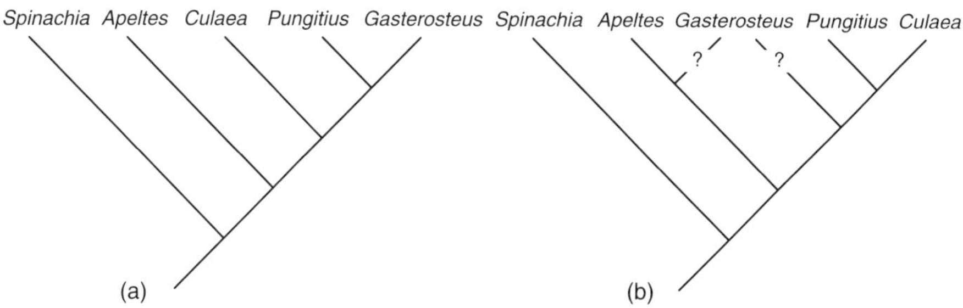

FIGURE 1.2 (a) Relationships proposed by Hall (from Hall, M.F., A Comparative Study of the Reproductive Behaviour of the Sticklebacks (Gasterosteidae), D.Phil. thesis, Department of Zoology, Oxford, 1956). (b) Relationships proposed by Reisman and Cade (from Reisman, H.M. and Cade, T.J., *Am. Midl. Nat.*, 77, 257, 1967. With permission).

Hall[167] performed an extensive comparative study of the behaviour of gasterosteids. Based on the number of differences between species, she hypothesized that *Pungitius* was most closely related to *Gasterosteus* because they shared permanent breeding colour, elaborate threat behaviour, use of filamentous nesting material, creeping through nest, zigzag courtship, and presence of the female head-up display. Hall placed *Culaea* as the sister group to *Pungitius* + *Gasterosteus* based on the single pelvic plate, two ventral fin rays, territorial females, insertion gluing, male quivering, postfertilization fanning peak, and use of a nursery. *Apeltes* and *Spinachia* represented successive sister groups to this clade (Figure 1.2a). Reisman and Cade[168] reinterpreted some of the behavioural data and agreed that *Spinachia* was the basal member of the family, but proposed that *Pungitius* and *Culaea* were sister groups based on the presence of a nursery, nesting in plants, and black nuptial colouration. They were unsure if *Gasterosteus* constituted the sister group to *Apeltes* (based on the red nuptial colouration and separated branchial slits) or *Pungitius* + *Culaea* (based on the [zigzag] dance and presence of a nest entrance) (Figure 1.2b).

The first molecular look into interfamilial relationships came with the chromosomal data of Chen and Reisman.[169] Although their study lacked information on *Spinachia* and they used *Apeltes* as a functional outgroup, their results agreed with Hall,[167] placing *Culaea* as a close relative of *Pungitius* + *Gasterosteus* (Figure 1.3).

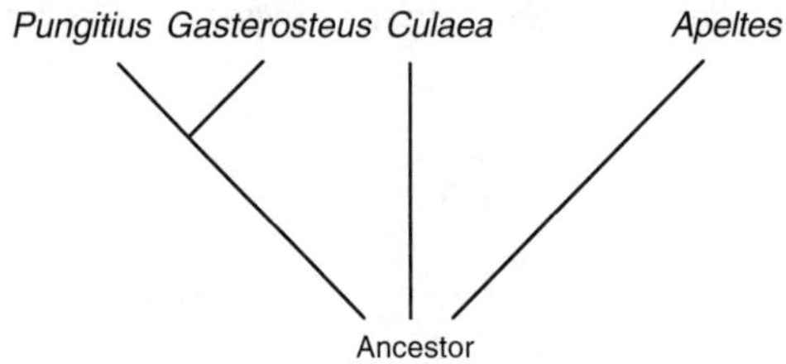

FIGURE 1.3 Relationships proposed by Chen and Reisman based on chromosomal data. (From Chen, T.R. and Reisman, H.M., *Cytogenetics*, 9, 321, 1970. With permission.)

Nelson[170] agreed that most characters place *Spinachia* as the basal member of this family and supported *Gasterosteus, Pungitius*, and *Culaea* as having shared a common ancestor. However, he considered each character separately and did not provide an overall picture of gasterosteid relationships, concluding that only a complete fossil record would provide the data necessary to correctly place each genus. Finally, Wootton examined all the available data and concluded that the information supported *Culaea* as the sister group to *Pungitius* + *Gasterosteus* (Figure 1.2a).[84] He felt, however, that this hypothesis was amenable to change based on new data, stating that "this scheme is provided more as a target for informed criticism than as a definitive statement … " (p. 335). The first attempt to reconstruct the phylogeny for the Gasterosteidae using modern phylogenetic systematic methodology came when Paepke polarized 20 morphological, behavioural, and ecological characters using the Aulorhynchidae and Hypoptychidae as the outgroups.[171] The results of that analysis supported Reisman and Cade's hypothesis, placing *Gasterosteus* as the sister group of *Culaea* + *Pungitius*. The sister group relationship between *Culaea* and *Pungitius* was supported by three characters: living wholly or mostly in freshwater, presence of a nursery, and black nuptial colouration (Figure 1.4). One year later,

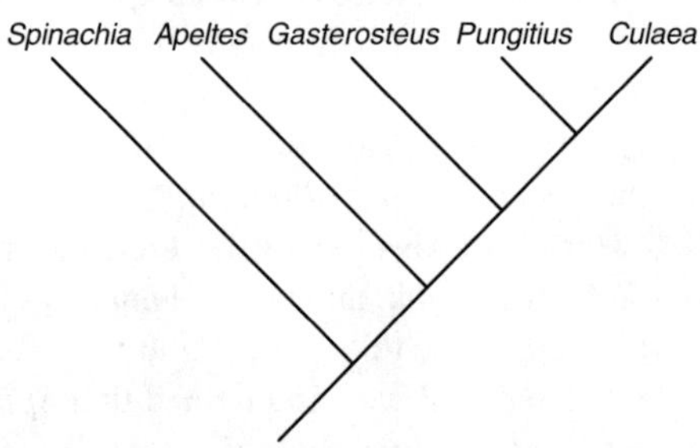

FIGURE 1.4 Phylogenetic relationships proposed by Paepke (from Paepke, H.-J., *Die Stichlinge*, Vol. 10, Westarp Wissenschaften, Magdeburg, 1996; Paepke, H.-J., *Die Stichlinge*, Vol. 10, A. Ziemsen Verlag, Wittenberg, Lutherstadt, 1983. With permission), McLennan (from McLennan, D.A. et al., *Can. J. Zool.*, 66, 2177, 1988; McLennan, D.A., *Copeia*, 318, 1993. With permission), McLennan and Mattern (from McLennan, D.A. et al., *Cladistics*, 17, 11, 2001. With permission), and the total evidence phylogeny of the Gasterosteidae based on morphological, behavioural, and molecular data (from Mattern, M.Y. et al., *Cladistics*, 20, 14, 2004. With permission).

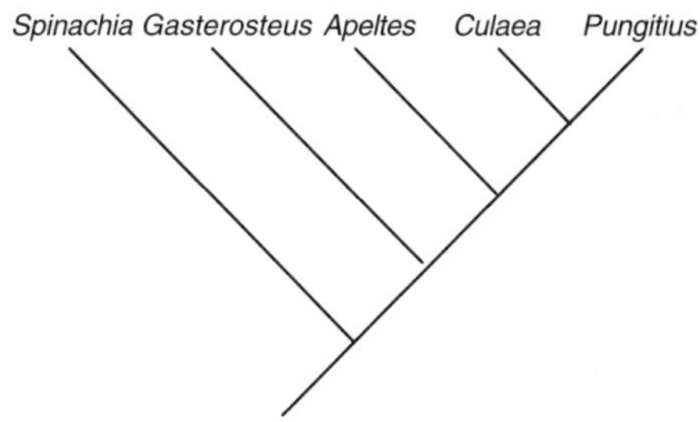

FIGURE 1.5 Relationships proposed by Bowne (from Bowne, P.S., The Systematic Position of Gasterosteiformes, Ph.D. thesis, University of Alberta, Edmonton, 1985) and phylogenetic relationships resulting from the analysis of five mitochondrial genes (12S rRNA, 16S rRNA, cytochrome *b*, ATPase 6, and control region) (from Mattern, M.Y., *Mol. Phylogenet. Evol.*, 30, 366, 2004. With permission).

Hudon and Guderley[172] investigated the relationships among four gasterosteid taxa (*Apeltes*, *Pungitius*, *G. wheatlandi*, and *G. aculeatus*) employing electrophoretic data. Using *Apeltes quadracus* as the functional out-group, they hypothesized that *G. aculeatus* and *G. wheatlandi* were more closely related to each other than either was to *Pungitius*, which is congruent with Paepke's result. Bowne[173] then undertook the arduous task of attempting to resolve the relationships among the Gasterosteiformes. She examined 347 morphological characters for 19 gasterosteiform and sygnathiform taxa, including 15 outgroups. As part of that analysis, she paid special attention to relationships within the Gasterosteidae. Unfortunately, she used successive members of the ingroup to polarize characters, so it is not surprising that her results, which were not obtained by any known phylogenetic methodology, were somewhat unclear. However, her shortest tree indicated that *Culaea* + *Pungitius* was the sister group to *Apeltes* with *Gasterosteus* and *Spinachia* forming successive sisters to that group (Figure 1.5), differing from Paepke only in the placement of *Apeltes*.

1.7.2 Recent Phylogenetic-Based Studies

McLennan et al.[174] and McLennan[175] returned the focus of gasterosteid systematics to phylogenetic systematic and behavioural studies. Fifty-one behavioural characters, including nuptial colouration, courtship, and parental care, produced a single tree that was identical to Paepke's hypothesis (Figure 1.4) and congruent with Reisman and Cade's proposal (Figure 1.2b). Bowne[176] responded by combining data from 50 osteological and external morphological characters (almost none of which were from her original thesis) to produce three equally parsimonious diagrams, all of which supported the basal placement of *Spinachia* and *Apeltes* but differed in the relationships among *Culaea* + *Pungitius* + *Gasterosteus*. These trees were extracted from MacClade, which does not have a tree-building algorithm, so once again the results were not robust. McLennan and Mattern[177] reviewed the status of the morphological and behavioural databases, combining them for the first time into a phylogenetic analysis based on all the available data. Their analysis of 89 morphological and 47 behavioural characters resulted in one tree with a consistency index of 82.5%

(excluding uninformative characters) that was identical to the behaviour-based tree (Figure 1.4).

I conducted the first family-wide molecular systematic study of gasterosteids, sequencing 2879 base pairs from five mitochondrial genes, 12S and 16S rRNA, cytochrome *b*, ATPase 6, and control region.[178] I selected these five genes (although all mitochondrial), because they are thought to evolve at different rates, therefore providing support for nodes of different ages. Although each gene presented a different hypothesis of relationships, the combined molecular analysis produced one tree of with a consistency index of 72.5%, which differed from the morphology + behaviour tree only in the placement of *Apeltes* ([*Spinachia*], [*Gasterosteus*], [*Apeltes*], [*Pungitius*], [*Culaea*]) (Figure 1.5). Combining my molecular data with the previous morphological and behavioural database produced a total data set of 3011 characters (48 behavioural characters, 84 morphological characters, and 2879 mtDNA base pairs), of which 805 were parsimony informative. This total evidence matrix produced one tree, identical to the morphology + behaviour topology (Figure 1.4) of 2241 steps (59 behavioural, 127 morphological, and 2055 molecular) with a CI of 0.735.[179] Clades were supported by the following unambiguous (nonhomoplasious) synapomorphies:

1. *Pungitius* + *Culaea*: Thirty base pair changes; morphological: parasphenoid not pierced by carotid foramen, small symplectic dorsal flange, and dorsal spines depressible into shallow dorsal groove; Behavioural: insertion gluing, snout above nest in nest show display, fanning present in nest show display, male dances towards the nest, and male moves the nursery.
2. *Pungitius* + *Culaea* + *Gasterosteus*: Seventeen base pair changes; morphological: anterior end of median ethmoid dorsal plate bent anteriorly, second ventral foramen present in the exoccipital, large cleithral extension to the coracoid fan, short and separated pelvic anterior processes, long pelvic posterior processes, pelvic branch ascends dorsally, and pelvic spines originate under the second dorsal spine; Behavioural: circle fighting, dorsal roll submission, female weather-vaning, male dances in front of female, male ventrolateral nuptial colouration, melanin in dorsolateral cells contracts during courtship, male nudges female flank with snout while female is in the nest, nursery formation, fry retrieval, and nest raiding.
3. *Pungitius* + *Culaea* + *Gasterosteus* + *Apeltes*: Seventy-eight base pair changes; morphological: lateral ethmoid confined to anterior orbit wall, no anterolateral extension of lateral ethmoid, lamina extending ventrally from lateral edge of nasal bending medially toward vomer and touching lateral ethmoid, short lachrymal, short premaxillary, medium symplectic dorsal flange, urohyal flange posteriorly entire, lateral foramen in pelvic plate, united pelvic anterior processes, extent of neural arch development along precaudal vertebra more than half the length of each vertebrum, and transverse processes arising from centre and ventral edges of centra; Behavioural: first stage of nest building is to collect vegetation, male constructs a nest entrance, head-down threat, broadside threat, courting female's initial response to male is stop and hold head up, male performs

nest maintenance during courtship, male ventilates nest during courtship, male quivers the female while she is in the nest, female is below male during follow to nest, male erects fins and spines during lead to nest, egg deposition in nest, and male extends nest after first fertilisation, male pulls ventilation holes in nest during egg development.

4. Gasterosteidae: Two hundred twenty-six base pair changes; morphological: suborbital and preoperculum touch, no lateral canal in dermopterotic, dermopterotic overlaps sphenotic by less than half, no exoccipital condyles, dentary and angular touch, no preoperculamandibular canal on articular, central lamina present in symplectic, operculum posterior margin rounded, no fourth pharyngobranchial, enlarged fourth actinost, cleithral lateral flange does not extend to ventral end of cleithrum, cleithral extension to coracoid fan is a low crest, pelvic and pectoral girdles touch each other, melanocytes present in connecting membrane of dorsal spines, and reduced cephalic lateral line system; Behavioural: pushing and boring during nest building, nest materials include both glue and collected plant materials, snout into nest during showing, male ventilates eggs via fanning, male removes decaying and dead eggs from nest, and male increases the number and duration of ventilation bouts across egg-guarding cycle.[179]

Keivany and Nelson[13] attempted to elucidate the relationships within *Pungitius* using 33 osteological characters. They reported that their matrix yielded 32 equally parsimonious trees of 71 steps; however, a reanalysis of the matrix presented in their paper yields only 11 trees of 62 steps. This discrepancy aside, the strict consensus of these trees supports Mattern and McLennan's total evidence analysis,[179] with one exception: these traits placed *Culaea* within *Pungitius* as the sister species to *P. hellenicus* (Figure 1.6). My analysis of five *Culaea* and two *Pungitius* populations,

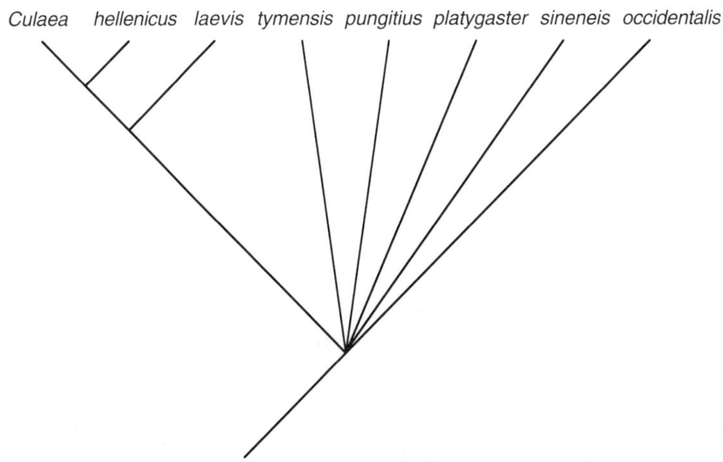

FIGURE 1.6 Phylogenetic relationships resulting from the reanalysis of Keivany and Nelson's dataset. (From Keivany, Y. et al., *Behaviour*, 141, 1485, 2004. With permission.) Tree represents the strict consensus of 11 equally parsimonious trees.

one from North America (*P. occidentalis*) and one from Europe (representing either *P. pungitius* or *P. laevis*) indicated that the two *Pungitius* populations were each other's closest relatives. The monophyly of *Pungitius* was supported by 30 molecular traits. Breaking that monophyly to incorporate *Culaea* into *Pungitius* was supported by only six morphological characters in Keivany and Nelson's study.[13] If the morphological data are telling us something real about relationships, then one of two patterns should have been recovered from my molecular data. Depending on which species was represented by the European specimen in my study, the analysis should have produced either: (1) ([*Culaea* + *P. laevis*] *P. occidentalis*) or (2) a polytomy (*Culaea*, *P. pungitius*, *P. occidentalis*). Because neither of these patterns was recovered, this means that the molecular and morphological data are in conflict. Unfortunately, there is no molecular data available for *P. hellenicus, P. laevis*, or *P. platygaster* to investigate whether the molecular data would produce a similar picture. Combining the data without the availability of these sequences is a pointless exercise, as *P. hellenicus* would still group with *Culaea* in the absence of any molecular evidence. Incidentally, when the mtDNA control region data available for multiple populations of *P. tymensis*, *P. sinensis,* and Japanese *P. pungitius*[71] were included in a quick reanalysis of the molecular data, the genus *Pungitius* still forms a monophyletic clade, which is the sister group to *Culaea*. Clearly, a more rigorous, holistic analysis of the evidence is needed to resolve the debate about the origins of *C. inconstans.*

1.8 GENERAL CONCLUSIONS

It is clear from this chapter that although extensive work has been done to resolve the relationships among the different genera of sticklebacks, the relationships within *Pungitius* and *Gasterosteus* are still left unresolved. To improve the state of knowledge in this area, we will need to investigate multiple molecular markers for a widespread sample of populations and species, using rigorous phylogenetic methodologies. Not only will this resolve the relationships among the species, it will also help identify the various species in a more definitive way. A rigorous phylogeny will also allow us to investigate how much of the morphological and behavioural variability is the result of evolution, and how much is the result of ecology.

REFERENCES

1. McPhail, J.D., Speciation and the evolution of reproductive isolation in the sticklebacks (*Gasterosteus*) of south-western British Columbia, in *The Evolutionary Biology of the Threespine Stickleback*,. Bell, M.A. and Foster, S.A., Eds., Oxford Science Publications, Oxford, 1994, p. 399.
2. Schluter, D., Experimental evidence that competition promotes divergence in adaptive radiation, *Science*, 266, 798, 1994.
3. Schluter, D., Adaptive radiation in sticklebacks: size, shape, and habitat use efficiency, *Ecology*, 74, 699, 1993.
4. Pelkwijk, J.J. ter and Tinbergen, N., Eine reizbiologische analyse einiger Verhaltensweisen von *Gasterosteus aculeatus* L., *Z. Tierpsychologie*, 1, 193, 1937.

5. Milinski, M. and Bakker, T.C.M., Female sticklebacks use male coloration in mate choice and hence avoid parasitized males, *Nature*, 344, 331, 1990.
6. Rowland, W.J., Proximate determinants of stickleback behaviour: an evolutionary perspective, in *The Evolutionary Biology of the Threespine Stickleback*, Bell, M.A. and Foster, S.A., Eds., Oxford Science Publications, Oxford, 1994, p. 297.
7. Kunzler, R. and Bakker, T.C.M., Female preferences for single and combined traits in computer animated stickleback males, *Behav. Ecol.*, 12, 681, 2001.
8. Largiader, C.R., Fries, V., and Bakker, T.C.M., Genetic analysis of sneaking and egg-thievery in a natural population of the three-spined stickleback (*Gasterosteus aculeatus* L.), *Heredity*, 86, 459, 2001.
9. Fitzgerald, G.J., Egg cannibalism by sticklebacks — spite or selfishness, *Behav. Ecol. Sociobiol.*, 30, 201, 1992.
10. Fitzgerald, G.J., Filial cannibalism in fishes — why do parents eat their offspring, *Trends Ecol. Evol.*, 7, 7, 1992.
11. Fitzgerald, G.J., The role of cannibalism in the reproductive ecology of the threespine stickleback, *Ethology*, 89, 177, 1991.
12. Brooks, D.R. and McLennan, D.A., *The Nature of Diversity: An Evolutionary Voyage of Discovery*, University of Chicago Press, Chicago, IL, 2002, p. 668.
13. Keivany, Y. and Nelson, J.S., Phylogenetic relationships of sticklebacks (Gasterosteidae), with emphasis on ninespine sticklebacks (*Pungitius* spp.), *Behaviour*, 141, 1485, 2004.
14. Croy, M.I. and Hughes, R.N., The influence of hunger on feeding behaviour and on the acquisition of learned foraging skills by the fifteen-spined stickleback, *Spinachia spinachia* (L.), *Anim. Behav.*, 41, 161, 1991.
15. Croy, M.I. and Hughes, R.N., The role of learning and memory in the feeding behaviour of the fifteen-spined stickleback, *Spinachia spinachia* (L.), *Anim. Behav.*, 41, 149, 1991.
16. Croy, M.I. and Hughes, R.N., Hierarchical response to prey stimuli and associated effects of hunger and foraging experience in the fifteen-spined stickleback, *Spinachia spinachia* (L.), *J. Fish Biol.*, 38, 599, 1991.
17. Croy, M.I. and Hughes, R.N., Effects of food supply, hunger, danger and competition on choice of foraging location by the fifteen-spined stickleback, *Spinachia spinachia* (L.), *Anim. Behav.*, 42, 131, 1991.
18. Croy, M.I. and Hughes, R.N., The combined effects of learning and hunger in the feeding behaviour of the fifteen-spined stickleback (*Spinachia spinachia*), *NATO ASI Ser. G. Ecol. Sci.*, 20, 215, 1990.
19. Hughes, R.N. and Croy, M.I., An Experimental-analysis of frequency-dependent predation (switching) in the 15-spined stickleback, *Spinachia spinachia*, *J. Anim. Ecol.*, 62, 341, 1993.
20. Kaiser, M.J., The ontogeny of predatory mechanisms in the 15-spined stickleback, *Spinachia spinachia* (L.), *J. Fish Biol.*, 40, 485, 1992.
21. Kaiser, M.J. and Croy, M.I., Population-structure of the 15-spined stickleback, *Spinachia spinachia* (L.), *J. Fish Biol.*, 39, 129, 1991.
22. Kaiser, M.J., Gibson, R.N., and Hughes, R.N., The effect of prey type on the predatory behavior of the 15-spined stickleback, *Spinachia spinachia* (L.), *Anim. Behav.*, 43, 147, 1992.
23. Kaiser, M.J. et al., Are digestive characteristics important contributors to the profitability of prey — a study of diet selection in the 15-spined stickleback, *Spinachia spinachia* (L.), *Oecologia*, 90, 61, 1992.

24. Ostlund, S., Female 15-spined sticklebacks detect males with empty nests by non-visual cues, *J. Fish Biol.*, 47, 1106, 1995.
25. Ostlund-Nilsson, S., Does paternity or paternal investment determine the level of paternal care and does female choice explain egg stealing in the fifteen-spined stickleback? *Behav. Ecol.*, 13, 188, 2002.
26. Ostlund-Nilsson, S., Fifteen-spined stickleback (*Spinachia spinachia*) females prefer males with more secretional threads in their nests: an honest-condition display by males, *Behav. Ecol. Sociobiol*, 50, 263, 2001.
27. Ostlund-Nilsson, S., Are nest characters of importance when choosing a male in the fifteen-spined stickleback (*Spinachia spinachia*)? *Behav. Ecol. Sociobiol.*, 48, 229, 2000.
28. Rademacher, K. and Kils, U., Predator prey dynamics of fifteen-spined stickleback (*Spinachia spinachia*) and the mysid (*Neomysis integer*), *Arch. Fish. Mar. Res.*, 43, 171, 1996.
29. Gross, H.P., Observations on the geographic variation of the marine coastal fish *Spinachia spinachia*, *Mar. Biol.*, 47, 297, 1978.
30. Scott, W.B. and Crossman, E.J., Freshwater fishes of Canada, *Bull. Fish. Res. Board Can.,* 184, 1973.
31. Scott, W.B. and Crossman, E.J., The freshwater fishes of New Brunswick: a checklist with distributional notes, *Contrib. ROM Div. Zool. Palaeontol.*, 51, 1, 1959.
32. Livingstone, D.A., Fresh water fishes of Nova Scotia, *Proc. Nova Scotia Inst. Sci.*, 23, 1, 1953.
33. Coad, B.W. and Power, G., Life history notes and meristic variation in the freshwater fourspine stickleabck, *Apeltes quadracus* (Mitchell), near Sept-Iles, Quebec, *Le Nat. Can.*, 100, 247, 1973.
34. Nelson, J.S., Salinitiy tolerance of brook sticklebacks, *Culaea inconstans*, freshwater ninespine sticklebacks, *Pungitius pungitius*, and freshwater fourspine sticklebacks, *Apeltes quadracus*, *Can. J. Zool.*, 46, 663, 1968.
35. Cooper, E.L., *Fishes of Pennsylvania and the Northeastern United States*, Pennsylvania State University Press, University Park, 1983, p. 243.
36. Scott, W.B. and Crossman, E.J., *Fishes Occurring in the Fresh Waters of Insular Newfoundland*, Department of Fisheries, Canada, Queen's Printer, 1964.
37. Dadswell, M.J., New records of freshwater fishes from the northwest coast of insular Newfoundland, *Can. Field Nat.*, 86, 289, 1972.
38. Rombough, P.J., Barbour, S.E., and Kerekes, J.J., Freshwater fishes from northern Newfoundland, *Can. Field Nat.*, 95, 359, 1981.
39. Stephenson, S.A. and Momot, W.T., Threespine, *Gasterosteus aculeatus*, and Fourspine, *Apeltes quadracus*, sticklebacks in the Lake Superior basin, *Can. Field Nat.*, 114, 211, 2000.
40. Holm, E. and Hamilton, J.G., Range extension for the fourspine stickleback, *Apeltes quadracus*, to Thunder Bay, Lake-Superior, *Can. Field Nat.*, 102, 653, 1988.
41. Campbell, C.E., Fourspine stickleback, *Apeltes quadracus*, from a freshwater lake on the Avalon Peninsula of eastern Newfoundland, *Can. Field Nat.*, 106, 400, 1992.
42. Mitchell, S.L., The fishes of New York described and arranged, *Trans. Lit. Philos. Soc. N.Y.*, 1, 355, 1815.
43. Krueger, W.H., Meristic variation in fourspine stickleback, *Apeltes quadracus*, *Copeia*, 442, 1961.
44. Cox, P., Regional variation of the fourspined stickleback, *Apeltes quadracus,* Mitchell, *Can. Field Nat.*, 37, 146, 1923.

45. Blouw, D.M. and Hagen, D.W., The adaptive significance of dorsal spine variation in the fourspine stickleback, *Apeltes quadracus*. 1. Geographic variation in spine number, *Can. J. Zool.*, 62, 1329, 1984.
46. Hagen, D.W. and Blouw, D.M., Heritability of dorsal spines in the fourspine stickleback (*Apeltes quadracus*), *Heredity*, 50, 275, 1983.
47. Kirtland, J.P., Descriptions of four new species of fishes, *Boston J. Nat. Hist.*, 3, 273, 1840.
48. Hansen, D.F., Variation in the number of spines and rays in the fins of the brook stickleback, *IL State Acad. Sci. Trans.*, 32, 207, 1939.
49. Moodie, G.E.E., Meristic variation, asymmetry, and aspects of habitat of *Culaea inconstans* (Kirtland), brook stickleback, in Manitoba, *Can. J. Zool.*, 55, 398, 1977.
50. Edge, T.A. and Coad, B.W., Reduced dorsal spine numbers in two isolated populations of the brook stickleback (*Culaea inconstans*) from eastern Canada, *Le Nat. Can.* (Quebec), 110, 99, 1983.
51. Lawler, G.H., Variation in number of dorsal spines in the brook stickleback, *Eucalia inconstans*, *Can. J. Zool.*, 36, 127, 1959.
52. Nelson, J.S., Evidence of a genetic basis for absence of pelvic skeleton in brook stickleback, *Culaea inconstans*, and notes on geographical distribution and origin of loss, *J. Fish. Res. Board. Can.*, 34, 1314, 1977.
53. Nelson, J.S. and Atton, F.M., Geographic and Morphological variation in presence and absence of pelvic skeleton in brook stickleback, *Culaea inconstans* (Kirtland), in Alberta and Saskatchewan, *Can. J. Zool.*, 49, 343, 1971.
54. Nelson, J.S., Geographic variation in brook stickleback, *Culaea inconstans*, and notes on nomenclature and distribution, *J. Fish. Res. Board. Can.*, 26, 2431, 1969.
55. Andraso, G.M. and Barron, J.N., Unusually long spines in brook stickleback (*Culaea inconstans*) from the Mad River drainage, Ohio, *Am. Midl. Nat.*, 147, 162, 2002.
56. Burks, D.J. et al., Geographic variation in agnostic responses of territorial male brook stickleback, *Culaea inconstans*, *Ohio J. Sci.*, 85, 23, 1985.
57. Andraso, G.M., A comparison of startle response in two morphs of the brook stickleback (*Culaea inconstans*): further evidence for a trade-off between defensive morphology and swimming ability, *Evol. Ecol.*, 11, 83, 1997.
58. Gach, M.H., Geographic variation in mitochondrial DNA and biogeography of *Culaea inconstans* (Gasterosteidae), *Copeia*, 563, 1996.
59. Koster, W.J., *Guide to the Fishes of New Mexico*, University of New Mexico Press, Albuquerque, New Mexico, 1957.
60. Mattern, M.Y., The Phylogeny of the Gasterosteidae with Emphasis on the Relationships within *Culaea inconstans* (Kirtland), Ph.D. thesis, University of Toronto, Toronto, 2006.
61. Münzing, J., Variabilität, Verbreitung und Systematik der Arten une Unterarten in der Gattung *Pungitius* Coste, 1848 (Pisces, Gasterosteidae), *Z. Zool. Syst. Evolutionsforsch.*, 7, 208, 1969.
62. McPhail, J.D., Geographic variation in North American ninespine sticklebacks, *Pungitius pungitius*, *J. Fish. Res. Board. Can.*, 20, 27, 1971.
63. Gross, H.P., Geographic variation in European ninespine sticklebacks, *Pungitius pungitius*, *Copeia*, 1979, 405, 1979.
64. Tanaka, S., Variations in ninespine sticklebacks, *Pungitius pungitius* and *P. sinensis* in Honshu, Japan, *Jpn. J. Ichthyol.*, 29, 203, 1982.
65. Takata, K., Goto, A., and Hamada, K., Geographic distribution and variation of three species of ninespine sticklebacks (*Pungitius tymensis, P. pungitius*, and *P. sinensis*) in Hokkaido, *Jpn. J. Ichthyol.*, 31, 312, 1984.

66. Takahashi, H., Takata, K., and Goto, A., Phylogeography of lateral plate dimorphism in the freshwater type of ninespine stickleback, genus *Pungitius*, *Ichthyol. Res.*, 48, 143, 2001.
67. Yang, S.Y. and Min, M.S., Genetic variation in sticklebacks, *Korean J. Zool.*, 33, 499, 1990.
68. Tanaka, S., Ten new species of Japanese fishes, *Dobutsugaku Zasshi* [*Zool. Mag. Tokyo*], 27, 565, 1915.
69. Masuda, H. et al., *The Fishes of the Japanese Archipelago*, Vol. 1, Tokai University Press, Tokyo, 1984.
70. Haglund, T.R., Buth, D.G., and Lawson, R., Allozyme variation and phylogenetic relationships of Asian, North American, and European populations of the ninespine stickleback, *Pungitius pungitius*, in *Systematics, Historical Ecology and North American Freshwater Fishes*, R.L. Mayden, Ed., Stanford, CA, Stanford University Press, 1992, p. 438.
71. Takahashi, K. and Goto, A., Evolution of East Asian ninespine stickleback as shown by mitochondrial DNA control region sequences, *Mol. Phylogenet. Evol.*, 21, 135, 2001.
72. Takahashi, H. and Takata, K., Multiple lineages of the mitochondrial DNA introgression from *Pungitius pungitius* (L.) to *Pungitius tymensis* (Nikolsky), *Can. J. Fish. Aquat. Sci.*, 57, 1814, 2000.
73. Takata, K., Genetic differences of ninespine sticklebacks genus *Pungitius*, in *Freshwater Fishes of Japan*, N. Mizuno and Goto, A., Eds., Tokai University Press, Tokyo, 1987, p. 134.
74. Niwa, T., Comparison of the gene frequency between sympatric population of ninespine sticklebacks, genus *Pungitius*, in Hokkaido, *Jpn. J. Ichthyol.*, 34, 184, 1987.
75. Kobayashi, H., Some new information found in the sticklebacks of Hokkaido, *J. Hokkaido Gakugei University (Sect. B)*, 8, 44, 1957.
76. Kobayashi, H., Cross experiments with three species of sticklebacks, *Pungitius pungitius* (L.), *Pungitius tymensis* (Nikolskii), and *Pungitius sinensis* (Guichenot), with special reference to their sympatric relationship, *J. Hokkaido Gakugei University (Sect. B)*, 10, 363, 1959.
77. Ziuganov, V.V. and Gomeluk, V.Y., Hybridization of two forms of ninespine stickleback, *Pungitius pungitius* and *Pungitius* platygaster, under experimental conditions and an attempt to predict the consequences of their contact in nature, *Environ. Biol. Fish.*, 13, 241, 1985.
78. Keivany, Y. and Nelson, J.S., Taxonomic review of the genus *Pungitius*, ninespine sticklebacks (gasterosteidae), *Cybium*, 24, 107, 2000.
79. Hubbs, C.L., The Atlantic American species of the fish genus *Gasterosteus*, *Occasional Papers Mus. Zool., Univ. Mich.*, 200, 1, 1929.
80. Putnam, F.W., Remarks on a supposed non-descript species of *Gasterosteus* from Massachusetts, *Proc. Commun. Essex Inst. Salem*, 5, 4, 1867.
81. Kendall, W.C., Description of a new stickleback, *Gasterosteus gladiunculus*, from the coast of Maine, *Proc. U.S. Nat. Mus.*, 18, 623, 1896.
82. Goode, G.B. and Bean, T.H., A list of the fishes of Essex County, including those of Massachusetts Bay according to the latest results of the works of the U.S. Fish Commission, *Bull. Essex Inst.*, 38, 1879.
83. Coad, B.W. and Power, G., Observations on ecology and meristic variation of ninespine stickleback, *Pungitius pungitius* (L. 1758) of Matamek-river system, Quebec, *Am. Midl. Nat.*, 90, 498, 1973.
84. Wootton, R.J., *The Biology of the Sticklebacks*, Academic Press, London, 1976.

85. Sargent, R.C. et al., A lateral plate cline, sexual dimorphism, and phenotypic variation in the black-spotted stickleback, *Gasterosteus-wheatlandi*, *Can. J. Zool.*, 62, 368, 1984.
86. Perlmutter, A., Observations of fishes of the genus *Gasterosteus* in the waters of Long Island, New York, *Copeia*, 1963, 168, 1963.
87. Reisman, H.M., Reproductive isolating mechanisms of blackspotted stickleback *Gasterosteus wheatlandi*, *J. Fish. Res. Board Can.*, 25, 2703, 1968.
88. McInerney, J.E., Reproductive behaviour of blackspotted btickleback, *Gasterosteus wheatlandi*, *J. Fish. Res. Board Can.*, 26, 2061, 1969.
89. Ayvazian, S.G., Observations of asymmetric reproduction along a morphocline of the blackspotted stickleback, *Gasterosteus* wheatlandi, *Can. J. Zool.*, 71, 1477, 1993.
90. Sauvage, H.E., Révision des espèces du groupe des Épinoches, *Nouv. Arch. Mus. Hist. Nat. Paris*, 10, 5, 1874.
91. Jordan, D.S. and Evermann, B.W., *The Fishes of North and Middle America*, T.F.H. Publ., NJ, 1896.
92. Cuvier, G. and Valenciennes, A., *Hist. Nat. Poissons*, 4, 1829.
93. Hagen, D.W. and Moodie, G.E.E., Polymorphism for plate morphs in *Gasterosteus aculeatus* on the east coast of Canada and an hypothesis for their global distribution, *Can. J. Zool.*, 60, 1032, 1982.
94. Bertin, L., Recherches bionomiques, biométriques et systématiques sur les Epinoches (Gastèrostéidés), *Ann. Inst. Oceanogr. Monaco*, 2, 1, 1925.
95. Hagen, D.W. and Gilbertson, L.G., Geographic variation and environmental selection in *Gasterosteus aculeatus* L. in the Pacific Northwest, America, *Evolution*, 26, 32, 1972.
96. Bakker, T.C.M. and Sevenster, P., Plate morphs of *Gasterosteus aculeatus* Linnaeus (Pisces: Gasterosteidae): comments on terminology, *Copeia*, 1988, 659, 1988.
97. Igarashi, K., Observations on the development of the scutes in the landlocked form of three-spined stickleback, *Gasterosteus aculeatus aculeatus* Linnaeus, *Bull. Japan. Soc. Sci. Fish.*, 30, 95, 1964.
98. Igarashi, K., On the variation of the scute in the three-spined stickleback, *Gasterosteus aculeatus* (Linnaeus) from Nasu Area, Tochigi-Ken, *Annot. Zool. Jpn.*, 43, 43, 1970.
99. Igarashi, K., Formation of the scutes in the marine form of the three-spined stickleback, *Gasterosteus aculeatus aculeatus* (Linnaeus), *Annot. Zool. Jpn.*, 43, 34, 1970.
100. Bell, M.A., Lateral plate polymorphism and ontogeny of the complete plate morph of threespine sticklebacks (*Gasterosteus aculeatus*), *Evolution*, 35, 67, 1981.
101. Hagen, D.W. and Gilbertson, L.G., Selective predation and the intensity of selection acting upon the lateral plates of threespine sticklebacks, *Heredity*, 30, 273, 1973.
102. Moodie, G.E.E., McPhail, J.D., and Hagen, D.W., Experimental demonstration of selective predation in *Gasterosteus aculeatus*, *Behaviour*, 47, 95, 1973.
103. Reimchen, T.E., Predator-induced cyclical changes in lateral plate frequencies of *Gasterosteus*, *Behaviour*, 132, 1079, 1995.
104. Reimchen, T.E., Injuries on stickleback from attacks by a toothed predator (*Oncorhynchus*) and implications for the evolution of lateral plates, *Evolution*, 46, 1224, 1992.
105. Giles, N., The possible role of environmental calcium levels during the evolution of phenotypic diversity in Outer Hebridean populations of the three-spined stickleback, *Gasterosteus aculeatus*, *J. Zool.*, 199, 535, 1983.
106. Baumgartner, J.V. and Bell, M.A., Lateral plate morph variation in California populations of the threespine stickleback, *Gasterosteus aculeatus*, *Evolution*, 38, 665, 1984.

107. Heuts, M.J., Experimental studies on adaptive evolution in *Gasterosteus aculeatus* L, *Evolution*, 1, 89, 1947.
108. Münzing, J., Biologie, Variabilität und Genetik von *Gasterosteus aculeatus* L. (Pisces), Untersuchungen im Elbegebeit, *Int. Rev. Hydrobiol.*, 44, 317, 1959.
109. Hagen, D.W. and Gilbertson, L.G., The genetics of plate morphs in freshwater threespine sticklebacks, *Heredity*, 31, 75, 1973.
110. Avise, J.C., Genetics of plate morphology in an unusual population of threespine sticklebacks (*Gasterosteus aculeatus*), *Genet. Res.*, 27, 33, 1976.
111. Ziuganov, V.V., Genetics of osteal plate polymorphism and microevolution of threespine stickleback (*Gasterosteus aculeatus* L.), *Theor. Appl. Genet.*, 65, 239, 1983.
112. Banbura, J., A new model of lateral plate morph inheritance in the threespine stickleback, *Gasterosteus aculeatus*. *Theor. Appl. Genet.*, 88, 871, 1994.
113. Colosimo, P.F. et al., The genetic architecture of parallel armor plate reduction in threespine sticklebacks, *PLoS Biol.*, 2, 109, 2004.
114. Ziuganov, V.V., Reproductive isolation among lateral plate phenotypes (low, partial, complete) of the threespine stickleback, *Gasterosteus aculeatus*, from the White Sea basin and the Kamchatka Peninsula, Russia, *Behaviour*, 132, 1173, 1995.
115. Ross, S.T., The systematics of *Gasterosteus aculeatus* (Pisces: Gasterosteidae) in central and southern California, *Contrib. Sci.*, 243, 1, 1973.
116. Hagen, D.W., Isolating mechanisms in threespine sticklebacks (*Gasterosteus*), *J. Fish. Res. Board Can.*, 24, 1637, 1967.
117. Kynard, B., Nest habitat preference of low plate number morphs in threespine sticklebacks (*Gasterosteus aculeatus*), *Copeia*, 1979, 525, 1979.
118. Mayr, E., *Animal Species and Evolution*, Harvard University Press, Cambridge, MA, 1963.
119. Miller, R.R. and Hubbs, C.L., Systematics of *Gasterosteus aculeatus*, with particular reference to intergradation and introgression along the Pacific coast of North America: a commentary on a recent contribution, *Copeia*, 1969, 52, 1969.
120. Hagen, D.W. and McPhail, J.D., The species problem within *Gasterosteus aculeatus* on the Pacific coast of North America, *J. Fish. Res. Board. Can.*, 27, 147, 1970.
121. Bell, M.A., Evolution of phenotypic diversity in *Gasterosteus aculeatus* superspecies on the Pacific coast of North America, *Syst. Zool.*, 25, 211, 1976.
122. Mayr, E., *Principles of Systematic Zoology*, McGraw-Hill, New York, 1969.
123. Bell, M.A., Lateral plate evolution in the threespine stickleback: getting nowhere fast, *Genetica*, 112, 445, 2001.
124. Bell, M.A., Palaeobiology and evolution of the threespine stickleback, in *The Evolutionary Biology of the Threespine Stickleback*, M.A. Bell and Foster, S.A., Eds., Oxford Science Publications, Oxford, 1994, p. 438.
125. Larson, G.L., Social behavior and feeding ability of two phenotypes of *Gasterosteus aculeatus* in relation to their spatial and trophic segregation in a temperate lake, *Can. J. Zool.*, 54, 107, 1976.
126. Bentzen, P. and McPhail, J.D., Ecology and evolution of sympatric sticklebacks (*Gasterosteus*) — specialization for alternative trophic niches in the Enos Lake species pair, *Can. J. Zool.*, 62, 2280, 1984.
127. McPhail, J.D., Ecology and evolution of sympatric sticklebacks (*Gasterosteus*) — morphological and genetic evidence for a species pair in Enos Lake, British Columbia, *Can. J. Zool.*, 62, 1402, 1984.
128. McPhail, J.D., Ecology and evolution of sympatric sticklebacks (*Gasterosteus*) — evidence for a species-pair in Paxton Lake, Texada Island, British Columbia, *Can. J. Zool.*, 70, 361, 1992.

129. Lavin, P.A. and Mcphail, J.D., The evolution of freshwater diversity in the threespine stickleback (*Gasterosteus aculeatus*) — site-specific differentiation of trophic morphology, *Can. J. Zool.*, 63, 2632, 1985.
130. Lavin, P.A. and McPhail, J.D., Adaptive divergence of trophic phenotype among freshwater populations of the threespine stickleback (*Gasterosteus aculeatus*), *Can. J. Fish. Aquat. Sci.*, 43, 2455, 1986.
131. Schluter, D. and McPhail, J.D., Ecological character displacement and speciation in sticklebacks, *Am. Nat.*, 140, 85, 1992.
132. Withler, R.E. and McPhail, J.D., Genetic variability in freshwater and anadromous sticklebacks (*Gasterosteus aculeatus*) of southern British Columbia, *Can. J. Zool.*, 63, 528, 1985.
133. Taylor, E.B. and McPhail, J.D., Historical contingency and ecological determinism interact to prime speciation in sticklebacks, *Gasterosteus*, *Proc. R. Soc. Lond. B*, 267, 2375, 2000.
134. Taylor, E.B. and McPhail, J.D., Evolutionary history of an adaptive radiation in species pairs of threespine sticklebacks (*Gasterosteus*): insights from mitochondrial DNA, *Biol. J. Linn. Soc.*, 66, 271, 1999.
135. Ridgway, M.S. and McPhail, J.D., Ecology and evolution of sympatric sticklebacks (*Gasterosteus*) — mate choice and reproductive isolation in the Enos Lake species pair, *Can. J. Zool.*, 62, 1813, 1984.
136. Nagel, L. and Schluter, D., Body size, natural selection, and speciation in sticklebacks, *Evolution*, 52, 209, 1998.
137. Rundle, H.D. and Schluter, D., Reinforcement of stickleback mate preferences: sympatry breeds contempt, *Evolution*, 52, 200, 1998.
138. Lavin, P.A. and Mcphail, J.D., Morphological divergence and the organization of trophic characters among lacustrine populations of the threespine stickleback (*Gasterosteus aculeatus*), *Can. J. Fish. Aquat. Sci.*, 44, 1820, 1987.
140. Hatfield, T. and Schluter, D., A test for sexual selection on hybrids of two sympatric sticklebacks, *Evolution*, 50, 2429, 1996.
140. Schluter, D., Adaptive radiation in sticklebacks — trade-offs in feeding performance and growth, *Ecology*, 76, 82, 1995.
141. Vamosi, S.M., Hatfield, T., and Schluter, D., A test of ecological selection against young-of-the-year hybrids of sympatric sticklebacks, *J. Fish Biol.*, 57, 109, 2000.
142. Brooks, D.R. and McLennan, D.A., Species: turning a conundrum into a research program, *J. Nematol.*, 31, 117, 1999.
143. Kraak, S.B.M., Mundwiler, B., and Hart, P.J.B., Increased number of hybrids between benthic and limnetic three-spined sticklebacks in Enos Lake, Canada; the collapse of a species pair? *J. Fish Biol.*, 58, 1458, 2001.
144. McPhail, J.D., Predation and evolution of a stickleback (*Gasterosteus*), *J. Fish. Res. Board. Can.*, 26, 3183, 1969.
145. Semler, D.E., Some aspects of adaptation in a polymorphism for breeding colours in threespine stickleback (*Gasterosteus aculeatus*), *J. Zool.*, 165, 291, 1971.
146. Moodie, G.E.E., Morphology, life-history, and ecology of an unusual stickleback (*Gasterosteus aculeatus*) in Queen-Charlotte Islands, Canada, *Can. J. Zool.*, 50, 721, 1972.
147. Bell, M.A., Differentiation of adjacent stream populations of threespine sticklebacks, *Evolution*, 36, 189, 1982.
148. Hagen, D.W., Moodie, G.E.E., and Moodie, P.F., Territoriality and courtship in the Olympic mudminnow (Novumbra hubbsi), *Can. J. Zool.*, 50, 1111, 1972.

149. Hagen, D.W. and Moodie, G.E.E., Polymorphism for breeding colors in *Gasterosteus aculeatus*. I. Their genetics and geographic distribution, *Evolution*, 33, 641, 1979.
150. Reimchen, T.E., Loss of nuptial color in threespine sticklebacks (*Gasterosteus aculeatus*), *Evolution*, 43, 450, 1989.
151. Scott, R.J. and Foster, S.A., Field data do not support a textbook example of convergent character displacement, *Proc. R. Soc. Lond. B*, 267, 607, 2000.
152. Scott, R.J., Sensory drive and nuptial colour loss in the three-spined stickleback, *J. Fish Biol.*, 59, 1520, 2001.
153. Scott, R.J., Assortative mating between adjacent populations of threespine stickleback (*Gasterosteus aculeatus*), *Ecol. Freshwater Fish.*, 13, 1, 2004.
154. Blouw, D.M. and Hagen, D.W., The adaptive significance of dorsal spine variation in the fourspine stickleback, *Apeltes quadracus*. 3. Correlated traits and experimental evidence on predation, *Heredity*, 53, 371, 1984.
155. Jamieson, I.G., Blouw, D.M., and Colgan, P.W., Parental care as a constraint on male mating success in fishes — a comparative study of threespine and white sticklebacks, *Can. J. Zool.*, 70, 956, 1992.
156. Jamieson, I.G., Blouw, D.M., and Colgan, P.W., Field observations on the reproductive biology of a newly discovered stickleback (*Gasterosteus*), *Can. J. Zool.*, 70, 1057, 1992.
157. Blouw, D.M., Evolution of offspring desertion in a stickleback fish, *Ecoscience*, 3, 18, 1996.
158. Paepke, H.-J., *Die Stichlinge*, Vol. 10, Westarp Wissenschaften, Magdeburg, 1996.
159. Haglund, T.R., Buth, D.G., and Blouw, D.M., Allozyme variation and the recognition of the white stickleback, *Biochem. Syst. Ecol.*, 18, 559, 1990.
160. Orti, G. et al., Global survey of mitochondrial DNA sequences in the threespine stickleback — evidence for recent migrations, *Evolution*, 48, 608, 1994.
161. Haglund, T.R., Buth, D.G., and Lawson, R., Allozyme variation and phylogenetic-relationships of Asian, North-American, and European populations of the threespine stickleback, *Gasterosteus aculeatus*, *Copeia*, 432, 1992.
162. Yamada, M., Higuchi, M., and Goto, A., Extensive introgression of mitochondrial DNA found between two genetically divergent forms of threespine stickleback, *Gasterosteus aculeatus*, around Japan, *Environ. Biol. Fish.*, 61, 269, 2001.
163. Linnaeus, C., *Systema Naturae*, 10th ed., Vol. Pisces in Vol. 1, 1758.
164. Walbaum, J.J., *Petri Artedi sueci genera Piscium in quibus systema totum ichthyologiae proponitur cum classibus, ordinibus, generum characteribus, specierum differentiis, observationibus plurimis*. Redactis speciebus 242 ad genera 52. Ichthyologiae Pars III. Ant. Ferdin. Rose, Grypeswaldiae [Greifswald]. Artedi Piscium Pt. 3 [i-viii] and 1-723, Pls. 1-3.
165. Jordan, D.S., *Manual of the Vertebrates of the Northern United States, Including the District East of the Mississippi River and North of North Carolina and Tennessee, Exclusive of Marine Species*, Chicago, IL, Jansen, McClurg and Company, 1876.
166. Leiner, M., Die drei europaischen Stichlinge (*Gasterosteus aculeatus*, *Gasterosteus pungitius*, und *Gasterosteus spinachia*) un ihre Kreuzungsprodukte, *Z. Morphol. Okol. Tiere*, 28, 107, 1934.
167. Hall, M.F., A Comparative Study of the Reproductive Behaviour of the Sticklebacks (Gasterosteidae), D.Phil. thesis, Department of Zoology, Oxford, 1956.
168. Reisman, H.M. and Cade, T.J., Physiological and behavioral aspects of reproduction in brook stickleback *Culaea inconstans*, *Am. Midl. Nat.*, 77, 257, 1967.
169. Chen, T.R. and Reisman, H.M., A comparative chromosome study of the North American species of sticklebacks (Teleostei: Gasterosteidae), *Cytogenetics*, 9, 321, 1970.

170. Nelson, J.S., Comparison of the pectoral and pelvic skeletons and of some other bones and their phylogenetic implications in the Aulorhynchidae and Gasterosteidae (Pisces), *J. Fish. Res. Board. Can.*, 28, 427, 1971.
171. Paepke, H.-J., *Die Stichlinge*, Vol. 10, A. Ziemsen Verlag, Wittenberg, Lutherstadt, 1983.
172. Hudon, J. and Guderley, H., An electrophoretic study of the phylogenetic relationships among four species of sticklebacks (Pisces: Gasterosteidae), *Can. J. Zool.*, 62, 2313, 1984.
173. Bowne, P.S., The Systematic Position of Gasterosteiformes, Ph.D. thesis, University of Alberta, Edmonton, 1985.
174. McLennan, D.A., Brooks, D.R., and McPhail, J.D., The benefits of communication between comparative ethology and phylogenetic systematics: a case study using gasterosteid fishes, *Can. J. Zool.*, 66, 2177, 1988.
175. McLennan, D.A., Phylogenetic relationships in the Gasterosteidae: an updated tree based on behavioral characters with a discussion of homoplasy, *Copeia*, 318, 1993.
176. Bowne, P.S., Systematics and morphologu of the Gasterosteiformes, in *The Evolutionary Biology of the Threespine Stickleback*, Bell, M.A. and Foster, S.A., Eds., Oxford Science Publications, Oxford, 1994, p. 28.
177. McLennan, D.A. and Mattern, M.Y., The phylogeny of the Gasterosteidae: combining behavioral and morphological data sets, *Cladistics*, 17, 11, 2001.
178. Mattern, M.Y., Molecular phylogeny of the Gasterosteidae: the importance of using multiple genes, *Mol. Phylogenet. Evol.*, 30, 366, 2004.
179. Mattern, M.Y. and McLennan, D.A., Total evidence phylogeny of Gasterosteidae: combining molecular, morphological and behavioral data, *Cladistics*, 20, 14, 2004.

APPENDIX: SYNONYMY

Spinachia Cuvier 1816

Etymology — From the Latin *spina* meaning "spine" or "prickle"
Spinachia Cuvier 1816: 320
Gastraea Sauvage 1874: 7

Spinachia spinachia (Linnaeus 1758)

Etymology — From the Latin *spina* meaning "spine" or "prickle"
Common name — Sea stickleback, 15-spined stickleback
Distribution — Coastal seas in northwestern Europe
Gasterosteus spinachia Linnaeus 1758: 296, Europe
Spinachia spinachia Cuvier 1816: 320, Baltic and northern seas
Spinachia vulgaris Fleming 1828: 219, English seas
Aulostoma polycanthus spinachia Swainson 1839: 175, no locality
Gasterosteus marinus Gronow (in Gray) 1854: 168, northern seas
Gastraea spinachia Sauvage 1874: 7, North pole, Berghem, Norway and Brest, Morlaix, and Bretagne, France (also erroneously identified from Newfoundland, Canada)
Spinachia linnei Malm 1877: 373, Sweden

Apeltes Dekay 1842

Etymology — From the Greek α meaning "lack of" and πελτη meaning "shield"
Apeltes Dekay 1842: 67

Apeltes quadracus (Mitchell 1815)

Etymology — From the Latin *quatuor* meaning "four" and *acus* meaning "needles" or "pins"
Common name — Four-spined stickleback
Distribution — Eastern coast of North America from the Gaspé Basin of Quebec to Virginia; also several freshwater lakes (see text for details)
Gasterosteus quadracus Mitchell 1815: 430, New York, U.S.
Gasterosteus apeltes Cuvier and Valenciennes 1829: 505, no locality
Apeltes quadracus Dekay 1842: 67, no locality
Gasterosteus millepunctatus Ayres 1842: 294, New York and Connecticut, U.S.

Culaea Whitley 1950

Etymology — Coined name to replace *Eucalia* [from the Greek ευ meaning "well" and καλια meaning "nest"]
Eucalia Jordan 1876: 248 (name already used for genus of butterfly)
Culaea Whitley 1950: 44, replacement name

Culaea inconstans (Kirtland 1840)

Etymology — From the Latin *inconstans* meaning "variable"
Common name — Brook stickleback, five-spined stickleback
Distribution — Bodies of freshwater in North America from the east coast to the Rockies and the Mackenzie River drainage south to Nebraska
Gasterosteus inconstans Kirtland 1840: 273, Mahoning County, Ohio, U.S. (usually cited as 1841 but the publication date is actually 1840 – N. W. Eschmeyer, Catalog of Fishes)
Gasterosteus pygmaeus Agassiz 1850: 314, Michipicotin on the northeast shore of Lake Superior, Ontario, Canada
Gasterosteus gymnetes Dawson 1859: 321, Montreal, Canada
Gasterosteus micropus Cope 1865: 81, Platte River near Fort. Riley, Kansas, U.S.
Eucalia inconstans Jordan 1876: 248, no locality
Eucalia inconstans cayuga Jordan 1876: 259, Cayuga Lake, Ithaca, New York, U.S.
Eucalia inconstans pygmaea Jordan 1876: 249, Lake Superior, Ontario, Canada

Pungitius Coste 1848

Etymology — From the Latin *pungi* meaning "prick," "puncture," or "mark with points"

Pungitius Coste 1848: 588

Pygosteus Gill (ex Brevoort) 1861: 39

Gasterostea Sauvage 1874: 7, 29

Pungitius pungitius (Linnaeus 1758)

Etymology — From the Latin *pungi* meaning "prick," "puncture," or "mark with points"

Common name — Nine-spined stickleback, ten-spined stickleback

Distribution — Atlantic, Arctic and Pacific coasts, and inland waters of Europe, Asia, and Japan

Gasterosteus pungitius Linnaeus 1758: 296, Europe

Gasterosteus laevis Cuvier 1829: 170, Somme and Bobigny near Paris, France

Gasterosteus vulgaris Mauduyt 1849–1851, La Verge near Poitiers, Bergue near Gençay, les Aiffes near St-Maurice, and Séguinièrer near St-Julien-Lars, Vienne, France

Gasterosteus lotharingus Blanchard 1866: 242–244, Lotharingen, Meuse River, France

Gasterosteus burgundianus Blanchard 1866: 244, Burgundy, France

Gasterosteus breviceps Blanchard 1866: 245, Caen, Normandy, France

Gasterostea pungitia Sauvage 1874: 29, Lille, Oise, and Abbeville, France and Westfalia, Germany

Gasterosteus burgundiana Sauvage 1874: 30, Dijon, France

Gasterostea laevis Sauvage 1874: 34, Bourg d'Ault, Oise, Seine, Bobigny, and Sarthe, France

Gasterostea lotharingus Sauvage 1874: 34, Meuse R. near St-Mihiel, France

Gasterostea breviceps Sauvage 1874: 34, Caen and Anjou, France

Gasterostea globiceps Sauvage 1874: 35, North America (Eschmeyer in Catalogue of Fishes notes this may be erroneous locality)

Gasterosteus sternus Kessler 1876: 6, Hu-lun Lake (Dalai-nor), Mongolia, China

Gasterosteus wosnesenjenskyi Kessler 1876: 9, west coast of Kamchatka, Russia

Gasterostea pungitia var. *laevis* Moreau 1881: 170, no locality

Gasterostea pungitia var. *breviceps* Moreau 1881: 171, no locality

Pygosteus pungitius Berg 1907: 451, Kamchatka, Russia, Alaska, U.S.

Pygosteus pungitius var. *hologymna* Bertin 1925: 122, no locality

Pygosteus pungitius var. *trachura* Bertin 1925: 122, no locality

Pygosteus pungitius var. *semiarmata* Bertin 1925: 122, no locality

Pygosteus pungitius var. *carinata* Bertin 1925: 164, no locality
Pungitius pungitius Berg 1932: 169, Europe
Pungitius pungitius pungitius Keivany and Nelson 2000: 117, Atlantic, Arctic, and Pacific coasts, and inland waters of Europe, Asia, and Japan
Pungitius pungitius laevis Keivany and Nelson 2000: 118, Ireland, southern England, and southern France

Pungitius hellenicus Stephanidis 1971

Etymology — Ελληνικσς after the Greek distribution of this fish
Common name — Greek nine-spined stickleback
Distribution — Sperchios drainage system, Greece
Pygosteus pungitius (non Linnaeus, 1758), Stephanidis 1943: 200–210, Sperchios Valley, Greece
Pungitius pungitius hellenicus Stephanidis 1971: 228–231, Sperchios Valley, Greece
Pungitius platygaster hellenicus Paepke 1983: 57–58, Sperchios drainage system, Greece
Pungitius hellenicus Paepke 1996: 75, Spechios drainage, Greece

Pungitius platygaster (Kessler 1859)

Etymology — From the Greek ρλατψ meaning "flat" and γαστηρ meaning "belly"
Common name — Ukrainian stickleback
Distribution — Black Sea, Sea of Azov, Aral Sea, and Caspian Sea basins
Gasterosteus pungitius (non Linnaeus, 1758), Nordmann 1840: 379–382, Crimea and Black Sea, Ukraine
Gasterosteus platygaster Kessler 1859: 202, Odessa and Aleshki on the Dnepr River, Ukraine
Gasterosteus pungitius var. *kessleri* Yakovlev 1870: 110, Astrakhan and Caspian Sea, Russia
Gasterosteus pungitius var. *niger* Yakovlev 1870: 110, Astrakhan and Caspian Sea, Russia
Gasterosteus platygaster var. *caucasias* Kessler 1877: 3, Transcaucus, Russia and Georgia
Gasterosteus platygaster var. *aralensis* Kessler 1877: 4, Aral Sea and Amu Darya River, Uzbekistan
Gasterosteus platygaster var. *danubica* Steindachner 1899: 542, Danube and Sava Rivers near Belgrade, Serbia
Pygosteus platygaster var. *nuda* Berg 1905: 218, Lake Charkal, Russia
Pygosteus platygaster var. *aralensis* Berg 1905: 218, Aral Sea
Pygosteus nudus Berg 1916 Lake Charkal, Ural River basin, Russia
Pungitius platygaster platygaster Paepke 1996: 73, Caspian, Azov, and Black seas

Pungitius platygaster aralensis Paepke 1996: 74, Aral Sea
Pungitius platygaster Keivany and Nelson 2000: 116, Black Sea, Sea of Azov, Aral Sea, and Caspian Sea basins

Pungitius tymensis (Nikolskii 1889)

Etymology — After the River Tym on Sakhalin Island, Russia
Common name — Sakhalin nine-spined stickleback
Distribution — Sakhalin Island, Russia and Hokkaido Island, Japan.
Gasterosteus tymensis Nikolskii 1889: 293, Tym River on Sakhalin Island, Russia
Pygosteus undecimalis Jordan and Starks 1902: 62, Chitose and Hokkaido, Japan
Pygosteus tymensis Berg 1907: 452, Sakhalin Island, Russia
Pungitius tymensis Berg 1949: 968, Sakhalin Island, Russia and Hokkaido, Japan

Pungitius occidentalis (Cuvier 1829)

Etymology — From the Latin *occidentalis* meaning "from the west"
Common name — North American nine-spined stickleback
Distribution — North America along the northern coastline from Cook Inlet, east of Aleutian Islands Alaska to New Jersey, U.S.; penetrates inland from Fort Nelson, British Columbia to western Quebec, and extends south to Minnesota and northern Indiana, U.S.
Gasterosteus occidentalis Cuvier 1829: 509, Newfoundland, Canada
Gasterosteus concinnus Richardson 1836: 57, Great Bear Lake and Saskatchewan River, Canada
Gasterosteus mainensis Storer 1837: 465, Kennebeck County, Maine, U.S.
Gasterosteus dekayi Agassiz 1850: 311, Lake Superior, Ontario, Canada
Gasterosteus nebulosus Agassiz 1850: 310–314, Lake Superior, Ontario, Canada
Gasterostea blanchardi Sauvage 1874: 32, New York, U.S.
Gasterostea occidentalis Sauvage 1874: 30–31, North America
Gasterostea dekayi Sauvage 1874: 31 New York, U.S.
Gasterostea mainensis Sauvage 1874: 33, Maine, U.S.
Gasterostea concinnas Sauvage 1874: 35, Saskatchewan to Great Bear Lake, Canada
Gasterosteus pungitius brachypoda Bean 1879: 129, Oosooadlin Mountain, Cumberland Gold, Greenland
Pungitius pungitius occidentalis Keivany and Nelson 2000: 118, North America along the northern coastline from Cook Inlet east of Aleutian Islands Alaska to New Jersey, U.S.; penetrates inland from Fort Nelson, British Columbia to western Quebec and extends south to Minnesota and northern Indiana, U.S.

Pungitius sinensis (Guichenot 1869)

Etymology — From the Latin for Chinese, *sino*
Common name — Chinese nine-spined stickleback
Distribution — China, Eastern Russia, Mongolia, Korea, and Northern Honshu Island and Hokkaido Island, Japan

Gasterosteus sinensis Guichenot 1869: 204, Yangtze River, Jiangxi Province, China

Gasterostea sinensis Sauvage 1874: 33, China

Pygosteus stenurus Kessler 1876: 6, Dalai-nor near Bost Mongolia

Gasterosteus wosnesenjenskyi Kessler 1876: 9, west coast of Kamtschatka, Russia

Gaterosteus japonicus Steindacher 1881: 264, Gulf of Strielok near Vladivostok, Russia, Sea of Japan (junior subjective homonym of *Gasteroteus japonicus* Houttuyn, 1782: 329; a Monocentridae)

Gasterosteus bussei Warpakchow 1887, Ilistaya River and Khanka Lake, Russia

Pygosteus steindachner Jordan and Synder 1901: 747 (replacement name for *Gasterosteus japonicus* Steindachner 1881)

Pungitius brevispinosus Otaki 1908: 87, Sapporo, Hokkaido, Japan

Pygosteus kaibarae Tanaka 1915: 565, Kichisho-in, southwest of Kyoto, Japan

Pungitius pungitius sinensis Berg 1949: 967, Eastern Manchuria, south to the Yangtze River, Northeast Korea, Hondo and Hokkaido, Sachalin Island, Primorsky region, Russia, Amur River drainage, Dalai-nor, Kuril Islands, and Sea of Okhotsk

Gasterosteus Linnaeus 1758

Etymology — From the Greek γαστηρ meaning "belly" and οστιυος meaning "made of bone"

Gasterosteus Linnaeus 1758: 295

Gasteracanthus Pallas 1814: 229

Gladiunculus Hubbs 1929: 1-9

Gasterosteus wheatlandi Putnam 1867

Etymology — Named after R. H. Wheatlandi who in 1859, collected the first specimens from Nahant, Massachusetts, U.S.
Common names — Black-spotted stickleback
Distribution — East coast of North America from Newfoundland to Long Island, New York

Gasterosteus wheatlandi Putnam 1867: 4, Nahant, Massachusetts, U.S.

Gasterosteus gladiunculus Kendall 1896: 623, Coast of Maine, U.S.

Gladiunculus wheatlandi Hubbs 1929: 2

Gasterosteus aculeatus Linnaeus 1758

Etymology — From the Latin *aculeatus* meaning "spiny" or "prickly"
Common name — Threespine or three-spined stickleback
Distribution — Atlantic, Arctic and Pacific coastal waters and inland waters of Europe, North America, and Asia

Gasterosteus aculeatus Linnaeus 1758: 295
Gasterosteus bispinosus Walbaum 1792: 450, North America
Gasterosteus teraculeatus Lacepède 1801: 295, no locality
Gasterosteus biaculeatus Shaw 1803: 608, New York, U.S. (Identified as *G. wheatlandi* by Jordan and Evermann 1896)
Gasteracanthus cataphractus Pallas 1814: 229, Kamtschatka, Russia
Gasterosteus trachurus Cuvier and Valenciennes 1829: 481, Europe
Gasterosteus leiurus Cuvier and Valenciennes 1829: 481, Europe
Gasterosteus gymnurus Cuvier and Valenciennes 1829: 170, England and France
Gasterosteus semiarmatus Cuvier and Valenciennes 1829: 493, Havre and Braie River near Abbeville, France
Gasterosteus semiloricatus Cuvier and Valenciennes 1829: 494, Somme, Oise, Rochelle, the coast of Normandy, Caen, Hable d'Ault, France and the environs of Berlin, Germany
Gasterosteus argyropomus Cuvier and Valenciennes 1829: 498, Florence, Italy
Gasterosteus brachycentrus Cuvier and Valenciennes 1829: 499, Florence, Italy
Gasterosteus tetracanthus Cuvier and Valenciennes 1829: 499, Florence, Italy
Gasterosteus obolarius Cuvier and Valenciennes 1829: 500, Kamtschatka, Russia
Gasterosteus noveboracensis Cuvier and Valenciennes 1829: 502, New York, U.S.
Gasterosteus niger Cuvier and Valenciennes 1829: 503, Newfoundland, Canada
Gasterosteus spinulosus Jenyns 1835, Edinburgh, Scotland
Gasterosteus loricatus Reinhardt 1837: 114, Greenland
Gasterosteus dimidiatus Reinhardt 1837: 114, Greenland
Gasterosteus ponticus Nordmann 1840: 380, Black Sea, Tauria, Ukraine
Gasterosteus biarmatus Nordmann 1840: 381, Tarkanckut, Crimea, Ukraine
Gasterosteus neoboracensis Dekay 1842: 66, New York, U.S.
Gasterosteus quadrispinosa Crespon 1844, Nîmes, France
Gasterosteus nemausensis Crespon 1844, Nîmes, France
Gasterosteus cuvieri Girard (in Storer) 1850, Labrador, Canada
Gasterosteus williamsoni Girard 1854: 133, Santa Clara River, California, U.S.

Gasterosteus microcephalus Girard 1854: 133, Four Creek, inlet of Tule Lake, California, U.S.

Gasterosteus plebius Girard 1854: 147, Presidio, Petaluma, California, U.S.

Gasterosteus inopinatus Girard 1854: 147, Presidio, California, U.S.

Gasterosteus insculptus Richardson 1855: 356, Northumberland Sound, Greenland

Gasterosteus serratus Ayres 1855: 2, San Francisco, California, U.S.

Gasterosteus dekayi Ayres 1855: 48, no locality

Gasterosteus intermedius Girard 1856: 135, Cape Flattery, Washington, U.S.

Gasterosteus pugetti Girard 1856: 135, Fort Steilacoom, Puget Sound, Washington, U.S.

Gasterosteus neustrianus Blanchard 1866: 220, Hafleur and Gournay, France

Gasterosteus bailloni Blanchard 1866: 231, Abbeville, France

Gasterosteus argentatissimus Blanchard 1866: 232, Avignon, la Sorgue, France

Gasterosteus elegans Blanchard 1866: 234, Cadillac and Langan, Gironde, France

Gasterosteus islandicus Sauvage 1874: 20, Iceland

Gasterosteus blanchardi Sauvage 1874: 32, New York, U.S.

Gasterosteus suppositus Sauvage 1874: 11, New York, U.S.

Gasterosteus texanus Sauvage 1874: 15, Texas, U.S.

Gasterosteus algeriensis Sauvage 1874: 17, Algeria

Gasterosteus atkinsii Bean 1879: 67, Schoodic Lake, Maine, U.S.

Gasterosteus bispinosus cuvieri (Girard) Jordan and Evermann 1896–1900: 749

Gasterosteus hologymnus Regan 1909: 435, Rome, Italy

Gasterosteus santaeannae Regan 1909: 437, Santa Anna River, California, U.S.

Gasterosteus williamsoni japonicus Franz 1910: 19, Misaki, Japan

Gasterosteus bispinosus johanseni Cox 1923: 147, Westbay, Port au Port Bay, Newfoundland, Canada

Gasterosteus aculeatus messinicus Stephanidis 1971: 202 Pamissos River near Messini, Greece

2 The Molecular Genetics of Evolutionary Change in Sticklebacks

David M. Kingsley and Catherine L. Peichel

CONTENTS

2.1 INTRODUCTION: THE GROWTH OF GENOMIC RESOURCES FOR THREE-SPINED STICKLEBACKS

The three-spined stickleback has long been used as a vertebrate model system for studies of behaviour, morphology, physiology, evolution, and ecology. A diverse and active research community has generated thousands of papers and several full-length books detailing the biology of different populations, their remarkable specializations to different environments, the ecological significance of different traits, and the tempo of evolutionary change in both existing and fossil populations.[1–7]

Despite the breadth, depth, and rich history of stickleback research, for many years the fish attracted relatively little attention from molecular biologists. When the first edition of *The Evolutionary Biology of the Threespine Stickleback*[5] was published in 1994, not a single nucleic acid sequence entry existed for *Gasterosteus aculeatus* in the public sequence databases. In the last 10 years, the situation has changed dramatically. Figure 2.1 shows the explosive growth of DNA sequence

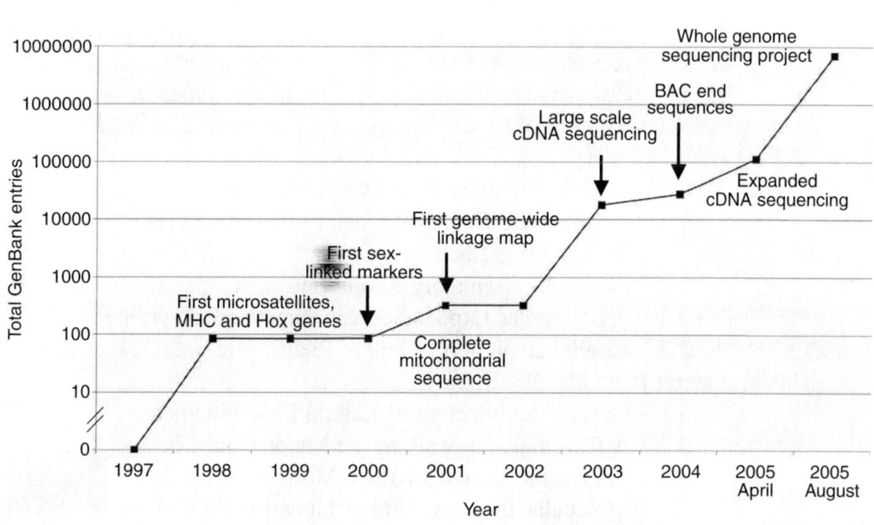

FIGURE 2.1 Exponential growth of sequence data in GenBank for *Gasterosteus aculeatus*. The total number of *Gasterosteus aculeatus* sequence entries in GenBank in each year since 1997, when no entries were present, through 2005, when the whole genome sequencing project was completed, is plotted on a log scale.

information available for *Gasterosteus aculeatus* over time. This progress began with the identification of several highly polymorphic microsatellite repeats,[8–10] followed by isolation of individual genes of particular interest to immunologists and developmental biologists, including members of the MHC[11] and *Hox* gene families.[12] Early studies of mitochondrial control region sequences provided pioneering information on the phylogenetic relationships of populations in different areas.[13–15] The complete sequence of the mitochondrial genome became available in 2001.[16]

The flood of information in recent years (Figure 2.1) reflects a concerted effort to develop *Gasterosteus aculeatus* as a new model system for studying the molecular basis of vertebrate evolution. Distinctive wild populations may show morphological, behavioural, and physiological differences as large as those seen between different species or genera, and reproductive isolation can occur between different forms in the wild. However, the stickleback radiation is so recent that most of the prezygotic isolating mechanisms found in the wild can be overcome using artificial fertilization in the laboratory. This key advantage makes it possible to begin with any trait of interest that differs between populations, establish crosses between distinct forms to generate F1 and F2 offspring, and measure how much of the variation in any given phenotype can be attributed to different chromosome regions inherited from the parents.

In many wild animals, studies of the molecular basis of evolution have been limited to theoretical models or analyzing the sequence and expression patterns of a few genes of particular interest to developmental or systematic biologists. Development of large-scale genetic and genomic resources for sticklebacks will make it possible to carry out an unbiased forward genetic analysis to address many long-standing questions in evolutionary biology. For example: Do evolutionary changes occur through a few genes of large effect or many genes of small effect? Do these genes act in a dominant, recessive, additive, or epistatic manner? Are new traits the result of changes in master regulatory genes long studied by developmental biologists in laboratory animals? Or do major regulatory genes act in so many developmental hierarchies that any mutation within them would likely reduce fitness and viability in natural populations? What types of DNA mutations generate new adaptive traits? Are such changes primarily in the coding regions of proteins or in regulatory sequences that control where and when genes are expressed? Do the key mutations that underlie new traits arise *de novo* following colonisation of new environments, or do these mutations already exist as standing genetic variants within the founding population? Finally, are there many different mechanisms to reach a given phenotypic endpoint, or are particular genetic mechanisms used over and over again when similar traits evolve in widely separated populations?

To begin a forward genetic analysis of evolutionary change in sticklebacks, a large set of highly polymorphic microsatellite markers was developed for monitoring the inheritance of different chromosome regions in crosses between different wild populations. The first genomewide linkage map for *G. aculeatus* showed that many classic armour and trophic traits could be mapped to particular chromosome regions.[17] Based on these encouraging initial results, a proposal was submitted to the U.S. National Institutes of Health (NIH) to develop a large set of genomic resources in sticklebacks to identify the actual genes and mutations responsible for evolutionary change in natural populations. An NIH Centre of Excellence in Genomic Science (CEGS) was established at Stanford in 2002, which has since generated publicly available large-insert genomic libraries, cDNA libraries, a physical map of the stickleback genome, and over 300,000 sequencing reads for stickleback genes and clones.[18] In November 2003, following discussions at the Fourth International Stickleback Conference in Sweden, *G. aculeatus* was nominated as a target for complete genome sequencing to the NIH National Human Genome Research Institute.[19] Sticklebacks were subsequently selected as a high-priority target for sequencing, and a survey to examine levels of polymorphism in various stickleback populations was performed in 2004 to select a population for complete genome sequencing. In the summer of 2005, millions of new sequence reads were generated at one of the major sequencing centres in the U.S. (the Broad Institute, Cambridge, MA). This amount of sequence data is equivalent to every base pair of the stickleback genome having been sequenced an average of six times. The veritable flood of new genetic and genomic information opens up many exciting possibilities for future studies. Here, we will summarise some of the most important new molecular tools now available for sticklebacks and how these tools can be used to study the genetic architecture and molecular basis of evolutionary change in different stickleback populations around the world.

2.2 DEVELOPING A TOOLKIT FOR MOLECULAR ANALYSIS

2.2.1 Expressed Gene Resources

For many years, examining the expression of particular genes of interest in sticklebacks required several months of laboratory work to isolate the *Gasterosteus* homolog of any gene. One of the major advantages of working with established model organisms is the existence of large sets of immediately and widely available clones of expressed genes with known sequences.

Several laboratories have reported the construction of libraries containing cloned copies of messenger RNA molecules (cDNA) that are present at particular developmental stages or in specific adult tissues.[17,20] The Stanford CEGS Centre has prepared cDNA libraries from adult mixed adult tissues, skin, gills, brain, eyes, and whole larval sticklebacks. Large numbers of clones were then randomly isolated from each library and sequenced from both ends.[18] Over 270,000 cDNA sequence entries have been deposited in public sequence databases (Figure 2.1), and the corresponding libraries and individual clones have been archived at Open Biosystems (http://www.openbiosystems.com/) for distribution to interested researchers.

To search for a stickleback cDNA clone for any gene of interest, one simple approach is to use the NCBI BLAST Web site to perform a similarity search with gene sequences already identified in other animals.[21] Paste the nucleic acid or protein sequence of the known gene into the "nucleotide-nucleotide BLAST" page or the "protein query vs. translated" database at the NCBI Web site (http://www.ncbi.nlm.nih.gov/blast/). Set the target database to "EST other" to target the search at large-scale gene collections in nonmammalian organisms. To limit the search to sequences from three-spined sticklebacks, paste the words "*Gasterosteus aculeatus* [Organism]" into the options field labelled "limit by entrez query" on the search page. After submitting the search, you will be taken to a new Web page listing any significant matches in existing databases. Clicking on individual entries will take you to sequence records with more information about the name of the clone, its sequence, and the stickleback tissue of origin. To confirm the identity of the clone, copy and paste the stickleback cDNA sequence into the NCBI BLAST window, search with this sequence against the "nonredundant (nr) database" and examine the top-ranking matches in other well-annotated genomes. The name of the clone can also be used to recover the corresponding sequence of the other end of the clone, and to order the actual cDNA clone from Open Biosystems (http://www.openbiosystems.com/).

Individual cDNA clones can be transcribed and used for *in situ* hybridization experiments to examine the spatial and temporal expression of the corresponding gene during development.[12,22–26] Alternatively, large sets of cDNA clones can be arrayed on glass slides. The microarrayed clones can then be hybridized to RNA samples from particular tissues to develop complex molecular signatures of both gene expression in specific tissues and expression differences between different populations or treatment conditions.[27] Large sets of stickleback cDNA clones have already been provided to Andrew Cossins' laboratory in Liverpool, and to Kevin Chipman's laboratory in Birmingham, U.K. Both labs are generating stickleback microarrays for expression profiling. With the recent completion of a stickleback genome assembly, it will be possible to generate additional microarrays containing the nearly complete set of stickleback transcription units for comprehensive analysis of gene expression changes under a variety of conditions.

2.2.2 Linkage Maps and Genetic Markers

Monitoring the inheritance of different chromosome regions in genetic crosses provides a general method of investigating whether particular chromosome regions are consistently associated with particular phenotypes. The most widely used markers for linkage mapping in sticklebacks are microsatellite markers, which are developed by designing polymerase chain reaction (PCR) primers to sequences that flank small di-, tri-, or tetranucleotide repeat sequences found throughout the genome. These markers are highly polymorphic both within and between populations and can be easily typed using PCR from small amounts of DNA. More than 80% of the markers are typically informative in any three-spined stickleback population or cross.[17] In addition, microsatellite markers are easy to distribute between laboratories, requiring only electronic files listing the sequences of primers that flank particular microsatellites and a description of PCR conditions that can be used to amplify a given locus with the primers. Furthermore, if the same set of microsatellite markers is used to genotype independent crosses, the locations of chromosome regions that underlie trait differences in different populations can be directly compared.

A set of 219 markers originally was ordered into 26 linkage groups.[17] Subsequently, 128 additional markers were added to the stickleback linkage map in the course of studies of particular linkage groups and regions. Twenty-five markers developed by

Rick Taylor's[9] and Theo Bakker's laboratories[10,28] can be found by searching the public NCBI nucleic acid database with the keywords "*Gasterosteus* microsatellite," and primers and reaction conditions for 345 microsatellite markers developed at Stanford can be found by searching the public NCBI nucleic acid database with the keywords "*Gasterosteus* Stn*" (see http://www.ncbi.nlm.nih.gov/entrez/query.fcgi).

The current density of microsatellite markers on the linkage map is approximately one marker every 5 cM. This is a sufficient density, in that virtually 100% of newly typed markers show significant linkage to other markers already present on the map. However, a few gaps must remain in the map, because cytological studies suggest there are 21 chromosomes,[29] and the initial microsatellite linkage map shows 26 stickleback linkage groups.[17]

Research is under way to increase the overall number and density of both microsatellite and other genetic markers in sticklebacks. An initial sequence survey identified 24 microsatellites in approximately 192 kb of random genomic sequence, suggesting that the three-spined stickleback genome of 670 Mb,[30,31] is likely to contain over 80,000 microsatellite sequences.[17] The abundance of microsatellites provides a simple means of developing new genetic markers in any region. For example, Shapiro et al.[25] identified new microsatellite sequences in stickleback genomic clones for several genes known to be involved in hindlimb development (*Pitx1*, *Pitx2*, and *Tbx4*). Mapping of the new microsatellites provided a simple way to compare the genetic location of candidate genes with the positions of the major chromosome regions already implicated in pelvic reduction by genetic mapping studies.[25] Colosimo et al.[32] used contiguous sequence information from genomic clones around the stickleback lateral plate locus to develop new microsatellite markers located every 10 kb throughout a 400-kb region. The high density of markers was key to identifying a much smaller region in linkage disequilibrium with the major locus controlling lateral plate morph.[32]

In human and mouse genetics, markers based on single base pair differences within a defined region are now being widely used in addition to microsatellite markers.[33] Single-nucleotide polymorphism (SNP) markers are even more abundant in stickleback genomic DNA than microsatellites, occurring at an average frequency of approximately one base pair difference every 1000 bp in limited initial surveys. SNP markers are somewhat more difficult to type than microsatellites and often have lower rates of heterozygosity in populations and crosses. However, they can be designed to virtually any sequence and are less prone to mutation than microsatellites, and so they are particularly useful for association mapping, haplotype, and phylogenetic studies.[34–37] As this chapter is being written, 2000 SNP-based genetic markers are being developed by the Stanford CEGS Center. To generate an SNP marker, PCR primers are designed to amplify either the 3 end of an expressed gene sequence from the cDNA project or a defined genomic fragment from large-scale genomic sequencing. The primers are used first to amplify and sequence the same region from multiple individuals to identify useful SNPs, and then to type a given SNP on progeny from genetic crosses. All 2000 SNPs are being typed on a single mapping panel to compare their segregation patterns with each other and with commonly used microsatellite markers. This project will increase the number of genetic markers on the *Gasterosteus aculeatus* linkage map by approximately tenfold by the time this book is published. The high density of markers will make it possible to compare the large-scale arrangement of genes in sticklebacks with their arrangement in other organisms and will also provide a key tool for long-range sequence assembly of the stickleback genome. In addition, the dense set of markers will provide ready entry points for detailed studies of particular chromosome regions.

The existing linkage map of sticklebacks has already made it possible to map chromosome regions that influence a large number of interesting traits related to skeletal armour, feeding adaptations, respiration, and sex determination.[17,25,38–41] Each chromosome region typically controls only a fraction of the phenotypic variance in a given trait. However, this fraction can vary anywhere from 6 to 100%, and some chromosome regions clearly act as nearly Mendelian factors (Table 2.1). These results suggest that a relatively small number of chromosome regions can control a substantial percentage of the variance in many interesting traits. Furthermore, these results are consistent with recent theoretical studies, and with experimental results from crosses in other animals.[42–44] Of course, chromosome regions with substantial phenotypic effects may actually harbour multiple genes or molecular changes that contribute to the overall effects seen. Nonetheless, the mapping results in sticklebacks thus far raise the exciting possibility that in many cases it eventually will be possible to track evolutionary differences to particular chromosome regions, genes, and mutations.

2.2.3 Physical Maps

Linkage maps provide a rapid means of detecting the number and location of chromosome regions that have a significant effect on a given trait. To identify the genes within a chromosome region of interest, methods are needed for recovering and analyzing the actual DNA sequences within that region. Given the total size of the stickleback genetic map (approximately 1500 cM), and the estimated size of the three-spined stickleback genome (670 Mb),[30,31] one centimorgan on the genetic map (corresponding to a 1% frequency of recombination between markers during meiosis) will typically contain approximately 500,000 base pairs of DNA.

Several DNA libraries have been made that can be used to clone and study large fragments of DNA from particular fish populations.[18,24,45] Libraries made in bacterial artificial chromosome (BAC) vectors typically have individual cloned fragments of the stickleback genome that range between 100,000 and 200,000 base pairs. Although the sizes of these clones are still substantially smaller than the average distance between markers on the stickleback genetic map, sets of overlapping clones can be assembled into larger sets that together span a larger physical region.

A first-generation physical map of the stickleback genome has been built by restriction mapping over 100,000 individual clones from a BAC library made from anadromous fish from the Salmon River in British Columbia.[18] A computer-based comparison of the internal fragments contained in each clone was used to detect overlaps between clones. This method automatically assembles individual clones into larger contiguous sets (called *contigs*). Contig data for over 100,000 stickleback BAC clones can now be searched using a freely available software tool called Internet Contig Explorer[46] (http://www.bcgsc.ca/ice/). Displayed information includes the

TABLE 2.1
Location and Magnitude of Effect for QTL Mapped to Date in Three-Spined Sticklebacks

Trait	Cross	LG	Marker	LOD	PVE (%)
Sex	Paxton[a]	19	*Idh*	N/A	Mendelian
	Priest[a]	19	*Idh*	N/A	Mendelian
Complete vs. reduced pelvis	Paxton[b]	7	*Pitx1*	72.6	83.0
	Alaska[c]	7	*Stn82*	N/A	Mendelian
Pelvic spine length	Paxton[b]	7	*Pitx1*	82.8	65.3
		2	*Stn21*	4.9	7.6
		4	*Gac4174*	4.9	5.8
	Priest[d]	8	*Stn94*	4.5	24.6
Pelvic girdle length	Paxton[b]	7	*Pitx1*	50.0	46.8
		1	*Stn7*	4.6	5.6
		2	*Stn21*	7.6	11.1
		4	*Gac4174*	4.7	5.6
Ascending branch height	Paxton[b]	7	*Pitx1*	45.1	44.5
		10	*Stn119*	5.3	6.6
Asymmetry	Paxton[b]	7	*Pitx1*	28.0	31.5
Lateral plate morph	Paxton[e]	4	*Gac4174*	116.9	77.6
	Friant, CA[e]	4	*Gac4174*	N/A	Mendelian
	Alaska[c]	4*	*Stn183*	N/A	Mendelian
Lateral plate number (Aa)	Paxton[e]	7	*Stn71*	4.6	11.3
		10	*Stn211*	8.7	20.3
		26	*Stn219*	12.8	28.6
Lateral plate number (aa)	Paxton[e]	26	*Stn218*	5.2	23.2
	Priest[d]	13	*Stn152*	5.5	25.6
		26	*Stn208*	4.6	22.2
Lateral plate width	Paxton[e]	4	*Gac4174*	10.7	12.9
		7	*Stn71*	9.2	11.1
		25	*Gac1125*	26.3	28.9
Lateral plate height	Paxton[e]	4	*Gac4174*	7.4	11.5
		7	*Stn71*	6.1	10.3
		25	*Gac1125*	10.9	17.9
Dorsal spine 1 length	Priest[d]	1	*Stn9*	4.7	20.9
		2	*Stn26*	3.6	17.2
Dorsal spine 2 length	Priest[d]	8	*Stn96*	4.5	22.4
		11	*Stn130*	3.4	17.0
Gill raker number	Priest[d]	11	*Stn131*	5.5	25.5
		16	*Stn178*	6.8	37.4
Opercle shape/size	Alaska[f]	19		6	30

TABLE 2.1 (CONTINUED)
Location and Magnitude of Effect for QTL Mapped to Date in Three-Spined Sticklebacks

Note: For each of the published stickleback QTL, the linkage group (LG), the name of the most closely linked marker, the LOD score at the most closely linked marker, and the percentage variance explained (PVE) by the QTL are indicated. *Stn183 is mapped to LG18 in Cresko et al. 2004 but coassembles with LG4 in Colosimo et al. 2004.

Sources: Superscripts indicate the publications from which the data were taken: [a]Peichel, C.L. et al., *Current Biology,* 14(16), 1416, 2004; [b]Shapiro, M.D. et al., *Nature,* 428(6984), 717, 2004; [c]Cresko, W.A. et al., *Proceedings of the National Academic Sciences U.S.A.,* 101, 6050, 2004; [d]Peichel, C.L. et al., *Nature,* 414(6866), 901, 200; [e]Colosimo, P.F. et al., *PLoS Biology,* 2, 635, 2004; [f]Kimmel, C.B. et al., *Proceedings of the National Academy of Sciences of the United States of America,* 102, 5791, 2005. All with permission.

names, sizes, and estimated overlaps of all clones in a given contig, and the restriction fragment digestion pattern of each individual clone.

High-density screening filters for BAC libraries from both marine and freshwater populations are available from Children's Hospital Oakland Research Institute (CHORI) BACPAC resources (http://bacpac.chori.org/). These filters can be hybridized with a probe sequence derived from any gene of interest to identify specific BAC clones that come from a given genomic region. In addition, the Stanford centre has determined the DNA sequence of both ends of the two clones that map to the far ends of every stickleback contig with three or more clones. All of the short-sequence reads have been deposited in Genbank and can be searched for matches to a gene or region of interest using the NCBI blast server (http://www.ncbi.nlm.nih.gov/blast/). Choose "nucleotide-nucleotide BLAST" and be sure to set the database option to "gss" for genome survey sequence. Once an initial clone has been found by library screening or by BLAST homology searches, sets of overlapping clones in the corresponding region can be recovered by entering clone names into Internet Contig Explorer.

Large-insert libraries and physical maps of the stickleback genome have already proved extremely useful for recovering much of the DNA in chromosome regions of particular biological interest. For example, the number and diversity of MHC alleles clearly play an important role in parasite resistance, innate immunity, and mate choice in sticklebacks.[47–51] Isolation and sequencing of BAC clones containing MHC genes has recently been used to identify gene duplication and gene conversion events that contribute to diversity of stickleback MHC class II genes.[45] Further genomic studies will help identify how the overall system of MHC alleles is inherited and evolves in different populations.

2.2.4 Large-Scale Stickleback Genome Sequencing

In the last 15 years, nearly complete genome sequences have been determined for many different organisms, including humans.[52] The large size of vertebrate genomes has required the development of new methods for both generating and assembling billions of base pairs of primary DNA sequence.[53] However, information on a genomewide scale has vastly accelerated research across many different fields.[54,55] Having nearly complete genome sequences has provided the first global pictures of gene content, sequence conservation, and chromosome organisation. The human genome project has led to the identification of the molecular basis of many human genetic diseases and the development of useful genetic markers for predicting risk and aiding in the diagnosis and treatment of cancer and other diseases. Complete genome information has also made it possible for many individual investigators to concentrate on questions of biomedical interest, rather than on the time-consuming process of identifying sequences each time research begins on a particular trait, chromosome region, or gene family.

Despite the incredible progress in genetics and genomics, it has only been possible to assign functions to a small percentage of the human genome. One of the best methods for learning how to read and interpret much of the remaining sequence is a detailed comparison between the sequences of multiple organisms with shared and different traits.[56] Many of the funding agencies and sequencing centres that contributed to large-scale sequencing of the human genome have thus been applying similar large-scale methods to other carefully chosen model systems (http://www.genome.gov/page.cfm?pageID=10002189). In 2003, sticklebacks were nominated for large-scale sequencing as a vertebrate model that had been extensively studied at all levels, including behaviour, physiology, morphology, ecology, and paleontology.[19] The ability to cross divergent forms offered an unusual opportunity to identify the genes and mutations that underlie complex traits in natural environments, without making any assumptions about the particular genomic features that may underlie phenotypic change. Sticklebacks were chosen as a high-priority target for sequencing in 2004, and assigned to one of the major sequencing centres of the human genome project, the Broad Institute in Cambridge, MA.

In 2004 and 2005, the Broad Institute carried out initial polymorphism surveys to identify a particular stickleback individual from which to begin large-scale sequencing. Experience with other genome projects had previously shown that the level of polymorphism present in the genome being sequenced has a large effect on the quality of the resulting genome assembly. Therefore, every effort was made to eliminate possible sources of heterogeneity for the stickleback genome sequencing project. All sequences would be derived from a single fish rather than a pool of individuals. All candidate fish would be females instead of males, to avoid heterozygosity for sequence variations present on the proto X and proto Y chromosome (see Section 2.3.1.3). The least polymorphic individual would be chosen by direct sequencing of several genomic regions from 16 candidate fish from populations thought to have been through genetic bottlenecks or forced brother–sister mating in the laboratory. Several stickleback laboratories provided individual fish from candidate populations. The lowest level of sequence heterogeneity (approximately one sequence difference per 2800 base pairs) was found in a female derived from Bear Paw Lake in Alaska, generated by four generations of brother–sister matings in the laboratory by Bill Cresko and John Postlethwait at the University of Oregon.

In 2005, a series of new DNA libraries were made from the Bear Paw Lake female, including deep-coverage BAC libraries generated for this individual by

Chris Amemiya at the Benaroya Research Institute. The new BAC libraries (VMRC26 and VMRC28) are publicly available, along with other stickleback libraries from CHORI (http://bacpac.chori.org/). The Broad Institute carried out approximately 7 million sequencing reads from May to July of 2005, generating a total of 6x high-quality base pair coverage for the stickleback genome, including the end sequences of most clones in the VMRC26 and VMRC28 BAC libraries. All the primary sequence reads are currently available and searchable from the NCBI trace sequence archives (http://www.ncbi.nlm.nih.gov/blast/mmtrace.shtml, set database to: *Gasterosteus aculeatus*).

The first whole genome assembly of stickleback sequences was released in February 2006. The entire sequence can be downloaded from http://www.broad.mit.edu/ftp/pub/assemblies/fish/stickleback/gasAcu1/. Particular regions of interest can also be found using the search page at NCBI (http://www.ncbi.nlm.nih.gov/BLAST/). Choose either a nucleotide (blastn) or translated database (tblastn) search. Then set "choose database" to "wgs" and "limit by entrez query" to "*Gasterosteus aculeatus* [ORGN]" to confine your search to the stickleback genome. The initial assembly has been combined with high resolution genetic maps and EST collections from Stanford, and detailed gene prediction work at Ensembl has produced an annotated version of the stickleback genome that can be searched, browsed, and compared to other organisms (http://www.ensembl.org). Further refinements are obviously required. The progress to date, however, should be very gratifying to all those who have long worked on the biology of *Gasterosteus*. The rich history of previous studies was key to the rationale for developing molecular resources for sticklebacks. The completion of a stickleback genome sequence will provide an exciting new starting point for additional studies, including the molecular basis of many complex traits in natural populations.

2.3 APPLICATIONS TO SPECIFIC TRAITS

Many classic morphological, physiological, and behavioural traits in sticklebacks are now being studied by a combination of linkage mapping and genomic cloning approaches (Figure 2.2). To illustrate how the new genetic and genomic resources can be applied to a variety of biological problems, we will review recent studies of sex determination, pelvic reduction, and lateral plate patterning. For each of these characters, genetic mapping studies have identified major chromosome regions that control much of the variation of the trait. The corresponding chromosome regions are now being intensively studied, and in some cases it has already been possible to identify specific genes and alleles that contribute to substantial phenotypic transformations in wild fish populations.

2.3.1 Sex Determination

2.3.1.1 Genetic Architecture of Sex Determination

Three-spined sticklebacks do not have heteromorphic sex chromosomes either by karyotype or analysis of synaptonemal complexes.[29,57] There was some evidence to suggest that sex ratios in sticklebacks are affected by environmental conditions such

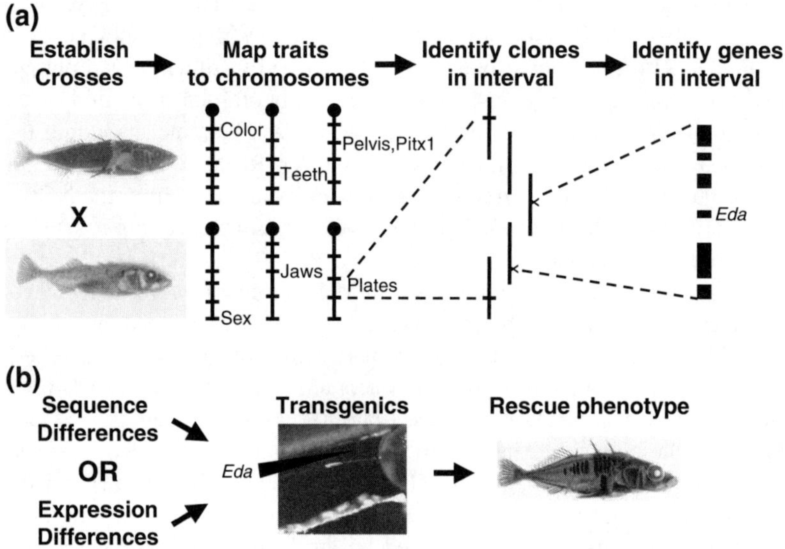

FIGURE 2.2 Genetic approach to identifying the genes and mutations that underlie evolutionary change. Sticklebacks offer a unique opportunity to cross natural populations that have evolved very different morphological, physiological, and behavioural traits. By comparing the distribution of traits and genetic markers in intercross and backcross progeny, it is possible to identify the particular chromosome regions that control genetic variation in a trait of interest. These key regions can then be isolated in overlapping DNA clones, using the large-insert DNA libraries and physical maps that have been developed for the stickleback genome. Comparative sequencing and expression studies can identify interesting genes and sequence variants in the cloned region that are strong candidates for controlling a particular trait. If the correct gene has been identified, it should be possible to introduce a new version of that gene into an evolved population, and show that the transfer of specific DNA sequences alters the development of the corresponding trait. A transgenic fish with extra lateral plates following introduction of the EDA gene is shown.

as rearing temperature and density.[58] However, protein and DNA markers with sex-specific alleles have been found,[59–62] suggesting that there might be a genetic basis for sex determination in three-spined sticklebacks. To determine if phenotypic sex in three-spined sticklebacks is determined primarily by genetic or environmental factors, the morphology of the adult gonads was examined in 92 offspring of a cross between benthic and limnetic individuals from Priest Lake, British Columbia,[17] and in 385 F2 offspring of a cross between a Japanese marine female and a benthic male from Paxton Lake, British Columbia.[25,38] A major locus on the distal end of linkage group (LG) 19 was found that segregated nearly perfectly with the male phenotype in both crosses.[40] This analysis showed that sex determination in three-spined sticklebacks is genetic and is controlled by a single major chromosome region.

An isocitrate dehydrogenase (IDH) allozyme was found to be sexually dimorphic in multiple populations in California and British Columbia.[59–61] The *Idh* gene was later identified in one of the clones from a cDNA library made from adult stickleback tissues.[17] PCR primers flanking a region of the 3′ untranslated region (UTR) of *Idh* amplify a single band from female genomic DNA, but two bands from male. This polymorphism segregates with phenotypic sex in both the Priest and Paxton crosses[40] as well as in stickleback populations from around the world,[63] suggesting that the master sex determination locus is very tightly linked to the *Idh* gene. The *Idh* marker, as well as other DNA markers that map near the sex determination region,[62] provides a simple and efficient method for determining the genotypic sex of individual sticklebacks. Such markers are already proving useful in environmental monitoring and toxicology studies examining the effects of endocrine-disrupting chemicals on the phenotypic sex of exposed individuals.[64,65]

2.3.1.2 Cloning the Sex-Determining Region

To identify the molecular nature of the sex-determining gene, a chromosome walk was initiated in the sex determination region. A chromosome walk relies on the genetic map to orient and order DNA markers relative to the trait of interest. To increase the power to detect rare recombination events near the sex determination region, a total of 699 F2 animals from the Paxton benthic cross were phenotyped by inspection of the gonads and genotyped with all the markers on LG 19.

Recombination events between markers can be traced to meiotic events that occurred in the F1 mother, which represent recombination between the two X chromosomes, or to meiotic events that occurred in the F1 father, which represent recombination between the X and the Y chromosomes. Most of the recombination events near the sex determination region have occurred during female meiosis, and so are not useful for defining the position of the sex determination locus. Rare recombinants in the male F1 suggest that the sex determination locus maps distal to the *Idh* locus (Figure 2.3). The sex determination locus is currently the most distal marker on the entire linkage group, suggesting it may map near the telomere. Further cytogenetic analysis and physical mapping will be required to confirm this.

To clone DNA sequences that never recombine with the sex determination locus and thus are likely in the same chromosome region, the most closely linked markers on the genetic map (*Idh*, *Stn191*, *Stn192*, and *Stn194*) were used to screen the Salmon River BAC library made from a combination of male and female fish.[18] The *Idh* probe identified 21 BACs, *Stn194* identified 16 BACs, *Stn191* identified 13 BACs, and *Stn192* identified 5 BACs (Figure 2.3). Additional BACs in the *Stn194*, *Stn191* and *Stn192* contigs were identified *in silico* using Internet Contig Explorer. The total number of BACs identified for each probe was roughly consistent with the *20x* coverage of the genome provided by this BAC library. The sex-specific polymorphism in the 3UTR of the *Idh* gene was used to genotype 15 *Idh* BACs as originating from the female (X) chromosome and 6 *Idh* BACs as originating from the male (Y) chromosome. The chromosome of origin (X vs. Y) has not yet been determined for the BACs in the *Stn194*, *Stn191*, or *Stn192* contigs. Each BAC end was sequenced, and PCR primers were designed to end sequences from each BAC to make new genetic markers that could be used for further experiments. Each of the new end markers was typed on all the BAC clones from the interval to confirm the overlaps between clones, to orient them with respect to each other, and to identify which BAC clones extended the furthest in each direction from the starting point.

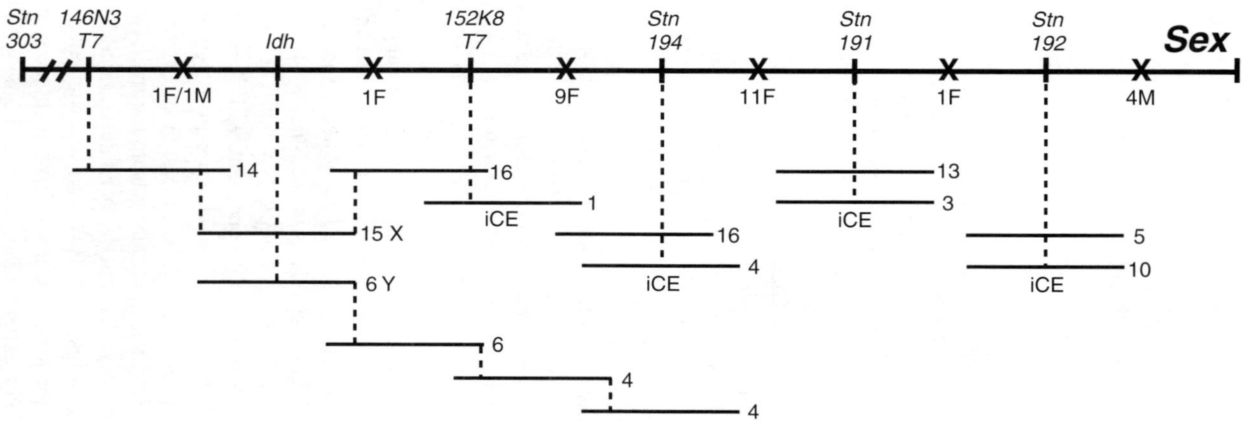

FIGURE 2.3 Genetic and physical map of the sex determination locus in three-spined sticklebacks. The top line shows the genetic map of the sex determination locus on linkage group 19. The names of the genetic markers mapped relative to the sex determination locus are indicated on the top of the line. The Xs on the line indicate recombination events that have occurred between the flanking markers. The numbers below refer to the number of recombinant F2 animals and (F) indicates that the recombination event occurred in the F1 female meiosis between two X chromosomes, while (M) indicates that the recombination event occurred in the F1 male meiosis between the X and the Y chromosome. The BAC contigs, which were isolated by screening the BAC library with a particular genetic marker, are indicated by a single line, and the total number of BACs within a contig is shown to the right of the line. Additional BACs that were identified by screening the iCE database are indicated with a separate line. X- and Y-chromosome-specific BACs were isolated using the *Idh* marker, and BACs specific to the X and Y chromosomes were isolated in subsequent walking steps. (*Note*: This figure is not drawn to scale.)

The chromosome walk around the sex determination locus was originally extended from the *Idh* BACs because these BACs could be distinguished as X or Y chromosome specific based on the sequence differences in the *Idh* gene. The Salmon River BAC library was screened again with markers from the ends of the initial *Idh* contig. Because primers from the right end of the male BAC contig did not amplify from any of the female BACs and primers from the right end of the female BAC contig did not amplify the male BACs, probes from both male and female BACs were used as probes to rescreen the BAC library. The two probes isolated completely different sets of BACs, suggesting that this region is highly divergent between male and female, and that there may be large rearrangements present that differentiate the X and Y chromosomes. Markers on the left side of the *Idh* contig are present on both X- and Y-specific BACs; therefore, only one end was used to probe the BAC library (Figure 2.3).

One of the BACs identified by the screen with the left end of the *Idh* contig contained a microsatellite marker (*146N3T7*). When this microsatellite marker was mapped on F2 animals from the Paxton benthic cross, rare animals were found that had inherited a chromosome with a recombination breakpoint between *146N3T7* and the *Idh*/sex determination region. Because the genotype at *Idh* was concordant with the phenotypic sex in this animal, the *146N3T7* marker must map proximal of the *Idh*/sex determination region. This genetic information therefore orients the walk and provides a proximal boundary in the search for the sex determination locus. The probe from the right end of the female *Idh* contig yielded 16 BACs. Using Internet Contig Explorer, BACs from this contig could be connected with BACs from the *Stn194* contig. However, the *Stn191* and *Stn192* contigs have still not been connected with each other or with the *Stn194* contigs.

None of the BACs isolated by the right end of the male *Idh* contig overlap with any of the female-specific BACs. Multiple rounds of chromosome walking have been performed using the male-specific BACs (Figure 2.3); this process has presumably isolated Y-chromosome-specific BACs. However, it is not yet clear how large the Y-specific region will be. Cytogenetic analysis and physical mapping of the region using telomere-specific probes will help define the actual size of the sex determination interval. To determine which of these BACs contain the sex determination locus, a transgenic approach can be used.[66] If a BAC contains the sex determination genes, it should transform an XX individual into a male. Any BAC that converts an XX individual into a male can be sequenced to completion to find the genes present. Specific candidate genes can then be analyzed to determine which is the master sex determination gene.

2.3.1.3 A Snapshot of Evolving Sex Chromosomes

Although the identity of the sex determination gene is not yet known, the genetic and physical mapping studies have already provided the first molecular glimpse of sex chromosome evolution in sticklebacks. Large sex-specific differences in recombination rate are clearly seen along the linkage group containing the sex determination region.[40] Markers close to the sex determination locus show a 4- to 20-fold reduction in recombination rate in male meiosis when compared to female meiosis.

However, this is not due to a lower general rate of recombination in males, because markers that are not closely linked to the sex determination locus show a twofold increase in recombination rate in male meiosis when compared to female meiosis. Suppression of recombination around a sex determination locus is a hallmark of an evolving Y chromosome and has been hypothesized to occur in order to reduce recombination between the sex determination locus and linked genes with sex-specific fitness effects.[67–69] This suppression of recombination leaves one chromosome in a consistently heterozygous state, which ultimately results in the degeneration of sex-linked loci in the heterogametic sex.[67,69,70]

To examine the effects of reduced recombination between the X and Y chromosomes at the molecular level, two X-specific BACs and two Y-specific BACs of the *Idh* contigs (Figure 2.3) were sequenced to completion. Comparative analysis of these sequences revealed that there is very poor overall homology between the X and Y chromosome, with only 63.7% sequence identity over the entire length of the alignment.[40] Although there are regions of very high homology between the X and Y chromosomes, the overall low sequence identity is the result of intervening gaps with virtually no homology between the X and Y chromosomes. These gaps in homology are mostly due to insertions of repetitive DNA elements and local duplications on the Y chromosome.[40] Many of these local duplications appear to be novel, stickleback-specific repetitive elements.[40]

Taken together, these data suggest that three-spined sticklebacks have a nascent Y chromosome that contains a single major sex determination region. Suppression of recombination, accumulation of repetitive DNA, and sequence divergence between the X and the Y chromosome has also been observed in other plant,[71–73] and animal[74–77] systems with evolving Y chromosomes. Further analysis of the nonrecombining region of the three-spined stickleback sex determination locus will lead to important insights into the early events that contribute to sex chromosome evolution and degeneration.

2.3.1.4 Comparative Evolution of Sex Chromosomes in Other Stickleback Groups

Previous cytogenetic analysis of the different stickleback species[29] demonstrated that *G. aculeatus*, *Pungitius pungitius*, and *Culaea inconstans* have no apparent sex chromosomes, whereas *G. wheatlandi* has heteromorphic chromosomes in males and *Apeltes quadracus* has heteromorphic chromosomes in females. This striking difference in sex chromosome complement, combined with the relatively young age of this family,[5] suggests that the different stickleback species have rapidly evolved independent mechanisms of sex determination. Future genetic and cytogenetic analyses of sex determination and sex chromosomes in these other gasterosteid species will therefore yield important insights into the evolution of sex chromosomes.

2.3.2 Pelvic Reduction

2.3.2.1 Genetic Architecture of Pelvic Reduction

Three-spined sticklebacks have a pelvic skeleton consisting of bilateral pelvic spines that articulate with an underlying pelvic girdle. The girdle covers part of the ventral surface and extends up the lateral side of the fish in an ascending branch that articulates with the lateral plates. The pelvic skeleton is protective against gape-limited, soft-mouthed predators.[78,79] However, several freshwater populations found throughout the wide distribution of three-spined sticklebacks have reduced or absent pelvic structures.[23,25,80–87] Several hypotheses have been put forward to explain the loss of pelvic structures, including the absence of predatory fish, reduced levels of calcium availability, or predation by macroinvertebrates.[79,87–92]

To examine the location and number of chromosome regions that control pelvic reduction in natural populations, 375 F2 progeny from a cross between Japanese marine female (complete pelvis) and Paxton benthic male (no pelvis) were analysed using genomewide linkage mapping.[25] All progeny were scored for length of the pelvic spines, length of the pelvic girdle, height of the ascending branch, and pelvis asymmetry. The progeny were also genotyped using the genomewide set of micro-satellite markers.

When pelvic reduction was simply scored as a qualitative trait (normal pelvic structures vs. any form of size reduction, loss or asymmetry), a near 3:1 Mendelian ratio of unaffected:affected fish was observed. A locus that explains much of the variation (83%) in this phenotype maps to the end of the linkage group 7. When different aspects of the trait were considered as quantitative traits (i.e., length of the pelvic spines, length of the pelvic girdle, height of the ascending branch), the same major chromosome region was again detected and found to control anywhere from 31.5 to 65.3% of the variance in particular characters. In addition, several loci of smaller effect are also observed on LG 1 (pelvic girdle length), LG 2 and LG 4 (pelvic spine length and pelvic girdle length), and LG 10 (ascending branch height). Both additive and epistatic interactions were observed between the major and minor loci that underlie pelvic reduction.[25]

2.3.2.2 A Candidate Gene Approach to Identifying the Major Pelvic Locus

Identification of the gene underlying pelvic reduction took advantage of the wealth of knowledge from developmental genetic studies on the molecular basis of limb development in traditional model organisms such as mouse and chicken. The pelvic spine in sticklebacks is a modified pelvic fin, and the pelvic fins of fish are homologous to the hindlimbs of terrestrial vertebrates. Therefore, genes that are specifically involved in development of the hindlimb in vertebrates were considered candidate genes for loss of the pelvic structures during evolution in sticklebacks. Several genes have been described that are specifically expressed in hindlimbs but not forelimbs or are required for normal hindlimb development in traditional vertebrate model systems, including the transcription factors *Pitx1*, *Pitx2*, and *Tbx4*.[93–99] Genes that are normally expressed in the hindlimb region, including *Pitx1* and its downstream target *Tbx4*, were not expressed in a Scottish population with severe pelvic reduction.[23] These results suggest that the pelvic loss arises through a failure to initiate fin development rather than by normal initiation with subsequent interruption of the normal limb development program. Furthermore, these results suggest that the genetic alterations leading to pelvic reduction must act at least as early as the establishment of *Pitx1* expression but could disrupt either *Pitx1* itself or any gene upstream of *Pitx1* that is required for normal expression and development of pelvic structures.

To compare the location of candidate genes with the major chromosome regions that control pelvic reduction in genetic crosses, Shapiro et al.[25] isolated stickleback homologues of the *Pitx1*, *Pitx2*, and *Tbx4* genes and mapped them in the same F2 progeny in which pelvic reduction had been mapped. The *Tbx4* and *Pitx2* genes map to positions on LGs 1 and 4, which clearly excludes them as candidate genes for the major locus controlling pelvic reduction. In contrast, the *Pitx1* gene mapped to the end of LG 7, in the same region as the major chromosome region that controls more than 65% of the variance in pelvic measurements. All F2 fish with bilateral loss of pelvic structures were homozygous for Paxton benthic alleles at the *Pitx1* locus. The complete absence of genetic recombination between *Pitx1* and the severe pelvic reduction phenotype suggests that *Pitx1* must be located extremely close to the pelvic reduction locus (within less than 0.5 centimorgans). The tight genetic linkage, combined with the known function of *Pitx1* in hindlimb development, suggest that *Pitx1* is an outstanding candidate for the major locus controlling pelvic reduction in three-spined sticklebacks.

2.3.2.3 Pitx1 and Regulatory Evolution

The entire open reading frame and the intron/exon junctions of *Pitx1* were sequenced; there were no changes that would result in amino acid differences between marine and pelvic-reduced fish.[25] These data are consistent with strong purifying selection that normally preserves the coding region of a major developmental regulator. To test for possible regulatory changes in the *Pitx1* locus, expression of *Pitx1* transcripts during normal development was analysed by *in situ* hybridization at stages when the pelvic structures are just starting to develop. Although similar patterns of *Pitx1* expression were observed in several soft tissues of larvae from both a complete-pelvis and a reduced-pelvis population, *Pitx1* expression was missing in the prospective pelvic region of Paxton benthic fish.[25] These results show that site-specific regulatory changes have occurred in Paxton benthic fish to alter the level of *Pitx1* expression at some locations while preserving its expression, splicing, and coding potential at other locations in the body (Figure 2.4). The mapping data show that the key genetic changes occur at or near the *Pitx1* locus itself, presumably by altering *cis*-acting regulatory information around *Pitx1* that controls its expression at specific locations.[25]

It has been proposed that morphological diversity is more likely to result from mutations in *cis*-regulatory elements of genes because these mutations would only affect expression in a specific structure and would not affect expression of the gene in structures required for the survival of the organisms.[100–103] The results of several

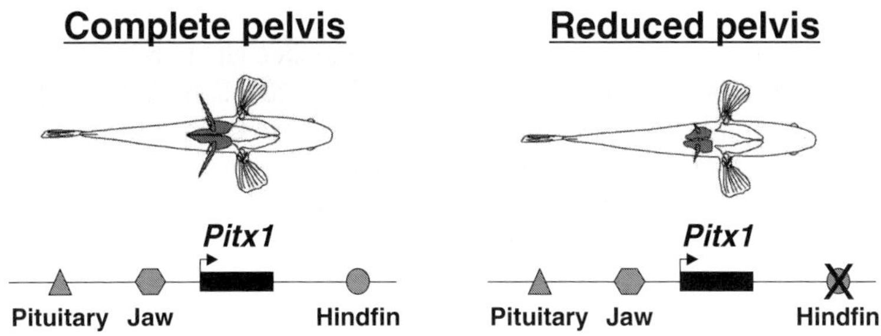

FIGURE 2.4 Model for *cis*-regulatory evolution at the *Pitx1* locus in pelvic-reduced sticklebacks. The *Pitx1* protein coding sequence is the same in complete and pelvic-reduced sticklebacks. *Pitx1* is expressed at several anatomical sites in both complete and pelvic-reduced sticklebacks, but is not expressed in the developing hindfin of pelvic-reduced sticklebacks. The loss of a modular *cis*-regulatory element that drives *Pitx1* expression specifically in the developing hindfin provides a mechanism to evolve new morphologies at one anatomical site, while preserving the function of the gene in other sites that are necessary for the viability of the fish in the wild.

genetic studies in a variety of taxa now provide support for the hypothesis that *cis*-regulatory evolution underlies morphological variation in natural populations.[104–112] However, in most of these studies, the molecular changes in the *cis*-regulatory elements have not yet been identified. In future, it will be particularly exciting to find the particular DNA sequences that normally control expression of key genes such as *Pitx1* in particular tissues, and determine how these sequences have changed in natural populations that have evolved major changes in anatomy.

2.3.2.4 A Molecular Explanation for Directional Asymmetry

Nearly all stickleback populations with pelvic reduction, including fossil sticklebacks, show directional asymmetry, in that the right side is more often reduced than the left (Figure 2.4).[23,25,80–82,113] Interestingly, mice with a null mutation of the *Pitx1* gene have a reduction in hindlimb structures, with a greater reduction of hindlimb structures on the right side than on the left.[97,99] This result is likely explained by partial compensation for loss of *Pitx1* by the closely related *Pitx2* gene, which is preferentially expressed on the left side during development in mammals, frogs, and fish.[99,114] When asymmetry in pelvic reduction was scored as a genetic trait in the Japanese marine by Paxton benthic cross, the trait mapped to the *Pitx1* locus,[25] suggesting that there is a conserved mechanism underlying directional asymmetry in sticklebacks and mammals.

2.3.2.5 Parallel Evolution of Pelvic Reduction

Pelvic reduction has evolved in multiple stickleback populations, including fish from Paxton Lake,[80] the Queen Charlotte Islands in British Columbia,[82,86] Quebec,[85] the Cook Inlet region of Alaska,[87] southern California,[81] the Outer Hebrides in Scotland,[83,84] and Iceland.[25] In addition, a fossil deposit of *Gasterosteus doryssus* shows pelvic reduction.[80] Finally, pelvic reduction has been observed in other gasterosteid fish, such as Canadian, Alaskan, Irish, Japanese, Russian, and Greek populations of the ninespine stickleback, *Pungitius pungitius,*[113,115–118] and in Canadian populations of the brook stickleback, *Culaea inconstans.*[119,120] This repeated evolution of pelvic reduction may be the result of the same underlying genetic and developmental mechanisms or could use different genetic and developmental mechanisms to reach the same endpoint.

The developmental basis of pelvic reduction has been compared across several populations of three-spined sticklebacks and two modes of reduction have been observed: paedomorphosis and gradual truncation of distal elements.[81] In the paedomorphic mode of pelvic reduction, the elements that are the last to appear during development are the first to be lost. This occurs in both the southern California populations and the fossil three-spined stickleback, *Gasterosteus doryssus.*[81] In the truncation of distal elements mode of pelvic reduction, the most distal elements (such as the pelvic spine) are lost before more proximal elements (the ascending branch and anterior process). Pelvic morphologies resulting from distal truncation do not resemble stages of normal pelvic development and therefore are not due to paedomorphosis. Most three-spined stickleback populations follow this mode of pelvic reduction, with the exception of the Boulton Lake population from the Queen Charlotte Islands.[81]

Cresko et al.[39] used genetic mapping and complementation studies to compare the genetic basis of pelvic reduction in three populations of pelvic-reduced fish from the Cook Inlet region of Alaska. Their results suggest that the same major locus likely underlies pelvic loss in all three nearby populations, although different minor loci may be present in some populations. Mapping experiments in the Alaskan fish showed linkage of the major locus to microsatellite markers on LG 7, the same chromosome that contains the major locus controlling pelvic reduction in the Paxton benthic cross. Shapiro et al.[25] used complementation crosses to compare the genetic basis of pelvic reduction in two populations that evolved in separate ocean basins thousands of miles apart. The F1 hybrid fish failed to develop a pelvis, again suggesting that similar genetic changes have occurred in widely separated populations. Pelvic-reduced fish from Paxton Lake, Cook Inlet, Iceland, as well as the Scottish fish previously used for marker gene expression studies all show directional asymmetry, with larger pelvic rudiments on the left- than right-hand side.[23,25] The combined mapping, complementation, expression, and morphological data suggest that *Pitx1* is the major locus underlying pelvic reduction in most stickleback populations around the world. Further work will reveal whether repeated use of the same major locus is due to a shared genetic variant already present in a common ancestor of multiple pelvic-reduced populations or due to repeated mutation of the same locus.

2.3.3 Lateral Plate Morphs

2.3.3.1 Genetic Architecture of Lateral Plate Number

One of the most prominent morphological changes seen in sticklebacks is major reduction of lateral plate number in freshwater populations.[121] Commonly observed lateral plate patterns are divided into three main morphs.[122] The complete morph has a full set of lateral plates and is usually found in marine populations. The low morph has only retained the anterior plates, while the partial morph has anterior plates and a posterior keel; these latter two morphs are usually found in freshwater populations. Although the reasons for the loss of lateral plates in freshwater are unknown, several hypotheses have been proposed, including low calcium levels,[91] salinity tolerance,[123] stream gradients,[124] parasite susceptibility,[125] increased body flexibility and changes in swimming performance,[126,127] and changes in predation regime[128–132] and climate.[133] Previous genetic studies have shown that plate morphs are reproducibly inherited in the laboratory and that crosses between different morphs result in relatively simple ratios of the plate morph phenotypes. Several models were proposed to explain these results, usually involving a one or two locus system.[59,134–137]

To directly analyse the number and position of genetic loci that control plate morph phenotypes, genomewide linkage mapping has been used with crosses between completely plated Japanese marine fish and low-plated benthic fish from Paxton Lake,[38] completely and low-plated fish from a single dimorphic site in California,[38] and completely plated marine and low-plated lake fish from Alaska.[39] In the California and Alaska crosses, plate morph segregates as a nearly Mendelian trait, and the low allele is largely recessive to the complete allele so that heterozygous fish are completely plated. These results both confirm a simple genetic model that was proposed by Avise in the same California site[59] and map a single major locus controlling plate morph to a particular chromosome region.

In contrast, in the Paxton benthic cross the same major chromosome region interacts with several modifier genes to determine the overall plate morph phenotype.[38] As in the California and Alaska crosses, almost all fish that carry two marine alleles (AA) at the major locus are completely plated, and fish that carry two Paxton benthic alleles (aa) are low plated. However, fish heterozygous for one marine and one Paxton benthic allele (Aa) develop either as complete or partially-plated fish, suggesting that the dominance relationship of alleles at the major locus varies in fish from different geographic locations.

Plate number was also analysed as a quantitative trait in the Paxton benthic cross, by scoring the number of lateral plates that develop in fish with different genotypes at the major locus. The major locus accounts for approximately 75% of the overall variance in plate number in F2 progeny. However, three unlinked modifier loci have substantial additive effects on the number of plates that develop in fish that are heterozygous at the major locus.[38] The total number of freshwater alleles at all three modifier loci can shift plate numbers between 15 and 30 plates per side and largely explain why fish heterozygous at the major locus (Aa) can develop as either complete or partial morphs. Increasing substitution of freshwater alleles at all three modifier genes has a smaller additive effect on the number of plates that develop in fish with two Paxton benthic alleles at the major locus (aa) and decreases the number of plates in low-morph animals from six to three plates per side. The same modifiers have very little effect on the number of plates in fish with two marine alleles at the major locus (AA), which always develop as completely plated fish.

These results provide excellent examples of both additive and epistatic effects on plate number development in the Paxton benthic cross.

The size of lateral plates also can vary significantly between different stickleback populations.[59,121,138] Three quantitative trait loci (QTL) were detected that act additively to affect plate size.[38] Two of these loci map to the same chromosome regions that affect plate morph or plate number (LG 4 and LG 7), suggesting that the pattern, number, and size of plates may be controlled by the same or linked genes on LGs 4 and 7.

2.3.3.2 A Chromosome Walk to the Major Plate Locus

To clone the major locus controlling plate development, Colosimo et al.[32] used many of the new genomic tools described earlier. Five hundred and eighty-three F2 fish from the Paxton benthic cross were phenotyped for lateral plate morph and genotyped with the molecular markers flanking the plate morph locus: *Gac4174* and *Stn183*. This analysis defined a genetic region of approximately 3.3 cM (39 recombinants in 1166 meioses), which is still a very large interval for positional cloning of a specific gene. To find markers that were more closely linked to the plate morph locus, the amplified fragment length polymorphism (AFLP) technique[139] was used to screen for markers that differed in allele size or frequency in pools of low-plated and completely plated progeny from the cross. Two more closely linked markers (*Stn345* and *Stn346*) were isolated, defining a new interval of 0.68 cM (8 recombinants in 1166 meioses).

These two markers were then used to screen the Salmon River BAC library, which was derived from completely plated marine fish.[18] Three rounds of chromosome walking led to the isolation of six overlapping BAC clones spanning the plate morph interval. One of the BAC ends was used to develop a new genetic marker (*Stn347*), which mapped one recombination event distal of the plate morph locus. In contrast, the *Stn345* marker mapped two recombination events proximal of the plate morph locus in the high-resolution-mapping cross. These two markers are approximately 539 kb from each other on the BAC contig and define the minimal physical interval in which the plate morph locus must reside. Two BAC clones within this interval were sequenced to completion, and a contiguous sequence of 407,051 bp was assembled.

To find the locus controlling plate morph phenotypes within this region, microsatellite markers were generated at 12 kb intervals throughout the sequence assembly and genotyped on a sample of 46 completely plated and 45 low-plated individuals from a single interbreeding population from Friant, CA.[59] This analysis takes advantage of historical meioses within natural populations; these meiotic events should homogenize allele frequencies at all markers in interbreeding completely and low-plated fish, except for those extremely closely linked to the actual mutations that causes the difference in plate phenotype.

This linkage-disequilibrium-mapping approach identified a small 16-kb region with significant allele frequency differences in completely and low-plated fish. Gene prediction in the surrounding region showed that the marker at the peak of linkage disequilibrium was located in intron 2 of the stickleback *Ectodysplasin* (*Eda*) gene.[32]

This gene is a member of a family of secreted signalling molecules that is required for development of a number of ectodermal derivatives in mammals, such as teeth, hair, and sweat glands, as well as dermal bones.[140,141] Furthermore, a mutation in the *Ectodysplasin receptor* (*Edar*) gene in medaka fish (*Oryzias latipes*) leads to a loss of scales, which are elements of the dermal skeleton.[142,143] The scales of medaka and dermal lateral plates of sticklebacks (which lack scales) have likely evolved from a common ancestral element.[143] The position of *Eda* at the peak of linkage disequilibrium in the chromosome walk and the known function of the gene in other animals suggest that molecular changes at the *Eda* locus are likely to underlie plate morph evolution in sticklebacks.

To determine if changes in the amino acid sequence of the *Eda* gene contribute to the evolution of the low-morph phenotype, the *Eda* locus was completely sequenced in BAC clones derived from the completely plated Salmon River population and low-plated Paxton benthic population. There were four amino acid changes in the EDA protein between the two populations; however, none of these amino acid changes occurred at sites that are highly conserved between mammals and fish nor in residues that were previously shown to be associated with defects in humans.[144,145] Numerous changes were found in the noncoding regions flanking *Eda* exons, and in surrounding regions.[32]

2.3.3.3 Transgenic Rescue of Plate Morph

Although the established role of EDA signalling in dermal bone and scale development made *Eda* a compelling candidate for the gene controlling lateral plate phenotypes, several other genes were present in the region. Furthermore, *Eda* expression in the completely and low-plated populations could not be directly assessed. To test directly whether changing levels of EDA signalling could alter plate development in sticklebacks, transgenic sticklebacks were generated with a full-length mouse *Eda* cDNA under the control of a broadly expressed promoter. The same construct had previously been used to restore the development of teeth, hair, and sweat glands in mice with a null mutation in the *Eda* locus.[146] Owing to the mosaic integration of injected DNA constructs in transgenic sticklebacks,[66] injected fish and uninjected control siblings were examined for mosaic patches of ectopic lateral plate formation after raising animals to a minimum of 30-mm standard length, at which point plate development should be complete.[147] Because the transgene was injected into low-plated embryos, extra lateral plates should only appear in fish carrying the transgene in regions that are normally unplated. In 3 of the 14 fish that carried the transgene, there were ectopic lateral plates, and no extra plates ever developed in uninjected control siblings or in fish injected with the same vector and a green fluorescent protein (GFP) insert instead of the *Eda* cDNA, confirming that EDA signalling is sufficient to trigger lateral plate formation and that *Eda* transgenes can partially rescue the low-plated phenotype of freshwater sticklebacks.[32] Future work will be required to determine the actual molecular changes at the *Eda* locus that are responsible for the difference between the complete and the low morphs.

2.3.3.4 Molecular Basis of Parallel Lateral Plate Reduction

Genetic evidence suggests that the same major genetic locus underlies the evolution of the low morph in populations from California, Alaska, British Columbia, and Japan.[38,39,59,148] Direct sequencing of *Eda* in a large survey of 10 completely plated populations and 15 low-plated populations from around the world showed that most of the latter shared many of the same base-pair changes.[32] In contrast, a phylogenetic analysis of these same populations using 25 random nuclear genes strongly rejects a common geographic origin of all low-plated populations.[32] Together with previous mitochondrial phylogenies,[13–15] these data suggest that the existing low-plated populations are not derived from a single low-plated population that has colonised numerous freshwater environments.

The striking difference in topology between the phylogenetic tree of *Eda* sequences and the phylogenetic tree of mitochondrial and control nuclear genes could be explained if an ancient low-plated *Eda* allele is present at a low frequency in marine populations and can spread to new locations by migration of completely plated fish. A direct survey of completely plated fish at the ocean outlets of two North American streams confirmed this prediction. The low-plated *Eda* allele is present at a frequency of 3.8% in an anadromous population from California and at a frequency of 0.2% in an anadromous population from British Columbia.[32] This low frequency is likely to be maintained by low levels of hybridization between completely plated anadromous fish and resident low-plated freshwater populations. Gene flow, together with the annual migration of anadromous fish between marine and freshwater environments, could thus provide a simple mechanism for introducing alleles for the low-plated phenotype into new populations (Figure 2.5).

Selection on the low-plated morph allele is likely to be strong and occur very quickly following colonisation of new streams and lakes. Several studies have reported rapid reduction of lateral plate numbers when marine fish are introduced into new freshwater environments.[149–152] Such strong selection could lead to a rapid increase in the frequency of a preexisting *Eda* allele and the rapid emergence of the alternative low-plated phenotype. It is still not clear why the low-plated *Eda* allele is favoured so strongly in freshwater. This could be due to change in predation regimes favouring different plate patterns in fresh water or selection on other pleiotropic effects of *Eda* itself. In addition, three other genes are usually inherited along with *Eda* in an ancient haplotype block that is shared between many low-plated populations. These closely linked genes may control correlated changes in salt tolerance, parasite resistance, or other factors that contribute to the selective advantage of the low-plate haplotype.[32] It will be fascinating to determine the environmental factors that contribute to rapid and repeated selection for the low-plate morph phenotype in freshwater populations of three-spined sticklebacks.

Although most low-plated stickleback populations are homozygous for an ancient low *Eda* haplotype, low-morph fish from Nakagawa Creek, Japan, do not share this haplotype.[32] However, crossing low-plated fish from this population with Canadian fish carrying the low-plated *Eda* allele results in only low-plated fish.[148] This failure to complement suggests that the low-plated morph in the Nakagawa Creek population is also due to a mutation at the *Eda* locus, but has resulted from

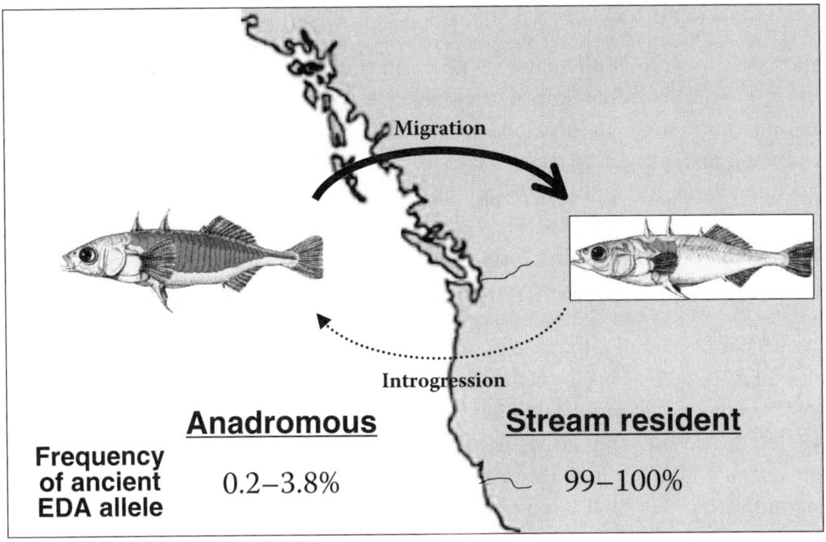

FIGURE 2.5 Changing skeletal morphology using genetic variants present at low frequencies in ancestral marine fish. Mapping, sequencing, and transgenic experiments have shown that the major locus controlling stickleback lateral plate morph is the *ectodysplasin* (*EDA*) gene. Most low-plated populations around the world have fixed an ancient genetic variant of *EDA*. Phylogenetic analysis suggests that an ancient *EDA* allele controlling the low-plated phenotype has existed for several million years. Marine fish harbour the ancient variant at low levels, where it is present in heterozygous form in completely plated individuals. Migration of anadromous sticklebacks can introduce the genetic variant into new freshwater environments. There, the *EDA* variant is rapidly swept to fixation in stream- or lake-resident populations. Low-level hybridization between marine and freshwater populations may help maintain the ancient allele at low frequencies in the ocean population, with repeated rounds of hybridization, introgression, and migration leading to worldwide spread of the genetic information for the alternative low-plated morphology. (Based on data and discussion in Colosimo, P.F. et al., *Science,* 307, 1928, 2005.)

an independent mutation. Therefore, parallel evolution of stickleback lateral plate morphs has occurred both by selection on existing genetic variation as well as by new mutation at the same locus.

Other stickleback species also show variation in plate number and pattern. The blackspotted stickleback (*G. wheatlandi*) is the sister species to the three-spined stickleback. Although *G. wheatlandi* is primarily a marine species, it is generally characterised by a low-plated morph phenotype[1]; this contrasts with the situation in *G. aculeatus*. However, the complete, partial, and low-plated morphs can be found in blackspotted sticklebacks from the Long Island Sound.[153] There is also variation in plate pattern found in the nine-spined stickleback, *Pungitius pungitius*.[154,155] It will be interesting to determine if plate morph variation is due to the *Eda* locus or an independent locus in these other stickleback species.

2.4 DISCUSSION

In the last 5 years, a full range of new molecular genetic and genomic tools has been developed for three-spined sticklebacks. Although detailed molecular studies of evolutionary traits in sticklebacks are still in their infancy, the results to date already suggest intriguing answers to several long-standing questions about how new traits evolve in natural populations.

2.4.1 How Many Genetic Changes Are Required to Achieve Major Phenotypic Change in Natural Populations?

Darwin, Wallace, and other early evolutionary biologists emphasised a view of evolutionary change based on the slow, cumulative selection of very small differences.[156] Genetic variants could lead to phenotypic variation in almost any direction, and selection could generate virtually any new character by steady accumulation of small variants. Around the turn of the century, the rediscovery of Mendel's laws led to a contrasting view that large phenotypic differences were often created by either major mutations or a small number of Mendelian factors.[157,158] Mutationists argued that the types of mutations that were generated in organisms largely determined the direction of evolutionary change. In this view, natural selection still played the important role of determining which contrasting phenotypes survived. However, the contrasting phenotypes themselves were built by mutations of large effect, and any detailed understanding of evolutionary changes would therefore require a better understanding of the genetic and mutational constraints that generated new phenotypes.

The neo-Darwinian synthesis of Mendelian genetics and population genetics showed how evolutionary change could arise from changes in allele frequency within populations. Fisher made a very influential argument that genetic variants with infinitesimally small phenotypic effects had the highest probability of being advantageous in complex animals evolving from one adapted state to a new adaptive state.[159] This "infinitesimal" view had many mathematical advantages. It also reemphasised the important role of natural selection in building the phenotypic traits seen in animals while shifting the focus away from any need to understand the details of particular genes or mutations. As Orr and Coyne have emphasised, the infinitesimal view was widely accepted for over 50 years despite a paucity of experimental data that had actually measured the size and number of mutations underlying real phenotypic traits in natural populations.[42–44]

Sticklebacks now provide one of the best available systems for studying the genetic architecture of evolutionary change in vertebrates. The ability to cross natural populations with major phenotypic differences and raise large numbers of F1 and F2 hybrid progeny makes it possible to experimentally measure the number and location of chromosome regions contributing to any trait of interest. Mapping studies of several different traits have clearly shown that particular chromosome regions can explain a substantial fraction of the variance in many morphological characters (Table 2.1). For sex determination, pelvic reduction, and plate reduction, a single genetic region can explain two thirds or more of the variance in crosses. For other traits, two major chromosome regions can account for much of the phenotypic variance seen in crosses (Table 2.1). For example, two chromosome regions together can explain over 50% of the phenotypic variance in gill raker number, and two independent chromosome regions account for up to a tenfold change in the length of the first dorsal spine (Table 2.1).[17] For most traits, multiple modifier loci with smaller phenotypic effects can also be found (Table 2.1). Thus, even traits that are strongly influenced by a major locus may be substantially modified by the influence of other genes.

It is possible that a large number of additional regions with very small effects also exist. The relatively small crosses performed to date do not have the power to detect genes of very small effect and may overestimate the percentage variance explained by some of the regions that have been detected.[160] The number of genetic factors that contribute to traits will continue to grow as larger and larger crosses are scored by increasingly quantitative methods. For example, the genetics of lateral plate number and pelvic reduction may look nearly Mendelian when scored semi-quantitatively or by qualitative categories in small families.[39] When the same traits are scored quantitatively in larger crosses, multiple modifier genes can be detected in addition to the major loci.[25,38] Although each of the modifier loci has a relatively small individual effect, their summed effects can be nearly as large as that of the major locus.[38]

Treating initial QTL as genetic cofactors may improve the detection of additional modifiers. Several of the modifier QTL identified in the plate genetic studies would not have passed the significance threshold for genomewide linkage mapping if the F2 progeny had not been subdivided into different categories based on whether they carried two marine alleles, one marine and one freshwater allele, or two freshwater alleles at the major plates locus.[38] Similarly, recent studies in yeast suggest that treating initial QTL as cofactors can nearly triple the number of overall QTL detected in data from a single cross.[161]

It is already clear from the initial studies that a substantial proportion of many phenotypic traits are controlled by QTL of much larger effect than predicted by the classic "infinitesimal" view of the neo-Darwinian synthesis. An important challenge for the future is to score enough traits in enough large crosses to get a better estimate of the overall distribution of number and size of genetic effects underlying new traits in stickleback populations. The major technical hurdles are the time and effort required to generate large crosses, phenotype large numbers of animals, and carry out genomewide linkage mapping studies to detect both major and minor QTL. As more and more laboratories begin applying the new genetic and genomic tools to the study of different traits in sticklebacks, there should be a rapid growth in detailed genetic information for a variety of different morphological, physiological, and behavioural traits that have long been studied in this classic system.

2.4.2 What Types of Genes Underlie New Phenotypes in Nature?

Over the last 25 years, geneticists and developmental biologists have identified a number of fundamental signalling pathways and transcription factor families that are used by many different animals to control the formation and patterning of body tissues. Many of these signals and transcription factors were first identified by laboratory mutations in flies, worms, and mice, or by disease-causing mutations in humans. Transformation of one body part into another, loss of entire organs or body regions, and ectopic formation of tissues in new regions are all dramatic phenotypes that highlight the importance of particular genes in normal development.

Are such major developmental control genes the same ones used to generate new phenotypes in wild populations? The dramatic effects of these mutations in laboratory animals often have obvious parallels to the anatomical differences that distinguish different species.[162,163] However, most of the laboratory mutations studied by developmental biologists also have deleterious pleiotropic effects that would reduce the overall fitness of animals in the wild.[156,159,164] The most extreme critics of this research have sharply questioned whether mutations in developmental control genes could ever generate useful new phenotypes in nature given the typical fertility and viability problems usually associated with such mutations in the laboratory.[165]

Only two of the QTL found by genetic mapping in sticklebacks (Table 2.1) have already been traced to particular genes. Nevertheless, it is striking that the first concrete examples of genes controlling major morphological changes in wild sticklebacks are both major developmental control genes that have long been studied based on major gene mutations in the laboratory. *Pitx1* is a homeodomain containing a transcription factor involved in development of multiple tissues, including the pituitary gland, jaw, and limbs.[94,97,98] *Eda* is a secreted signalling molecule required for induction or growth of many different structures, including hair, teeth, secretory glands, and dermal bones in the skull.[166] Although it is true that mutations in these genes cause extensive pleiotropic defects in laboratory mice and would be predicted to reduce overall viability or fitness in the wild, sticklebacks provide very strong evidence that evolution in natural populations can occur by changes in these same kinds of genes.

2.4.3 What Types of Mutations Control Evolutionary Change?

Regulatory mutations may help resolve the old debate about the types of genes that underlie evolution of new traits in wild populations.[25,100–103] Mutations in the protein-coding regions of genes typically alter the function of that gene in all tissues where it is normally expressed. In contrast, mutations in regulatory control regions may alter the function of a gene at a particular time and location while preserving its functions in other tissues.

Most of the previously studied laboratory mutations in either *Pitx1* or *Eda* are changes that alter the encoded protein, including deletions, stop codons, or splicing mutations that would reduce or eliminate the function of the gene in all tissues. In contrast, sticklebacks from pelvic-reduced and low-plated populations have *Pitx1* and *Eda* coding sequences that can be identical to those of marine fish.[25,32] The lack of coding-region changes in these populations, in combination with the genetic and expression evidence for tissue-specific changes in gene expression,[23,25] strongly argue that evolution has proceeded by making regulatory changes in the stickleback *Pitx1* and *Eda* genes.

A major challenge for the future is to identify the actual DNA sequence changes that have altered the tissue-specific expression of genes such as *Pitx1* or *Eda* in populations that have evolved new skeletal morphologies. The regulatory regions surrounding major developmental control genes are often highly modular, with separate DNA-binding sequences that control expression at different times and places during development. These *cis*-acting regulatory modules may be located at large distances from the coding regions of the gene. Although we do not yet have the ability to recognise and decode this regulatory information from simple inspection of the primary DNA sequence, a variety of methods have been established in other organisms to screen for regulatory information using functional assays in transgenic animals. The recent development of gene transfer and reporter methods for sticklebacks[66] should make it possible to carry out a detailed dissection of the DNA sequence differences responsible for altered function of key genes such as *Pitx1* or *Eda* in different sticklebacks. Isolation of such sequences will make it possible to test whether the key regulatory alterations consist of single or multiple lesions and whether the same or different molecular events have occurred in independent populations.

2.4.4 Are There Few or Many Ways of Evolving Particular Traits?

One of the most striking aspects of the evolutionary radiation of sticklebacks is the repeated evolution of similar traits in many widely separated freshwater environments. The consistent association of particular phenotypes with particular ecological conditions provides a strong argument that the corresponding traits are adaptive and have been repeatedly selected in different locations.[167] Widespread parallel evolution of sticklebacks provides an outstanding opportunity to ask whether there are few or many molecular mechanisms to evolve a particular phenotype in response to natural selection in the wild.

A variety of recent genetic experiments show that pelvic reduction and lateral-plate-patterning changes are based on the same major genes in many different populations. This striking conclusion is now supported by failure of phenotypes to complement in crosses between different populations,[25,38,39,59] by independent genetic mapping of the same trait to similar chromosome regions,[25,38,39] and finally, by direct sequence analysis of the *Eda* gene in many different low-plated populations around the world.[32]

It remains to be seen whether similar trends will hold for both the major and the minor QTL controlling evolutionary differences between populations. Mapping studies show that some QTL controlling minor quantitative variations in pelvic size map to different locations in the Priest Lake and Paxton Lake populations.[17,25] Crosses of Alaskan pelvic-reduced populations show partial restoration of pelvic development, suggesting that minor modifier genes may differ between populations more often than major loci.[39] On the other hand, modifier QTL controlling quantitative variation in lateral plate number clearly map to similar chromosome locations in at least two different stickleback populations.[17,38]

Clearly, more data are needed to determine the overall frequency with which the same genetic mechanisms are used in different populations. Given the vast range of the stickleback radiation, it would not be surprising if some populations have used different genetic mechanisms from the majority of populations. Nonetheless, the existing data for pelvic reduction and lateral plate patterning already suggest that widespread parallel evolution often occurs using very similar genetic mechanisms. Similar results have recently been reported for parallel evolution of denticle and pigmentation patterns in *Drosophila,*[110,168] melanism in mammals and birds,[169–172] and tetrodotoxin resistance in natural populations of garter snakes.[173]

Why might some genes be preferential substrates for evolutionary change in natural populations? Some mutations may be able to persist as standing variants within populations, making it possible for old mutations to be repeatedly selected when populations encounter new environments. Some genes may be more susceptible to *de novo* mutation than others, owing to sequence content, overall gene size, presence of repeats, or presence of sequences prone to recombination or transposition. Alternatively, if morphological evolution is typically based on regulatory rather than coding-region mutations, genes containing many independent regulatory modules may be altered in more specific ways than genes with a simple promoter driving expression in all tissues.

The overall linkage relationships of genes along a chromosome could also influence which regions are more likely to be used for evolutionary change. Many different phenotypes may be coselected when animals colonise a new environment. For example, marine fish are exposed to very different salinities, predators, parasites, and types of defensive cover when they colonise freshwater streams and lakes. If some chromosome regions contain multiple genes that can contribute to coselected phenotypes, those regions may be favoured substrates for evolutionary change.

Finally, some evolutionary constraints may act at levels much higher than individual genes or gene clusters. Recent work on systems biology has emphasised the degree to which some developmental pathways incorporate multiple feedback loops. These loops stabilise the overall output of signalling systems, producing robust circuits that show little change in output even in the face of perturbation of any individual component.[174] Perhaps those points in the network where changes in single components do produce significant changes in overall function will turn out to be preferential substrates for evolution of new phenotypes.

We do not yet have sufficient data to determine which of these mechanisms may be contributing to the striking reuse of particular genes when similar phenotypes evolve repeatedly in sticklebacks or other organisms. Nevertheless, the recent cloning of the stickleback lateral plates locus provides a good example of the way these issues can now be formulated in much more concrete form for future study. In mice and humans, mutations in three different components of the ectodysplasin signalling pathway can produce very similar phenotypes.[140,141] These genes encode the ectodysplasin signalling molecule (EDA), its receptor (EDAR), and an intracellular adapter molecule that binds to the cytoplasmic tail of the receptor (EDARRAD). A range of mutations has been identified in all three genes in human medical clinics and laboratory mice. In contrast to the diversity of mutations that give similar phenotypes in mammals, the widespread evolution of the low-plated phenotype in sticklebacks is largely based on mutations in only one of the three genes (EDA).

Why has EDA been used repeatedly as the major locus instead of the other known components of the pathway? All but one of the low-plated populations surveyed share a single 16-kb haplotype that arose several million years ago. This ancient haplotype is present at low but detectable levels in fully plated anadromous sticklebacks, corresponding to frequencies of 1 in 50 to 500 chromosomes.[32] These data confirm Lindsey's 40-year-old prediction that rapid evolution of stickleback traits may be based on repeated selection of variants already present in the ancestral marine population.[58]

Although selection from standing variation is clearly a key mechanism underlying parallel phenotypic evolution of low-plated populations, it is still not clear why variants of EDA predominate rather than variants of EDAR or EDARRAD. Perhaps variants at the three different loci may differ in their effects in heterozygotes, altering their ability to be spread back through the marine population following low-frequency hybridization between fully plated anadromous fish and freshwater low-plated stream-resident populations. The different signalling components may have different copy numbers in the stickleback genome, an issue that can be directly addressed using the complete stickleback sequence now being generated. It is also possible that the regulatory sequences surrounding EDA are more modular than the regulatory control sequences surrounding the other genes. The presence of a separate *cis*-acting regulatory module driving expression in developing lateral plates would make it possible to generate specific plate phenotypes while avoiding the pleiotropic consequences of general loss of ectodysplasin signalling associated with coding-region mutations found in mice and humans. Recent development of transgenic methods in sticklebacks should make it possible to test whether specific *cis*-acting modules for lateral plate development surround the EDA locus or other genes. Finally, the coding and regulatory regions of several other genes are present along with the EDA locus in the conserved 16-kb ancient haplotype block controlling repeated evolution of lateral plates. The neighbouring genes encode products previously implicated in immune functions, nervous system functions, and potential salt regulation in other animals. The shared haplotype present in many different populations may provide a suite of coselected genes in sticklebacks, with the neighbouring genes contributing to other phenotypes that also change when sticklebacks evolve in freshwater.[32]

2.5 CONCLUDING REMARKS

We are still at a very early stage in the use of sticklebacks for studying the molecular genetics of evolutionary change. Only a few traits have been mapped, only two of the mapped traits have been traced to specific genes, and the causative DNA sequence changes are still not known for either the *Pitx1* or *Eda* genes. Nevertheless, the progress since the publication of *The Evolutionary Biology of the Threespine Stickleback* in 1994[5] has been remarkable. For the first time, we now have comprehensive genetic and genomic tools that can be utilised in the study of any interesting phenotype in sticklebacks. Classic traits have been traced to specific genes, and in the case of the lateral plate morph, to specific alleles whose frequency can be measured in both ancestral marine and derived freshwater populations. This progress is already making it possible to reframe classic issues in evolution, developmental biology, and population genetics in terms of specific chromosome regions, genes, and alleles that clearly underlie major phenotypic changes in natural populations. One of the most exciting prospects is that the same methods can in principle be extended to a large range of other interesting anatomical, physiological, and behavioural phenotypes that have evolved in different stickleback populations. Sticklebacks are one of the very few vertebrates that have already been studied in depth by many different approaches, including anatomy, behaviour, physiology, ecology, and paleontology. The recent developments in stickleback genomics now bring a full complement of molecular genetic methods to this well-studied fish. The recent availability of a stickleback genome sequence will further accelerate research and make it possible to examine the molecular basis of evolutionary change across the entire genome. Although long considered a "nonmodel" organism, sticklebacks are clearly emerging as a new "supermodel" organism for bridging disparate fields and revealing a detailed picture of how many different traits evolve in natural populations of vertebrates.[175]

ACKNOWLEDGMENTS

Our interest in sticklebacks was originally stimulated by an outstanding review of stickleback biology, evolution, genetics, and collecting techniques by Mike Bell in the book *Evolutionary Genetics of Fishes*. When we found additional papers by Don McPhail, Don Hagen, and Tom Reimchen, we knew we wanted to become stickleback researchers too. We have special thanks to Dolph Schluter for our long-standing and productive collaboration on stickleback genetics. The "Big Cross" sprang forth following our very first visit to Vancouver in September 1998 and has been a joy to analyse together ever since. When we began molecular studies of sticklebacks, we thought it might take 10 or 20 years to identify the molecular basis of particular traits. It has gone faster than we ever dared hope because of the wonderful group of research assistants, graduate students, postdocs, and faculty colleagues who have brought their own talents and enthusiasm to the project. To Kris Nereng, Kenny Ohgi, Bonnie Cole, Ben Blackmann, Pam Colosimo, Melissa Marks, Mike Shapiro, Kim Hosemann, Brian Summers, Craig Miller, Frank Chan, Anne Knecht, Clint Matson, Joe Ross, Brian Fritz, Jun Kitano, Tiffany Malek, Amanda Bruner, Anna Greenwood, Abby Wark, Rick Myers, Jane Grimwood, Jeremy Schmutz, and Will Talbot: thank you and happy fishing!!!

REFERENCES

1. Wootton, R.J., *The Biology of the Sticklebacks*, Academic Press, London, 1976.
2. Coad, B.W., A bibliography of sticklebacks, *Syllogeus,* 35, 1, 1981.
3. Wootton, R.J., *A Functional Biology of Sticklebacks*, Croom Helm, London, 1984.

4. Ziuganov, V.V., *Stickleback Family (Gasterosteidae) of the World Fauna*, Nauka, Leningrad, 1991.
5. Bell, M.A. and Foster, S.A., *The Evolutionary Biology of the Threespine Stickleback*, Oxford University Press, Oxford, 1994.
6. Paepke, H.-J., *Die Stichlinge*, 2nd ed., Westarp Wissenschaften, Germany, 1996.
7. Keivany, Y., http://www.geocities.com/CapeCanaveral/Hall/1345/stickbibl.html, 2002.
8. Rico, C. et al., Characterization of hypervariable microsatellite loci in the threespine stickleback *Gasterosteus aculeatus*, *Molecular Ecology,* 2(4), 271, 1993.
9. Taylor, E.B., Microsatellites isolated from the threespine stickleback *Gasterosteus aculeatus*, *Molecular Ecology,* 7, 925, 1998.
10. Largiader, C.R. et al., Isolation and characterization of microsatellite loci from the three-spined stickleback (*Gasterosteus aculeatus* L.), *Molecular Ecology,* 8(2), 342, 1999.
11. Sato, A. et al., Cloning of major histocompatibility complex (MHC) genes from threespine stickleback, *Gasterosteus aculeatus*, *Molecular Marine Biology and Biotechnology,* 7(3), 221, 1998.
12. Ahn, D.G. and Gibson, G., Expression patterns of threespine stickleback *Hox* genes and insights into the evolution of the vertebrate body axis, *Development Genes and Evolution,* 209, 482, 1999.
13. Ortí, G. et al., Global survey of mitchondrial DNA sequences in the threespine stickleback: evidence for recent migrations, *Evolution: International Journal of Organic Evolution,* 48(3), 608, 1994.
14. Deagle, B.E., Reimchen, T.E., and Levin, D.B., Origins of endemic stickleback from the Queen Charlotte Islands: mitochondrial and morphological evidence, *Canadian Journal of Zoology,* 74(6), 1045, 1996.
15. Taylor, E.B. and McPhail, J.D., Evolutionary history of an adaptive radiation in species pairs of threespine sticklebacks (*Gasterosteus*): insights from mitochondrial DNA, *Biological Journal of the Linnean Society,* 66, 271, 1999.
16. Miya, M., Kawaguchi, A., and Nishida, M., Mitogenomic Exploration of higher Teleostean phylogenies: a case study for moderate-scale evolutionary genomics with 38 newly determined complete mitochondrial DNA sequences, *Molecular Biology and Evolution,* 18(11), 1993, 2001.
17. Peichel, C.L. et al., The genetic architecture of divergence between threespine stickleback species, *Nature,* 414(6866), 901, 2001.
18. Kingsley, D.M. et al., New genomic tools for molecular studies of evolutionary change in sticklebacks, *Behaviour,* 141, 1331, 2004.
19. Kingsley, D.M., Sequencing the genome of threespine sticklebacks (*Gasterosteus aculeatus*), National Human Genome Research Institute White Paper, 2003, http://www.genome.gov/12512292.
20. Jones, I. et al., Molecular cloning and characterization of spiggin: an androgen-regulated extraorganismal adhesive with structural similarities to von willebrand factor-related proteins, *Journal of Biological Chemistry,* 276, 17857, 2001.
21. Altschul, S.F. et al., Basic local alignment search tool, *Journal of Molecular Biology,* 215, 403, 1990.
22. Ahn, D.G. and Gibson, G., Axial variation in the threespine stickleback: relationship to *Hox* gene expression, *Development Genes and Evolution,* 209, 473, 1999.
23. Cole, N.J. et al., Expression of limb initiation genes and clues to the morphological diversification of threespine sticklebacks, *Current Biology,* 13, R951, 2003.
24. Cresko, W.A. et al., Genome duplication, subfunction partitioning, and lineage divergence: *Sox9* in stickleback and zebrafish, *Developmental Dynamics,* 228, 480, 2003.

25. Shapiro, M.D. et al., Genetic and developmental basis of evolutionary pelvic reduction in threespine sticklebacks, *Nature,* 428(6984), 717, 2004.
26. Tanaka, M. et al., Developmental genetic basis for evolution of pelvic fin loss in the pufferfish *Takifugu rubripes*, *Developmental Biology,* 281, 227, 2005.
27. Brown, P.O. and Botstein, D., Exploring the new world of the genome with DNA microrrays, *Nature Genetics,* 21, 33, 1999.
28. Heckel, G. et al., Microsatellite markers for the three-spined stickleback (*Gasterosteus aculeatus* L.) and their applicability in a freshwater and an anadromous population, *Conservation Genetics,* 3(1), 79, 2002.
29. Chen, T.-R. and Reisman, H.M., A comparative chromosome study of the North American species of sticklebacks (Teleostei: Gasterosteidae), *Cytogenetics,* 9, 321, 1970.
30. Hinegardner, R., Evolution of cellular DNA content in teleost fishes, *American Naturalist,* 102(928), 517, 1968.
31. Hinegardner, R. and Rosen, D.E., Cellular DNA content and evolution of teleostean fishes, *American Naturalist,* 106(951), 621, 1972.
32. Colosimo, P.F. et al., Widespread parallel evolution in sticklebacks by repeated fixation of ectodysplasin alleles, *Science,* 307, 1928, 2005.
33. Sachidanandam, R. et al., A map of human genome sequence variation containing 1.42 million single nucleotide polymorphisms, *Nature,* 409, 928, 2001.
34. Lander, E.S., The new genomics: global views of biology, *Science,* 274, 536, 1996.
35. Risch, N. and Merikangas, K., The future of genetic studies of complex human diseases, *Science,* 273, 1516, 1996.
36. Collins, F.S., Guyer, M.S., and Chakravarti, A., Variations on a theme: cataloging human DNA sequence variation, *Science,* 278, 1580, 1997.
37. Kruglyak, L., The use of a genetic map of biallelic markers in linkage studies, *Nature Genetics,* 17(1), 21, 1997.
38. Colosimo, P.F. et al., The genetic architecture of parallel armor plate reduction in threespine sticklebacks, *PLoS Biology,* 2, 635, 2004.
39. Cresko, W.A. et al., Parallel genetic basis for repeated evolution of armor loss in Alaskan threespine stickleback populations, *Proceedings of the National Academy of Sciences of the United States of America,* 101, 6050, 2004.
40. Peichel, C.L. et al., The master sex-determination locus in threespine sticklebacks is on a nascent Y chromosome, *Current Biology,* 14(16), 1416, 2004.
41. Kimmel, C.B. et al., Evolution and development of facial bone morphology in threespine sticklebacks, *Proceedings of the National Academy of Sciences of the United States of America,* 102, 5791, 2005.
42. Orr, H.A. and Coyne, J.A., The genetics of adaptation — a reassessment, *American Naturalist,* 140(5), 725, 1992.
43. Orr, H.A., The genetics of species differences, *Trends in Ecology and Evolution,* 16, 343, 2001.
44. Orr, H.A., The genetic theory of adaptation: a brief history, *Nature Reviews Genetics,* 6, 119, 2005.
45. Reusch, T.B.H., Schaschl, H., and Wegner, K.M., Recent duplication and inter-locus gene conversion in major histocompatibility class II genes in a teleost, the three-spined stickleback, *Immunogenetics,* 56, 427, 2004.
46. Fjell, C.D. et al., Internet Contig Explorer (iCE) — a tool for visualizing clone fingerprint maps, *Genome Research,* 13, 1244, 2003.
47. Reusch, T.B.H. et al., Female sticklebacks count alleles in a strategy of sexual selection explaining MHC polymorphism, *Nature,* 414, 300, 2001.

48. Wegner, K.M., Reusch, T.B.H., and Kalbe, M., Multiple parasites are driving major histocompatibility complex polymorphism in the wild, *Journal of Evolutionary Biology,* 16(2), 224, 2003.
49. Kurtz, J. et al., Major histocompatibility complex diversity influences parasite resistance and innate immunity in sticklebacks, *Proceedings of the Royal Society of London Series B,* 271, 197, 2004.
50. Milinski, M. et al., Mate choice decisions of stickleback females predictably modified by MHC peptide ligands, *Proceedings of the National Academy of Sciences of the United States of America,* 102(12), 4414, 2005.
51. Reusch, T.B. and Langefors, A., Inter- and intralocus recombination drive MHC class IIB diversification in a teleost, the three-spined stickleback *Gasterosteus aculeatus*, *Journal of Molecular Evolution,* 61, 531, 2005.
52. International Human Genome Sequencing Consortium, Finishing the euchromatic sequence of the human genome, *Nature,* 431, 931, 2004.
53. Collins, F.S., Morgan, M., and Patrinos, A., The human genome project: lessons from large-scale biology, *Science,* 300, 286, 2003.
54. International Human Genome Sequencing Consortium, Initial sequencing and analysis of the human genome, *Nature,* 409, 860, 2001.
55. Venter, J.C. et al., The sequence of the human genome, *Science,* 291, 1304, 2001.
56. Waterston, R.H. et al., Initial sequencing and comparative analysis of the mouse genome, *Nature,* 420, 520, 2002.
57. Cuñado, N. et al., Synaptonemal complex analysis in oocytes and spermatocytes of threespine stickleback *Gasterosteus aculeatus* (Teleostei, Gasterosteidae), *Genetica,* 114, 53, 2002.
58. Lindsey, C.C., Experimental study of meristic variation in a population of threespine stickleback, *Gasterosteus aculeatus*, *Canadian Journal of Zoology,* 40, 271, 1962.
59. Avise, J.C., Genetics of plate morphology in an unusual population of threespine sticklebacks (*Gasterosteus aculeatus*), *Genetical Research,* 27, 33, 1976.
60. Withler, R.E. and McPhail, J.D., Genetic variability in freshwater and anadromous sticklebacks (*Gasterosteus aculeatus*) of Southern British Columbia, *Canadian Journal of Zoology,* 63, 528, 1985.
61. Withler, R.E., McPhail, J.D., and Devlin, R.H., Electrophoretic polymorphism and sexual dimorphism in the freshwater and anadromous threespine sticklebacks (*Gasterosteus aculeatus*) of the Little Campbell River, British Columbia, *Biochemical Genetics,* 24, 701, 1986.
62. Griffiths, R. et al., DNA sex identification in the three-spined stickleback, *Journal of Fish Biology,* 57, 1331, 2000.
63. Peichel, C.L., Unpublished data.
64. Hahlbeck, E., Griffiths, R., and Bengtsson, B.-E., The juvenile three-spined stickleback (*Gasterosteus aculeatus* L.) as a model organism for endocrine disruption I. Sexual differentiation, *Aquatic Toxicology,* 70, 287, 2004.
65. Hahlbeck, E. et al., The juvenile three-spined stickleback (*Gasterosteus aculeatus* L.) as a model organism for endocrine disruption II — kidney hypertrophy, vitellogenin and spiggin induction, *Aquatic Toxicology,* 70, 311, 2004.
66. Hosemann, K.E. et al., A simple and efficient microinjection protocol for making transgenic sticklebacks, *Behaviour,* 141, 1345, 2004.
67. Bull, J.J., *Evolution of Sex Determining Mechanisms*, Benjamin Cummings, Menlo Park, 1983.

68. Rice, W.R., The accumulation of sexually antagonistic genes as a selective agent promoting the evolution of reduced recombination between primitive sex chromosomes, *Evolution: International Journal of Organic Evolution,* 41(4), 911, 1987.
69. Charlesworth, B., The evolution of sex chromosomes, *Science,* 251, 1030, 1991.
70. Rice, W.R., Genetic hitchhiking and the evolution of reduced genetic activity of the Y sex chromosome, *Genetics,* 116(1), 161, 1987.
71. Filatov, D.A. et al., Low variability in a Y-linked plant gene and its implications for Y-chromosome evolution, *Nature,* 404, 388, 2000.
72. Okada, S. et al., The Y chromosome in liverwork *Marchantia polymorpha* has accumulated unique repeat sequences harboring a male-specific gene, *Proceedings of the National Academy of Sciences of the United States of America,* 98(16), 9454, 2001.
73. Liu, Z. et al., A primitive Y chromosome in papaya marks incipient sex chromosome evolution, *Nature,* 427, 348, 2004.
74. Steinemann, M. and Steinemann, S., Enigma of Y chromosome degeneration: Neo-Y and Neo-X chromosomes of *Drosophila miranda* a model for sex chromosome evolution, *Genetica,* 102/103, 409, 1998.
75. Nanda, I. et al., Amplification of a long terminal repeat-like element on the Y chromosome of the platy fish, *Xiphophorous maculatus*, *Chromosoma,* 109, 173, 2000.
76. Volff, J.-N. and Schartl, M., Sex determination and sex chromosome evolution in the medaka, Oryzias latipes, and the platyfish, Xiphophorous maculatus, *Cytogenetic and Genome Research*, 99, 170, 2002.
77. Bachtrog, D., Adaptation shapes patterns of genome evolution on sexual and asexual chromosomes in *Drosophila*, *Nature Genetics,* 34(2), 215, 2003.
78. Hoogland, R.D., Morris, D., and Tinbergen, N., The spines of sticklebacks (*Gasterosteus* and *Pygosteus*) as a means of defense against predators (*Perca* and *Esox*), *Behaviour,* 10, 205, 1957.
79. Reimchen, T.E., Structural relationship between spines and lateral plates in threespine stickleback (*Gasterosteus aculeatus*), *Evolution: International Journal Organic Evolution,* 37, 931, 1983.
80. Bell, M.A., Reduction and loss of the pelvic girdle in *Gasterosteus* (Pisces): a case of parallel evolution, *Natural History Museum of Los Angeles County Contributions in Science,* 257, 1, 1974.
81. Bell, M.A., Interacting evolutionary constraints in pelvic reduction of threespine sticklebacks, *Gasterosteus aculeatus* (Pisces: Gasterosteidae), *Biological Journal of the Linnean Society,* 31, 347, 1987.
82. Moodie, G.E.E. and Reimchen, T.E., Phenetic variation and habitat differences in *Gasterosteus* populations of the Queen Charlotte Islands, *Systematic Zoology,* 25, 49, 1976.
83. Campbell, R.N., Morphological variation in the three-spined stickleback (*Gasterosteus aculeatus*) in Scotland, *Behaviour,* 93, 161, 1985.
84. Campbell, R.N., Sticklebacks (*Gasterosteus aculeatus* [L.], and *Pungitius pungitius* [L.]) in the Outer Hebrides, Scotland, *Hebridean Naturalist,* 3, 8, 1979.
85. Edge, T.A. and Coad, B.W., Reduction of the pelvic skeleton in the three-spined stickleback *Gasterosteus aculeatus* in 2 lakes of Quebec Canada, *Canadian Field-Naturalist,* 97, 334, 1983.
86. Reimchen, T.E., Status of unarmoured and spine-deficient populations (Queen Charlotte Unarmoured Stickleback) of threespine stickleback, *Gasterosteus* sp., on the Queen Charlotte Islands, British Columbia, *Canadian Field-Naturalist,* 98, 120, 1984.

87. Bell, M.A. et al., Evolution of pelvic reduction in threespine stickleback fish: a test of competing hypothesis, *Evolution: International Journal of Organic Evolution,* 47, 906, 1993.
88. Reimchen, T.E., Spine deficiency and polymorphism in a population of *Gasterosteus aculeatus*-an adaptation to predators, *Canadian Journal of Zoology,* 58, 1232, 1980.
89. Reist, J.D., Selective predation upon pelvic phenotypes of brook stickleback, *Culaea inconstans*, by northern pike, *Esox lucius*, *Canadian Journal of Zoology,* 58, 1245, 1980.
90. Reist, J.D., Predation upon pelvic phenotypes of brook stickleback, *Culaea inconstans*, by selected invertebrates, *Canadian Journal of Zoology,* 58, 1253, 1980.
91. Giles, N., The possible role of environmental calcium levels during the evolution of phenotypic diversity in Outer Hebridean populations of three-spined stickleback, *Gasterosteus aculeatus*, *Journal of Zoology,* 199, 535, 1983.
92. Ziuganov, V.V. and Zotin, A.A., Pelvic girdle polymorphism and reproductive barriers in the ninespine stickleback *Pungitius pungitius* (L.) from northwest Russia, *Behaviour,* 132, 1095, 1995.
93. Gibson-Brown, J.J. et al., Evidence of a role for T-box genes in the evolution of limb morphogenesis and the specification of forelimb/hindlimb identity, *Mechanisms of Development,* 56(1–2), 93, 1996.
94. Shang, J., Luo, Y., and Clayton, D., Backfoot is a novel homeobox gene expressed in the mesenchyme of developing hind limb, *Developmental Dynamics,* 209, 242, 1997.
95. Logan, M., Simon, H.G., and Tabin, C., Differential regulation of T-box and homeobox transcription factors suggests roles in controlling chick limb-type identity, *Development,* 125(15), 2825, 1998.
96. Logan, M. and Tabin, C.J., Role of Pitx1 upstream of Tbx4 in specification of hindlimb identity, *Science,* 283(5408), 1736, 1999.
97. Lanctôt, C. et al., Hindlimb patterning and mandible development require the *Pitx1* gene, *Development,* 126, 1805, 1999.
98. Szeto, D.P. et al., Role of the Bicoid-related homeodomain factor Pitx1 in specifying hindlimb morphogenesis and pituitary development, *Genes and Development,* 13, 484, 1999.
99. Marcil, A. et al., Pitx1 and Pitx2 are required for development of hindlimb buds, *Development,* 130, 45, 2003.
100. King, M. and Wilson, A.C., Evolution at two levels in humans and chimpanzees, *Science,* 188, 107, 1975.
101. Stern, D.L., Evolutionary developmental biology and the problem of variation, *Evolution: International Journal of Organic Evolution,* 54, 1079, 2000.
102. Davidson, E.H., *Genomic Regulatory Systems: Development and Evolution*, Academic Press, San Diego, CA, 2001.
103. Carroll, S.B., Grenier, J.K., and Weatherbee, S.D., *From DNA to Diversity: Molecular Genetics and the Evolution of Animal Design*, 2nd ed., Blackwell Sciences, Malden, MA, 2004.
104. Belting, H.G., Shashikant, C.S., and Ruddle, F.H., Modification of expression and *cis*-regulation of Hoxc8 in the evolution of diverged axial morphology, *Proceedings of the National Academy of Sciences of the United States of America,* 95, 2355, 1998.
105. Stern, D.L., A role of *Ultrabithorax* in morphological differences between *Drosophila* species, *Nature,* 396, 463, 1998.
106. Wang, R.L. et al., The limits of selection during maize domestication, *Nature,* 398, 236, 1999.

107. Sucena, E. and Stern, D.L., Divergence of larval morphology between *Drosophila sechellia* and its sibling species caused by cis-regulatory evolution of *ovo/shaven-baby*, *Proceedings of the National Academy of Sciences of the United States of America,* 97, 4530, 2000.
108. Wang, X. and Chamberlin, H.M., Multiple regulatory changes contribute to the evolution of the *Caenorhabditis lin-48 ovo* gene, *Genes and Development,* 16, 2345, 2002.
109. Wittkopp, P.J., Vaccaro, K., and Carroll, S.B., Evolution of *yellow* gene regulation and pigmentation in *Drosophila*, *Current Biology,* 12, 1547, 2002.
110. Sucena, E. et al., Regulatory evolution of *shavenbaby/ovo* underlies multiple cases of morphological parallelism, *Nature,* 424, 935, 2003.
111. VanLaere, A.-S. et al., A regulatory mutation in *IGF2* causes a major QTL effect on muscle growth in the pig, *Nature,* 425, 832, 2003.
112. Gompel, N. et al., Chance caught on the wing: *cis*-regulatory evolution and the origin of pigment patterns in *Drosophila*, *Nature,* 433, 481, 2005.
113. Bell, M.A., Francis, R.C., and Havens, A.C., Pelvic reduction and its directional asymmetry in threespine sticklebacks from the Cook Inlet region, Alaska, *Copeia,* 1985, 437, 1985.
114. Campione, M. et al., The homeobox gene *Pitx2*: mediator of asymmetric left-right signaling in vertebrate heart and gut looping, *Development,* 126(6), 1225, 1999.
115. Nelson, J. S., Absence of pelvic complex in ninespine sticklebacks, *Pungitius pungitius*, collected in Ireland and Wood Buffalo National Park region, Canada, with notes on meristic variation, *Copeia* (4), 707, 1971.
116. Takata, K. and Goto, A., Pelvic spine deficiency of *Pungitius tymensis* in Hokkaido, *Japanese Journal of Ichthyology,* 32, 100, 1985.
117. Zyuganov, V.V. and Rozanov, A.S., Genetics of the structural polymorphism of the pelvic skeleton of the ninespine stickleback, *Doklady Akademii nauk SSSR,* 293, 459, 1987.
118. Keivany, Y. and Nelson, J.S., Comparative osteology of the Greek Ninespine stickleback, *Pungitius hellenicus* (Teleostei, Gasterosteidae), *Journal of Ichthyology,* 38(6), 430, 1998.
119. Nelson, J.S. and Atton, F.M., Geographical and morphological variation in the presence and absence of the pelvic skeleton in the brook stickleback, *Culaea inconstans* (Kirkland), in Alberta and Saskatchewan, *Canadian Journal of Zoology,* 49, 343, 1971.
120. Nelson, J.S., Evidence of a genetic basis for absence of pelvic skeleton in brook stickleback, *Culaea inconstans*, and notes on geographical distribution and origin of loss, *Journal of the Fisheries Research Board of Canada,* 34(9), 1314, 1977.
121. Cuvier, G.L. and Valenciennes, M.A., *Histoire naturelle des poissons: Tome quatrieme*, Chez F.G. Levrault, Paris, 1829.
122. Hagen, D.W. and Gilbertson, L.G., Geographic variation and environmental selection in *Gasterosteus aculeatus* L in the Pacific Northwest, America, *Evolution: International Journal of Organic Evolution,* 26(1), 32, 1972.
123. Heuts, M.J., Experimental studies on adaptive evolution in *Gasterosteus aculeatus* L., *Evolution: International Journal of Organic Evolution,* 1, 89, 1947.
124. Baumgartner, J.V. and Bell, M.A., Lateral plate morph variation in California populations of the threespine stickleback, *Gasterosteus aculeatus*, *Evolution: International Journal of Organic Evolution,* 38, 665, 1984.
125. MacLean, J., Ecological genetics of threespine sticklebacks in Heisholt Lake, *Canadian Journal of Zoology,* 58, 2026, 1980.

126. Taylor, E.B. and McPhail, J.D., Prolonged and burst swimming in anadromous and fresh-water threespine stickleback, *Gasterosteus aculeatus*, *Canadian Journal of Zoology,* 64(2), 416, 1986.
127. Bergstrom, C.A., Fast-start performance and reduction in lateral plate number in threespine stickleback, *Canadian Journal of Zoology,* 80, 207, 2002.
128. Hagen, D.W. and Gilbertson, L.G., Selective predation and the intensity of selection acting upon lateral plates of threespine sticklebacks, *Heredity,* 30(June), 273, 1973.
129. Moodie, G.E.E., McPhail, J.D., and Hagen, D.W., Experimental demonstration of selective predation on *Gasterosteus aculeatus*, *Behaviour,* 47, 95, 1973.
130. Reimchen, T.E., Injuries on sticklebacks from attacks by a toothed predator (*Oncorhyncus*) and implications for the evolution of lateral plates, *Evolution: International Journal of Organic Evolution,* 46, 1224, 1992.
131. Reimchen, T.E., Predator-induced cyclical changes in lateral plate frequencies of *Gasterosteus*, *Behaviour,* 132, 1079, 1995.
132. Reimchen, T.E., Predator handling failures of lateral plate morphs in *Gasterosteus aculeatus*: implications for stasis and distribution of the ancestral plate condition, *Behaviour,* 137, 1081, 2000.
133. Hagen, D.W. and Moodie, G.E.E., Polymorphism for plate morphs in *Gasterosteus aculeatus* on the east coast of Canada and an hypothesis for their global distribution, *Canadian Journal of Zoology,* 60(5), 1032, 1982.
134. Münzing, J., Biologie, variabilität und genetik von *Gasterosteus aculeatus* L. (Pisces) untersuchugen im elbegebiet, *Internationale Revue der Gesamten Hydrobiologie,* 44, 317, 1959.
135. Hagen, D.W. and Gilbertson, L.G., The genetics of plate morphs in freshwater threespine sticklebacks, *Heredity,* 31(August), 75, 1973.
136. Ziuganov, V.V., Genetics of osteal plate polymorphism and microevolution of threespine stickleback (*Gasterosteus aculeatus* L.), *Theoretical and Applied Genetics,* 65, 239, 1983.
137. Banbura, J., A new model of lateral plate morph inheritance in the threespine stickleback, *Gasterosteus aculeatus*, *Theoretical and Applied Genetics,* 88, 871, 1994.
138. Miller, R.R. and Hubbs, C.L., Systematics of Gasterosteus aculeatus, with particular reference to intergradation and introgression along the Pacific coast of North America: a commentary on a recent contribution, *Copeia,* 1969, 52, 1969.
139. Ransom, D.G. and Zon, L.I., Mapping zebrafish mutations by AFLP, *Methods in Cell Biology,* 60, 195, 1999.
140. Thesleff, I. and Mikkola, M.L., Death receptor signaling giving life to ectodermal organs, *Science's STKE: Signal Transduction Knowledge Environment,* 131, PE22, 2002.
141. Kangas, A.T. et al., Nonindependence of mammalian dental characters, *Nature,* 432(7014), 211, 2004.
142. Kondo, S. et al., The medaka *rs-3* locus required for scale development encodes *ectodysplasin-A* receptor, *Current Biology,* 11, 1202, 2001.
143. Sire, J.Y. and Huysseune, A., Formation of dermal skeleton and dental tissues in a fish: a comparative and evolutionary approach, *Biological Reviews,* 78, 219, 2003.
144. Paakonen, K. et al., The mutation spectrum of the EDA gene in X-linked anhidrotic ectodermal dysplasia, *Human Mutation,* 17(4), 349, 2001.
145. Vincent, M.C. et al., Mutational spectrum of the ED1 gene in x-linked hypohidrotic ectodermal dysplasia, *European Journal of Human Genetics,* 9, 355, 2001.
146. Srivastava, A.K. et al., Ectodysplasin-A1 is sufficient to rescue both hair growth and sweat glands in Tabby mice, *Human Molecular Genetics,* 10(26), 2973, 2001.

147. Bell, M.A., Lateral plate polymorphism and ontogeny of the complete plate morph of threespine sticklebacks (*Gasterosteus aculeatus*), *Evolution: International Journal of Organic Evolution,* 35, 67, 1981.
148. Schluter, D. et al., Parallel evolution and inheritance of quantitative traits, *American Naturalist,* 163, 809, 2004.
149. Klepaker, T., Morphological changes in a marine population of threespined stickleback, *Gasterosteus aculeatus*, recently isolated in freshwater, *Canadian Journal of Zoology,* 71, 1231, 1993.
150. Bell, M.A., Lateral plate evolution in the threespine stickleback: getting nowhere fast, *Genetica,* 112-113, 445, 2001.
151. Kristjánsson, B.K., Skúlason, S., and Noakes, D.L.G., Rapid divergence in a recently isolated population of threespine stickleback (*Gasterosteus aculeatus*), *Evolutionary Ecology Research,* 4, 659, 2002.
152. Bell, M.A., Aguirre, W.E., and Buck, N.J., Twelve years of contemporary armor evolution in a threespine stickleback population, *Evolution: International Journal of Organic Evolution,* 58, 814, 2004.
153. Sargent, R.C. et al., A lateral plate cline, sexual dimorphism, and phenotypic variation in the black-spotted Stickleback, *Gasterosteus wheatlandi*, *Canadian Journal of Zoology,* 62(3), 368, 1984.
154. McPhail, J.D., Geographic variation in North American ninespine sticklebacks, *Pungitius pungitius*, *Journal of the Fisheries Research Board of Canada,* 20, 27, 1963.
155. Münzing, J., Variabilität, Verbreitung und Systematik der Arten und Unterarten in der Gattung Pungitius Coste, 1848 (Pisces, Gasterosteidae), *Zeitschrift fur Zoologische Systematik und Evolutionforschung,* 7, 208, 1969.
156. Darwin, C., *The Origin of Species*, J. Murray, London, 1859.
157. Bateson, W., Heredity and variation in modern lights, in *Darwin and Modern Science: Essays in Commemoration of the Centenary of the Birth of Charles Darwin and of the Fiftieth Anniversary of the Publication of "The Origin of Species,"* Seward, A.C., Ed., Cambridge University Press, Cambridge, 1909.
158. de Vries, H., *The Mutation Theory; Experiments and Observations on the Origin of Species in the Vegetable Kingdom*, Open Court Publishing Company, Chicago, IL, 1909.
159. Fisher, R.A., *The Genetical Theory of Natural Selection*, Oxford University Press, Oxford, 1930.
160. Beavis, W.D., QTL analyses: power, precision, and accuracy, in *Molecular Dissection of Complex Traits*, Paterson, A.H., Ed., CRC Press, Boca Raton, FL, 1998, p. 145.
161. Brem, R.B. et al., Genetic interactions between polymorphisms that affect gene expression in yeast, *Nature,* 436, 701, 2005.
162. Goldschmidt, R., *The Material Basis of Evolution*, Yale University Press, New Haven, CT, 1940.
163. Lewis, E.B., A gene complex controlling segmentation in Drosophila, *Nature,* 276, 565, 1978.
164. Lande, R., Microevolution in relation to macroevolution, *Paleobiology,* 6, 233, 1980.
165. Wells, J., *Icons of Evolution: Science or Myth?: Why Much of What We Teach About Evolution is Wrong*, Regnery Publishing, Washington, DC, 2000.
166. Mikkola, M.L. and Thesloff, I., Ectodysplasin signaling in development, *Cytokine Growth Factor Reviews,* 14, 211, 2003.
167. Schluter, D., *The Ecology of Adaptive Radiation*, Oxford University Press, New York, 2000.

168. Gompel, N. and Carroll, S.B., Genetic mechanisms and constraint governing the evolution of correlated traits in drosophilid flies, *Nature,* 424, 931, 2003.
169. Theron, E. et al., The molecular basis of an avian plumage polymorphism in the wild: a *melanocortin-1-receptor* point mutation is perfectly associated with the melanic plumage morph of the bananaquit, *Coereba flaveola*, *Current Biology,* 11, 550, 2001.
170. Eizirik, E. et al., Molecular genetics and evolution of melanism in the cat family, *Current Biology,* 13, 448, 2003.
171. Nachman, M.W., Hoekstra, H.E., and D'Agostino, S.L., The genetic basis of adaptive melanism in pocket mice, *Proceedings of the National Academy of Sciences of the United States of America,* 100, 5268, 2003.
172. Mundy, N.I. et al., Conserved genetic basis of a quantitative plumage trait involved in mate choice, *Science,* 303, 1870, 2004.
173. Geffeney, S.L. et al., Evolutionary diversification of TTX-resistant sodium channels in a predator-prey interaction, *Nature,* 434, 759, 2005.
174. Lander, A.D., A calculus of purpose, *PLoS Biology,* 2, e164, 2004.
175. Gibson, G., The synthesis and evolution of a supermodel, *Science,* 307, 1890, 2005.

3 Speciation in Sticklebacks

Janette Wenrick Boughman

CONTENTS

3.1 INTRODUCTION

The three-spined stickleback (*Gasterosteus aculeatus* spp.) has risen to prominence as a model system for understanding the mechanisms by which new species are formed in nature. The pioneering work of McPhail and his colleagues set the stage for the recent interest in the system, and its success in revealing how speciation proceeds. This review builds on McPhail's earlier review of stickleback speciation,[1] extending it by broadening the geographic scope of populations considered, and incorporating new theory and data. We focus especially on work published subsequent to another fine review of the topic.[2] We cover several topics that have received recent attention, including the role of predation, phenotypic plasticity, the genetics of species differences, and conservation concerns. Part of our objective is to point out areas of research that are likely to be especially fruitful and where we are sorely lacking data.

One reason research on sticklebacks has made such a contribution to our understanding of speciation is the existence of several sets of phenotypically differentiated and reproductively isolated pairs of sticklebacks found throughout their holarctic range (Figure 3.1). These pairs are composed of sympatric or parapatric populations that maintain some level of reproductive isolation (Table 3.1). Members of these pairs differ in behavioural, morphological, and physiological traits that have been shown in many cases to have a heritable basis. In addition, the adaptive significance of the traits is often known or can be inferred from studies of selection on them. In most cases, these pairs are replicated in that multiple, independent populations of each member of the pair exist. This evolutionary replication provides a unique ability to test hypotheses about mechanisms that drive the divergence and maintenance of these pairs. Combine this with their experimental tractability and the wealth of information on their behaviour, morphology, ecology, and genetics, and you have an emerging model system.

We begin this review by setting the stage for speciation in sticklebacks — laying out their phylogeography, historical factors, geographic distribution, and describing

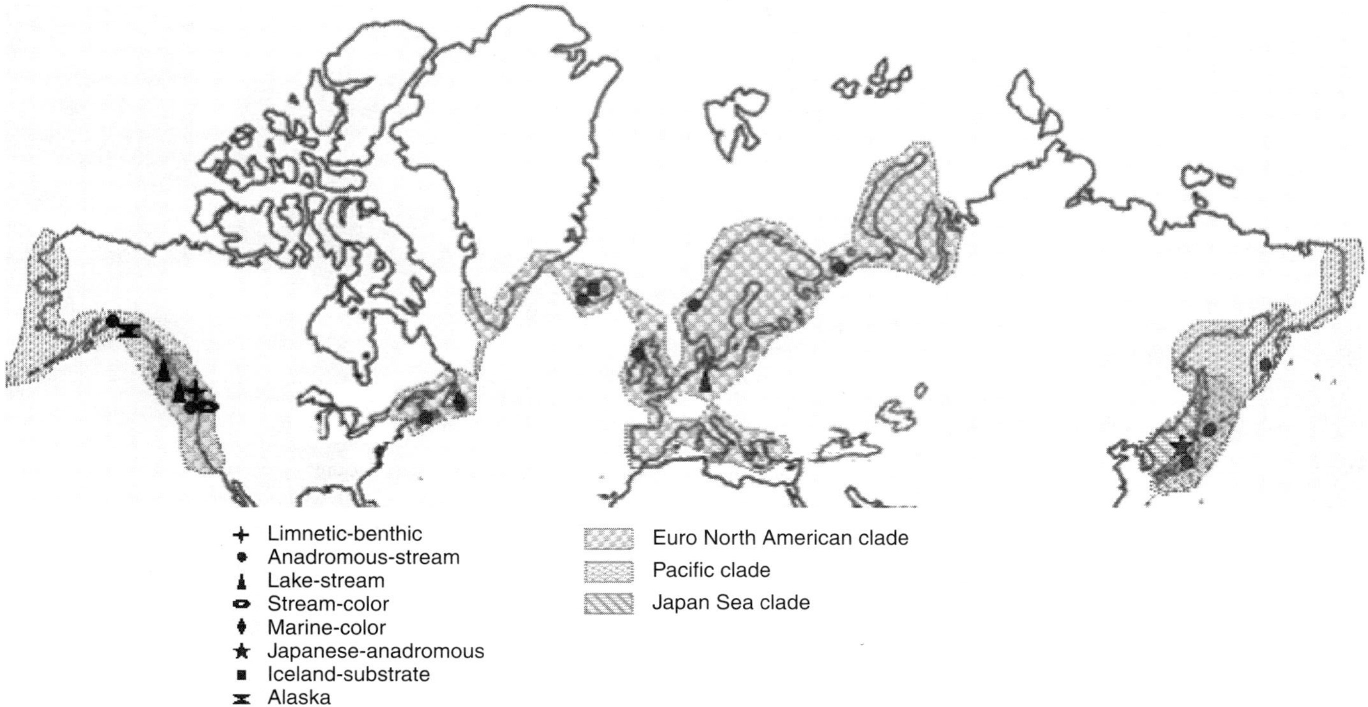

FIGURE 3.1 Worldwide distribution of stickleback clades and divergent pairs. Filled areas indicate the known distribution of three different clades. Further work may extend these distributions. Symbols indicate reported locations of pairs that have been shown to differ in morphological and behavioural traits and are reproductively isolated. The symbol for Alaska indicates the location of lakes and streams where two different ecotypes were trapped. Ongoing work will test if these are good biological species.

TABLE 3.1
Descriptive Data for Stickleback Pairs with Clade, Genetic Differentiation, Major Phenotypic Differences, and Mechanisms of Reproductive Isolation

Pair	Clade	Number Pairs	Genetic Distance	Size	Colour	Shape	Trophic	Armour	Court-ship	Preference	Habitat	Premating RI Traits	RI Mechanisms	References
Limnetic-benthic	EN	4	0.018*; 0.27 (Fst)	Y	Y	Y	Y	Y	Y	Y	Y	Habitat, size, colour, odour	Divergent ns, comp, pred, divergent ss, ss hybrids, postmating extrinsic	1,19,59,60, 62,63,92, 95,96,124
Anadromous-stream	EN, P, J	100s	0–0.008*; 0.193–1.978~; 0.037~	Y	Y sm	Y	Y	Y	Y	U	Y	Size, court?, others	Divergent ns, pred?, ss hybrids?	5,8,39
Stream-colour	EN	1	U	N	Y	N	Y	Y	N	Y?	Y	Colour	Comp, pred, divergent ss?, postmating intrinsic?	46,74,81, 82,118, 158

Lake-stream	EN, P	2	0.007–0.426~	Y	Y	Y	Y	Y	U	U	Y	Habitat, colour?	Divergent ns?	84,85,106,107
Marine-colour	EN	1	0–0.003*	Y	Y	Y	Y	Y	Y	Y	Y	Habitat, colour, court?	Divergent ss	65,86,87,176
Japanese-anadromous	P, J	1	0.428*; 0.735~	Y	Y	N	Y	Y	Y	Y	U	Size, court?, others	Postmating intrinsic, divergent ss?	4,5,8,29,73
Iceland	EN	2 +	U	Y	Y sm	Y	Y	Y	U	U	Y	U	Divergent ns, comp?, pred	57,108,110

Note: Clades are indicated by EN = Euro North American; P = Pacific; J = Japan Sea. Genetic distances are estimated by Nei's D on allozymes* or microsatellites~. Y indicates that the pair differs in the trait; N indicates the pair does not differ; U indicates that data are unavailable; and "?" indicates that a difference or mechanism is probable based on inference but has not been demonstrated directly. Abbreviations used include RI = reproductive isolation; ns = natural selection; ss = sexual selection; pred = predation; comp = competition; sm = small.

features of their biology that seem to promote rapid speciation. This is followed with a description of the basis for reproductive isolation in the pairs. Then we move on to the main body of the chapter — the mechanisms of speciation. Here we cover forms of natural selection that have been implicated in the divergence of traits underlying reproductive isolation, including divergent natural selection, competition, predation, and the process of reinforcement. We also evaluate the evidence for sexual selection, including sexual selection against hybrids. We briefly cover the implications of recent work on the genetics of adaptation for our understanding of speciation (see Chapter 2 for more information). Finally, we conclude with a discussion of the tenuous nature of stickleback species pairs, and implications for their conservation.

Throughout, we begin by describing what is known for limnetic-benthic pairs because they have been so extensively studied, followed by data on other pairs when available. It becomes obvious that we need far more information on these other pairs to evaluate the generality of results for limnetic-benthic pairs, and to round out our understanding of stickleback diversity.

3.2 SETTING THE STAGE

3.2.1 Phylogeographic Relationships

Genetic data give us some information on the phylogeographic relationships among various stickleback populations spread throughout the world. Present mitochondrial DNA (mtDNA) data suggest two clades, the Pacific and Euro-North American. A third clade, the Japan Sea, is supported by allozyme[3,4] and microsatellite data,[5] but not by mtDNA data.[6,7] It is thought that the lack of genetic differentiation in mtDNA data is due to introgressive hybridization between Japan Sea and Pacific clades, and that the Japan Sea is, in fact, a distinct third clade. Discrepancies between mitochondrial and nuclear DNA are fairly common in phylogeographic studies, especially when there has been a history of gene flow, to which mtDNA is particularly susceptible. Additional studies that employed nuclear markers would complement those using mitochondrial DNA, and using a larger number of markers might allow us to resolve relationships among populations with greater certainty. The Pacific and Euro-North American clades are estimated to have diverged 0.9 to 1.3 mya.[8] The Pacific and Japan Sea clades are estimated to have diverged 1.5 to 2 mya based on microsatellite[5] and allozyme data.[4] Thus, both of these deep divergences precede the extensive post-*Pleistocene* diversification of freshwater populations.

The distribution of these clades gives us some insight into the historical nature and likely geographic context of stickleback diversification. The Pacific clade is found in the Pacific basin: on the East coast of Japan,[4] throughout coastal Alaska, in Haida Gwai,[8] and in northern Vancouver Island, British Columbia.[9] The Japan Sea clade is found on both coasts of Japan, Korea, and western Russia. Sometimes it is sympatric with the Pacific clade, especially on Hokkaido Island in Japan.[4] The Euro-North American clade is found in the Pacific and Atlantic basins on both coasts of North America.[8] This includes the Pacific coast of North America from Alaska south to California, and on the Atlantic coast from Long Island north through coastal

Canada including Nova Scotia and the St. Lawrence Seaway. This clade is also found in Iceland, and the coasts of Northern Europe.[8]

3.2.2 Historical and Geographical Features of Stickleback Speciation

Much of the present day phenotypic diversity among populations has arisen post-glacially. The exception to this is the Japanese anadromous pair.[4,5,8] The distribution of species pairs gives us some insight into the likely geographic nature of speciation. There is currently no strong evidence that any pair arose strictly in sympatry. Instead, various patterns of allopatry and parapatry are likely. Divergence that began in allopatry has continued in sympatry.

3.2.2.1 Limnetic-Benthic Pairs

Because at present their populations are fully sympatric and early mtDNA data suggested that limnetics and benthics in several lakes were sister taxa, limnetic-benthic pairs are often cited as an example of sympatric speciation.[10] The available data argue against such a conclusion. Instead, the evidence we have so far supports the double invasion hypothesis proposed by McPhail.[11] He suggested that two separate invasions of marine or anadromous sticklebacks into recently created low-elevation lakes allowed for a period of allopatric divergence and the evolution of some reproductive isolation, setting the stage for further diversification. (Anadromous fish spend much of their life in the ocean but move into estuarine or fresh water to reproduce.) Several lines of evidence support this hypothesis. Physiological data suggest that limnetics are the descendants of the second invasion.[12] Both mtDNA and microsatellite data also argue that limnetics and benthics within a lake have descended directly from marines and not each other, and that limnetics and benthics in separate drainages are the result of independent colonisations.[13,14] Limnetic-benthic pairs within each lake have unique assemblages of mtDNA haplotypes, most of which differ from common marine and anadromous haplotypes by a single restriction site. In contrast, limnetics and benthics from different lakes always differ by more than one site, arguing against a single origin for all limnetics or all benthics. In addition, a phylogeny based on microsatellite data provides little support for a single origin of limnetics and of benthics.[13] Microsatellite data also show that limnetics and benthics within a lake are not sister taxa, arguing against sympatric speciation.[14] The nuclear and mtDNA patterns are in conflict, but the mtDNA pattern could easily result from past hybridization.

Taylor and McPhail[14] argue for a combination of chance and determinism in the evolution of limnetic-benthic species pairs. Chance plays a role through lake elevation and location, which either allowed for two separate colonisations or did not. Therefore, low-elevation lakes close to the ocean can host pairs, whereas high-elevation lakes and those far from the ocean cannot.[9] These chance events act in concert with deterministic processes. In particular, the mechanisms of natural and sexual selection that caused reproductive isolation to evolve are deterministic (see sections 3.4 and 3.5). The combination of these processes produced species pairs in some lakes, but not others.

The importance of both factors is highlighted because not all lakes that probably had two invasions of anadromous sticklebacks host species pairs. A lack of competitors and predators appears to be crucial for the diversification of sticklebacks. Lakes with species pairs are distinct in having only one other species of fish present, namely, cutthroat trout.[15] Other low-elevation lakes in the region contain several fish species. The lack of species pairs in these lakes suggests that the presence of these other fishes inhibited diversification of the sticklebacks. The presence of other fishes is historical; their effect on stickleback diversification is likely deterministic. Either predation or competition could be responsible.

There are two major ways in which predation could factor in: (1) restricting access to certain niches, and (2) changing the strength or nature of divergent selection.[16] First, many small fishes find refuge from predators in the vegetated littoral zone and avoid the dangerous pelagic zone.[17] The vegetated littoral habitat occupied by benthics may be better habitat, and suggests at least one reason why the first colonists evolved to occupy that niche. The pelagic zone may not be available to sticklebacks when predators are present. Therefore, limnetics may have been able to exploit the open water, planktivorous niche only in lakes without open-water fish predators. Some indirect evidence that predators exclude sticklebacks from the open water comes from the covariation between planktivorous trophic morphology and the presence of open-water fish predators. Many, long gill rakers are a specialisation for planktivory.[18] In lakes with many fish species, solitary populations (those that are not part of a species pair) have few gill rakers, suggesting that they do not exploit the open water, planktivorous niche. In contrast, in lakes with only cutthroat trout, solitaries have many gill rakers, suggesting they do use the open water habitat.[15,19] A second way in which predators could affect diversification is by altering the strength or nature of divergent selection. When high predation risk limits habitat segregation by excluding individuals from risky habitats,[20] it also reduces the strength of divergent selection. In addition, high predator-induced mortality is predicted to decrease the strength of divergent selection, hindering diversification.[16] Therefore, in lakes with multiple predators and high predation risk, selection may not have been strongly divergent, so diversification did not proceed.

Alternatively, competition between sticklebacks and these other fish species may have prevented diversification of sticklebacks. Besides cutthroat trout, the two most common fish species found sympatrically with sticklebacks in the region are rainbow trout (*Oncorhynchus mykiss*) and prickly sculpin (*Cottus asper*).[15] Juvenile salmonids may compete with sticklebacks for zooplankton,[21] whereas sculpin may compete for benthic invertebrates.[22] The presence of these competitors in the littoral and pelagic habitats may have prevented limnetics and benthics from specialising on the resources found there. The consequence may have been that when the second colonists entered the lake, competition with other fish species prevented divergence from the resident population. Instead, they became extinct or merged with those residents.

Present data indicate that a combination of chance (geography, history, preconditions, and fish community) and determinism (adaptation for resource exploitation and predator avoidance) influence speciation in limnetic and benthic sticklebacks. Geographic and historical particulars influence the possibility of speciation by setting up an ecological context that could promote diversification, but selection causes the divergence. We have insufficient data to determine whether this is also likely for other pairs. However, we do have data on geographic context and the likelihood of independent evolution for some. A period of allopatric divergence brought about by geological processes is likely for Japanese, lake-stream, and Washington stream-colour pairs, but not as likely for anadromous-stream, marine-colour, or Iceland pairs. We now turn to describing these other pairs.

3.2.2.2 Other Pairs

Allopatric divergence in glacial refugia may have played a role in the divergence of lake-stream pairs found on Haida Gwai and northern Vancouver Island, British Columbia. These areas and some areas of coastal Alaska are purported to have been glacial refugia.[8,23] Glacial refugia may have allowed a period of allopatric divergence for freshwater populations.[24] Once the glaciers receded, marines probably migrated into the area, and may have encountered refugial freshwater populations from which they were at least partly reproductively isolated. Consistent with different times of colonisation, Misty Lake and Misty inlet populations on Northern Vancouver Island show deep mtDNA divergence: the lake population is the Euro-North American clade and the inlet stream population the Pacific clade with approximately 2.7% sequence divergence.[25] Both the Pacific and Euro-North American clades are found in different lakes on Haida Gwai.[24] However, not all lake-stream pairs on Haida Gwai are split into separate mtDNA clades,[25] so refugial divergence may play a role for only some pairs. In these other pairs, population boundaries and clade boundaries are not congruent, nor does morphological variation covary with mtDNA clade.[26] Moreover, large-scale sampling suggests that refugial populations probably fused with new migrants, as many populations contain haplotypes from both clades.[26] Therefore, the geographic context for lake-stream pairs may be quite complex.

A different role for geography and history is implied for a set of lakes and rivers in Germany near the Baltic Sea. In contrast to the steep, mountainous topography of coastal British Columbia that restricts dispersal between drainages, in northern Germany the watersheds are shallow, and therefore, major river flow patterns and lake boundaries could have been altered over time and provided colonisation and dispersal routes that are no longer present.[27] Reusch et al.[28] found that lake, river, and estuarine populations from three drainages form distinct clades with mean F_{ST} values on the order of 0.14 to 0.18 for these ecotypes. Within habitat type, there was increasing isolation by distance, suggesting dispersal and neutral evolution shape genetic differentiation. It seems unlikely that these lake-stream populations evolved by parallel evolution in a manner similar to the British Columbia lake-stream pairs. Instead, the data are consistent with a single origin of each type followed by dispersal. However, the authors sampled multiple populations from the same drainages and lakes, so the populations may not be evolutionarily independent; hence, these results must be interpreted with caution.

A substantial period of allopatry is very likely for the Japanese anadromous pair because low sea levels cut off the Sea of Japan from the Pacific Ocean about 2 mya and again during the last glacial period 10 to 70 tya.[5] This time frame corresponds closely to the estimated divergence time for this pair.[4,5] Adaptation in these isolated areas appears to have resulted in substantial phenotypic and genetic differentiation.[5] This pair has evolved both pre- and postmating reproductive isolation[5,29] and shows the greatest genetic differentiation and the most intrinsic postzygotic isolation.

In contrast, anadromous-stream resident pairs probably did not experience a period of allopatry, but instead have been parapatric throughout their evolutionary divergence.[1] Therefore, gene flow may have occurred throughout differentiation of types and the evolution of reproductive isolation. If gene flow was ongoing, divergent selection must have been strong enough to counter its effects, given the repeated evolution of so many pairs. These pairs are extremely widespread, and are found throughout most of the range of sticklebacks. Vicariant events are less likely because of the broad distribution. However, one possibility is that the stream-resident fish were landlocked during the Pleistocene glaciation, and came into secondary contact with anadromous populations once water levels rose. We have no evidence to test this hypothesis at present. Neither do we have data for most populations on whether the anadromous and stream-resident forms derive from the same or different ancestors (e.g., different mtDNA clades). Yet, it seems safe to say that the evolution of pairs in western North America is independent of pairs in northern Europe or Asia. Thus there is the potential for massively parallel evolution of phenotypic differentiation and reproductive isolation for anadromous-stream pairs.

Anadromous and lake-resident sticklebacks have been found in the same location in several lakes and streams in Alaska.[30] Given the number of lakes and abundance of sticklebacks in coastal Alaskan waters, the potential for additional species pairs exists and deserves further study.

Iceland substrate pairs have probably been sympatric throughout their divergence. Icelandic lakes formed following the Pleistocene glaciation, and the lava substrate formed subsequent to that due to nearby volcanic activity. These lava flows provide a second distinct habitat, without which diversification into mud- and lava-associated populations would not be possible.

Unfortunately, we know very little about the likely geographic context of speciation in the Washington stream-colour pairs. Currently, red and black populations are parapatric, but particulars of the timing of colonisation of headwater regions relative to downstream regions, and the possibility of changing watersheds is not well known. We also have little direct information on the marine-colour pair of Nova Scotia, although white and marine populations may have been sympatric throughout their divergence. At least, no obvious geographic barrier separated these populations.

3.2.3 Rapid Ecological Speciation

Speciation in sticklebacks is often rapid. Is there something unique about sticklebacks that predisposes them to rapid speciation? Three factors seem likely to be important. These are the opportunity to enter novel and enemy-free environments, to survive and reproduce there, and to adapt rapidly to these new environments.[1,31–33]

Adaptive diversification is facilitated when organisms enter "empty" habitats with few competitors, predators, and parasites. This leaves multiple niches available for colonists to evolve into, with little ecological constraint. The reduction of antagonistic forces allows for unhindered adaptation to local conditions. Vast numbers of streams and lakes were formed all over the northern hemisphere after the last glaciers receded. Early colonising species would have found these streams and lakes to be enemy-free habitat. The anadromous lifestyle and broad distribution of sticklebacks would have enabled them to be among those early colonists.

After colonising these novel habitats, individuals must survive and reproduce to establish a viable population. Phenotypic plasticity may have played a role here. Phenotypic plasticity occurs when the environment induces changes in an individual's morphology, physiology, or behaviour. Plastic responses may allow individuals to survive and reproduce in novel habitats, thus allowing sufficient time for populations to adapt to novel selection pressures imposed by those habitats.[34,35] Selection can then act on those plastic phenotypes and their underlying genetics, causing evolutionary (genetic) change. The derived population is likely to be better adapted to the new habitat.[36] Phenotypic plasticity may complement divergent selection when it brings a population closer to the local optimum.

A number of traits show phenotypic plasticity in sticklebacks, including trophic morphology,[37,38] body size,[39] nuptial colour,[40–42] and many forms of behaviour.[43] Some of these are the same traits that differ substantially between populations and underlie premating or postmating reproductive isolation. In many cases, the differences between populations or species have a genetic basis, indicating evolutionary change.[44–48] Although the available evidence is suggestive, the role of phenotypic plasticity in diversification and speciation for sticklebacks has not been directly tested. This would be well worth doing both for its potential to increase our understanding of stickleback speciation and to test fundamental theories of adaptive evolution and speciation.

Colonists who adapted rapidly to the new environment furthered their chances of persisting and of diversifying into available niches. Sticklebacks show incredible geographic variation in morphological, physiological, and behavioural traits.[49] Some traits are phenotypically plastic,[37,43] but others are likely to have a substantive genetic basis.[46–48,50–55] Much of this diversity is adaptive, and much of it has arisen over a brief period of time (since the last glaciation). In some cases adaptive evolution has been extremely rapid. There are several remarkable cases of substantive phenotypic change over very few generations. One of these is rapid evolution of lateral plates in Loberg Lake, Alaska.[56] The lake was fish free in 1982 and recolonised by anadromous sticklebacks between 1983 and 1988. The first sticklebacks were observed in 1990 at which point 96% of these were fully plated and 0% low plated. By 2001 only 11% were fully plated while 75% were low plated. This corresponds to evolutionary rates for the fully plated morph of 0.19 haldanes and for the low-plated morph of 0.12 haldanes. Other traits may show similar rates of evolutionary change (M. A. Bell, personal communication, December 2004). Another example of extremely rapid evolution comes from a small Icelandic lake created in 1987. In just 12 years, the lake population diverged from anadromous sticklebacks in armour characters (spine length and plate number), body size, and shape.[57] This corresponds to evolutionary rates from 0.19 haldanes[58] to as high as 0.8 haldanes (dorsal spine length). The lake has two distinct habitats consisting of mud flats and lava rubble. Fish sampled in each of these habitats also showed other phenotypic differences, including shape, trophic structures, and number of armour plates. It appears that divergence of freshwater fish from anadromous ancestors occurred first, followed by further differentiation into populations inhabiting distinct habitats, resulting in a sympatric pair.

These rapid evolutionary events are extraordinary examples of anagenic change, as we currently have no data suggesting these populations are reproductively isolated from their anadromous ancestors. However, the facility for extremely rapid adaptation almost certainly plays a role in stickleback speciation because, as we will see later, adaptive divergence of populations produces reproductive isolation. The more (and faster) the adaptive divergence, the more (and faster) the opportunity for reproductive isolation. With the combination of empty habitat to fill and the capacity for rapid evolutionary change, the stage is set for sticklebacks to diversify — even into new species. Therefore, understanding what leads to this diversification illuminates what causes reproductive isolation to evolve.

3.3 THE BASIS OF REPRODUCTIVE ISOLATION

3.3.1 Premating Isolation

3.3.1.1 Limnetic-Benthic Pairs

Premating (sexual) isolation for limnetics and benthics depends on a combination of habitat segregation, assortative mating on body size, and asymmetric isolation due to nuptial colour and colour preference.[59] Sexual isolation depends on the same traits in multiple pairs, supporting the hypothesis that ecologically dependent natural or sexual selection are involved.

Both types nest in the littoral zone, but prefer different microhabitats for breeding.[1] Limnetics nest in shallow, open areas around fallen logs, whereas benthics nest in deeper, vegetated areas. This habitat segregation reduces encounter rates between ecotypes, but does not eliminate encounters completely because these two habitat types form a mosaic in the shallow littoral zone of the lakes where breeding occurs. Other forms of premating isolation also play a role.

Foremost among these is assortative mating on body size between species.[59] Heterospecific spawnings are seen most often between the smallest benthics and largest limnetics.[60] Both female and male size-based mating behaviour is likely to be involved in size-assortative mating between species. In choice tests, Paxton limnetic males preferred smaller limnetic females as mates over larger benthic females, and the strength of this preference increased as the difference in size between potential mates increased.[61] In contrast, solitary populations of limnetic-like males preferred the larger benthic females, and more so when benthic females were much larger than limnetic females. Both kinds of evidence suggest that limnetic male size preferences have been altered in sympatry with benthics, and this contributes to sexual isolation. Given that strong size-assortative mating between types contributes to sexual isolation, it is surprising that we lack information on size preferences and size assortative mating within species for both males and females.[59] Without this information we cannot evaluate the relative importance of sexual selection or natural selection on body size evolution and size-based sexual isolation.

Nuptial colour and colour preference also play a role, but affect premating isolation in an asymmetric manner.[59] Limnetic females have strong preferences for red colour[62] and reject benthic males who tend to have reduced colour (area and intensity of red), so differences in colour preference and colour display contribute to isolation in this direction. However, benthic females retain weakly positive preferences for red colour[62] and are more likely to mate with limnetic males with high than low colour.[59] Therefore, colour preference decreases sexual isolation in this direction.

Odour and odour preferences also contribute to premating isolation in an asymmetric manner. Benthic females use olfactory cues to recognize and reject heterospecific males, whereas limnetic females do not.[63] This suggests that odour differs between species and that odour perception may also differ, likely due to differences in ecology. Therefore, both species appear to use two mechanisms for mate recognition. Limnetics use colour and size, whereas benthics use odour and size.

Differences in courtship behaviour do not appear to contribute significantly to sexual isolation.[59] This is surprising, given the variation among populations in courtship,[43,64,65] evolutionary change in courtship,[66] and the very interactive nature of stickleback courtship. Nonetheless, current evidence suggests that it is primarily body size, colour, and odour differences that isolate limnetic-benthic pairs.

Other traits, such as differences in body shape may also contribute to sexual isolation, but have not yet been tested. In addition, learning through social experience may play a role in shaping social interactions and mate recognition. This could occur because of social interactions or imprinting. Shoaling preferences appear to be influenced in other fish species by social experience and phenotype matching,[67–70] and this may influence mate preferences. At present, there is no evidence for imprinting, but the one study published to date had low power and may not be conclusive.[71] Further work on social influences is warranted.

Further evidence that body size and nest location may have contributed to sexual isolation in the limnetic-benthic species pairs comes from a study on limnetic-like and benthic-like allopatric populations.[72] Body size exerted the strongest effect. Limnetic-like females mated assortatively on body size and selected the smaller, usually limnetic-like males. Size was also important to benthic-like females, who mated with larger, usually benthic-like males irrespective of their own body size. Nesting habitat may also have played a role. Benthic-like males nested in deeper water, and limnetic-like males tended to nest in open areas although not significantly so. These nesting preferences would contribute to assortative mating by ecotype if limnetic-like and benthic-like females shared these habitat preferences, but this was not directly tested.

3.3.1.2 Other Pairs

Much less information is available on the basis of sexual isolation in other pairs, but in some cases it appears that the same traits confer reproductive isolation. Research on premating isolation for other pairs is sorely needed.

Anadromous and stream-resident sticklebacks from around the world differ substantially in size, and pairs are isolated largely by body size.[39,73] Populations that differ most in size show the strongest sexual isolation. An experimental manipulation of size demonstrates its importance in assortative mating.[39] Anadromous females were manipulated to a small size, and stream females to a large size by varying the length of the growing period. Males were wild-caught and their size was unmanipulated. Mating was size assortative, even between ecotypes. For example, experimentally small anadromous females mated readily with small stream males but rejected large anadromous males. Other currently unknown ecotypic differences also make a small contribution to sexual isolation; however, colour does not appear to play a large role.[39] Despite the extensive data showing within-population colour preferences for marine and anadromous fish[66,74] and strong sexual selection on colour, we have little information on colour and colour preference for stream populations. Hence, colour and colour preferences may not differ substantially between these forms, and it appears that body size differences override any colour effect.[39]

Studies of several Pacific anadromous and stream-resident populations in Japan found substantial differences in male courtship behaviour and life history traits that may contribute to isolation.[5,73,75] In particular, Mori and collaborators[76–80] have collected data on behaviour and reproductive ecology on several freshwater populations within Japan, including stream-resident and spring-resident populations. These freshwater populations appear to be substantially reproductively isolated from both the Japan Sea and Pacific anadromous populations.[73] This is true despite the fact that these freshwater populations descend from the Pacific anadromous populations, show few behavioural differences, and are not genetically differentiated from them. Body size and life histories do vary, and probably contribute more to sexual isolation than do behavioural traits.[5,73]

Japan Sea males show more extensive divergence in courtship behaviour than other populations or ecotypes.[5] Japan Sea males have lost the zigzag display, and instead perform a rolling lateral display. Otherwise, courtship is composed of similar elements, but the frequency and sequence of behaviours vary. Japan Sea males court conspecific females more vigorously than heterospecific females. Pacific anadromous males appear to be indiscriminate in courtship behaviour.[5] Heterospecific matings occur despite these conspicuous differences in courtship behaviour, but in one direction only. Pacific anadromous females reject Japan Sea males, but Japan Sea females mate readily with both males.[5,73] Body size appears to play a role, but because females of both species prefer larger males, it contributes in an asymmetric manner. Other traits such as shape, odour, and colour may also contribute, but further work is needed to identify and assess the relative importance of such traits. Sexual isolation is weakened by being asymmetrical. In this system more than any other, postmating reproductive isolation appears to play an essential role in reproductive isolation.

Colour plays a more important role in Washington stream-colour pairs, and may contribute asymmetrically as it does in limnetic-benthic pairs. Both red and black populations prefer colourful males (bright red or deep black) over dull males.[74] A study on Connor Creek fish found that females from allopatric red populations show reduced interest in black males, which could contribute to sexual isolation.[81] Black females from within the hybrid zone have lost their preference for red — they are as likely to mate with black as with red males. However, females from allopatric black populations actually prefer red males, which would counter sexual isolation if these populations came into contact. A later study showed that red and black fish appear to mate assortatively in the Chehalis River system, and that colour may be the primary cue.[82] Despite partially conflicting results, both studies show that female colour preference has evolved in sympatric populations, and suggest it plays a role in sexual isolation. Surprisingly, we lack good data regarding how sexual selection acts on colour in these systems. These studies implicate colour in sexual isolation; however, they did not investigate other traits that are also likely to differ among these populations, such as size, shape, and behavioural traits. Given that size differences are an important trait isolating other pairs,[39,59,60] their potential role here should be investigated.

The apparent lack of lake-stream hybrids in nature,[1,83] coupled with the extent of phenotypic and genetic differentiation (both mtDNA and 28S rRNA[25,26]), suggests that lake and stream fish are reproductively isolated from one another. Although experimental data are not extensive, there is evidence for strong assortative mating in the Mayer Lake pair[84] and weak premating isolation in the Drizzle pair.[85] Yet, we have little information on the basis of reproductive isolation for these pairs. In Mayer Lake, habitat appears to be an important premating isolation factor, as fish appear to breed in their respective environments, although some breed in the intervening inlet area. In Drizzle Lake, lake males slightly but significantly prefer lake females, probably based on female colour and size.[85]

Marine-colour pairs appear to be completely reproductively isolated.[86] Several factors are likely to play a role in sexual isolation. The first and probably most important is habitat segregation. Most white stickleback males nest in filamentous algae, further offshore, or in deeper water than sympatric typical marine sticklebacks, which nest on the mud or rock substrate in shallow water near to shore.[65] Some populations of white sticklebacks nest in the rocky intertidal.[87] Occasionally, courtship occurs between typical white males and females, but these courtships invariably break off once the female reaches the nest,[65,86] further evidence that nest site characteristics, possibly including odour, and habitat influence assortative mating between the types. White sticklebacks are smaller than typical marine fish,[86] and body size may contribute to sexual isolation as it does for so many other stickleback pairs, although this has not been directly tested. In addition, male colour probably plays a role. Both types of females respond equally strongly to the white males, but white females respond only weakly to the ancestral typical marine males.[86] This may be due to both colour and courtship behaviour differences, as white stickleback males seem to have lost several behaviours, including dorsal pricking and meandering leads.[65] Therefore, it appears that both male mating traits and female preferences have diverged in white sticklebacks, but manipulative experiments to identify the factors that underlie sexual isolation have not been done.

There is some evidence of assortative mating between plate morphs in two widely separated Russian lakes, but the resulting sexual isolation is asymmetric and incomplete.[88,89] In both the White Sea (northern Russia) and Lake Azabachije (Kamchatka peninsula), mating in choice tests was observed most often between males and females of the same plate morph (complete, partial, or low) from the same body of water. Plate morphology is not the only cue however, as low-plated females were more likely to spawn with their own males than with low-plated males from the other lake.[89] Whether the basis of sexual isolation is plate number has not been established.

Ongoing work is attempting to identify the factors that confer sexual isolation on a number of pairs for which these data are currently lacking, including the Alaska populations and Iceland substrate pairs.

In many cases, we lack information on how the traits that confer sexual isolation between populations are acted on within populations. For example, body size is known to be an important component of sexual isolation in anadromous-stream pairs and limnetic-benthic pairs, yet there is very little information on whether body size is under sexual selection within populations. We do not know the answer to basic questions such as: Are there preferences for large[90] or small partners? Do those preferences differ between populations? We know that colour preferences differ between limnetics and benthics[62] and marine-colour pairs,[86] but lack similar data on most other pairs. Investigating such questions will give us insight into the processes involved in the evolution of sexual isolation.

3.3.2 Postmating Isolation

Postmating isolation is almost entirely extrinsic for nearly all pairs, meaning that hybrids suffer because they are poorly adapted, not because of developmental problems. Extrinsic postmating reproductive isolation has been well studied in limnetic-benthic pairs. These studies have found that selection against hybrids seems to be strong and that much of this selection is ecologically based.[91,92] Hybrid phenotypes are intermediate between the divergent phenotypes of the parental species, so hybrids fall between parental niches and have reduced foraging efficiency relative to both parental species.[93,94] This reduces growth rate and presumably affects survival to maturity. Even if some individuals survive to adulthood, they have low mating success.[95,96] We know more about the mechanisms of postmating isolation and less about the traits that underlie the low fitness of hybrids; however, it appears that trophic, antipredator, and mating traits are involved.[16,93–98] Details of how natural and sexual selection act against hybrids are discussed in sections 3.4 and 3.5.

There are only three pairs for which appreciable intrinsic postzygotic isolation has been found: Japanese anadromous-stream, Japanese anadromous, and Washington stream-colour pairs. In each case the data are from crosses with a single set of populations, so we do not know if the pattern is a general one. In crosses between a single Pacific anadromous and freshwater population derived from the Pacific anadromous, F1 hybrid males and females were sterile,[29,99] although there is some indication that rare backcross and F2 hybrids are formed.[29] A cross between a Japan Sea female and a Pacific anadromous male produced sterile F1 hybrid males, whereas the reciprocal cross produced fertile F1 hybrid males. Both crosses produced fertile F1 hybrid females.[5] Interestingly, the direction of postmating isolation is opposite to that for premating isolation in the Japanese anadromous pair.

One study found that Washington stream F1 hybrid clutches had slightly reduced fertility, and about 50% reduced viability in backcrosses and F2 hybrids.[81] Reduced viability may occur because of aberrant male behaviour — hybrid males show reduced fanning (5% compared to 35% of time spent fanning), so eggs were insufficiently oxygenated. In contrast to these results, other crosses showed little intrinsic postzygotic isolation in the Chehalis River system.[46] We have little information on the possibility of extrinsic postzygotic isolation in Japanese or Washington stream-colour pairs.

3.4 MECHANISMS OF SPECIATION: NATURAL SELECTION AND REPRODUCTIVE ISOLATION

The remainder of this chapter will focus on addressing the question of how the traits that confer reproductive isolation came to differ. We turn to the mechanisms of speciation.

Stickleback speciation seems to be primarily the result of ecological processes. In ecological speciation, reproductive isolation evolves as a consequence of divergent selection between environments.[100] Divergent selection can arise from ecological differences between populations, including abiotic factors like salinity and temperature, and biotic factors like competition and predation. Both pre- and postmating isolation can be involved. Evidence in support of this hypothesis is extensive and includes ecological, performance, and behavioural data from field and laboratory studies on pre- and postmating isolation. Additional genetic data are emerging. A substantial body of work provides evidence that natural selection plays a fundamental role in the evolution of reproductive isolation for sticklebacks. Much of this work has been recently reviewed.[1,2,33,101] Consequently, here we will briefly describe some of the major findings and describe new work on the role of competition and predation.

3.4.1 Divergent Natural Selection

Several types of evidence suggest strongly that divergent natural selection has been important in the evolution of reproductive isolation in sticklebacks. This evidence comes from parallel speciation experiments that test natural selection's influence on premating isolation,[39,102] reciprocal transplant experiments that test its role for postmating isolation,[94,103,104] and experiments with ecologically differentiated allopatric populations that are sexually isolated.[72]

Parallel speciation occurs when reproductive isolation evolves in parallel for populations that have colonised similar environments that differ from their common ancestor's environment.[105] The hypothesis predicts that populations in similar environments will mate freely, whereas populations in different environments will be reproductively isolated. This pattern is predicted if reproductive isolation arises as a by-product of adaptation to different niches.

3.4.1.1 Limnetic-Benthic Pairs

The first test of parallel speciation found that limnetic and benthic ecotypes reject each other as mates, but that limnetics mate freely with limnetics from any lake and benthics mate freely with benthics from any lake.[102] Thus, mating compatibility is determined by ecotype.

Reciprocal transplant experiments document extensive fitness trade-offs for parentals and selection against hybrids because of their intermediate phenotypes.[94] F1 hybrids had reduced growth rates in both parental habitats, demonstrating that despite high fitness in the laboratory, hybrids suffer reduced fitness in the wild.[103] A recent study controlled for the possibility that intrinsic genetic incompatibilities contributed to the low fitness of hybrids by comparing fitness of parental species and backcrosses in alternate environments.[104] The backcross hybrids show the expected trade-offs in growth (the fitness measure). Benthic backcrosses grew well in the littoral habitat and poorly in the pelagic habitat, whereas limnetic backcrosses grew well in the pelagic habitat and poorly in the littoral habitat. However, trade-offs for parentals were as expected in the littoral habitat but not the pelagic habitat, where both parental species grew poorly, so the findings were mixed.

Selection against hybrids was also found in the Priest Lake pair by tracking the proportion of hybrids at different life stages using species diagnostic microsatellite markers.[92] A low but substantive hybrid frequency in juveniles (8%) fell to less than half that amount in adults (3%). Although the source of selection cannot be identified in this study, selection clearly acts before sexual maturity and is likely to be ecologically driven. Moreover, estimates of gene flow were an order of magnitude lower (0.16%) than hybridization rates, suggesting that either sexual selection or natural selection acts against adult hybrids.

To test the importance of ecological divergence without the possibility of reinforcement, sexual isolation between limnetic-like and benthic-like allopatric populations was assessed in seminatural pond enclosures.[72] Mating interactions were not observed directly, but females spawned disproportionately with ecologically similar males. These results further support the primacy of divergent natural selection in sexual isolation for sticklebacks and argue that it probably contributed to initial divergence before secondary contact.

3.4.1.2 Other Pairs

A study of parallel speciation in anadromous-stream pairs from around the northern hemisphere (British Columbia, Alaska, Iceland, Scotland, Norway, and Japan) found support for the hypothesis and identified the traits that underlie reproductive isolation.[39] Here too, similar ecotypes mated freely whereas different ecotypes did not. This is true for allopatric and parapatric populations tested with populations in their own or from distant regions. Neither geographic nor genetic distance contributes to reproductive isolation, despite long periods of allopatry and the opportunity for the accumulation of substantial genetic differences. The primary trait conferring reproductive isolation was body size, with a smaller role for unidentified traits such as body shape or odour that may differ systematically among ecotypes.

There are few direct tests of parallel speciation in other pairs. Additional tests performed on other pairs with the goal of establishing the basis for isolation would be valuable in order to test whether the same traits are used repeatedly, or if different pairs are isolated with different traits. We do have indirect evidence consistent with divergent selection for a few pairs, because parallel evolution occurs in the traits that are known to underlie reproductive isolation. A pattern of parallel evolution of traits has long been taken as evidence for the role of selection.

Stream-colour pairs show parallel patterns of variation in several phenotypic traits, consistent with the action of divergent selection. They obviously differ in colour, but also show parallel evolution in trophic, antipredator, and physiological traits. For example, black populations from multiple localities have few gill rakers, short spines, and low salinity tolerance, and these differences have a genetic basis.[81]

Lake-stream pairs show parallel evolution of several phenotypic traits.[1,106] Genetic data indicate that several lake-stream pairs have evolved independently of one another, and in parallel.[25] Typically, stream fish are smaller, deeper bodied, wider mouthed, and have fewer gill rakers than the parapatric lake fish. They also differ in colour — all three lake populations are black, whereas the stream populations are mottled brown. Body shape, colour, and meristic differences between lake and inlet populations in Misty Lake have a genetic basis.[106] As with limnetic-benthic pairs, parallel evolution implies that divergent selection has given rise to differentiation of lake-stream pairs. There is also some data from reciprocal transplant studies[107] that suggests that each type does best in its native environment. However, almost all fish actually lost weight, so the data are somewhat difficult to interpret.

Mud and lava populations in four Iceland lakes have diverged in morphology[108] and behaviour,[109] suggesting that divergent selection plays a role. However, the pattern of divergence is not strictly in parallel; that is, the particular traits that differ between populations vary from lake to lake and the magnitude of divergence is not constant. Complicating matters is the fact that there is substantial sexual dimorphism in morphology. In fact, differences between males and females are frequently larger than differences between habitats. Sex differences in ecology could underlie sexual dimorphism, but that has not yet been tested. The current data support the role of divergent selection, but suggest that mud and lava habitats are not similar across all lakes studied. For example, predation intensity varies among lakes as does intraspecific competition.[110,111] Both are likely to affect the traits under divergent selection and the extent of divergence. In addition, lakes vary in how much lava habitat they have and whether it is vegetated.[108] Small areas of distinct habitat may not provide a sufficient resource base to support a population of sticklebacks.

3.4.2 Competition

In addition to divergent selection arising because of differences in environment, selection can also arise owing to ecological interactions between populations, such as competition.[112,113] This can also cause divergence in ecological traits. The connection to reproductive isolation has not been clearly made, but the connection to phenotypic divergence is supported by several lines of evidence. A large body of work has investigated how competition for resources has affected divergence in trophic traits for sticklebacks. Much of this work has been reviewed recently,[33,100,101] so here we focus primarily on work not covered in those reviews.

3.4.2.1 Limnetic-Benthic Pairs

One of the first experiments to test the role of competition showed that the presence of a competitor alters natural selection and favours divergence in trophic traits.[114] Selection acted most strongly against individuals with similar phenotypes to the competitor. The measured strength of selection would produce the difference in trophic traits observed in natural populations in about 500 generations.

Further evidence that competition is involved comes from a study testing how the strength of competition changes with the amount of phenotypic divergence between competitors.[115] If competition drives character displacement, one predicts that competition strength will wane as character divergence proceeds. Pritchard and Schluter's study found stronger competition between anadromous and intermediate sticklebacks (which are similar morphologically and ecologically) than between anadromous and benthic sticklebacks (which differ more extensively). Competition theory also predicts that selection will be frequency dependent — changing in intensity with the frequency of phenotypes.[116] This prediction was tested and confirmed by Schluter[117] by comparing the fitness of intermediates in the presence of either limnetics or benthics, while controlling for density. The most limnetic-like intermediates suffered when competing with limnetics, whereas the most benthic-like intermediates suffered when competing with benthics. Thus, selection acted most strongly against the intermediate phenotype resembling the competitor. The phenotype that was in high frequency (i.e., limnetic) relative to the other phenotype appears to have experienced stronger competition and, hence, stronger divergent selection.

3.4.2.2 Other Pairs

Washington stream-colour pairs may have adapted to compete with another species of fish, the Olympic mudminnow, *Novumbra hubbsi*, which are also black, although this hypothesis is somewhat controversial.[46,81] Several mechanisms might favour black sticklebacks, including competition with mudminnows, predation by them, and ecologically dependent sexual selection. Black males may compete more effectively against mudminnows than red males for breeding territories.[81,118] There is nearly perfect geographical correspondence between mudminnow and black stickleback distributions,[46] supporting the hypothesis that mudminnows are selective agents. However, this geographical pattern may also arise because mudminnows may be responding to similar environmental features as sticklebacks. The importance of competition with mudminnows was challenged by Scott and Foster,[119] who argue that encounter rates between stickleback males and mudminnows are insufficient to drive such divergence in colour. They also argue that males do not express full melanic colouration until they enter the parental phase. If so, black colouration is unlikely to play a role in establishment of nesting sites.

Other relevant data are for solitary populations. To directly test the prediction that frequency-dependent competition causes disruptive selection, Bolnick[120] surveyed lakes with solitary populations and asked if selection is disruptive on trophic traits. He used body size and relative gonad mass as proxies for fitness, and found some evidence for weak disruptive selection in Cedar and Muskeg Lakes, both part of the Amor de Cosmos watershed on Vancouver Island, British Columbia. He followed this survey with an experimental manipulation of density in natural populations, positing that disruptive selection should be stronger in high-density treatments because they should experience stronger competition. Although experiments failed in several lakes and the pattern was not consistent across the remaining lakes, he found some support for the prediction that disruptive selection is stronger with higher competition. This result lends support to the idea that competitive interactions can drive diversification in natural systems.

Robinson[121] also found support for the role of competition in a solitary population in Cranby Lake, Texada Island, British Columbia. Morphology varies from very limnetic-like to benthic-like, and he found a correspondence with capture location and morphology. Foraging efficiency experiments showed that fish with limnetic-like traits foraged most efficiently on plankton, whereas those with benthic-like traits foraged most efficiently on benthic prey. These patterns mirror those found in the limnetic-benthic species pairs, and suggest that individual fish specialise on prey types that their morphology makes them most suited to exploit.

3.4.3 Predation

Predation has been implicated as an additional factor underlying the divergence in phenotypes, including behaviour and morphology. There are four ways that predation may have played a role: (1) different suites of predators in distinct habitats may generate divergent selection on antipredator traits, (2) adaptation to one set of predators may generate trade-offs, (3) predators may alter the strength of divergent selection through competition, and (4) predation may select against migrants between habitats. Predation has been shown to be important between limnetic-benthic pairs. Its role in the other pairs is less clear. We need more work on the role of predation in reproductive isolation and divergence.

3.4.3.1 Limnetic-Benthic Pairs

First, habitat specialisation brought about by competition may expose limnetics and benthics to different suites of predators or levels of predation risk. These predator assemblages then may select for different defensive traits, especially if the predators have different modes of attack. In this scenario, resulting divergence in antipredator traits is a by-product of ecological character displacement. The pelagic habitat used primarily by limnetics is thought to be dominated by diving birds and piscivorous fishes, whereas the littoral habitat used primarily by benthics is thought to be dominated by invertebrate predators such as dragonfly nymphs and backswimmers.[122] These predators have different attack behaviour, so it seems reasonable that traits that provide protection against diving birds might not provide protection against dragonfly nymphs,[123] and these different predators could select for different traits. Differences in predation regimes should primarily affect the evolution of antipredator traits.

Divergence in antipredator traits provides evidence in support of this hypothesis, as limnetics and benthics differ in armour traits, escape, and schooling behaviour.

Benthics have fewer plates, shorter spines, and a reduced pelvic girdle relative to limnetics. Armour traits differ largely because of evolutionary reduction in benthic armour, whereas limnetics do not differ from solitary populations in armour traits.[124] Schooling has been inferred to be a defence against diving birds and piscivorous fishes,[125] and may be less effective against insect predators. The density of birds and fish predators is higher in the pelagic habitat where limnetics are found, and limnetics show a stronger tendency to school.[126,127] Also, escape behaviour differs between limnetics and benthics.[58,128] Limnetics quickly dart away while remaining suspended in the water column and school tightly. In contrast, benthics dive to the bottom and attempt to hide under or behind cover. Limnetics and benthics have diverged in predator defences: benthics use refuges to reduce predation risk and encounter rates, whereas limnetics use schooling behaviour and rapid escape to reduce their per capita risk and protect themselves further with body armour. The overall reduction in armour and loss of schooling behaviour in benthics is consistent with the hypothesis that benthics experience reduced or different predation pressure. Studies on other taxa have shown that vegetation provides a refuge for small fishes.[17] This refuge is not available to limnetics.

A second way that predation could influence divergence and speciation is if populations experience survival trade-offs in different habitats. This would arise if adaptations to avoid one set of predators made individuals more susceptible to another set. The existence of such trade-offs would imply that different predation regimes impose divergent selection on antipredator traits. Predation could compound fitness trade-offs experienced through competition. One study found evidence for such trade-offs.[126] Limnetics and benthics were exposed to predators common in native and nonnative habitats. Each species experienced higher survival in the presence of predators common in its own habitat and reduced survival in the presence of predators common in the habitat of the other species. Predation may thus create steeper fitness trade-offs than competition alone.

Third, the presence of predators may change the strength of divergent selection through competition. This hypothesis does not require different suites of predators in alternative habitats, just that predation changes the nature of selection and competitive interactions. Changes in predation-induced divergent selection should primarily affect divergent evolution of trophic traits rather than antipredator traits. If predators lower overall density, they might decrease the strength of competitive interactions, which could decrease the strength of divergent selection. Alternatively, predation could increase the strength of divergent selection if it alters foraging behaviour to increase habitat segregation, thereby exposing groups of prey to different selection regimes associated with exploiting resources in those different habitats. Data support this latter hypothesis.[16] Divergent selection was stronger in the presence of predators than in their absence. In addition, higher mortality increased the strength of divergent selection, whereas greater competition decreased its strength. Two findings suggest that predators affected divergent selection through competition, but not through predation. Selection was strong on trophic traits (number of gill rakers) but not on antipredator traits[58] in this experiment, arguing against the hypothesis that divergent selection acted on armour traits. In addition, there was no pattern of consistent selection on survival in the presence of predators.

Measures of the way that selection acts through predators provide additional evidence. A pond experiment shows that the intensity and nature of selection via predation differs between habitats and may thus have been a factor favouring divergence of limnetics and benthics.[98] Trout-induced mortality was higher for limnetic than for benthic fish, suggesting selection acts more strongly on limnetics. Limnetics had lower survival than benthics in the presence of cutthroat trout, but higher survival in their absence. This suggests that limnetics have higher encounter rates with trout than benthics.

Predation does not appear to be a component of ecological selection against hybrids, because F1 hybrids experienced equally low survival in the presence or absence of trout.[98] Hybrids did have lower survival than the mean for limnetics and benthics, but this appears to be not due to increased vulnerability to predators relative to parental species but, rather, decreased competitiveness.

And lastly, predation may select against migrants between habitats.[129] If limnetics enter the littoral zone, they will be exposed to predators to which they are not well adapted, and are likely to suffer higher mortality. The same is likely to be true for benthics. Selection against these migrants reduces their frequency in the new population, and so reduces encounter rates between migrants and residents. In this way, predation may reduce gene flow between species or populations and also will favour strong habitat segregation and fidelity. Currently, we have no data to test this hypothesis for limnetic-benthic pairs.

3.4.3.2 Other Pairs

Predation may play a role in the diversification of sticklebacks in Icelandic lakes. Populations from lake to lake differ in various morphological, behavioural, diet, and armour traits.[57,108–110] In addition, habitat-specific morphs are found sympatrically in several lakes.[108] Trait variation among these populations corresponds to the intensity of predation. Antipredator behaviour varies substantially among lakes that differ in the intensity of predation, with the strongest antipredator responses in the population with the highest predation pressure, which is the population inhabiting mud substrates in Thingvallavatn.[109,110] Divergence between sympatric morphs in morphology and behaviour is greater in those lakes with high predator density than in lakes with low predator density.[108,110] This finding supports the hypothesis that predation increases the strength of divergent selection. This probably occurs by increasing habitat segregation, as suggested by Sandlund et al.,[111] who found that sticklebacks were confined to the vegetated areas of mud substrate in Thingvallavatn, where they are partially protected from predators. Competition and predation probably both play a role in this system, and the relative importance of each may vary across lakes, which is suggested by the fact that lake pairs differ in the traits that show most divergence.[108,109] Unfortunately, we know very little about competition in these Icelandic populations.

Predation has been proposed as another agent favouring the evolution of black colouration in Washington stream-colour pairs. Moodie[83] showed reduced predation by trout on black males and predicted that black males would be found in habitats with intense trout predation. McPhail[81] presented evidence that mudminnow predation on stickleback nests and juveniles may favour black males, who are better able to ward off these predators. They also have higher reproductive output.[118] The likely independent evolution of black colouration in four drainages on the Olympic peninsula,[46] almost complete correspondence between the distribution of black males and mudminnows,[46] and the putative selective pressure caused by mudminnows has been taken as good evidence that this species has played a role in diversification of red and black populations. Similar to the limnetic-benthic pairs, both competition and predation appear to be important, because mudminnows both prey on juveniles and compete with males for breeding territories.

Sometimes, divergent selection acts to cause divergence between the sexes within a species, producing sexual dimorphism rather than sexual isolation between diverging populations. Evidence for this comes from sexual dimorphism in spine number in a solitary population in Boulton Lake, Haida Gwai.[130] The sexes are segregated by habitat and show fairly extensive ecological differentiation. Thus, they are exposed to different suites of predators. Males use the littoral habitat, whereas females use the pelagic habitat. Habitat segregation by sex is similar to that found between limnetics and benthics. The primary predators in the littoral zone are dragonfly naiads, who appear to select for reduced spine length and number,[131] especially during summer when they are active.[123] In contrast, piscivorous birds are predominant predators in the pelagic zone and appear to select for increased spine length and number,[132] especially in winter.[123] Selection differentials sometimes vary between the sexes, providing direct evidence that divergent natural selection contributes to sexual dimorphism in spine number.[130] Sexual dimorphism in Icelandic populations may result from a similar process. This process is similar to the one that produces distinct species, although because males and females share the same alleles, genetic correlations between the sexes change the evolutionary response to divergent selection.[133]

Indirect evidence that predation may be involved in other pairs, particularly anadromous-stream pairs, comes from the parallel loss of armour in freshwater fish. This certainly implies that selection on armour traits differs in the marine and stream environments. The source of this selection is probably a combination of reduced availability of calcium for armour development, and reduced predation pressure in freshwater habitats. If so, the costs of armour production are higher in freshwater habitats and the benefits of armour are reduced. Reimchen[131,134] has worked out the function of plates and the pelvic girdle, so we have some understanding of how predators select for these traits. However, we lack a good understanding of the role of predation and armour loss in speciation. For example, there appears to be little sexual isolation between fully plated and low-plated forms where it has been studied.[89] Therefore, despite the consistent differences between anadromous sticklebacks and many freshwater forms, plate number may not play an important role in sexual isolation. Yet, there may be some effect on postmating isolation. One possibility is that partially plated hybrids may suffer reduced fitness through a combination of increased predation and increased cost of armour production; however, this has not been studied. Armour loss is also seen in solitary lake populations in many regions (Alaska, Norway, Scotland, etc.). The repeated pattern of armour loss suggests that divergent selection has contributed to differentiation in armour traits among populations. Yet, this parallel evolution is probably facilitated by parallel genetic mechanisms for loss of the lateral plates[53,54] and the pelvic girdle.[52,135] One of the most exciting results to come out of recent genetic studies is the implication that single genes of large effect control the expression of these traits, and that similar genomic regions underlie differences in multiple populations. Whether this parallel genetic evolution results from some mutational bias in certain genetic regions, or selection acting repeatedly on the same genes is currently unknown. Continued work on the role of predation and the genetics underlying armour traits will certainly yield insight into rapid adaptive evolution and speciation in sticklebacks.

3.4.4 Reinforcement

Reinforcement has been implicated in stickleback speciation by several studies.[61,136–138] Reinforcement occurs when selection against hybrids favours mechanisms to increase prezygotic isolation.[139] The predicted pattern is reproductive character displacement, which is the greater difference in mating traits between two closely related species in sympatry than allopatry, or when sympatric populations show greater heterospecific mate discrimination.[140] Such reproductive character displacement facilitates recognition of conspecifics in sympatry, reducing wasteful heterospecific matings. Character displacement in mating traits is the predicted outcome of reinforcement, but can also occur due to direct selection on mate discrimination.[141]

3.4.4.1 Limnetic-Benthic Pairs

Limnetic-benthic pairs show both stronger premating isolation and character displacement in mating traits than do solitary (allopatric) populations. Ecological character displacement may strengthen premating isolation as a by-product of divergent selection, making it important to control for its effects. Two studies controlled for this factor by comparing sympatric populations to ecologically similar allopatric populations. This ensures that mate discrimination does not arise from ecological differences. Rundle and Schluter[137] found that benthic females discriminated against limnetic males, but females from benthic-like allopatric populations did not. These results show stronger mate discrimination in sympatry. Albert and Schluter[61] found reproductive character displacement of limnetic male size preference. Limnetic males preferred small limnetic females over large benthic females and the strength of mate discrimination increased with size difference. One population of allopatric males preferred large benthic females, whereas the other showed no significant preference. Preference for large females is expected because of their increased fecundity[142] and has been shown for several other allopatric freshwater and anadromous populations[143–145] suggesting it is the ancestral state. The findings of both studies are consistent with either reinforcement or direct selection on mate preferences. Albert and Schluter[61] suggest their results are more likely owing to direct selection because sympatric males were aggressive toward large benthic females, who are known to cannibalize eggs.[146] Distinguishing between reinforcement and direct selection in stickleback speciation deserves further study.

Recently, Albert et al.[138] tested for character displacement of colour by comparing limnetic-benthic pairs to solitary populations. They found evidence for character displacement in both throat colour (red) and body colour (blue). Red and blue reflectance (here reflectance is a measure of colour intensity or saturation) show significant differences between limnetics and benthics, whereas solitaries are intermediate. Two explanations for character displacement in colour were offered, both of which depend on habitat differences. The first is that differences in light quality between habitats selects for brighter colours (both red and blue) in limnetics than in benthics. This is the prediction from sensory drive. The second is that the distinct diets of the two species might differ in carotenoid content, and produce the pattern, at least for red. The limnetic diet may be rich in carotenoids, enabling them to produce a redder colour. Benthic diet may be poor in carotenoids, resulting in low red. And the less specialised solitary diet might have intermediate levels of carotenoids, resulting in intermediate expression.

3.4.4.2 Other Pairs

The only study of reinforcement on other pairs is one by Borland[136] on an anadromous-stream pair that found reproductive character displacement in male size preferences. Anadromous fish are larger than stream fish. The study found that allopatric stream-resident males preferred larger stream females, likely because fecundity increases with female size. In contrast, sympatric stream-resident males preferred smaller stream females, who differ most from the sympatric anadromous females. This finding is consistent with reinforcement, but alternative hypotheses were not directly considered or ruled out. Work on other pairs is required before we can draw any general conclusions about the role of reinforcement in stickleback speciation. Sticklebacks have given us some of the best tests of reinforcement to date.[147] Given the important recent advances in the theory of reinforcement,[141,148–153] this area deserves attention.

3.5 MECHANISMS OF SPECIATION: SEXUAL SELECTION AND REPRODUCTIVE ISOLATION

In addition to natural selection, sexual selection is implicated as an important cause of sexual isolation. Although not as extensively studied as natural selection, most of the evidence to date indicates that the type of sexual selection that contributes to sexual isolation is ecologically dependent. The data include tests of sensory drive[62] and parallel evolution of mating traits.[59] Here, we will focus on how sexual selection acts on traits that we know underlie sexual isolation — primarily size and colour — but will also discuss courtship behaviour. Courtship behaviour appears to be influenced by sexual selection independent of environment. More work on the role of sexual selection in speciation is certainly warranted.

3.5.1 Ecologically Dependent Sexual Selection

3.5.1.1 Limnetic-Benthic Pairs

The first line of evidence tests sensory drive as the mechanism of divergence in mating traits and preferences. This work has found that differences in mating habitat underlie differences in nuptial colour and colour preference in limnetic-benthic pairs.[62] Benthics nest and court in habitats where the light is red-shifted, whereas limnetics nest in habitats with more broad-spectrum light. The sensory drive hypothesis predicts that mating traits should be selected to be conspicuous in their mating habitat because conspicuous signals are easier to detect, and females can discriminate between strong and weak signals more easily.[154,155] As predicted, limnetics and benthics differ in nuptial colour. Limnetic males display large areas of intense red, whereas benthics have reduced red or are black.[62] These colours render the males conspicuous in their respective mating habitats.[42,164] However, limnetics and benthics achieve conspicuousness in different ways. Limnetics are brighter than the background, whereas benthics are darker. Limnetics also have high colour contrast.

Colour preferences also differ between limnetics and benthics. Limnetic females have strong preferences for red males, whereas benthics have weak or no preference.[62] These preferences depend partly on sensitivity to red light. Limnetic females are quite sensitive to red light and have strong preferences. In contrast, benthic females have low sensitivity and weak preferences. In addition, there is a strong correlation between the extent of red shift in the habitat and sensitivity to red. Therefore, colour sensitivity and colour preference seem to be determined by light environment. Essentially, colour and colour preference are adaptations to different mating habitats in the two species.

Differences in preference lead to differences in the strength of sexual selection on colour, and the importance of colour in sexual isolation. Sexual selection on colour is strong on limnetic males because limnetic females have strong colour preferences,[62] and limnetics rely on colour differences to avoid heterospecific matings.[59] In contrast, benthic females have weak or no preference for red males,[62] which reduces the strength of sexual selection on colour in this species and relegates colour to a minor role in sexual isolation for this species.[59]

3.5.1.2 Other Pairs

Ecologically dependent sexual selection has been implicated in Washington stream-colour pairs as well.[119] As an alternative to competition with mudminnows, Scott and Foster echo Reimchen's[156] suggestion that water colour may play a role in the evolution of black colouration, much as it does in lake populations.[62,156,157] A survey showed that red males are not found in heavily red-shifted habitats,[158] providing some support for this hypothesis, although black males are found in full-spectrum habitats, suggesting other factors may also be important, possibly including competition. Scott[158] found evidence of assortative mating based on colour in Connor Creek. Females from the headwater region and the contact zone chose melanic males. In contrast, females from the creek mouth chose typical red and blue males. The findings for females in the contact zone differ from earlier work[81] that found a lack of preference for colour. In either case, there appears to be evolutionary change in the tendency for females to mate with melanic or red males, and this corresponds to the extent of red shift in their habitat. Such assortative mating is likely to contribute to sexual isolation between red and black populations.

We have no data on whether sensory drive contributes to divergence in most other pairs, including the marine-colour pairs, lake-stream pairs, and Japanese anadromous pairs. Given substantial colour differences and distinct mating habitats in the marine-colour and lake-stream pairs, this topic deserves further study.

3.5.2 Parallel Divergence in Reproductive Isolation and Mating Traits

The second approach to exploring divergent sexual selection tests for parallel evolution in mate recognition and mating traits and gives us further evidence for the importance of ecology. Parallel evolution has long been used to infer the action of selection. Similar environments impose similar selection on traits, leading to their parallel evolution. Different environments impose different selection on traits, leading to their divergent evolution. In contrast, parallel evolution between independent evolutionary lineages is unlikely under drift because random changes should be uncorrelated with environment. Nor is it likely if sexual selection is not ecological but instead arises from interactions between the sexes. In this case, traits are likely to differ between populations, but not in correlation with environment and not in parallel. We can therefore use patterns of parallel evolution to test the extent to which sexually selected traits and the traits that confer sexual isolation evolve under the action of ecologically dependent selection.

3.5.2.1 Limnetic-Benthic Pairs

Data from limnetic-benthic pairs indicate substantial parallel evolution of sexual isolation.[59] Multiple pairs use the same two traits — body size and nuptial colour — in essentially the same way. We find strong assortative mating based on body size, and strong asymmetric isolation based on nuptial colour. Thus, the basis for sexual isolation has evolved in parallel, consistent with findings from Rundle et al.[102] who found that the patterns of sexual isolation were in parallel but did not identify the traits responsible.

In addition, the traits that confer sexual isolation have evolved in a parallel manner, as predicted if selection is ecologically dependent.[59] Differences in male and female reproductive and courtship traits, in body size, and size dimorphism are extensive. This divergence is nearly as extensive as that found for trophic traits[19] and armour traits.[124] The divergence is also very much in parallel.[59] All limnetics have evolved large areas of intense red colour, whereas benthics have evolved smaller areas of reduced colour. All limnetics have evolved small body size and benthics large body size. So two traits that confer sexual isolation in pairs show strong parallel evolution, suggesting that selection on them has been ecologically dependent.

Another difference in sexually selected traits between limnetics and benthics is in the extent of condition dependence in nuptial colour.[159] Here again, the pattern is one of parallel evolution. Three limnetic populations show strong condition-dependent expression of red, whereas three benthic populations do not. These results suggest that the particular form of sexual selection operates in parallel — it is similar for all limnetics but distinct from that in all benthics. If condition dependence has been important in the evolution of female preference, then the strength of sexual selection on colour is predicted to correlate with the degree of condition dependence. This prediction is borne out. Moreover, the extent of condition dependence correlates with the importance of colour differences in sexual isolation, confirming the connection between differences in sexual selection on colour and reproductive isolation in these pairs.[159]

In contrast, courtship behaviour shows only partly parallel evolution.[59] For example, bite rate has evolved in parallel (benthics bite more), but zigzag rate has not. This suggests two possibilities. Selection on courtship behaviour may arise not from the environment, but from interactions between the sexes over mating. This would cause the traits to evolve in arbitrary directions with respect to environment. Alternatively, some behavioural traits may be under ecologically dependent selection whereas others may not. The results for behavioural traits indicate that not all sexually selected traits evolve in an ecologically dependent manner, but that the traits that confer sexual isolation do.

3.5.2.2 Other Pairs

Very little information is available for other pairs regarding whether the basis for reproductive isolation or mating traits have evolved in parallel. The only exception to this are the anadromous-stream pairs, who show a strong pattern of parallel evolution of mate recognition, as described earlier.[39] Stream populations have repeatedly evolved smaller body size, reduced nuptial colour, reduced armour, and possibly changes in courtship behaviour. Body size consistently has the largest effect on reproductive isolation between anadromous and stream populations. That other phenotypic traits also show patterns of parallel evolution is consistent with ecologically dependent sexual selection.

3.5.3 Sexual Selection Against Hybrids

3.5.3.1 Limnetic-Benthic Pairs

Sexual selection typically plays a role in premating isolation, but can also contribute to postmating isolation if hybrids suffer a mating disadvantage relative to parental species. Estimates of gene flow in the pair from Priest Lake are consistent with sexual selection against hybrids.[92] In limnetic-benthic pairs, even this form of sexual selection appears to be ecologically dependent. A laboratory-based study found no mating disadvantage for F1 hybrid males that had opportunities to mate with both limnetic and benthic females.[95] Despite being discriminated against by both parentals, hybrid males did obtain some matings from both; thus, their average mating success was equal to that of parentals. However, in the field, hybrid males had reduced average mating success.[96] Hybrid males nest in the same mating habitat as limnetics (open areas) and are therefore likely to encounter limnetic females because gravid (reproductively receptive) limnetic females also prefer this habitat. Limnetic females discriminate against hybrid males and prefer to mate with limnetics, so hybrid males obtain few matings from limnetic females. In contrast, hybrid males are unlikely to nest near benthics (in vegetated areas), which is where they would encounter gravid benthic females. Therefore, hybrid males do not have the opportunity to mate with both types, which is required to make up for reduced success with each type. Thus, mating habitat preference of both parentals and hybrids,[1] coupled with reduced mating success with each parental species, results in sexual selection against hybrid males. At present, we have no data on mating success of hybrid females.

3.5.3.2 Other Pairs

We also do not have much data on sexual selection against hybrids in other pairs. The only study of which I am aware[89] focuses on anadromous-stream pairs in Russia, and suggests that completely plated females discriminate against partially plated males, which are presumed to be hybrids.[89] There appears to be little isolation between low-plated and partially plated morphs. Unlike the limnetic-benthic pairs, this isolation does not appear to depend on habitat or encounter rates. The actual cues have not yet been identified.

3.6 THE GENETICS OF PARALLEL EVOLUTION AND SPECIATION

There are many examples of parallel evolution in sticklebacks: of adaptive traits, sexual traits, and traits that underlie reproductive isolation. These patterns provide strong evidence for the parallel action of selection. But selection alone does not result in evolutionary change. Evolution occurs only when selection acts on available genetic variation. There are certain axes upon which evolutionary change is more likely to occur because there is more genetic variation on which selection can act.[160] To the extent that populations share biases in genetic variation, they are more likely to evolve in similar directions. The genetic variation present in the ancestral marine population (standing genetic variation) is shared among freshwater populations descended from those marines. Some genetic differences among populations may result from lineage sorting.[161,162] It is also possible that populations share biases in their production of new variation through mutations. These genetic biases predispose populations to respond in similar ways to selection. In addition, some genes or variants might be recruited by selection more often than others because they do not have negative pleiotropic effects that could counter any beneficial effects favoured by selection. This might lead to the repeated use of the same loci in multiple independent instances of adaptive evolution. An important question to address then is, to what extent does parallel phenotypic evolution involve the repeated use of the same loci?

Another important set of questions that relate to the rapid adaptation and speciation seen in sticklebacks has to do with the genetic architecture of adaptive traits. By genetic architecture, we mean the number of genes that control the expression of a trait, the magnitude of their effects, their location in the genome, and their gene action. Traits controlled by a few genes of major effect might respond more rapidly to selection than polygenic traits. Genes that show additive effects can be acted on more efficiently by selection, also allowing more rapid evolutionary change. Therefore, exploring the genetic architecture of traits of known adaptive value and those that confer reproductive isolation is an important avenue to further our understanding of the role selection plays in the diversification of sticklebacks.

Fortunately, new genetic techniques are infusing the study of adaptive divergence and speciation in sticklebacks with the ability to address some of these questions of long-standing and general interest.[163] Here, we review quantitative genetic studies and mapping studies that are beginning to reveal the genetic basis of adaptation and speciation. Many of the studies focus on armour traits, in part because they differ conspicuously between populations and are relatively easy to study. Fewer studies have focused on other adaptive traits such as trophic traits (gill rakers) and body shape, and studies are just beginning to emerge on the genetics of reproductive isolation, body size, colour, and behavioural traits, but more of these studies are under way. This is an area of increasingly active research and promises great advances over the next few years.

3.6.1 Quantitative Genetics Studies

Schluter et al.[55] used a quantitative genetics approach to investigate whether parallel evolution of plate number and body shape in independent stream populations depends on parallel genetic changes. They made three sets of crosses. The first crossed Japanese anadromous sticklebacks with stream sticklebacks from the same clade, both collected in Japan. The second crossed Euro-North American anadromous sticklebacks with stream sticklebacks from the same clade, both collected in British Columbia. These two sets are undoubtedly independent. The third crossed both stream populations in a complementation test.

For plate number, they found evidence for strong parallel evolution, similar gene action (partial dominance but not epistasis), and that similar small numbers of genes (one or two) underlie armour loss in both lineages. Using a complementation test, they found that changes in the same locus underlie parallel loss of plates in the stream fish. The pattern for body shape was also in parallel, but more complicated, perhaps because of its polygenic nature. They characterized body shape using partial warps and principal component analysis, and found strong parallel evolution in the first principal component that separated anadromous from stream fish. However, the second principal component showed that stream fish differed from one another; in fact, phenotypic evolution here was divergent rather than in parallel. Even so, both lineages show similar gene action (additive but no dominance or epistasis), indicative of substitutions causing similar phenotypic changes. They were unable to determine if the substitutions were at similar or different loci for body shape. These results are corroborated by molecular genetics and mapping studies of plate number.

Lewandowski and Boughman[42] focused on the genetics of colour differences between populations. They used a paternal half-sib, split-clutch design to estimate genetic and environmental effects on expression of male nuptial colour for red and black populations. They found genetic differences between black and red populations, as well as genetic variation within populations for colour expression. Both red and black colour also showed some phenotypic plasticity, responding to differences in light regime (full spectrum or red-shifted). Despite this phenotypic plasticity, red and black colour measures showed significant genetic correlation across environments, indicating significantly more genetic than environmental variation in colour expression. Genetic correlations among traits suggest different genetic mechanisms are at work in red as compared to black populations. Red populations had positive correlations among red and black colour traits, whereas black populations had negative correlations.

3.6.2 Mapping Studies

Mapping studies have located quantitative trait loci (QTL) and actual genes responsible for important adaptive traits that differ between populations, so we are beginning to understand the genetic basis of population or species differences.[165] Many of the traits that have been mapped are structural armour traits, such as spine length, pelvic girdle morph, and plate number, or structural trophic traits such as gill raker number.[50] Trophic traits certainly play a role in ecologically dependent postmating isolation.[93,94] Armour traits are also likely to be involved, although this is not as firmly established.[98,124,126] Hence, the QTL discovered may not contain "speciation genes," but they certainly code for traits that have played pivotal roles in the adaptive diversification of stickleback species. Given that Chapter 2 covers the results of these studies in some detail, let us focus on how these studies further our understanding of rapid parallel evolution, especially in traits that play a role in reproductive isolation.

A robust finding that is emerging from these studies is that a few loci of major effect control the differences between populations and species, and these loci seem to underlie parallel evolution in multiple independent freshwater populations. Evolution has repeatedly used the same loci. For example, differences in plate number between British Columbia anadromous and Paxton benthic fish seems to be determined by a single QTL that explains 75% of the variation among populations, with smaller contributions from three other QTL.[54] This same major QTL is also involved in a California stream population polymorphic for plate number. Perhaps more surprisingly, the same minor QTL are also involved in both populations, and the QTL have similar gene action. In both British Columbia and California populations, the major QTL shows partial dominance, where Aa individuals are either partial or complete morphs. Three minor QTL influence plate number and determine whether Aa individuals develop as complete or partial morphs. These modifier QTL show additive effects: the more benthic alleles, the fewer plates. The modifier QTL have little effect in the AA background, but reduce plate number by half in both Aa and aa individuals. Thus, there is epistasis between the major and minor QTL. A similar pattern was found in three Alaska freshwater populations.[53] This study found that the same major chromosomal region appears to be involved in the reduction in plate number for the three Alaska populations, and this region maps to the same location as the major QTL in the British Columbia and California fish. However, in these populations the fully plated alleles are fully dominant. The simple genetic architecture might facilitate rapid evolution, and partly explain the extremely rapid loss of plates seen in natural populations.[56,57]

A major QTL governs pelvic reduction in Paxton benthic fish and explains 13 to 44% of the variation between it and its anadromous ancestors, with contributions from four minor QTL.[52] As with lateral plate QTL, a combination of additive and epistatic effects were found. Increasing the number of Paxton benthic alleles at all QTL led to decreases in the size of all pelvic structures in an additive fashion. Yet, there appears to be epistasis among the minor QTL, and their effect depends on the genotype at the major QTL. All of these QTL map to different regions from the plate reduction QTL, so different genes control lateral plate and pelvic traits. A complementation test indicated that the same genes control pelvic reduction in a population from a small lake in Iceland. Not only that, but it appears that the same chromosomal region appears to underlie pelvic reduction in the three Alaska populations studied by Cresko et al.[53]; however, the complete morph alleles appear to be dominant in these populations. In contrast, it appears that different genetic changes have occurred at the modifiers in the three Alaska populations, as the pelvic phenotypes did not fully complement.

Genetic studies will undoubtedly continue to give us answers to fundamental and long-standing questions. Future work should focus on the traits that have been identified as pivotal in pre- and postmating isolation. There are some current studies in this area. Fortunately, the flurry of papers focused on this topic seems unlikely to abate any time soon.

3.7 PERSISTENCE AND CONSERVATION OF SPECIES

What are the prospects for persistence of these stickleback species? At the moment, not very good. Bell[31] and McPhail[1] have noted a pattern of long-term stasis and persistence of marine populations that give rise to multiple diversified freshwater populations. These freshwater populations seem not to persist over evolutionary time. For example, paleontological evidence demonstrates extinction and recolonisation by a phenotypically different population,[166] although the evidence can also be explained as a case of rapid evolution within a population (Bell, personal communication, December 2004). Bell suggests that the short persistence time of freshwater populations is because their habitats are in essence, temporary. Freshwater boreal lakes can silt up over time, streams can dry out or flood, and repeated glacial events are likely to wipe out both lakes and streams over longer time periods. These are geologic processes and threaten the long-term persistence of nearly all freshwater populations.

There is also evidence that biological and anthropogenic processes threaten the persistence of species, and these threats are far more immediate than geologic processes. The limnetic-benthic species pairs have been listed as endangered in Canada.[167,168] Reasons for their listing are their restricted distribution to only six lakes in four watersheds in British Columbia, and recent or ongoing loss of some pairs.[169–172] The primary threats appear to be introduced or invasive species, and water and land use practices adjacent to the lakes.[167] The Hadley Lake pair recently went extinct shortly after the introduction of brown bullhead catfish (*Ameirus nebulosus*).[169] Predation by bullheads effectively wiped out the pair within a very short time span. Signal crayfish (*Pacifastacus leniusculus*) were introduced in Enos Lake and have increased dramatically in population size (personal observation). Their population explosion corresponds closely in time to an increase in hybridization, implicating them as a primary cause. The breakdown of reproductive isolation threatens to obliterate the Enos limnetic-benthic pair and turn them into a hybrid swarm.[171,172] Estimates of hybridization for the Enos pair (24%) are nearly five times the rate for the Paxton and Priest pairs (5.2 and 4.5%, respectively). Although long-term gene flow estimates are an order of magnitude lower, gene flow in Enos Lake is twice the rate in Paxton and four times the rate in Priest Lake (0.0032, 0.0018, and 0.0007, respectively).[172] Introgression is largely unidirectional, with limnetic alleles introgressing into the benthic genome. No genetically pure limnetics were detected in a sample of 192 fish.[172]

The effects of crayfish on hybridization are not well known, but probably include direct effects such as nest predation and resource competition, and indirect effects such as alteration of habitat and resources. They could be interfering with both premating and postmating reproductive isolation. The possibility of future species introductions to other lakes poses a real threat to the other pairs. Nearly as threatening are water and land use. Licenses for water use permit the extraction of substantial volumes of water from several lakes, in some cases equal to or exceeding current lake volumes. Such drawdown would undoubtedly have negative impacts by destroying littoral habitat necessary for breeding for both species and for feeding by benthics, and reducing the area of pelagic habitat necessary for feeding by limnetics. In addition, logging, mining, and land development may increase turbidity or pollute lake waters. Increased turbidity could hamper visually based mate recognition,[155,173] resulting in a collapse of sexual isolation perhaps similar to the collapse of cichlid species in Lake Victoria, Africa.[174] Pollution may poison the fish themselves or their prey. The specific effects of these ecological disturbances on survival and hybridization of the pairs are not well known, and deserve further study.

Hope that conservation efforts can actually protect and recover the species pairs comes, in part, from evidence that the Paxton pair recovered from past hybridization.[172] For unknown reasons, this pair went through a period of increased hybridization; however, hybridization rates are currently quite low[1,172] and the pair is stable. This suggests that gene pools were not fully homogenized during the hybridization event, and that ecological conditions after the event prevented continued gene flow. Careful study of the Enos pair should be done to compare historical samples (pre-hybridization) to current samples (ongoing hybridization) and to follow changes as the pair (hopefully) recovers. This work should include investigation of genetics, morphology, ecology, and behaviour. Tracking these changes may give insight into the effects of hybridization at all levels, and allow us to describe the signature of hybridization.

Identifying the specific causes of increased hybridization in the Enos pair has obvious benefits for conserving this and other pairs. But an additional benefit is that we can gain insight into the factors essential for reproductive isolation. Because speciation is ecological and the pairs arose and are maintained by divergent selection, changing selection regimes can reverse the diversification process.[175] Thus, research into the causes of hybridization and how we might prevent it has dual benefits — protection of the species pairs and insight into fundamental processes of speciation. Such work will undoubtedly be fruitful on both fronts.

ACKNOWLEDGMENTS

Thanks to J. McKinnon, F. vonHippel, and the Boughman laboratory for comments on the manuscript. And thanks to all the scientists who have amassed copious amounts of intriguing data on stickleback diversity and speciation.

REFERENCES

1. McPhail, J.D., Speciation and the evolution of reproductive isolation in the sticklebacks (*Gasterosteus*) of south-western British Columbia, in *The Evolutionary Biology of the Threespine Stickleback,* Bell, M.A. and Foster, S.A., Eds., Oxford University Press, Oxford, 1994, p. 399.
2. McKinnon, J.S. and Rundle, H.D., Speciation in nature: the threespine stickleback model systems, *Trends in Ecology and Evolution,* 17, 480, 2002.
3. Haglund, T.R., Buth, D.G., and Lawson, R., Allozyme variation and phylogenetic relationships of Asian, North American, and European populations of the threespine stickleback, Gasterosteus aculeatus, *Copeia,* 2, 432, 1992.
4. Higuchi, M. and Goto, A., Genetic evidence supporting the existence of two distinct species in the genus Gasterosteus around Japan, *Environmental Biology of Fishes,* 47, 1, 1996.
5. Kitano, J., Mori, S., and Peichel, C.L., Phenotypic divergence and reproductive isolation between sympatric populations of anadromous threespine sticklebacks, *Biological Journal of the Linnean Society,* in press.
6. Yamada, M., Higuchi, M., and Goto, A., Extensive introgression of mitochondrial DNA found between two genetically divergent forms of threespine stickleback, Gasterosteus aculeatus, around Japan, *Environmental Biology of Fishes,* 61, 269, 2001.
7. Watanabe, K., Mori, S., and Nishida, M., Genetic relationships and origin of two geographic groups of the freshwater threespine stickleback, "hariyo," *Zoological Science,* 20, 265, 2003.
8. Orti, G., Bell, M.A., Reimchen, T.E., and Meyer, A., Global survey of mitochondrial DNA sequences in the threespine stickleback — evidence for recent migrations, *Evolution,* 48, 608, 1994.
9. Johnson, L.S. and Taylor, E.B., The distribution of divergent mitochondrial DNA lineages of threespine stickleback (Gasterosteus aculeatus) in the northeastern Pacific basin: post-glacial dispersal and lake accessibility, *Journal of Biogeography,* 31, 1073, 2004.
10. Johannesson, K., Parallel speciation: a key to sympatric divergence, *Trends in Ecology and Evolution,* 16, 148, 2001.

11. McPhail, J.D., Ecology and evolution of sympatric sticklebacks (Gasterosteus): origin of the species pairs, *Canadian Journal of Zoology,* 71, 515, 1993.
12. Kassen, R., Schluter, D., and McPhail, J.D., Evolutionary history of threespine sticklebacks (Gasterosteus spp) in British Columbia: insights from a physiological clock, *Canadian Journal of Zoology,* 73, 2154, 1995.
13. Taylor, E.B. and McPhail, J.D., Evolutionary history of an adaptive radiation in species pairs of threespine sticklebacks (Gasterosteus): insights from mitochondrial DNA, *Biological Journal of the Linnean Society,* 66, 271, 1999.
14. Taylor, E.B. and McPhail, J.D., Historical contingency and ecological determinism interact to prime speciation in sticklebacks, Gasterosteus, *Proceedings of the Royal Society of London Series B,* 267, 2375, 2000.
15. Vamosi, S.M., The presence of other fish species affects speciation in threespine sticklebacks, *Evolutionary Ecology Research,* 5, 717, 2003.
16. Rundle, H.D., Vamosi, S.M., and Schluter, D., Experimental test of predation's effect on divergent selection during character displacement in sticklebacks, *Proceedings of the National Academy of Sciences of the United States of America,* 100, 14943, 2003.
17. Werner, E.E., Gilliam, J.F., Hall, D.J., and Mittelbach, G.G., An experimental test of the effects of predation risk on habitat use in fish, *Ecology,* 64, 1540, 1983.
18. Langeland, A. and Nost, T., Gill raker structure and selective predation on zooplankton by particulate feeding fish, *Journal of Fish Biology,* 47, 719, 1995.
19. Schluter, D. and McPhail, J.D., Ecological character displacement and speciation in sticklebacks, *American Naturalist,* 140, 85, 1992.
20. Lima, S.L. and Dill, L.M., Behavioral decisions made under the risk of predation: a review and prospectus, *Canadian Journal of Zoology,* 68, 619, 1990.
21. O'Neill, S.M. and Hyatt, K.D., An experimental study of competition for food between sockeye salmon (Oncorhynchus nerka) and threespine sticklebacks (Gasterosteus aculeatus) in a British Columbia coastal lake, Smith, H.D., Margolis, L., and Wood, C.C., Eds., *Sockeye Salmon (Oncorhynchus nerka) Population Biology and Future Management*, Ottawa, Department of Fisheries and Oceans Canada, 1987, p. 143.
22. Rosenfeld, J., Effects of fish predation in erosional and depositional habitats in a temperate stream, *Canadian Journal of Fisheries and Aquatic Sciences,* 57, 1369, 2000.
23. Hocutt, C.H. and Wiley, E.O., *The Zoogeography of North American Freshwater Fishes*, John Wiley & Sons, New York, 1986.
24. O'Reilly, P., Reimchen, T.E., Beech, R., and Strobeck, C., Mitochondrial DNA in Gasterosteus and pleistocene glacial refugium on the Queen Charlotte Islands, British Columbia, *Evolution,* 47, 678, 1993.
25. Thompson, C.E., Taylor, E.B., and McPhail, J.D., Parallel evolution of lake-stream pairs of threespine sticklebacks (Gasterosteus) inferred from mitochondrial DNA variation, *Evolution,* 51, 1955, 1997.
26. Deagle, B.E., Reimchen, T.E., and Levin, D.B., Origins of endemic stickleback from the Queen Charlotte Islands: mitochondrial and morphological evidence, *Canadian Journal of Zoology,* 74, 1045, 1996.
27. Donner, J., *The Quaternary History of Scandinavia*, Cambridge University Press, Cambridge, 1995.
28. Reusch, T.B.H., Wegner, K.M., and Kalbe, M., Rapid genetic divergence in postglacial populations of threespine stickleback (Gasterosteus aculeatus): the role of habitat type, drainage and geographical proximity, *Molecular Ecology,* 10, 2435, 2001.

29. Honma, Y., Chiba, A., and Tamura, E., Fine structure of the sterile testis of hybrid threespine stickleback between marine and landlocked forms, *Japanese Journal of Ichthyology,* 33, 262, 1986.
30. Von Hippel, F.A. and Weigner, H., Sympatric anadromous-resident pairs of threespine stickleback species in young lakes and streams at Bering Glacier, Alaska, *Behaviour,* 141, 1441, 2004.
31. Bell, M.A., Evolutionary phenetics and genetics: the threespine stickleback, Gasterosteus aculeatus, and related species, Turner, B.J., Ed., *Evolutionary Genetics of Fishes,* New York, Plenum Press, 1984, p. 431.
32. Bernatchez, L., Chouinard, A., and Lu, G.Q., Integrating molecular genetics and ecology in studies of adaptive radiation: whitefish, Coregonus sp., as a case study, *Biological Journal of the Linnean Society,* 68, 173, 1999.
33. Schluter, D., *The Ecology of Adaptive Radiation*, Oxford University Press, Oxford, 2000.
34. Adams, C.E. and Huntingford, F.A., Incipient speciation driven by phenotypic plasticity? Evidence from sympatric populations of Arctic charr, *Biological Journal of the Linnean Society,* 81, 611, 2004.
35. Skulason, S., Snorrason, S.S., and Jonsson, B., Sympatric morphs, populations and speciation in freshwater fish with emphasis on Arctic charr, Magurran, A.E. and May, R., Eds., *Evolution of Biological Diversity*, Oxford, Oxford University Press, 1999, p. 70.
36. Price, T.D., Qvarnstrom, A., and Irwin, D.E., The role of phenotypic plasticity in driving genetic evolution, *Proceedings of the Royal Society of London Series B,* 270, 1433, 2003.
37. Day, T., Pritchard, J., and Schluter, D., A comparison of two sticklebacks, *Evolution,* 48, 1723, 1994.
38. Day, T. and McPhail, J.D., The effect of behavioural and morphological plasticity on foraging efficiency in the threespine stickleback (Gasterosteus sp.), *Oecologia,* 108, 380, 1996.
39. McKinnon, J.S., Mori, S., Blackman, B.K., David, L., Kingsley, D.M., Jamieson, L., Chou, J., and Schluter, D., Evidence for ecology's role in speciation, *Nature,* 429, 294, 2004.
40. Candolin, U., Changes in expression and honesty of sexual signalling over the reproductive lifetime of sticklebacks, *Proceedings of the Royal Society of London Series B,* 267, 2425, 2000.
41. Frischknecht, M., The breeding coloration of male three-spined sticklebacks (Gasterosteus aculeatus) as an indicator of energy investment in vigor, *Evolutionary Ecology,* 7, 439, 1993.
42. Lewandowski, E. and Boughman, J.W., Genetic and environmental effects of color expression in threespine sticklebacks, in review.
43. Foster, S.A., The geography of behaviour: an evolutionary perspective, *Trends in Ecology and Evolution,* 14, 190, 1999.
44. McPhail, J.D., Ecology and evolution of sympatric sticklebacks (Gasterosteus) — evidence for a species-pair in Paxton Lake, Texada Island, British Columbia, *Canadian Journal of Zoology,* 70, 361, 1992.
45. McPhail, J.D., Ecology and evolution of sympatric sticklebacks (Gasterosteus) — morphological and genetic evidence for a species pair in Enos Lake, British Columbia, *Canadian Journal of Zoology,* 62, 1402, 1984.
46. Hagen, D.W. and Moodie, G.E.E., Polymorphism for breeding colors in *Gasteroseus aculeatus* I. Their genetics and geographic distribution, *Evolution,* 33, 641, 1979.

47. Hatfield, T., Genetic divergence in adaptive characters between sympatric species of stickleback, *American Naturalist,* 149, 1009, 1997.
48. Girvan, J.R. and Braithwaite, V.A., Orientation behaviour in sticklebacks: modified by experience or population specific?, *Behaviour,* 137, 833, 2000.
49. Bell, M.A. and Foster, S.A., *The Evolutionary Biology of the Threespine Stickleback,* Oxford University Press, Oxford, 1994.
50. Peichel, C.L., Nereng, K.S., Ohgi, K.A., Cole, B.L.E., Colosimo, P.F., Buerkle, C.A., Schluter, D., and Kingsley, D.M., The genetic architecture of divergence between threespine stickleback species, *Nature,* 414, 901, 2001.
51. Shapiro, M.D., Schluter, D., and Kingsley, D.M., The genetic control of upper jaw morphology in threespine sticklebacks, *Integrative and Comparative Biology,* 42, 1310, 2002.
52. Shapiro, M.D., Marks, M.E., Peichel, C.L., Blackman, B.K., Nereng, K.S., Jonsson, B., Schluter, D., and Kingsley, D.M., Genetic and developmental basis of evolutionary pelvic reduction in threespine sticklebacks, *Nature,* 428, 717, 2004.
53. Cresko, W.A., Amores, A., Wilson, C., Murphy, J., Currey, M., Phillips, P., Bell, M.A., Kimmel, C.B., and Postlethwait, J.H., Parallel genetic basis for repeated evolution of armor loss in Alaskan threespine stickleback populations, *Proceedings of the National Academy of Sciences of the United States of America,* 101, 6050, 2004.
54. Colosimo, P.F., Peichel, C.L., Nereng, K.S., Blackman, B.K., Shapiro, M.D., Schluter, D., and Kingsley, D.M., The genetic architecture of parallel armor plate reduction in threespine sticklebacks, *PLos Biology,* 2, 635, 2004.
55. Schluter, D., Clifford, E.A., Nemethy, M., and McKinnon, J.S., Parallel evolution and inheritance of quantitative traits, *American Naturalist,* 163, 809, 2004.
56. Bell, M.A., Aguirre, W.E., and Buck, N.J., Twelve years of contemporary armor evolution in a threespine stickleback population, *Evolution,* 58, 814, 2004.
57. Kristjansson, B.K., Skulason, S., and Noakes, D.L.G., Rapid divergence in a recently isolated population of threespine stickleback (Gasterosteus aculeatus L.), *Evolutionary Ecology Research,* 4, 659, 2002.
58. Bergstrom, C.A., Fast-start swimming performance and reduction in lateral plate number in threespine stickleback, *Canadian Journal of Zoology,* 80, 207, 2002.
59. Boughman, J.W., Rundle, H.D., and Schluter, D., Parallel evolution of sexual isolation in sticklebacks, *Evolution,* 59, 361, 2005.
60. Nagel, L.M. and Schluter, D., Body size, natural selection, and speciation in sticklebacks, *Evolution,* 52, 209, 1998.
61. Albert, A.Y.K. and Schluter, D., Reproductive character displacement of male stickleback mate preference: reinforcement or direct selection?, *Evolution,* 58, 1099, 2004.
62. Boughman, J.W., Divergent sexual selection enhances reproductive isolation in sticklebacks, *Nature,* 411, 944, 2001.
63. Rafferty, N. and Boughman, J.W., Olfactory mate recognition in asympatric pair of three-spined sticklebacks, *Behavioral Ecology,* 17, 2006.
64. Ridgway, M.S. and McPhail, J.D., Ecology and evolution of sympatric sticklebacks (*Gasterosteus*): mate choice and reproductive isolation in the Enos Lake species pair, *Canadian Journal of Zoology,* 62, 1813, 1984.
65. Jamieson, I.G., Blouw, D.M., and Colgan, P.W., Field observations on the reproductive-biology of a newly discovered stickleback (Gasterosteus), *Canadian Journal of Zoology,* 70, 1057, 1992.

66. McLennan, D.A. and McPhail, J.D., Experimental investigations of the evolutionary significance of sexually dimorphic nuptial coloration in *Gasterosteus aculeatus* (L) — the relationship between male color and female behavior, *Canadian Journal of Zoology,* 68, 482, 1990.
67. Engeszer, R.E., Ryan, M.J., and Parichy, D.M., Learned social preference in zebrafish, *Current Biology,* 14, 881, 2004.
68. Barber, I. and Wright, H.A., How strong are familiarity preferences in shoaling fish?, *Animal Behaviour,* 61, 975, 2001.
69. Landeau, L. and Terborgh, J., Oddity and the confusion effect in predation, *Animal Behaviour,* 34, 1372, 1986.
70. Rosenthal, G.G. and Ryan, M.J., Assortative preferences for stripes in danios, *Animal Behaviour,* 70, 1063, 2005.
71. Griffin, A.S., Social learning about predators: a review and prospectus, *Learning and Behaviour,* 32, 131, 2004.
72. Vines, T.H. and Schluter, D., Strong assortative mating between allopatric sticklebacks as a by-product of adaptation to different environments, *Proceedings of the Royal Society of London Series B,* 273, 911, 2006.
73. Ishikawa, M. and Mori, S., Mating success and male courtship behaviors in three populations of the threespine stickleback, *Behaviour,* 137, 1065, 2000.
74. McKinnon, J.S., Video mate preferences of female three-spined sticklebacks from populations with divergent male coloration, *Animal Behaviour,* 50, 1645, 1995.
75. Mori, S., Divergence in reproductive ecology of the threespined stickleback, Gasterosteus aculeatus, *Japanese Journal of Ichthyology,* 34, 165, 1987.
76. Mori, S., Reproductive-behavior of the landlocked three-spined stickleback, Gasterosteus aculeatus microcephalus, in Japan. 1. The year-long prolongation of the breeding period in waterbodies with springs, *Behaviour,* 93, 21, 1985.
77. Mori, S., Geographical variations in fresh-water populations of the three-spined stickleback, Gasterosteus aculeatus, in Japan, *Japanese Journal of Ichthyology,* 34, 33, 1987.
78. Mori, S., The breeding system of the three-spined stickleback, Gasterosteus aculeatus (forma-leiura) with reference to spatial and temporal patterns of nesting activity, *Behaviour,* 126, 97, 1993.
79. Mori, S., Nest-site choice by the three-spined stickleback, Gasterosteus aculeatus (form leiurus), in spring-fed waters, *Journal of Fish Biology,* 45, 279, 1994.
80. Mori, S., Factors associated with and fitness effects of nest-raiding in the three-spined stickleback, Gasterosteus aculeatus, in a natural situation, *Behaviour,* 132, 1011, 1995.
81. McPhail, J.D., Predation and the evolution of a stickleback (Gasterosteus), *Journal of the Fisheries Research Board Canada,* 26, 3183, 1969.
82. Scott, R.J., Assortative mating between adjacent populations of threespine stickleback (Gasterosteus aculeatus), *Ecology of Freshwater Fish,* 13, 1, 2004.
83. Moodie, G.E.E., Predation, natural selection and adaptation in an unusual threespine stickleback, *Heredity,* 28, 155, 1972.
84. Moodie, G.E.E., Morphology, life history, and ecology of an unusual stickleback (Gasterosteus aculeatus) in Queen Charlotte Islands, Canada, *Canadian Journal of Zoology,* 50, 721, 1972.
85. Stinson, E.M., Threespine sticklebacks (Gasterosteus aculeatus) in drizzle lake and its inlet, Queen Charlotte Islands: ecological and behavioural relationships and their relevance to reproductive isolation, MS thesis, University of Alberta, 1983.

86. Blouw, D.M. and Hagen, D.W., Breeding ecology and evidence of reproductive isolation of a widespread stickleback fish (Gasterosteidae) in Nova Scotia, Canada, *Biological Journal of the Linnean Society,* 39, 195, 1990.
87. MacDonald, J.F., Bekkers, J., MacIsaac, S.M., and Blouw, D.M., Intertidal breeding and aerial development of embryos of a stickleback fish (Gasterosteus), *Behaviour,* 132, 1183, 1995.
88. Ziuganov, V.V., Golovatjuk, G.J., Savvaitova, K.A., and Bugaev, V.F., Genetically isolated sympatric forms of threespine stickleback, Gasterosteus aculeatus, in Lake Azabachije (Kamchatka Peninsula, USSR), *Environmental Biology of Fishes,* 18, 241, 1987.
89. Ziuganov, V.V., Reproductive isolation among lateral plate phenotypes (low, partial, complete) of the threespine stickleback, Gasterosteus aculeatus, from the White Sea Basin and the Kamchatka Peninsula, Russia, *Behaviour,* 132, 1173, 1995.
90. Kraak, S.B.M., Bakker, T.C.M., and Mundwiler, B., Sexual selection in sticklebacks in the field: correlates of reproductive mating, and paternal success, *Behavioral Ecology,* 10, 696, 1999.
91. Rundle, H.D. and Schluter, D., Natural selection and speciation in sticklebacks, in *Adaptive Speciation,* Dieckmann, U., Metz, H., Doebeli, M., and Tautz, D., Eds., Cambridge University Press, Cambridge, 2004, p. 1.
92. Gow, J.L., Peichel, C.L., and Taylor, E.B., Selection against hybrids in natural populations of sympatric threespine sticklebacks, in revision.
93. Schluter, D., Adaptive radiation in sticklebacks — size, shape, and habitat use efficiency, *Ecology,* 74, 699, 1993.
94. Schluter, D., Adaptive radiation in sticklebacks — trade-offs in feeding performance and growth, *Ecology,* 76, 82, 1995.
95. Hatfield, T. and Schluter, D., A test for sexual selection on hybrids of two sympatric sticklebacks, *Evolution,* 50, 2429, 1996.
96. Vamosi, S.M. and Schluter, D., Sexual selection against hybrids between sympatric stickleback species: evidence from a field experiment, *Evolution,* 53, 874, 1999.
97. Bentzen, P. and McPhail, J.D., Ecology and evolution of sympatric sticklebacks (Gasterosteus) — specialization for alternative trophic niches in the Enos Lake species pair, *Canadian Journal of Zoology,* 62, 2280, 1984.
98. Vamosi, S.M. and Schluter, D., Impacts of trout predation on fitness of sympatric sticklebacks and their hybrids, *Proceedings of the Royal Society of London Series B,* 269, 923, 2002.
99. Honma, Y. and Tamura, E., Anatomical and behavioral differences among threespine sticklebacks — the marine form, the landlocked form and their hybrids, *Acta Zoologica,* 65, 79, 1984.
100. Schluter, D., Ecology and the origin of species, *Trends in Ecology and Evolution,* 16, 372, 2001.
101. Rundle, H.D. and Nosil, P., Ecological speciation, *Ecology Letters,* 8, 336, 2005.
102. Rundle, H.D., Nagel, L.M., Boughman, J.W., and Schluter, D., Natural selection and parallel speciation in sympatric sticklebacks, *Science,* 287, 306, 2000.
103. Hatfield, T. and Schluter, D., Ecological speciation in sticklebacks: environment-dependent hybrid fitness, *Evolution,* 53, 866, 1999.
104. Rundle, H.D., A test of ecologically dependent postmating isolation between sympatric sticklebacks, *Evolution,* 56, 322, 2002.
105. Schluter, D. and Nagel, L.M., Natural selection and parallel speciation, *American Naturalist,* 146, 292, 1995.

106. Lavin, P.A. and McPhail, J.D., Parapatric lake and stream sticklebacks on northern Vancouver island — disjunct distribution or parallel evolution, *Canadian Journal of Zoology,* 71, 11, 1993.
107. Hendry, A.P., Taylor, E.B., and McPhail, J.D., Adaptive divergence and the balance between selection and gene flow: lake and stream stickleback in the Misty system, *Evolution,* 56, 1199, 2002.
108. Kristjansson, B.K., Skulason, S., and Noakes, D.L.G., Morphological segregation of Icelandic threespine stickleback (Gasterosteus aculeatus L)., *Biological Journal of the Linnean Society,* 76, 247, 2002.
109. Doucette, L.I., Variability in resource-based behavior of the threespine stickleback (Gasterosteus aculeatus) in Iceland, MS thesis, Institute of Biology, University of Iceland, 2001.
110. Doucette, L.I., Skulason, S., and Snorrason, S.S. Risk of predation as a promoting factor of species divergence in threespine sticklebacks (Gasterosteus aculeatus l.), *Biological Journal of the Linnean Society,* 82, 189, 2004.
111. Sandlund, O.T., Jonasson, P.M., Jonsson, B., Malmquist, H.J., Skulason, S., and Snorrason, S.S., Threespine stickleback *Gasterosteus aculeatus* in Thingvallavatn: habitat and food in a lake dominated by arctic charr *Salvelinus alpinus*, *Oikos,* 64, 365, 1992.
112. Taper, M.L. and Case, T.J., Models of character displacement and the theoretical robustness of taxon cycles, *Evolution,* 46, 317, 1992.
113. Abrams, P.A., Alternative models of character displacement and niche shift.1. Adaptive shifts in resource use when there is competition for nutritionally nonsubstitutable resources, *American Naturalist,* 41, 651, 1987.
114. Schluter, D., Experimental evidence that competition promotes divergence in adaptive radiation, *Science,* 266, 798, 1994.
115. Pritchard, J.R. and Schluter, D., Declining interspecific competition during character displacement: summoning the ghost of competition past, *Evolutionary Ecology Research,* 3, 209, 2001.
116. Abrams, P.A., The importance of intraspecific frequency-dependent selection in modeling competitive coevolution, *Evolutionary Ecology,* 3, 215, 1989.
117. Schluter, D., Frequency dependent natural selection during character displacement in sticklebacks, *Evolution,* 57, 1142, 2003.
118. Hagen, D.W., Moodie, G.E.E., and Moodie, P.F., Polymorphism for breeding colors in Gasterosteus aculeatus 2. Reproductive success as a result of convergence for threat display, *Evolution,* 34, 1050, 1980.
119. Scott, R.J. and Foster, S.A., Field data do not support a textbook example of convergent character displacement, *Proceedings of the Royal Society of London Series B,* 267, 607, 2000.
120. Bolnick, D.I., Can intraspecific competition drive disruptive selection? An experimental test in natural populations of sticklebacks, *Evolution,* 58, 608, 2004.
121. Robinson, B.W., Trade offs in habitat-specific foraging efficiency and the nascent adaptive divergence of sticklebacks in lakes, *Behaviour,* 137, 865, 2000.
122. Reimchen, T.E. Predators and morphological evolution in threespine stickleback, in *The Evolutionary Biology of the Threespine Stickleback,* Bell, M.A. and Foster, S.A., Eds., Oxford University Press, Oxford, 1994, p. 240.
123. Reimchen, T.E. and Nosil, P., Temporal variation in divergent selection on spine number in threespine stickleback, *Evolution,* 56, 2472, 2002.
124. Vamosi, S.M. and Schluter, D., Character shifts in the defensive armor of sympatric sticklebacks, *Evolution,* 58, 376, 2004.

125. Magurran, A.E., The adaptive significance of schooling as an antipredator defense in fish, *Annales Zoologici Fennici,* 27, 51, 1990.
126. Vamosi, S.M., Predation sharpens the adaptive peaks: survival trade-offs in sympatric sticklebacks, *Annales Zoologici Fennici,* 39, 237, 2002.
127. Odling-Smee, L.C., Boughman, J.W., and Braithwaite, V.A., Sympatric species of threespine stickleback differ in their performance in a spatial learning task, in revision.
128. Law, T.C. and Blake, R.W., Comparison of the fast-start performances of closely related, morphologically distinct threespine sticklebacks (Gasterosteus spp), *Journal of Experimental Biology,* 199, 2595, 1996.
129. Nosil, P., Vines, T.H., and Funk, D.J., Reproductive isolation caused by natural selection against immigrants from divergent habitats, *Evolution,* 59, 705, 2005.
130. Reimchen, T.E. and Nosil, P., Variable predation regimes predict the evolution of sexual dimorphism in a population of threespine stickleback, *Evolution,* 58, 1274, 2004.
131. Reimchen, T.E., Spine deficiency and polymorphism in a population of Gasterosteus aculeatus — an adaptation to predators, *Canadian Journal of Zoology,* 58, 1232, 1980.
132. Reimchen, T.E., Structural relationships between spines and lateral plates in threespine stickleback (Gasterosteus aculeatus), *Evolution,* 37, 931, 1983.
133. Bolnick, D.I. and Doebeli, M., Sexual dimorphism and adaptive speciation: two sides of the same ecological coin, *Evolution,* 57, 2433, 2003.
134. Reimchen, T.E., Predator handling failures of lateral plate morphs in Gasterosteus aculeatus: functional implications for the ancestral plate condition, *Behaviour,* 137, 1081, 2000.
135. Kawano, K., Character displacement in stag beetles (Coleoptera : Lucanidae), *Annals of the Entomological Society of America,* 96, 503, 2003.
136. Borland M., Size-Assortative Mating in Threespine Sticklebacks from Two Sites on the Salmon River, British Columbia, MS thesis, University of British Columbia, 1986.
137. Rundle, H.D. and Schluter, D., Reinforcement of stickleback mate preferences: sympatry breeds contempt, *Evolution,* 52, 200, 1998.
138. Albert, A.Y.K., Millar, N.P., and Schluter, D., Character displacement of a sexually selected trait in threespine sticklebacks, *Biological Journal of the Linnean Society,* in press.
139. Dobzhansky, T., *Genetics and the Origin of Species*, Columbia University Press, New York, 1951.
140. Brown, W.L. and Wilson, E.O., Character displacement, *Systematic Zoology,* 5, 49, 1956.
141. Servedio, M.R., Beyond reinforcement: the evolution of premating isolation by direct selection on preferences and postmating, prezygotic incompatibilities, *Evolution,* 55, 1909, 2001.
142. Wootton, R.J., Effect of size of food ration on egg-production in female three-spined stickleback, Gasterosteus aculeatus l, *Journal of Fish Biology,* 5, 89, 1973.
143. Sargent, R.C., Bell, M.A., Krueger, W.H., and Baumgartner, J.V., A lateral plate cline, sexual dimorphism, and phenotypic variation in the black-spotted stickleback, Gasterosteus wheatlandi, *Canadian Journal of Zoology,* 62, 368, 1984.
144. Rowland, W.J., The ethological basis of mate choice in male threespine sticklebacks, Gasterosteus aculeatus, *Animal Behaviour,* 38, 112, 1989.
145. Kraak, S.B.M. and Bakker, T.C.M., Mutual mate choice in sticklebacks: attractive males choose big females, which lay big eggs, *Animal Behaviour,* 56, 859, 1998.
146. Foster, S.A., Inference of evolutionary pattern — diversionary displays of three-spined sticklebacks, *Behavioural Ecology,* 5, 114, 1994.

147. Servedio, M.R. and Noor, M.A.F., The role of reinforcement in speciation: theory and data, *Annual Review of Ecology Evolution and Systematics,* 34, 339, 2003.
148. Servedio, M.R., The evolution of premating isolation: local adaptation and natural and sexual selection against hybrids, *Evolution,* 58, 913, 2004.
149. Servedio M.R. and Saetre, G.P., Speciation as a positive feedback loop between postzygotic and prezygotic barriers to gene flow, *Proceedings of the Royal Society of London Series B,* 270, 1473, 2003.
150. Servedio, M.R., Reinforcement and the genetics of nonrandom mating, *Evolution,* 54, 21, 2000.
151. Kirkpatrick, M. and Servedio, M.R., The reinforcement of mating preferences on an island, *Genetics,* 151, 865, 1999.
152. Kirkpatrick, M., Reinforcement during ecological speciation, *Proceedings of the Royal Society of London Series B,* 268, 1259, 2001.
153. Kirkpatrick, M., Reinforcement and divergence under assortative mating, *Proceedings of the Royal Society of London Series B,* 267, 1649, 2000.
154. Endler, J.A., Signals, signal conditions, and the direction of evolution, *American Naturalist,* 139, 125, 1992.
155. Boughman, J.W., How sensory drive can promote speciation, *Trends in Ecology and Evolution,* 17, 571, 2002.
156. Reimchen, T.E., Loss of nuptial color in threespine sticklebacks (*Gasterosteus aculeatus*), *Evolution,* 43, 450, 1989.
157. McDonald, C.G., Reimchen, T.E., and Hawryshyn, C.W., Nuptial colour loss and signal masking in *Gasterosteus*: an analysis using video imaging, *Behaviour,* 132, 963, 1995.
158. Scott, R.J., Sensory drive and nuptial colour loss in the three-spined stickleback, *Journal of Fish Biology,* 59, 1520, 2001.
159. Boughman, J.W., Condition dependent expression of red color differs between stickleback species, *Evolution,* in review.
160. Schluter, D., Adaptive radiation along genetic lines of least resistance, *Evolution*, 50, 1766, 1966.
161. Wu, C.I. Inferences of species phylogeny in relation to segregation of ancient polymorphisms, *Genetics,* 127, 429, 1991.
162. Takahashi, K., Terai, Y., Nishida, M., and Okada, N., Phylogenetic relationships and ancient incomplete lineage sorting among cichlid fishes in Lake Tanganyika as revealed by analysis of the insertion of retroposons, *Molecular Biology and Evolution,* 18, 2057, 2001.
163. Kingsley, D.M., Zhu, B., Osoegawa, K., de Jong, P.J., Schein, J., Marra, M., Peichel, C.L., Amemiya, C., Schluter, D., Balabhadra, S., Friedlander, B., Cha, Y.M., Dickson, M., Grimwood, J., Schmutz, J., Talbot, W.S., and Myers, R.M., New genomic tools for molecular studies of evolutionary change in sticklebacks, *Behaviour,* 141, 1331, 2004.
164. Boughman, J.W., Conspicuous nuptial color in sticklebacks: effects of environment on signal divergence, in preparation.
165. Orr, H.A., The genetics of species differences, *Trends in Ecology and Evolution,* 16, 343, 2001.
166. Bell, M.A., Baumgartner, J.V., and Olson, E.C., Patterns of temporal change in single morphological characters of a Miocene stickleback fish, *Paleobiology,* 11, 258, 1985.

167. Hatfield, T., Rosenfeld, J., Murray, C., Foote, C.J., Jesson, D., McPhail, J.D., Richardson, J., Schluter, D., Taylor, E.B., and Wood, P., National Recovery Strategy for Stickleback Species Pairs (*Gasterosteus spp.*) in British Columbia, prepared for British Columbia Ministry of Water, Land and Air Protection, and Fisheries and Oceans Canada, 2003.
168. Wood, P.M., Will Canadian policies protect British Columbia's endangered pairs of sympatric sticklebacks?, *Fisheries,* 28, 19, 2003.
169. Hatfield, T., Status of the stickleback species pair, *Gasterosteus* spp., in Hadley Lake, Lasqueti Island, British Columbia, *Canadian Field Naturalist,* 115, 579, 2001.
170. Kraak, S.B.M., Mundwiler, B., and Hart, P.J.B., Increased number of hybrids between benthic and limnetic three-spined sticklebacks in Enos Lake, Canada; the collapse of a species pair?, *Journal of Fish Biology,* 58, 1458, 2001.
171. Taylor, E.B., Boughman, J.W., Groenenboom, M., Sniatynski, M., Schluter, D., and Gow, J.L., Speciation in reverse: morphological and genetic evidence of the collapse of a three-spined stickleback (Gasterosteus aculeatus) species pair, *Molecular Ecology,* 15, 343, 2006.
172. Gow, J.L., Peichel, C.L., and Taylor, E.B., Contrasting hybridization rates between sympatric threespine sticklebacks highlight the fragility of reproductive barriers between evolutionarily young species, *Molecular Ecology,* 15, 739, 2006.
173. Stockner, J.G., Rydin, E., and Hyenstrand, P., Cultural oligotrophication: causes and consequences for fisheries resources, *Fisheries,* 25, 7, 2000.
174. Seehausen, O., van Alphen, J.J.M., and Witte, F., Cichlid fish diversity threatened by eutrophication that curbs sexual selection, *Science,* 277, 1808, 1997.
175. Coyne, J.A. and Allen, O.H., *Speciation*, Sinauer, Sunderland, MA, 2004.
176. MacDonald, J.F., Macisaac, S.M., Bekkers, H., and Blouw, D.M., Experiments on embryo survivorship, habitat selection, and competitive ability of a stickleback fish (Gasterosteus) which nests in the rocky intertidal zone, *Behaviour,* 132, 1207, 1995.

4 Antipredator Defences in Sticklebacks: Trade-Offs, Risk Sensitivity, and Behavioural Syndromes

Felicity Huntingford and Susan Coyle

CONTENTS

4.1 PREDATORS OF STICKLEBACKS

Being small, sticklebacks fall prey to a wide variety of predators. For example, in his comprehensive review of the effects of predation on stickleback biology,

TABLE 4.1
Ecological Details and Classification of Predation Regime in Three Icelandic Lakes

Site	Depth (m)	Vegetation	Special Features	Predators Present	Risk Category
Thingvallavatn, lava	1	Patchy green algae	Lava rubble, with cavities	Piscivorous (charr and trout)	High
Thingvallavatn, soft-bottom	17	Dense stands of *Nitella opaca*	No lava. Sticklebacks concentrated in Nitella stands	Piscivorous (charr and trout)	Very high
Frostastadavatn, lava	5	Minimal	Steep laval shoreline	Recently introduced charr and trout	Moderate
Frostastadavatn, soft bottom	1–1.5	*Nitella* common; other species patchy	No lava, shallow, sandy shoreline	Recently introduced charr and trout	Moderate
Sauravatn	<0.5	Dense, multispecies	Very shallow, marshy lake. Sticklebacks concentrated in vegetation at shoreline	No piscine predators	Minimal

Source: Modified after Doucette, L.I. et al., *Biol. J. Linn. Soc.*, 82, 189, 2004.

Reimchen[1] cites 6 species of invertebrates (including jellyfish, *Aurelia* sp., water scorpions, *Ranatra* sp., and backswimmers, *Notonecta* sp.), 22 species of fish, 1 species of reptile, 5 species of mammal, and 34 species of bird as predators of adult sticklebacks. Leeches, diving beetles (*Dytiscus*), and dragonfly nymphs also feed on eggs and larvae.[2] Thus, in any one location and depending on life history stage, sticklebacks may require protection from many different kinds of predators with distinct methods of hunting, and they also require different antipredator responses. Additionally, the literature abounds with examples describing how the predation regime experienced by sticklebacks at different sites varies in terms both of the absolute pressure of predation on sticklebacks and the relative pressure from different kinds of predator[1,3,4] (Table 4.1 gives one example). Both the range of predators to which sticklebacks fall prey and the spatial variability of predation regimes have been an important force in their evolution.

4.2 ANTIPREDATOR ADAPTATIONS IN STICKLEBACKS

Sticklebacks have a suite of traits (morphological and behavioural) that protect them against the wide range of predators to which they fall victim. These variously reduce their chances of encountering or being detected by a predator, and of being attacked successfully should an encounter occur. In addition, because survival is only one component of fitness, sticklebacks have mechanisms (often complex and subtle) for balancing the need for antipredator defence against other imperatives, such as the need to feed and to reproduce. In this chapter, we describe some of these adaptations and balancing acts, concentrating on work published since 1994, when Michael Bell and Susan Foster published their influential review of *The Evolutionary Biology of the Threespine Stickleback*.*

4.2.1 Morphological Adaptations

For most of the year, sticklebacks tend to be well camouflaged, with a dark dorsal surface, a (counter-shaded) lighter ventral surface, and diffuse vertical bars of darker skin along their flanks. They have some capacity to lighten or darken their colour in relation to their background, and the lateral stripes can intensify in response to perceived risk[5] (and personal observation). However, most work on morphological adaptations to predation risk has concentrated on the sticklebacks' protective armour, including lateral bony plates and the spines after which these fish are named.

Thus, as its name implies, the three-spined stickleback (*Gasterosteus aculeatus*) has three dorsal spines along its back, two pelvic spines, and several lateral plates running along its flanks; the dorsal and pelvic spines can be locked into an upright position. Hoogland, Morris, and Tinbergen showed that perch (*Perca fluviatilis*) and pike (*Esox lucius*) took fewer three-spined stickleback than the less armoured nine-spined stickleback (*Pungitius pungitius*), which in turn were eaten in lower frequencies than several species of cyprinid with no body armour.[6] More recent direct evidence for the effectiveness of spines as a defence comes from Mathis and Chivers,[7] who found that in mixed shoals of brook stickleback (*Culaea inconstans*) and fathead minnows (*Pimephales promelas*), the latter were captured more readily by yellow perch (*Perca flavicens*). The anterior lateral plates (running from behind the head to the second dorsal spine) provide structural support for the dorsal and pectoral spines, buttressing them against deflection or fracture. The posterior plates (running from the rear of the second dorsal spine and to the tail fin) interfere with swallowing of prey. The effects of lateral plates on handling times and probability of escape can be seen in Figure 4.1.[8]

Indirect evidence for the effectiveness of stickleback body armour comes from studies linking the development of armour to predation risk. For example, in several North American populations sympatric limnetic and benthic sticklebacks exist. The former, which live in open water where they are exposed to a greater risk of predation by vertebrates, are heavily armoured compared to the benthic form[9] (also see Chapter 3). A relationship between well-developed armour and relaxation of other defences provides further indirect evidence for the effectiveness of armour. Thus, the armoured brook stickleback is more likely to forage under threat of predation than the unarmoured fathead minnow.[10] Minnows (*Phoxinus phoxinus*) take longer to emerge from a refuge after being attacked than do three-spined sticklebacks.[11] Furthermore, three-spined sticklebacks with a pelvic girdle and spines sometimes take more risks when foraging in the presence of a predator than those without girdle and spines.[12]

---

* *Evolutionary Biology of the Threespined Stickleback*, Bell, M.A. and Foster, S.A., Eds., Oxford University Press, Oxford, 1994.

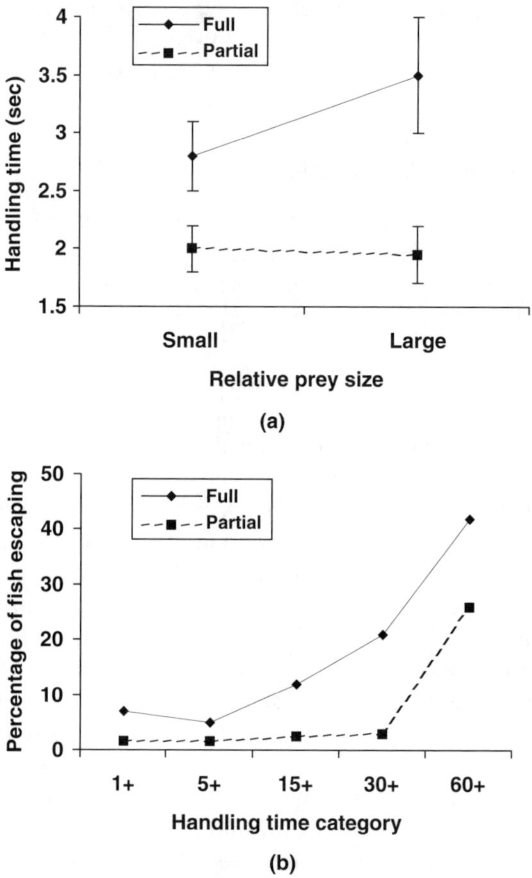

FIGURE 4.1 Armour as protection against piscivorous fish. (a) Handling time for piscivorous predators feeding on two populations of three-spined stickleback with different levels of armour (fully plated and partially plated) as a function of relative prey size (small: ratio of prey depth to predator mouth size = 0.5–0.8. Large: ratio of prey depth to predator mouth size = 0.8–1.0). (b) Percentage of escapes following capture as a function of handling time by three-spined sticklebacks with different levels of armour. (Modified after Reimchen, T.E., *Behaviour*, 137, 1081, 2000.)

4.2.2 Behavioural Adaptations

Notwithstanding the effectiveness of their armour, sticklebacks have a range of behavioural adaptations that also reduce their risk of predation (Figure 4.2).[13] One important form of predator avoidance in sticklebacks, as in other animals, is to make preferential use of areas in which they are less likely to encounter or be detected by a predator (habitat selection). For example, the nine-spined stickleback (*Pungiteus pungitius*), which has smaller spines giving less protection against predators,[6] makes greater use of weeded habitats than does the better-protected three-spined stickleback.[5,13] In the breeding season, male three-spined sticklebacks prefer to build their

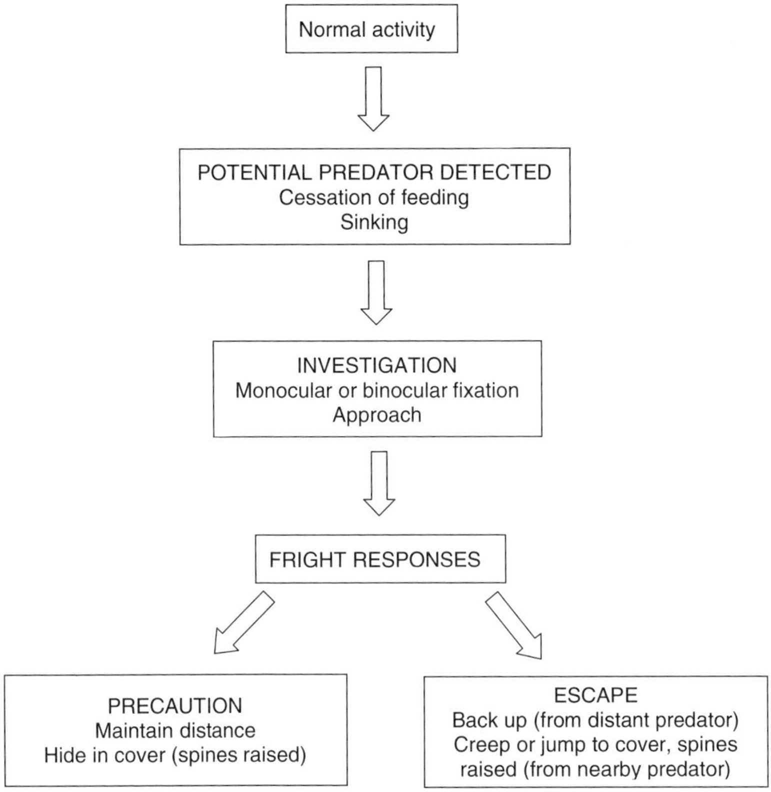

FIGURE 4.2 Schematic representation of responses in sticklebacks during encounters with potential predators. (Modified from Benzie, V.L., Some Aspects of the Anti-Predator Responses of Two Species of Stickleback, Thesis, 1965.)

nests in sites that are well hidden by macrophytes, where they are relatively protected from nest predators and attract more females.[14] When exploring a novel and potentially dangerous environment or on encountering a predator in a familiar environment, sticklebacks make use of various form of shelter (clumps of weed and spaces beneath stones) to avoid detection. The extent to which they do so relates to vulnerability. For example, nine-spined sticklebacks make greater use of weed during exploration than do three-spined sticklebacks, both when exploring a novel environment (Figure 4.3)[13] and when assessing feeding patches[15] (see the following pages).

Sticklebacks also protect themselves against predation by detecting predators early and taking evasive action before being detected themselves. For example, the brook stickleback *Culea inconstans* avoids traps marked with conspecific skin extract, which probably contains an alarm pheromone (see also Chapter 6). Larger (presumably older) fish are better able to make this discrimination, suggesting that this may be something that they have to learn.[16] Certainly, sticklebacks can learn to associate olfactory cues from a predator with a frightening experience (for example, induced by alarm pheromone[17,18]). Avoiding detection can also be achieved by

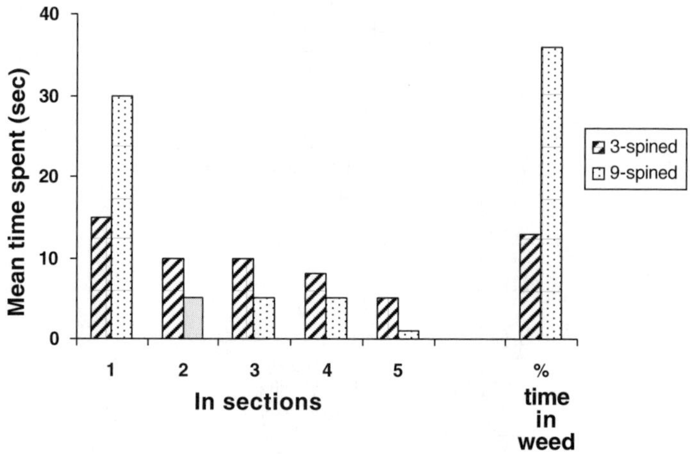

FIGURE 4.3 Relationship between armour and precautionary behaviour. Time spent at different distances from shelter (1 = close to shelter, 5 = far from shelter) in three- and nine-spined sticklebacks placed in an unfamiliar environment, together with percentage of time spent in weed during exploration. (Modified after Benzie, V.L., Some Aspects of the Anti-Predator Responses of Two Species of Stickleback, Thesis, 1965.)

general vigilance in unfamiliar and potentially dangerous environments. Many studies assess how responsive sticklebacks are to risk by measuring both time to emerge from shelter in a potentially dangerous environment and time spent swimming "cautiously" with spines raised after they do emerge (see the following pages).

Once prey and predator come into contact, prey can protect themselves through the initially surprising response of approaching a predator (slowly, and with frequent stops and starts), either alone, in pairs, or in groups. This behaviour is called *predator inspection* and was described for sticklebacks by Benzie[13] (Figure 4.2 to Figure 4.4). One function of predator inspection is to provide information about how much of a threat a potential predator poses (Does it belong to a predatory species? Is it the right size to eat this particular prey? Is it hungry?) and, therefore, whether it is safe to continue other activities such as feeding. Another possible function is pursuit deterrence, whereby information flows in the opposite direction; by facing and approaching a predator a stickleback conveys the information that it has detected the predator, is on the alert, and is not worth attacking.[19] However, Milinski et al.[20] found no evidence that predator inspection serves a pursuit deterrence function in three-spined sticklebacks. Once a potential predator does attack, sticklebacks show a range of responses to avoid capture, including rapid fast-start escape, creeping for cover and freezing (Figure 4.2),[13] and locking the spines in position in anticipation of capture. Differences in immediate response to an attacking predator are seen between species (for example, nine-spined sticklebacks make greater use of weed than do three-spined ones; Figure 4.3)[13] and between individuals within populations (for example, individual fish escape at different speeds[21]). Differences in response are also seen in the same individual on different occasions; for example, gravid females, which are

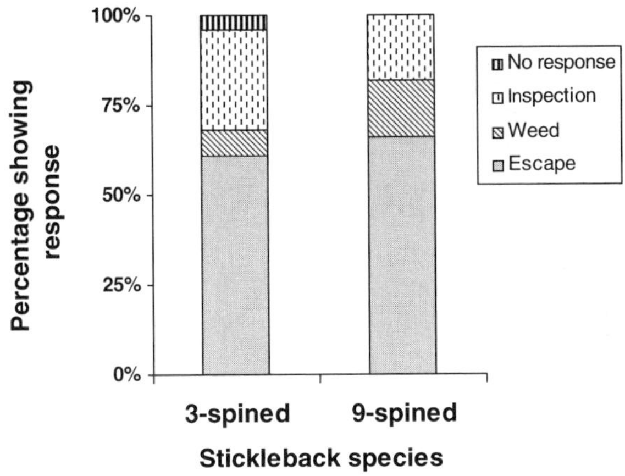

FIGURE 4.4 Relationship between armour and antipredator behaviour. Response on first interaction during exposure to a pike by laboratory-reared three- and nine-spined sticklebacks with no previous experience of a pike. (Modified after Benzie, V.L., Some Aspects of the Anti-Predator Responses of Two Species of Stickleback, Thesis, 1965.)

particularly vulnerable, incorporate unpredictable, protean elements into their escape responses, whereas nongravid females simply flee.[22]

Outside the breeding seasons, sticklebacks tend to form shoals (loose associations of groups of fish, brought together through mutual attraction), which provide protection against predators through safety in numbers, enhanced vigilance, and the confusion effect.[19] The behavioural processes involved in shoaling can be complex and subtle. For example, individual three-spined stickleback shoal preferentially with fish of their own size, probably because odd individuals within shoals experience higher rates of predation.[23,24] Given a choice, they also show a preference for shoals containing familiar siblings over those with nonfamiliar unrelated fish.[25] As a final example, in the presence of a predator, brook sticklebacks are more likely to shoal with unarmoured heterospecifics (fathead minnows) than with other sticklebacks, perhaps because predators find it easier to capture and eat minnows.[7]

4.3 EFFECTS OF LOCAL PREDATION REGIMES

Predation regime varies markedly between habitats and sites, and such variation is associated with differences in morphology and behaviour among populations of sticklebacks.[1,26] Walker[27] carried out an extensive analysis of foraging mode and predator avoidance as predictors of morphological variability among three-spined sticklebacks from 40 lakes. This study established that the presence of native predatory fish has selected directly for increased morphological protection, including deeper heads and bodies, more posteriorly placed dorsal spines, pelvic processes that extend further forward, and shorter medial fins. Populations of three-spined stickleback that coexist with perch show more predator inspection than do those from perch-free sites; they also make more complex adjustments to the detailed appearance of potential predators.[28] Several aspects of antipredator responses vary

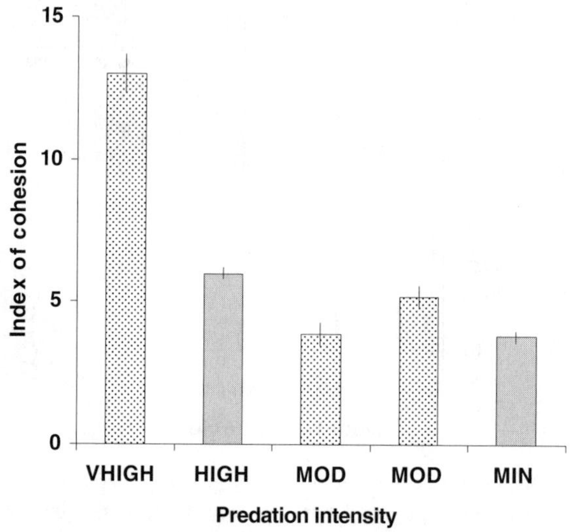

FIGURE 4.5 Relationship between shoaling tendency and predation risk; mean and 95% confidence intervals of index of cohesion (a measure of strength of shoaling) in three-spined sticklebacks from different areas of two lakes in Iceland classified according to predation risk. (Modified from Doucette, L.I. et al., *Biol. J. Linn. Soc.*, 82, 189, 2004; also see Table 4.1.)

between Icelandic populations of three-spined sticklebacks, including markedly stronger shoaling responses in fish from the site with the highest risk of predation[4] (Figure 4.5). On a finer scale, a study of sympatric three-spined sticklebacks showed that the benthic form (which inhabits areas relatively free from vertebrate predators) has poorly developed armour and does not shoal with conspecifics, whereas the limnetic form (inhabiting areas exposed to diving avian predators) has well-developed armour, including long spines, and readily forms shoals with conspecifics[29] (also see Chapter 3). Short-term changes in morphology in relation to immediate predation regime have also been described. For example, Reimchen and Nosil[30] studied a large sample of three-spined sticklebacks collected from a coastal lake over 15 years. They used stomach contents to assess relative use of benthic and pelagic foraging habitat by males and females and compared trait distributions before and after known periods of strong selection (winter and summer) to estimate the size and direction of selection. In adult fish, for example, selection favoured higher numbers of dorsal and pelvic spines in females than in males, associated with a greater use of pelagic habitat by females (and hence, greater exposure to vertebrate predators), and a greater use of benthic habitat (with associated exposure to invertebrates) in males. This effect was stronger at times when the differences in feeding ecology between the sexes were greatest (Figure 4.6).

4.4 COSTS, BENEFITS, AND TRADE-OFFS

The examples given earlier show that sticklebacks have a whole suite of adaptations (some very complex) that protect them against the risk of predation and that vary

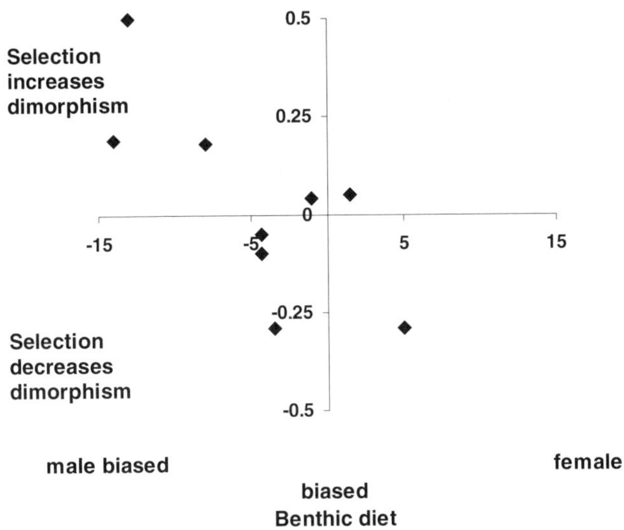

FIGURE 4.6 Short-term changes in morphology in relation to predation risk. Gender differences in spine number in three-spined stickleback (positive values indicate selection favours relatively longer spines in males compared to females) in relation to relative use of benthic foraging by females and males (positive values indicate that females make greater use of benthic foraging). (Modified after Reimchen, T.E. et al., *Evolution*, 58, 1274, 2004.)

in response to local predation regimes. Protective morphology and behaviour are not developed to their fullest extent in all sticklebacks, because such adaptations have their costs. The particular form taken in each case represents a compromise or trade-off between different selective forces, again sometimes complex and subtle. In this section, we examine such trade-offs, using as illustration studies from the last decade or so of literature on stickleback morphology and behaviour.

4.4.1 Conflicting Adaptations to Different Predators

The study by Reimchen and Nosil[30] is one of a series demonstrating how selection by vertebrate and invertebrate predators acts in opposite directions on the development of stickleback armour (also see Chapter 3). Well-developed spines and lateral plates strongly protect against predatory fish, making handling difficult and escape more likely (Figure 4.1), whereas they make sticklebacks more vulnerable to predation by invertebrates, which use the spines to hold on during attacks.[1] In sympatric populations of three-spined sticklebacks, limnetic fish (which have long spines) are particularly vulnerable to invertebrate predators, whereas benthic fish (which have short spines) are particularly vulnerable to avian predators.[29] Even among vertebrate predators, selection by piscivorous fish and birds may act in different directions, with trout predation being associated with an increased plate numbers and predation by diving birds being associated with reduced plate numbers.[1] Such variable selection regimes, depending on the precise combination of predators and on habitat selection, contribute to the maintenance of variability in armour development in sticklebacks.

4.4.2 Conflicting Morphological and Behavioural Adaptations

Well-developed armour clearly makes sticklebacks more difficult for piscivorous fish to handle and swallow. However, in some conditions, having armour may make them more rather than less vulnerable to such predators, owing to a trade-off between physical protection and the hydrodynamics of swimming. For example, during fast escapes, brook sticklebacks without defensive armour are able to bend their bodies more during the initial phase and can move further, faster, and with greater acceleration than can those with armour (Figure 4.7a).[21] Impaired escape in relation to the development of body armour has also been shown in a comparison between populations of brook sticklebacks (Figure 4.7b).[31] A similar trade-off has been demonstrated for three-spined sticklebacks, with lateral plate number being negatively correlated with the speed and extent of fast-start escape responses.[32]

4.4.3 Conflicting Adaptations for Avoiding Attack and Gathering Information

The importance of information gathering during predator inspection is demonstrated by the fact that sticklebacks that coexist with abundant predatory fish show particularly strong inspection.[28] However, during predator inspection, information is gathered at the expense of increased risk of attack. For example, using newly killed sticklebacks mounted on and moved by fine wire, Milinski et al.[20] showed that the probability of an "inspecting" fish being attacked by a pike increases markedly as it approaches the predator (Figure 4.8a). Inspecting in pairs, which is commonly seen in sticklebacks, substantially reduces the predation cost associated with predator inspection (Figure 4.8a).[20] Small sticklebacks from ponds containing piscivorous fish (perch) adjust their behaviour accordingly, inspecting more when in the company of a larger companion than when alone.[28] Pairwise inspection only protects the leading stickleback if its companion is close behind (less than 5 cm; Figure 4.8b).[20] This raises the question of why the following fish cooperates (by moving forward) rather than cheating (by hanging back) during pairwise predator inspection.

An important model of the evolution of mutual cooperation is the "prisoner's dilemma," in which both participants do best of all when both cooperate. However, if one player defects when the other cooperates, the defector does better than if it had cooperated whereas the latter does significantly worse than if both had defected. Axelrod and Hamilton[33] identified a successful strategy for resolving the prisoner's dilemma, called *tit-for-tat*, where a player cooperates on its first move and after that copies its opponent's previous move. Milinski[34] showed that a single three-spined stickleback moves toward a predator faster when it is next to a mirror angled so that its image appears to be moving forward with it (cooperating) than when the mirror is angled in such a way that the image appears to be moving away (defecting). Based on this result, Milinski made the intriguing suggestion that sticklebacks inspecting in pairs might be behaving according to the *tit-for-tat* strategy. Whether his finding rules out alternative explanations, such as an effect of safety in numbers, is controversial and has been the subject of a long academic exchange.[35–38]

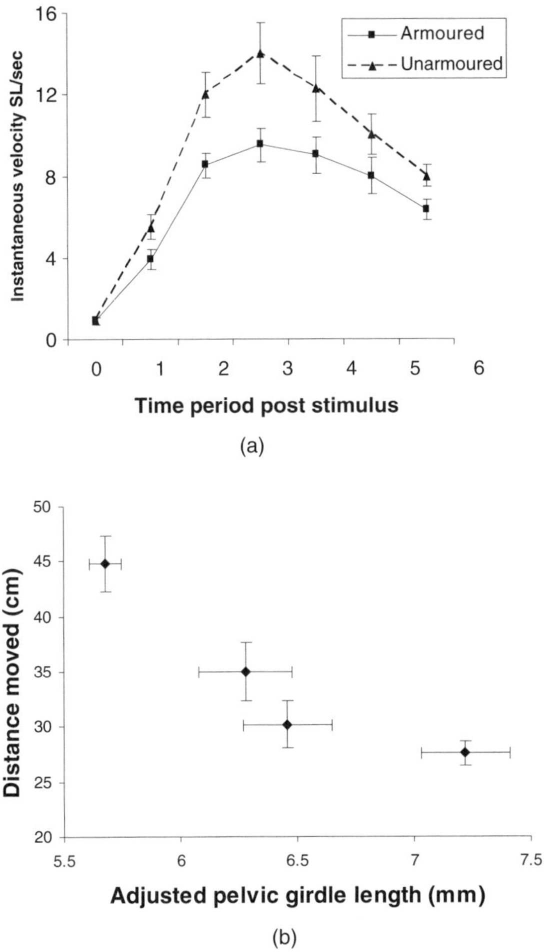

FIGURE 4.7 Effect of armour on escape in brook sticklebacks. (a) Instantaneous velocity (mean, SEM) at 6 points in the fast startle response (total duration = 0.1 sec) in sticklebacks with body armour (pelvic girdle and pelvic spines; closed squares) and without armour (closed triangles) (modified after Andraso, G.M., *Evol. Ecol.*, 11, 83, 1997). (b) Distance moved during a fixed time during fast start responses (corrected for body length. Mean, SEM) in relation to relative length of pelvic girdle (correlated for body length). (Modified after Andraso G.M. et al., *Can. J. Zool.*, 73, 1147, 1995.)

What we do know is that three-spined sticklebacks in the wild can be found shoaling with the same individual on several different occasions,[39] that within small groups inspecting a predator under laboratory conditions sticklebacks have preferred inspecting companions,[40] and that they base decisions about whether to join an inspecting companion on whether that companion cooperated or defected on previous occasions.[40] In addition, analysis of the dynamics of predator inspection showed that three-spined sticklebacks from a site with abundant predatory

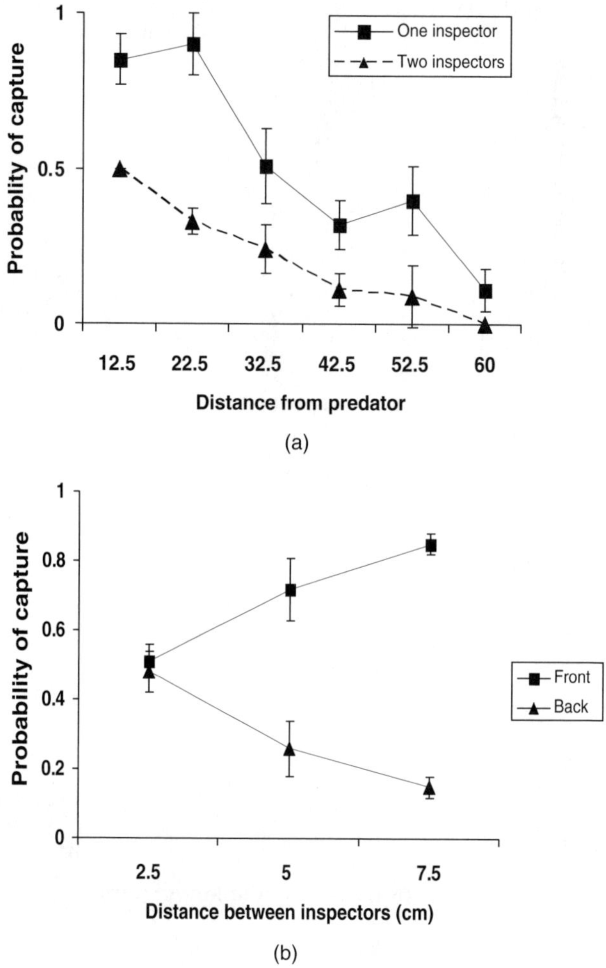

FIGURE 4.8 Costs and benefits of predator inspection in three-spined sticklebacks. (a) Probability of capture (Mean, SEM) for recently killed sticklebacks moved experimentally to fixed distances from a pike, for single fish and for paired fish at the same distance. (b) Probability of capture (Mean, SEM) for the front and back stickleback of a pair placed 10 cm from a predator, with different distances between them. (Modified after Milinski, M. et al., *Proc. R. Soc. Lond. B*, 264, 831, 1997.)

fish (but not those from a low risk site) were significantly more likely to reciprocate on a given move in cases where its companion had just reciprocated than when it had just defected (Figure 4.9).[41] Whether or not such behaviour corresponds to playing *tit-for-tat*, it is clear that during encounters with potential predators sticklebacks have complex behavioural mechanisms for reducing the cost of collecting information.

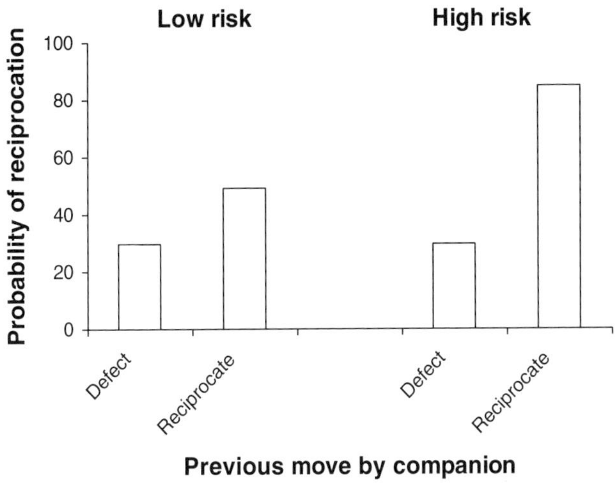

FIGURE 4.9 Effect of previous behaviour of companion on decisions during predator inspection. Probability of reciprocation (moving forward to accompany an inspecting partner, as opposed to hanging back or defection) by three-spined sticklebacks from a site with a low predation risk and from a site with a high predation risk, in relation to the behaviour of its companion partner at the previous opportunity for reciprocation. (Modified after Huntingford, F.A. et al., *Animal Behavior*, 47, 413, 1994.)

4.4.4 Conflicting Needs for Predator Avoidance and Feeding

The trade-off between predator avoidance and feeding is particularly well studied in behavioural ecology. Work on sticklebacks has been important in this context, because Milinski's classic experiment was one of the earliest demonstrations of the effects of predation risk on prey choice.[42] To give some more recent examples of a trade-off between predator avoidance and various aspects of foraging, rates of biting at live prey behind glass are lower in three-spined sticklebacks from sites nearer to the coast, where predators are abundant, than in those from inland rivers, where predation pressure is reduced.[43] Taking cover to avoid predation represents lost opportunity to feed (for example), and is finely traded off against both predation risk and food requirements. For example, three-spined sticklebacks emerge from shelter after a simulated predatory attack sooner than minnows. Minnows also lose weight more slowly during food deprivation than sticklebacks and so, besides being less well protected, can afford to go without food for longer. In addition, a period of food deprivation shortens emergence time in both species and, among sticklebacks, larger fish, which lose weight proportionally more slowly during deprivation, have longer emergence times, especially after a period of food deprivation (Figure 4.10).[11]

Shoaling provides various kinds of protection against predation, but because shoal mates may compete for food, there is a potential conflict between avoiding

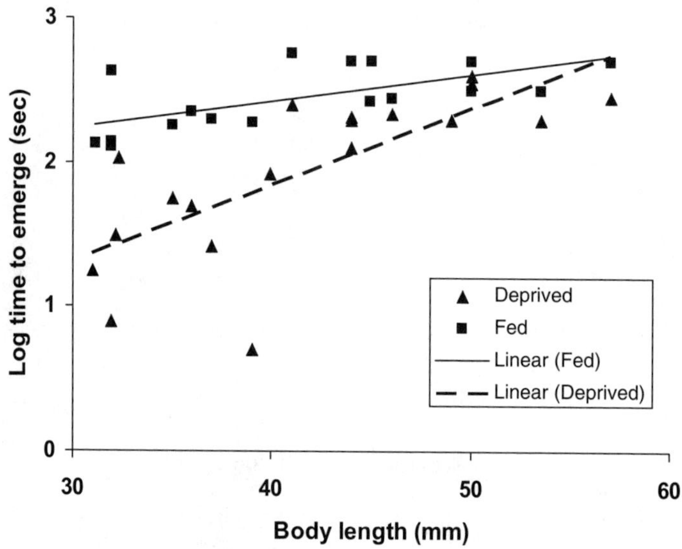

FIGURE 4.10 Trade-off between foraging and avoiding predation. Time (log) taken to emerge from a shelter in relation to body length in well-fed sticklebacks (fed = filled square) and in stickleback that had been deprived of food for 2 d (deprived = filled triangle). (Modified after Krause, J. et al., *Behaviour*, 137, 1113, 2000.)

predation and gaining food; this is reflected in shoaling behaviour. For example, three-spined sticklebacks spend less time with a shoal (and with the larger of two shoals when given a choice) when hungry than when satiated. In a freely moving shoal, hungry sticklebacks also spend more time near the front, where feeding is richer but predation risk greater.[44] Owing to the oddity effect, sticklebacks (like other small fish) are more vulnerable to predation when they differ in size from their shoal mates, especially if they are larger, and behaviour within shoals is adjusted to this risk. Large sticklebacks spend more time foraging (as opposed to maintaining vigilance, for example) as the number of other large fish in a shoal increases, and hence, the extent to which they stand out from the herd falls.[45] A trade-off between predation risk and an even more complex aspect of foraging has been demonstrated in a study of the use of social cues during foraging.[15] These researchers showed that nine-spined sticklebacks (but not three-spined) gather information from shoal mates about food patch quality by hiding in the weed and watching from a distance (making use of "public information"). It seems that three-spined sticklebacks use social information in the form of local enhancement (a tendency to go to where other fish are feeding and see what is up[46]) rather than using public information. Arguably, they can afford to do so because they are well armoured. Thus, these two species use different strategies when foraging socially, which reflects their vulnerability to predation.

4.4.5 Conflicting Needs for Predator Avoidance and Breeding

The requirements for successful reproduction (in the case of male sticklebacks, developing breeding colouration, moving to breeding grounds and establishing territories, building nests, courting females, and caring for young) impose an increase in predation risk. The conflicting requirements of breeding successfully and surviving at least to the end of the current breeding episode are reflected in all these aspects of reproductive behaviour.

Starting at the beginning of the reproductive periods, red colouration is an early sign of breeding in male three-spined sticklebacks and has an important role in aggressive interactions and during courtship. However, colouration makes breeding males more conspicuous, and it can be adjusted to reduce predation risk, for example, by being brightest only during courtship and fighting. Risk-related changes in breeding colouration can be extremely complex. For example, breeding male three-spined sticklebacks were given a 3-week period either with normal feeding or on low rations; they were then screened for the size of their red throat patch (corrected for fish size) after a bout of courtship. Half of the males in each group were exposed to sight of predators (a perch) just beforehand and the other half left undisturbed. In well-fed males, exposure to a predator had the effect of reducing the area of colouration (Figure 4.11). However, in food-deprived males, which had larger red areas than well-fed males in both conditions, patch area increased after exposure to a predator. It seems that poorly fed males (whose lipid reserves were very low) take greater risks (in terms of maintaining bright colouration) because with their depleted energy reserves they have little chance of a second breeding attempt.[47]

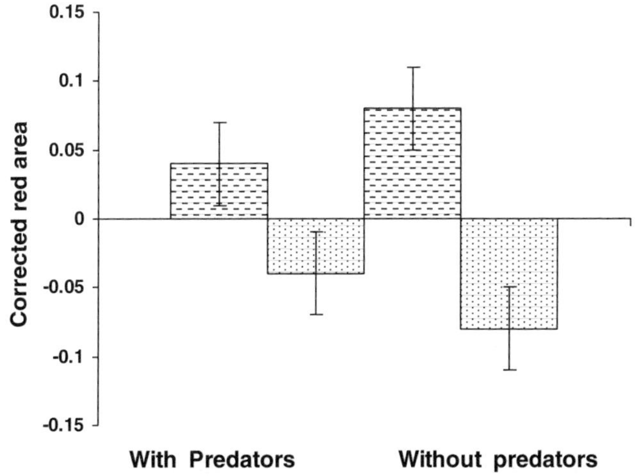

FIGURE 4.11 Complex trade-offs between reproducing and avoiding predators. Mean (SEM) area of red breeding colouration (corrected for body size) in sticklebacks that have been either food-deprived (hatched bars) or fed (stippled bars), and in the presence or absence of a predator. (Modified from Candolin, U., *Anim. Behav.*, 58, 1261, 1999.)

In terms of reaching the breeding grounds, small three-spined sticklebacks arrive before large ones. In doing so, they incur a predation risk (because there are more predators about early in the seasons, and there is less safety in numbers), but have a better chance of getting a territory, which they cannot do when larger males are about.[48] Once on the breeding grounds, male three-spined sticklebacks tend to build their nests close to macrophyte cover. Only in shallow water (where vulnerability to attack by avian predators is thought to be high) only, larger, redder males occupy the sites with macrophytes. This effect is probably not due to better competitive ability in large bright males, because not all concealed nest sites were used up. Instead, their choice of nest sites seems to be an adaptive response to their greater conspicuousness.[14] Toward the end of the season, male three-spined sticklebacks chose more exposed nest sites in shallow water, where they get more encounters with females, and their eggs develop faster because water temperature is higher. Thus, as the breeding season draws to a close, males can increase their chances of gaining and rearing eggs, but at the cost of a greater predation risk.[49]

Attack on a territorial intruder by male three-spined sticklebacks breeding in the laboratory is inhibited by exposure to a predator. This is adjusted to the value of the defended territory, with fish maintaining higher attack rates under predation risk when their nest is full of eggs compared to when it is empty (Figure 4.12).[50] In the field, such adjustments result in males (especially small ones) defending smaller territories when protective cover is removed.[51] Courtship is also modulated by predation risk. Thus, male three-spined sticklebacks from sites nearer to the coast, where predators are abundant, court at a slower rate than those from inland rivers,

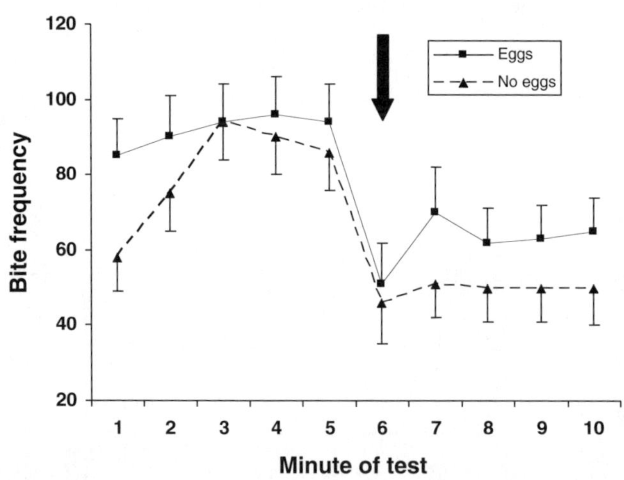

FIGURE 4.12 Trade-off between defending a territory and avoiding predators. Frequency of bites to a standard conspecific intruder by breeding male sticklebacks, before and during visual exposure to a predator (dashed line represents males with empty nests and solid line represents males with a clutch of eggs; ↓ indicates start of a 5-min exposure). (Modified after Ukegbu, A.A. et al., *Ethology*, 78, 72, 1988.)

where predation pressure is reduced.[43] Individual male three-spined sticklebacks show reduced courtship in the presence of predators, but with an interaction between levels of competition and habitat complexity. Thus, males breeding with competitors (but not those breeding alone) reduced their courtship more when nesting in open sites than when nesting in concealed sites.[52]

4.5 INDIVIDUAL VARIABILITY IN RISK TAKING

4.5.1 Behavioural Syndromes in Sticklebacks

Taken together these studies show that the suite of morphological and behavioural traits that defend sticklebacks against predators reflects numerous trade-offs between the need for protection and other fitness imperatives. As illustrated previously, such trade-offs can be complex and the effectiveness of any particular defence or combination of defences depends on the environment. As a result, it is hard (for sticklebacks and scientists alike) to predict the combination of traits that will maximise overall fitness. For this reason perhaps, different individuals strike a different balance between avoiding predators and other functions. For example, within the same population, individuals may differ consistently in the extent of development of both protective morphology and risk taking (Figure 4.13; Coyle, unpublished data[53]). Here we consider the nature and evolutionary significance of within-population variation in antipredator behaviour.

Studies of variation in risk taking in sticklebacks go back a long way. Benzie[13] compared the behaviour of three-spined and nine-spined sticklebacks in direct encounters with a predator and when exploring an unfamiliar and potentially dangerous environment. Whereas both species behave with caution in both contexts, the more robustly armoured three-spined stickleback takes greater risks, approaching a predator more frequently (Figure 4.4) and exploring an unfamiliar environment more quickly, making less use of cover in the process (Figure 4.3). The three-spined stickleback also takes greater risks during the breeding season, nesting on more exposed locations and showing strikingly higher levels of aggression (Figure 4.14).[54]

Following on from Benzie's work, in a study designed to elucidate the relationship between aggressive behaviour shown in different contexts, Huntingford[55] compared the performance of identified individual three-spined sticklebacks (collected from more than one site) in a variety of situations. In confrontations with a live pike, individual sticklebacks differed greatly in their response. Some stayed hidden throughout with their spines locked in the raised position, whereas others emerged from cover, approaching and inspecting the predator and often foraging with lowered spines. The same individual fish also differed consistently in the readiness with which they emerged from shelter to explore an unfamiliar and potentially dangerous environment (with several weeks elapsing between the two tests). The relative boldness or timidity of the same individuals in the two conditions (measured using multivariate analysis) were positively correlated, so differences in general response to risk influence how individual sticklebacks behave in various dangerous situations. When these same fish came into breeding condition (several months later), the males showed striking differences in how fiercely they would attack a variety of intruders,

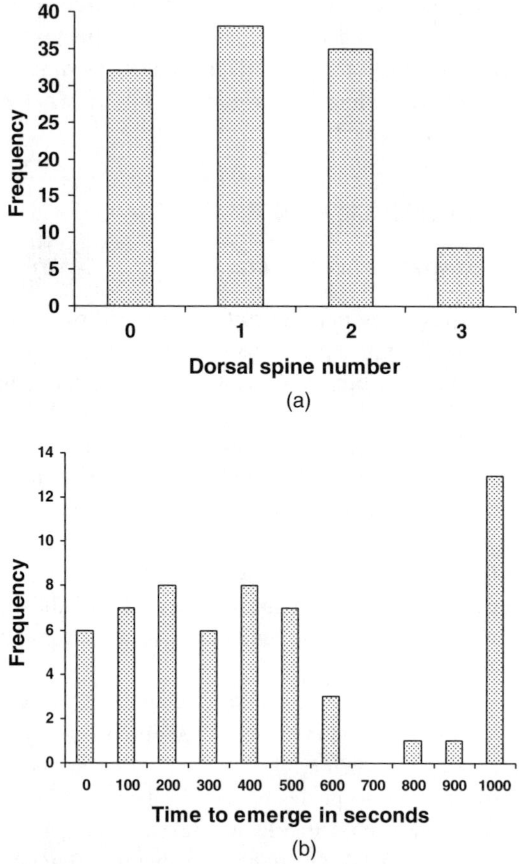

FIGURE 4.13 Variability in morphology and behaviour in three-spined sticklebacks from a site in the Outer Hebrides, Scotland. (a) Frequency distribution of fish with 0 to 3 dorsal spines. (b) Frequency distribution of time to emerge from cover into an unfamiliar tank. Test duration was 15 min and a notional value of 1000 given to fish that never emerged. (From Coyle, S., unpublished data, 2006. With permission.)

including rival males, females, males of a closely related species, and nest predators. Individual responses to these different kinds of intruder were all strongly correlated, so once again, general sensitivity to threat predicts how individuals behave when confronting different threats. Comparing overall levels of aggression during the breeding seasons to risk taking in dangerous situations outside the breeding season (both in the pike test and in the unfamiliar environment test) revealed a small but significant positive correlation. Here again, general responsiveness to risk influences how fiercely breeding sticklebacks fight, as well as of how they respond to more obvious threats. It seems that individual stickleback make different trade-offs between avoiding predation and other fitness imperatives. To borrow a phrase from Gosling,[56] they have "personalities," one dimension of which involves a shy–bold

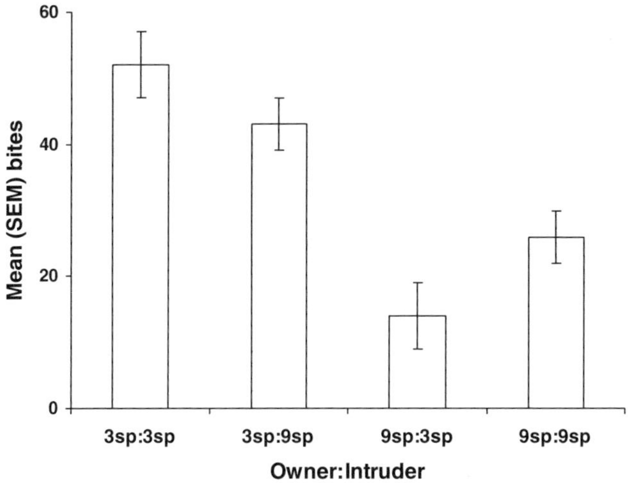

FIGURE 4.14 Variability in aggressiveness between stickleback species. Number of bites per 5 min given by breeding male three-spined (3sp) and nine-spined (9sp) sticklebacks toward territorial intruders (breeding male three-spined and nine-spined males) confined in a glass flask 5 cm from their (empty) nest. (From Huntingford, F.A., A Comparison of Anti-predator Behaviour and Aggression Towards Conspecifics in the Three-Spined Stickleback, *Gasterosteus aculeatus*, Thesis, 1973. With permission.)

continuum and an association between boldness toward predators and aggression toward conspecifics.

The existence of consistent individual differences in interrelated suites of correlated behaviours have been described for many vertebrates and also for some invertebrates.[57] Such suites of traits have been labelled as behavioural syndromes[58] or as different coping strategies.[59] Recent interest in such patterns of variability was stimulated by Wilson et al.,[60] who identified a "shy–bold continuum" among individuals from natural populations of juvenile pumpkinseed sunfish (*Lepomis gibbosus*). Their results suggested that variability along this continuum depended on environmental conditions rather than an innate tendency to be shy or bold.[61] A comparative approach to animal personalities has been used to answer various questions about the causes and consequences of such variation.[56] Sih, Bell, and Johnson[58] focused on whether behavioural syndromes are context specific (individuals may be shy in one context, e.g., predator avoidance, but bold in another, e.g., social interactions) or domain general (individuals may be bold [or timid] across a range of functional behavioural categories, e.g., feeding, mating, and exploration). This has implications for evolution, because domain general behaviour syndromes may act as constraints on evolution.

A number of studies have observed variable risk taking and associated behaviour in fish (see examples in Table 4.2). In the light of this burgeoning interest in behavioural syndromes, attention has focused on the topic of consistent individual differences in risk taking among three-spined sticklebacks, and whether these are reflected in different functional contexts. We review these briefly here, draw some

TABLE 4.2
Recent Articles Describing Variable Risk Taking (with and without behavioural syndromes) in Fish Other than Sticklebacks

Species	Behavioural Variable	Findings	Authors
Pumpkinseed sunfish (*Lepomis gibbosus*)	Entering versus not entering a trap	Shy–bold continuum Stable over time in field but not in lab	Wilson et al.[60]
Pumpkinseed sunfish (*Lepomis gibbosus*)	Time to approach potential threat stimulus and novel food	Shy–bold continuum Stable over time Not correlated across contexts	Coleman and Wilson[74]
Trinidad killifish (*Rivulus hartii*)	Rate of exploration of a novel environment	Shy–bold continuum Boldness in lab predicts dispersal in the field	Fraser et al.[75]
Bluegill carp (*Lepomis macrochirus*), Crucian carp (*C. langsdorfii*), goldfish (*C. auratus*)	Rate of exploration of a novel environment	Shy–bold continuum Species-specific levels of boldness (bluegill > goldfish > crucian)	Yoshida et al.[76]
Poeciliid (*Brachyraphis episcopi*)	Rate of exploration of a novel environment	Shy–bold continuum Boldness correlates negatively with size	Brown and Braithwaite[77]

Lion-head cichlid (*Steatocranus casuarius*)	Rate of exploration of a novel environment Aggression toward conspecific	Continuum of boldness and aggression Behaviours not consistent in juveniles, but stable in older fish	Budaev et al.[78]
Cichlid fish (*Nannacara anomala*)	Inspection of model predator Pairwise fights	Continuum of boldness and aggression Bold pairs fight more fiercely than timid pairs	Brick and Jakobsson[79]
Brown trout (*Salmo trutta*)	Response to a novel object Pairwise dominance tests	Shy–bold continuum Boldness not correlated with metabolism or size but correlated with dominance	Sundstrom et al.[80]
Red-spotted cherry salmon (*Oncorhynchus masou macrostomus*)	Responses to food Responses to unfamiliar object	A dimension of boldness, activity, reactivity, greediness, and carefulness identified Complex interactions between genetic and phenotypically plastic components of behaviour	Iguchi et al.[81]
Rainbow trout (*Oncorhynchus mykiss*)	Method of food capture Time spent out of cover Activity	Shy–bold continuum Bold fish learned faster than timid fish	Sneddon[82]

general conclusions, and highlight areas where more research would be particularly valuable.

Ward et al.[39] worked with nonbreeding sticklebacks (wild-caught, from a single population) during the winter. Fish were marked and given four tests designed to quantify boldness in different contexts; intervals of about 1 week elapsed between tests, during which the fish were kept in groups. These authors looked at latency to resume feeding after a simulated attack from a predator and time spent by solitary fish near a small shoal, testing each fish twice and finding both scores to be statistically repeatable. Comparing across these two contexts, fish that took a long time to recover from the predatory attack spent more time near the shoal. Fish classified as bold on the basis of their combined score in these two tests were more likely to be near the front of a small freely moving shoal than were timid fish (Figure 4.15a). Bold fish also captured a larger number of prey than a timid companion when feeding in competition (Figure 4.15b). No gender effects were found. Weight loss during a short period of starvation was similar in bold and timid fish, suggesting that they probably have similar metabolic rates, but the boldest fish grew faster during the experimental period. This study has demonstrated consistent individual differences in risk taking that are reflected in functionally different contexts, such as direct response to a predatory attack and competitive interactions with a conspecific. The shy–bold continuum is not related in a simple way to metabolic rate and, in the conditions in which these fish were held, bold fish enjoy faster growth.

Bell[62] added a comparative dimension by quantifying risk taking in different contexts in sticklebacks from two different populations captured during the breeding season in two sites in different drainage systems. At one site, sticklebacks coexisted with abundant predators and, probably as a consequence, had well-developed armour; at the other, there were few predators and the sticklebacks were less well armoured. Males and females were allowed to breed (providing 11 broods from each site for subsequent genetic studies) and then screened individually for aggression towards a same-sex intruder and for time to recover from a simulated predatory attack. The tests were always carried out in this sequence with intervals of 30 to 60 min between them. In this case, gender differences were found, with females being more willing to risk exposure to a predator to get food, presumably because they were growing up new batches of eggs and so were highly motivated to feed. In sticklebacks from the site with abundant predators, but not in the low-predation site, boldness under risk of predation was positively correlated with levels of aggression toward a conspecific (Figure 4.16). Genetic correlations between boldness and aggression were significantly higher in sticklebacks from the high-risk site than in those from the site with the lower predation risk (mean and 95% confidence intervals = 0.84 (0.28–0.99) and 0.26 (0.77–0.93), respectively).

Extending the comparative approach, Dingemanse et al.[63] screened fry (collected during the autumn from 12 sites) for risk taking in a number of different contexts, including aggression toward a size-matched conspecific, response to a novel environment, and following a change in a familiar tank (including visual exposure to a perch). Boldness in a novel environment and in the face of change tended to be correlated in all populations. The relationship between boldness and aggression ranged from significantly negative to significantly positive.[64] These detailed and

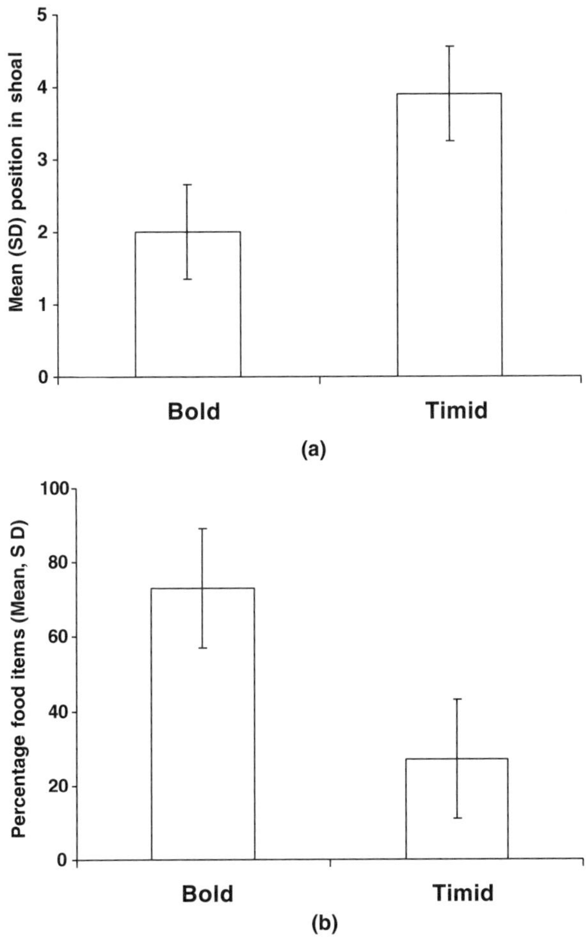

FIGURE 4.15 Behavioural syndromes in sticklebacks. (a) Mean (SD) position in a free moving shoal (1 = at the front) in stickleback previously classified as bold and shy. (b) Mean (SD) percentage of food items captured during competitive feeding. (Modified after Ward, A.J.W. et al., *Behav. Ecol. Sociobiol.*, 55, 561 2004.)

intriguing studies raise a number of areas in which further investigation would be valuable.

4.5.2 Causes of Behavioural Variation and Covariation

Studies on other vertebrates suggest that in individuals at the bold, aggressive end of the shy–bold continuum (or proactive copers, to use an alternative terminology), the physiological response to challenge centres on the adrenal-sympathetic system. In contrast, in those at the timid, nonaggressive end of the spectrum, physiological responses to challenge centre on the hypothalamic-cortisol system.[59] There is some evidence that variation in boldness and aggression in rainbow trout depends on an

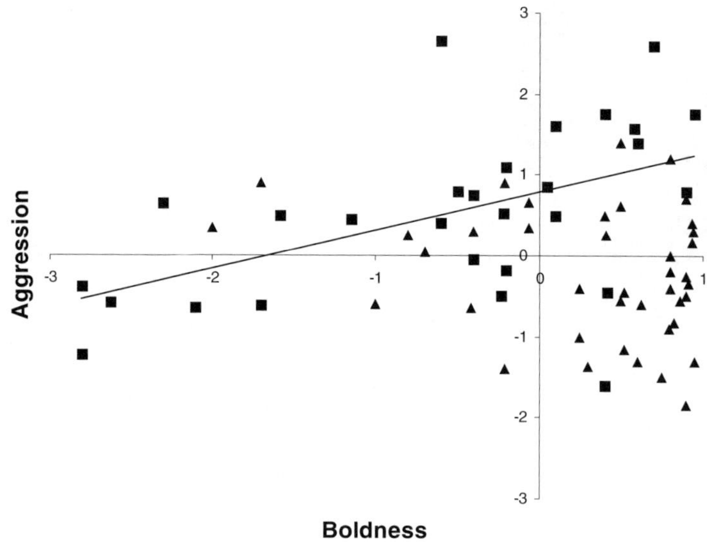

FIGURE 4.16 Variable expression of behavioural syndromes in sticklebacks. Relationship between multivariate scores of aggression toward a conspecific and boldness in a novel environment in individual sticklebacks from a site with abundant predators (filled square and solid line) and from a site with few predators (filled triangles). (Modified after Bell, A.M., *J. Evol. Biol.*. 18, 464, 2005.)

equivalent physiological distinction. Rainbow trout selectively bred for a high-cortisol response to stress are less bold and aggressive than those bred for low-cortisol responsiveness[65] and also show lower brain serotonergic activity and plasma catecholamine concentrations when exposed to confinement stress.[66] Their small size makes such studies more difficult to carry out on sticklebacks, but given how much we know about other aspects of the shy–bold continuum in these fish, information on its physiological bases would be particularly valuable.

4.5.3 Inheritance and Ontogeny of Boldness and Aggression

It is also clear from studies of other vertebrates that in some cases the different coping strategies have a heritable component, although the extent to which the underlying genetic mechanisms have been identified is variable. Thus, for mice we know exactly where on the Y chromosome are the alleles that turn peaceful males into fighters,[67] whereas for rainbow trout we only know that the physiological traits that underpin such behavioural differences respond to selection.[65] In three-spined sticklebacks, aggressiveness in breeding males responds to selective breeding for high and low levels, taking with it aggression among females and juveniles.[68] Family studies have also shown that differences in boldness and aggression in three-spined sticklebacks are heritable and that, in the high-risk population, there is significant genetic covariance between boldness and aggression.[62] Because the stickleback genome is becoming increasingly well characterised (see Chapter 2), it would be both valuable and possible to extend our knowledge of the genetic mechanisms that underlie the patterns of behavioural variation described in these studies, using the molecular tools of QTL mapping[69] and candidate gene analysis.[70]

In terms of ontogenetic processes, the short life span of sticklebacks means that the behaviour of identified individuals can be tracked from egg to adulthood. Such studies are rare in the literature on behavioural syndromes, though Sinn et al.[71] provide an exception, as do Bell and Stamps.[72] The latter authors carried out a developmental study of boldness and aggression and the relationship between them in three-spined sticklebacks from the two sites described previously. In this way, they followed the same individuals at three different life history stages (juvenile, subadult, and adult). Boldness was measured using an appropriate predator for each stage (sculpin, egret, and bass, respectively). Tracking individuals across developmental time produced complex results. In fish from the low-risk site, aggression scores in the juvenile phase were positively related to those in the subadult phase, as were boldness scores in subadult and adult phases. There were no other significant relationships and, for this population, a correlation between boldness and aggression was found only in juvenile fish. In fish from the high-risk site, there was a very weak tendency for individual aggressiveness to be correlated across stages (Spearman's rank order correlation coefficient, $R_s = 0.33$, 0.32, and 0.39, $P = 0.07$, 0.08, and 0.03 for the juvenile: subadult, subadult: adult, and juvenile: adult comparisons, respectively). However, no trace of a correlation across stages was found for boldness. This is in spite of the fact that significant associations were found between these two traits within each stage. As Bell and Stamps[72] point out, such complex developmental histories raise interesting questions about the extent to which behavioural syndromes depend on the effects of experience in one context being generalised to other contexts, and also about how key life history transitions influence both absolute levels and correlations of the relevant behavioural traits.

4.5.4 Ecological Correlates and Evolutionary Consequences of Behavioural Syndromes

Two issues raised by the literature on behavioural syndromes or coping strategies concern the mechanisms that maintain such striking variation in boldness and aggression within populations and the evolutionary consequences of the correlation between them. In general, it seems likely that a combination of spatial and temporal heterogeneity in selection regimes (with the different phenotypes flourishing in different conditions), and possibly frequency-dependent selection, explains the maintenance of within-population variation in boldness and aggression.[73] The evolutionary significance of behavioural syndromes lies in the fact that associations between boldness and aggression may act as a constraint on evolution. For example, selection favouring high levels of aggression may be compromised if this carries with it dangerously high levels of risk taking during encounters with predators.[58,62]

The studies of Bell[62] and Dingemanse et al.[63] show clearly that the linkage between boldness and aggression described by Huntingford[55] and Ward et al.[39] are not fixed. Instead, behaviour in these two contexts can be uncoupled, and in some cases the relationship between them is negative. Thus, they will not act as constraints on evolution, but rather respond adaptively to local selection regimes,[62] with variable patterns of predation being an influential factor. Thus, in Bell's study,[62] aggression and boldness were correlated in the high-predation site only and Dingemanse et al.[64] found positive correlations in fish from large ponds inhabited by piscivorous predators and negative correlations in small ponds lacking such predators. Further studies are needed on the fitness consequences of variable risk taking and the selective forces that favour coupling or uncoupling of boldness and aggression.

4.6 CONCLUSIONS

Because they have adapted in the postglacial period to such a wide variety of habitats, sticklebacks have proved to be excellent models for studying the influence of variable predation regimes on the evolution of morphological and behavioural antipredator adaptations. Studies by many researchers have generated accurate data for many populations to which powerful statistical tools can be applied (Walker's study provides an example[27]). They have also produced a growing number of long-term data sets that can be used in analyses of the action of selection on such traits (Reimchen and Nosil's study provides an example[30]). The small size and tractability of these fish make it possible to carry out highly controlled experimental studies of sophisticated behavioural adaptations. Such adaptations include cooperation during predator inspection (as demonstrated by Milinsk[20]) and use of "public information" in assessing resource quality (as demonstrated by Coolen et al.[15]), as well as complex trade-offs between different fitness imperatives (Condolin,[47] for example). It is also possible to create controlled conditions in which the individual variability in behaviour in various contexts can be expressed and measured. This is one reason why sticklebacks have proved a valuable model for studying behavioural syndromes, their development, and their ecological correlates (for example, Bell and Stamps[72] and Dingemanse et al.[63]). The short generation time found in sticklebacks has led to the development of a genomewide linkage map and sequencing of the stickleback genome. These have already given a molecular handle on the genetic process underlying the evolution of adaptive morphological traits (see Chapter 2), and they open up the real possibility of doing the same for behavioural ones.

ACKNOWLEDGMENTS

We would like to thank Alison Bell and Neils Dingemanse for their help, Tom Reimchem for reading the manuscript and for his useful comments, and finally, thanks to Lorna Kennedy.

REFERENCES

1. Reimchen, T.E., Predators and morphological evolution in threespine stickleback, in *The Evolutionary Biology of the Threespine Stickleback*, Bell, M.A. and Foster, S.A., Eds., Oxford Scientific Publications, Oxford, 1994.
2. Wootton, R.J., *The Biology of the Sticklebacks*, Academic Press, London, 1976.
3. Giles, N., Development of the overhead fright response in wild and predator-naive three-spined sticklebacks, *Gasterosteus aculeatus* L., *Anim. Behav.,* 32, 276, 1984.
4. Doucette, L.I., Skulason, S., and Snorrason, S.S., Risk of predation as a promoting factor of species divergence in threespine sticklebacks (*Gasterosteus aculeatus* L.), *Biol. J. Linn. Soc.,* 82, 189, 2004.
5. Wootton, R.J., *A Functional Biology of Sticklebacks*, Croom Helm, London, 1984.
6. Hoogland, R., Morris, D., and Tinbergen, N., The spines of sticklebacks (*Gasterosteus* and *Pygosteus*) as means of defense against predators (*Perca* and *Esox*), *Behaviour,* 10, 236, 1957.
7. Mathis, A. and Chivers, D.P., Overriding the oddity effect in mixed-species aggregations: group choice by armoured and nonarmoured prey, *Behav. Ecol.*, 14, 334, 2003.
8. Reimchen, T.E., Predator handling failures of lateral plate morphs in *Gasterosteus aculeatus*: functional implications for the ancestral plate condition, *Behaviour*, 137, 1081, 2000.
9. Vamosi, S.M. and Schluter, D., Character shifts in the defensive armor of sympatric sticklebacks, *Evolution*, 58, 376, 2004.
10. Abrahams, M.V., The interaction between antipredator behaviour and antipredator morphology: experiments with fathead minnows and brook stickleback, *Can. J. Zool.*, 73, 2209, 1995.
11. Krause, J. et al., Species-specific patterns of refuge use in fish: the role of metabolic expenditure and body length, *Behaviour*, 137, 1113, 2000.
12. Grand, T.C., Risk-taking by threespine stickleback (*Gasterosteus aculeatus*) pelvic phenotypes: does morphology predict behaviour?, *Behaviour*, 137, 889, 2000.
13. Benzie, V.L., Some Aspects of the Anti-Predator Responses of Two Species of Stickleback, Thesis, 1965.
14. Kraak, S.B.M., Bakker, T.C.M., and Hocevar, S., Stickleback males, especially large and red ones, are more likely to nest concealed in macrophytes, *Behaviour*, 137, 907, 2000.
15. Coolen, I. et al., Species difference in adaptive use of public information in sticklebacks, *Proc. R. Soc. Lond. B*, 270, 2413, 2003.
16. Chivers, D.P. and Smith, R.J.F., Intraspecific and interspecific avoidance of areas marked with skin extract from brook stickleback (*Cilaea inconstans*) in a natural habitat, *J. Chem. Ecol.*, 20, 1517, 1994.
17. Chivers, D.P., Brown, G.E., and Smith, J.F., Acquired recognition of chemical stimuli from pike, *Esox lucius*, by brook sticklebacks, *Culaea inconstans* (Osteichthyes, Gasterosteidae), *Ethology*, 99, 234, 1995.
18. Mirza, R.S. and Chivers, D.P., Learned recognition of heterospecific alarm signals: the importance of a mixed predator diet, *Ethology*, 107, 1007, 2001.
19. Pitcher T.J. and Parrish, J.K., Functions of shoaling and behavior in teleosts, in *Behavior of Teleost Fishes*, Reuter H. and Breckling, B., Eds., Chapman and Hall, London, 1993.
20. Milinski, M. et al., Cooperation under predation risk: experiments on costs and benefits, *Proc. R. Soc. Lond. B*, 264, 831, 1997.

21. Andraso, G.M., A comparison of startle response in two morphs of the brook stickleback (*Culaea inconstans*): further evidence for a trade-off between defensive morphology and swimming ability, *Evol. Ecol.*, 11, 83, 1997.
22. Rodewald A.D. and Foster, S.A., Effects of gravidity on habitat use and antipredator behaviour in three-spined sticklebacks, *J. Fish. Biol.*, 52, 973, 1998.
23. Peuhkuri, N., Ranta, E., and Seppa, P., Size-assortative schooling in free-ranging sticklebacks, *Ethology*, 103, 318, 1997.
24. Peuhkuri, N., Size-assorted fish shoals and the majority's choice, *Behav. Ecol. Sociobiol.,* 46, 307, 1999.
25. Frommen, J.G. and Bakker, T.C.M., Adult three-spined sticklebacks prefer to shoal with familiar kin, *Behaviour*, 141, 1401, 2004.
26. Huntingford, F.A., Wright, P.J., and Tierney, J.F., Adaptive variation in antipredator behaviour in threespine stickleback., in *Evolutionary Biology of the Threespined Stickleback*, Bell, M.A. and Foster, S.A., Eds., Oxford University Press, Oxford, 1994.
27. Walker, J.A., Ecological morphology of lacustrine threespine stickleback *Gasterosteus aculeatus* L. (Gasterosteidae) body shape, *Biol. J. Linn. Soc.,* 61, 3, 1997.
28. Walling, C.A. et al., Predator inspection behaviour in three-spined sticklebacks (*Gasterosteus aculeatus*): body size, local predation pressure and cooperation, *Behav. Ecol. Sociobiol.*, 56, 164, 2004.
29. Vamosi, S.M., Predation sharpens the adaptive peaks: survival trade-offs in sympatric sticklebacks, *Ann. Zool. Fenn.*, 39, 237, 2002.
30. Reimchen, T.E. and Nosil, P., Variable predation regimes predict the evolution of sexual dimorphism in a population of threespine stickleback, *Evolution*, 58, 1274, 2004.
31. Andraso G.M. and Barron, J.N., Evidence for a trade-off between defensive morphology and startle-response performance in the brook stickleback (*Culaea inconstans*), *Can. J. Zool.*, 73, 1147, 1995.
32. Bergstrom, C.A., Fast-start swimming performance and reduction in lateral plate number in threespine stickleback, *Can. J. Zool.*, 80, 207, 2002.
33. Axelrod, R. and Hamilton, W., The evolution of behaviour, *Science*, 211, 1390, 1981.
34. Milinski, M., Tit-for-tat in sticklebacks and the evolution of cooperation, *Nature*, 325, 433, 1987.
35. Master, W.M. and Waite, T.A., Tit-for-tat during predator inspection, or shoaling?, *Behaviour*, 39, 603, 1989.
36. Lazarus, J. and Metcalfe, N.B., Tit-for-tat cooperation in sticklebacks: a critique of Milinski, *Anim. Behav.*, 39, 987, 1990.
37. Milinski, M., No alternative to tit-for-tat cooperation in sticklebacks, *Anim. Behav.,* 39, 989, 1989.
38. Milinski, M., On cooperation in sticklebacks, *Anim. Behav.*, 40, 1190, 1992.
39. Ward, A.J.W. et al., Correlates of boldness in three-spined sticklebacks (*Gasterosteus aculeatus*), *Behav. Ecol. Sociobiol.*, 55, 561, 2004.
40. Milinski, M., Kulling, D., and Kettler, R., Tit-for-tat: sticklebacks (*Gasterosteus aculeatus*) trusting a cooperative partner, *Behav. Ecol.,* 1, 7, 1990.
41. Huntingford, F.A., Lazarus, J., Barrie, B., and Webb, S.A., A dynamic analysis of cooperative predator inspection in sticklebacks, *Animal Behavior*, 47, 413–419, 1994.
42. Milinski, M. and Heller, R., Influence of a predator on the optimal foraging behaviour of sticklebacks (*Gasterosteus aculeatus*), *Nature*, 275, 642, 1978.
43. Peeke, H.V.S. and Morgan, L.E., Behavioural differentiation of adjacent marine and fluvial populations of threespine stickleback in California: a laboratory study, *Behaviour*, 137, 1011, 2000.

44. Krause, J., The influence of hunger on shoal size choice by three-spined sticklebacks, *Gasterosteus aculeatus*, *J. Fish. Biol.*, 43, 775, 1993.
45. Peuhkuri, N., Shoal composition, body size and foraging in sticklebacks, *Behav. Ecol. Sociobiol.*, 43, 333, 1998.
46. Krause, J., Ideal free distribution and the mechanism of patch profitability assessment in three-spined sticklebacks (*Gasterosteus aculeatus*), *Behaviour*, 123, 27, 1992.
47. Candolin, U., The relationship between signal quality and physical condition: is sexual signalling honest in the three-spined stickleback?, *Anim. Behav.*, 58, 1261, 1999.
48. Candolin, U. and Voigt, H.R., Do changes in risk-taking affect habitat shifts of sticklebacks?, *Behav. Ecol. Sociobiol.*, 55, 42, 2003.
49. Candolin, U. and Voigt, H.R., Size-dependent selection on arrival times in sticklebacks: why small males arrive first, *Evolution*, 57, 862, 2003.
50. Ukegbu, A.A. and Huntingford, F.A., Brood value and life expectancy as determinants of parental investment in male three-spined sticklebacks, *Gasterosteus aculeatus*, *Ethology*, 78, 72, 1988.
51. Candolin, U. and Voigt, H.R., Correlation between male size and territory quality: consequences of male competition or predation susceptibility, *OIKOS*, 95, 225, 2001.
52. Candolin, U. and Voigt, H.R., Predator-induced nest site preference: safe nests allow courtship in sticklebacks, *Anim. Behav.*, 56, 1205, 1998.
53. Coyle, S., unpublished data, 2006.
54. Huntingford, F.A., A Comparison of Anti-Predator Behaviour and Aggression Towards Conspecifics in the Three-Spined Stickleback, *Gasterosteus aculeatus*, Ph.D. thesis, Oxford University, 1973.
55. Huntingford, F.A., The relationship between anti-predator behaviour and aggression among conspecifics in the three-spined stickleback, *Anim. Behav.*, 24, 245, 1976.
56. Gosling, S.D., From mice to men: what can we learn about personality from animal research, *Psychol. Bull.*, 127, 45, 2001.
57. Sinn, D.L. and Moltschaniwskyj, N.A., Personality traits in Dumpling squid (*Euprymna tasmanica*): context-specific traits and their correlation with biological characteristics, *J. Comp. Pyschol.*, 119, 99, 2005.
58. Sih, A., Bell, A., and Johnson, J.C., Behavioral syndromes: an ecological and evolutionary overview, *Trends Ecol. Evol.*, 19, 372, 2004.
59. Koolhaas, J.M. et al., Coping styles in animals: current status in behavior and stress-physiology, *Neurosci. Behav. Rev.*, 23, 925, 1999.
60. Wilson, D.S. et al., Shyness and boldness in humans and other animals, *Trends Ecol. Evol.*, 9, 442, 1994.
61. Wilson, D.S. et al., The shy-bold continuum: an ecological study of a psychological trait, *J. Comp. Pyschol.*, 107, 250, 1993.
62. Bell, A.M., Behavioural differences between individuals and two populations of stickleback (*Gasterosteus aculeatus*), *J. Evol. Biol.*, 18, 464, 2005.
63. Dingemanse, N.J. et al., The evolution of behavioural syndromes within and between stickleback populations, in press.
64. Dingemanse, N.J., personal communication, 2006.
65. Pottinger T.G. and Carrick, T.R., Modification of the plasma cortisol response to stress in rainbow trout by selective breeding, *Gen. Comp. Endocrinol.*, 116, 122, 1999.
66. Schjolden, J. et al., Brain serotonergic activity in rainbow trout divergent in stress responsiveness, in press.
67. Sluyter, F., van Oortmerssen, G.A and Koolhaas, J.M., Studies on wild house mice VI: differential effects of the Y chromosome on intermale aggression, *Aggressive Behav.*, 20, 379, 1994.

68. Bakker, T.C.M., Aggressiveness in sticklebacks (*Gasterosteus aculeatus* L.): a behaviour-genetic study, *Behaviour*, 98, 1, 1986.
69. Wright, D. et al., QTL analysis of behavioral and morphological differentiation between wild and laboratory Zebrafish (*Danio rerio*), *Behav. Genet.*, 1, 2006.
70. Fitzpatrick, M.J. et al., Candidate genes for behavioural ecology, *Trends Ecol. Evol.*, 20, 96, 2005.
71. Sinn, D.L. et al., Early temperamental traits in an octopus (*Octopus bimaculoides*), *J. Comp. Pyschol.*, 115, 351, 2001.
72. Bell A.M. and Stamps, J.A., Development of behavioural differences between individuals and populations of sticklebacks, *Gasterosteus aculeatus*, *Anim. Behav.*, 68, 1339, 2004.
73. Dingemanse, N.J. et al., Fitness consequences of avian personalities in a fluctuating environment, *Proc. R. Soc. Lond. B*, 271, 847, 2004.
74. Coleman K. and Wilson, D.S., Shyness and boldness in pumpkinseed sunfish: individual differences are context specific, *Anim. Behav.*, 56, 927, 1998.
75. Fraser, D.F. et al., Explaining leptokurtic movement distributions: intrapopulation variation in boldness and exploration, *Am. Nat.*, 158, 124, 2001.
76. Yoshida, M., Nagamine, M., and Uematsu, K., Comparison of behavioral responses to a novel environment between three teleosts, bluegill *Lepomis macrochirus*, crucian carp *Carassius langsdorfii*, and goldfish *Carassius auratus*, *Fish. Sci.*, 71, 314, 2005.
77. Brown C. and Braithwaite, V.A., Size matters: a test of boldness in eight populations of the peociliid *Brachyraphis episopi*, *Anim. Behav.,* 68, 1325, 2004.
78. Budaev, S.V., Zworykin, D.D., and Mochek, A.D., Consistency of individual differences in behaviour of the lion-head cichlid, *Steatocranus casuarius.*, *Behav. Process.*, 48, 49, 1999.
79. Brick O. and Jakobsson, S., Individual variation in risk taking: the effect of a predatory threat on fighting behavior in *Nannacara anomala*, *Behav. Ecol.*, 13, 439, 2002.
80. Sundström, L.F. et al., Hatchery selection promotes boldness in newly hatched brown trout (*Salmo trutta*): implications for dominance, *Behav. Ecol.*, 15, 192, 2004.
81. Iguchi, K.I., Matsubara, N., and Hakoyama, H., Behavioural individuality assessed from two strains of cloned fish, *Anim. Behav.*, 61, 351, 2001.
82. Sneddon, L.U., The bold and the shy: individual differences in rainbow trout, *J. Fish. Biol.,* 62, 971, 2003.

5 Reproductive Behaviour in the Three-Spined Stickleback

Sara Östlund-Nilsson

CONTENTS

5.1 INTRODUCTION

Being widely distributed in the northern hemisphere, where it inhabits virtually all types of waters, the three-spined stickleback shows considerable variability not only in morphology but also in its reproductive strategies and behaviour. Normally, it becomes reproductively active from late April until July, when the male leaves the shoal and settles on the bottom in shallow water to establish a territory. When he has completed the construction of a nest, he begins to attract females through characteristic courtship dances. The female responds with her own dance, but after she has laid her eggs in his nest, she leaves the male alone to attend to the eggs until they hatch.

The reproductive biology of sticklebacks has been studied intensively for many decades. Earlier literature on this subject is largely covered in the books of Wootton[1,2] and Bell and Foster,[3] and the primary aim of this chapter is to highlight the research done in recent years.

I will describe the reproductive behaviour in chronological order and focus on our present understanding of the mechanisms and selection pressures behind the activities that sticklebacks perform during the different phases of the breeding season. I will first discuss territoriality and the aggressive behaviour of the male and then continue with nest building, courtship, and parental behaviour. I will also consider the roles of male sneaking and egg-napping behaviour. Clearly, the literature on these activities reflects considerable variability in reproductive behaviour between different populations of this widespread species, although it should be kept in mind that some of the divergent data seen in the literature may partly be related to the use of different experimental or observational methods. This variability also provides us with numerous clues as to how the reproductive behaviour of sticklebacks has evolved, and through this variability we may soon be able to deduce the genetic background of stickleback behaviour, given the rapid progress now being made in stickleback genetics.

5.2 TERRITORIALITY

During spring, the male establishes a territory in which he builds a nest. Male aggression is tightly connected to the establishment and maintenance of the territory, and he will repeatedly defend his nest and territory against intruding males. Males also protect their nests and eggs from female raiders as well as from various egg predators. Male territorial aggression in sticklebacks was first described in 1855 by Warington[4] and has since provoked great interest among biologists. The early work on male aggression, as well as stickleback behaviour in general, was primarily done from a descriptive ethological perspective, whereas today the evolutionary background of the behaviour is much more in focus.

The level of male aggression does not always coincide with the size of a territory.[5–9] There are often more factors involved, one being male size. Bigger males are often dominant over smaller males, and the size of a male is an important determinant for the outcome of a combat. Size and aggressive behaviour in combination have often been found to correlate with territorial characteristics such as size, vegetation cover, or habitat complexity and depth. In a population studied by Candolin and Voigt,[10] the biggest males were found to defend the largest territories with low structural complexity and high female encounter rate. The authors experimentally manipulated competition intensity and habitat structure and showed that larger males increased the size of their territory more than smaller males when a neighbouring male was removed. Moreover, they reduced their territory less than smaller males when habitat complexity and cover from predators were reduced.

In the field, bigger males have been found to have higher reproductive success.[11] However, small males tend to arrive first to the breeding grounds and thus may compensate for their size disability and acquire larger territories than they would be able to get later in the reproductive season.[12] Smaller males arriving early may also gain a few more matings, but this is probably at the cost of higher predation risk.[12]

Although size will in many cases determine the outcome of territorial fighting, male territorial aggression may also be influenced by earlier social experiences. Bolyard and Rowland[13] showed that males that had social experience outside their territories were less aggressive toward neighbours compared to males without social experience. A large male may not only benefit from a large territory but also from other features of a territory. Kraak and coworkers[14] showed that in shallow water, larger and redder males, that is, males that could be presumed to have higher competitive ability and greater conspicuousness, choose to settle in habitats with a higher density of macrophytes. The depth at which a territory is established may be important in terms of predation threat. Males of the closely related blackspotted stickleback (*G. wheatlandi*) that built their nests in deeper areas returned sooner to the nests if scared away by a predator, compared to males that nested in more shallow waters.[15]

During spring, males develop blue eyes and red colouration on their cheek and under their belly. These colours have been shown to be important for mating success in the three-spined stickleback (see Section 5.5). Tinbergen[16] found that the red colouration of the males may also work as a signal during male–male encounters, and redder males may elicit attacks by other males. However, later studies have shown that the red intensity alone may not determine the attack response of a male, but that redness is a context-related signal that is dependent on the momentary situation of the male. In two studies, by Rowland and coworkers,[17,18] males were presented simultaneously with three video images of other males that had three levels of red colouration: bright, moderate, and dull. The results showed that males that were on neutral ground were most aggressive against the more moderately coloured video image. By contrast, when males were presented with the same video images in their own territory, they mostly attacked the brightest coloured male on the screen.

5.3 NEST BUILDING

Individual and population differences may be found in nest-building behaviour of members of the stickleback family (Gasterosteidae), but all species build elaborate nests (see Chapter 11 for other species). In the three-spined stickleback, the male generally begins nest building by digging a small pit in the bottom substrate into which he later puts the nesting material. He forms the nest material into a spherical form and creates a tunnel through it. The nest material is "glued" together with spider-web-like threads, in order to stabilise the nest construction. This "glue" is produced in the reproductively active male by kidney tubuli cells that have transformed into secretory cells during the breeding period and is stored in the urine bladder.[19,20] Jakobsson and coworkers[21] and later Jones and coworkers[22] described the molecular character of this glue and demonstrated that it is a protein complex. This protein was named *spiggin* after the Swedish name for the three-spined stickleback, *spigg*.[21]

Males commonly select pieces of soft filamentous algae as the main nest-building material. The nest may be built in the vegetation, but it is most commonly built on the bottom substrate in the small pit the male has dug. The bottom substrate varies a lot between different populations. Laboratory studies have shown that males prefer to build their nest on the substrate typical of their population, when offered a choice.[23] If there is no suitable nest material in the immediate area, he will gather it from the surroundings.

Visual stimuli play an important role in nest-building behaviour. Schüts[24] showed that presenting males with nest-building material under glass stimulated their nest-digging behaviour, and if presented to a sandy material under glass, they were stimulated to collect nest materials. Males are also selective in their choice of nest-building materials. Wunder,[25] Leiner,[26] and Morris[27] gave males nest-building materials of different colours and found that males placed material with brighter colours around their nest entrance. Morris found that male preference for colouration around the nest entrance shifted during the nest-building phase.[27] He showed that males ignored red colours early in their nest-building phase but chose red colours more often toward the completion of the nest. Morris also suggested that the contrasting colours could be used by the male himself as a landmark. In a more recent study by Östlund-Nilsson and Holmlund,[28] the males were given conspicuously looking material, such as shiny metallic sticks and spangles of different colours. The males were provided with a surplus of their normal nest-building material (algae) so that the sticks and the spangles were not needed for the nest construction, but their incorporation would reveal if males showed preference for particular colours and shapes. The results did show that males put colourful metallic sticks and spangles in their nest (Figure 5.1). They often decorated the nest entrance with these artefacts. Most often, the males chose the red colouration, and they preferred the shape of sticks to spangles. The study also investigated the female preference for conspicuous materials in the nests (see Section 5.5.2.1).

5.4 PATERNAL CARE

In teleost fish, parental care takes many forms: paternal care, maternal care, and biparental care. In 78% of all fish families, no care is provided, but among those who do provide care, paternal care prevails.[29–31] The possible reasons for why paternal care dominates have been reviewed by Gross and Sargent[32] and Clutton-Brock,[33] and they include:

1. When females lay their gametes first, only they can desert the brood, thereby "forcing" the male to stay with the offspring (the "gamete order hypotheses").[34,35]
2. When fish have external fertilisation, male confidence in paternity should be higher and paternal care should therefore evolve (the "paternity confidence hypothesis").[34,36]
3. Male territoriality and external fertilisation promote paternal care (the "association hypothesis").[37,38]

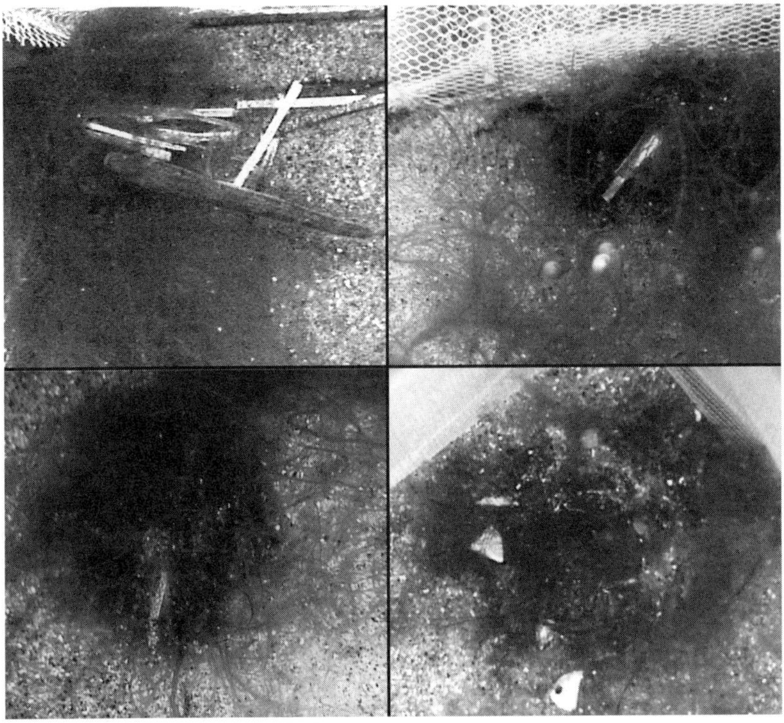

FIGURE 5.1 Video frames of males that have "decorated" their nests with artificial materials (foil sticks and spangles), or below to the right, with a mix of artificial and natural objects. (Photo by Sara Östlund-Nilsson.)

4. The benefits of parental care are similar to either sex, but the costs are typically larger for females, as their reproductive success more often is dependent on body size.[31,32,37,39]
5. Finally, one may of course make the more general suggestion that paternal care may be a sexually selected trait, attractive to females.

In the three-spined stickleback, a male in courtship phase may receive eggs from one to several females. He subsequently enters the paternal phase, and he is not sexually active during the time the eggs (i.e., embryos) develop. After the eggs have hatched, the male starts repairing his nest or builds a new nest, whereupon he again begins to attract females to his nest. The shifts between the sexual phase and the paternal phase are regulated hormonally (see Chapter 8). During the shift to the paternal phase the male also loses his nuptial colouration and becomes duller and, thus, more cryptic.[40] A male may collect up to 20 clutches of eggs from different females for each breeding cycle.[1] Males attend the eggs at least until they hatch, which takes around 5 to 10 d, depending on the water temperature.[1,41]

The care of the developing eggs includes providing them with a current of water, which he accomplishes by fanning the nest in intervals with his pectoral fins. Fanning probably serves at least two purposes: cleaning and oxygenating the eggs. The latter may be essential for a nest-building species such as the three-spined stickleback, where the nest itself reduces water circulation. Indeed, in the absence of fanning, the eggs will die.[25,42] While fanning, he positions himself with his snout into the nest and moves his pectoral fins to create a water current over and through the nest.[1,2,43–46] Not only does the male fan the eggs, he also protects them from being napped by neighbouring males (Section 5.4.3) or being predated upon by other organisms and other sticklebacks, independent of sex. When the eggs grow older and become more metabolically active, the male often makes several openings in the nest. This is probably to increase the water flow over the eggs.[47]

Not all sticklebacks have extensive paternal care. In the "white stickleback," the male builds a nest directly into filamentous algae and after the eggs have been fertilised, he removes them from the nest and disperses them over the surrounding area, where they develop without any care from the male. Allozyme data do not support the hypothesis that the white stickleback is a separate species from the three-spined stickleback,[48] although field and laboratory observations of mating behaviour reveal striking differences.[49,50] Indeed, it lives sympatric with the three-spined stickleback but is reproductively isolated from it.[49] The white stickleback either builds its nest in filamentous algae,[49,50] or, as seen in one population, settles for a nest built directly on the rock substrate, which may be in the intertidal zone where the eggs may occasionally be air-exposed.[51,52] In both cases, no parental care is provided. Experiments have shown that the desertion of the young is a heritable trait.[53] This apparent evolutionary reversal from parental care to desertion of the young has not been seen in other fishes. An obvious advantage of deserting the eggs after fertilisation is that the male can immediately spawn again.

5.4.1 Energetic Costs of Parental Care

Parental investment is the cost associated with parental care.[33] The cost of parental care may be seen from a life-history trade-off perspective,[54] and be measured as reduced survival, breeding rate, and future fitness.[33,55] Costs association with parental activities may also be measured directly as energy expenses. In sticklebacks, it has been shown that male parental activities are energetically costly (for a review, see Wootton[56]) and cause a depletion of lipid and glycogen reserves over the course of the breeding season.[57] The production of the nest-building protein, spiggin, used as a glue during the nest building, is energetically expensive. The synthesis of the corresponding protein (tangspiggin) in the 15-spined stickleback (*Spinachia spinachia*) accounts for at least 5 to 10% of the resting metabolic rate of the male, and the production rate has been shown to correlate with food intake[58] (see Chapter 11). Similar costs for glue production are likely to apply for other stickleback species.

In the three-spined stickleback, high brood-fanning activity can cause a reduction in body fat,[59] and territorial fighting has been found to cause a decline in somatic glycogen levels.[60] Moreover, starved males lose weight faster if they conduct parental activities,[59] and males fed a reduced ration are less successful in defending their territory and provide less parental care compared to well-fed males.[7] Finally, by using respirometry and ration manipulation, Smith and Wootton[61] found that the energy expenditure of parenting males was higher compared to nonparenting males.

Thus, there is ample evidence that parental care is energetically costly for sticklebacks.

The high energy expenditure during parental care may inflict costs not only on the parent but also on the brood. Males may eat some or even all of their own eggs, (i.e., filial cannibalism) to compensate for weight loss and reduced food intake during the paternal care period[62] (see Section 5.4.4).

5.4.2 Stealing Fertilisations

During the times a male establishes a territory, builds a nest, courts females, fertilises eggs and aerates them, and defends the nest against predators, he also participates in a whole range of other "side" activities. Commonly, he engages in "nest raidings" or "raidings," which are terms used to describe actions involving stealing fertilisations (sneaking), eggs, and nest material, and egg cannibalism. Not only male behaviour but also other characteristics of a male may affect raiding behaviour in sticklebacks. The frequency of raiding events may be dependent on the structure of the habitat. Surprisingly, Mori[63] found that the degree of raiding was higher in apparently well-concealed nests in dense vegetation. The reason for this could be that raiders were less easily detected by the nest owners when there was much vegetation for them to hide in.

I will start by describing sneaking, which sometimes may work as an alternative mating strategy for males. Thus, it may have important implications for the genetics and speciation in many species, sticklebacks included. Sneaking is when a male swims in and fertilises the eggs in another male's nest immediately after the female has laid her eggs, thereby partly or fully depriving the nest owner of his paternity. Sneaking behaviour has been intensively studied for many years in the three-spined stickleback.[5,9,63–75] Arguably, there should be a strong selection favouring this parasitic reproductive behaviour, especially in species with paternal care such as sticklebacks, where a sneaker does not have to pay costs for parental care. Mori[63] investigated a population in the field and noted that a male raider was often the victim of raiding himself and was not necessarily a subordinate individual that could neither build a nest nor hold a territory. In fact, the great majority of nest raidings were performed by territorial males with similar body size and territory size as males that were not observed raiding.[63] In contrast, De-Fraipont and coworkers,[71] who specifically studied sneaking interactions, found that older and younger males may have different reproductive tactics. Thus, young nonterritorial males were more likely to engage in sneaking compared to older and territorial males. They also found that young males produced larger quantities and more motile sperm that could make sneaking a particularly successful way of competing with older and bigger males for fitness.

Territorial males can protect themselves against the impact of sneaking by adjusting their behaviour in ways that may not necessarily involve aggression toward rival males, but instead involve reducing their rate of courtship toward females in the presence of rivals, thereby reducing the risk of attracting sneaker males to their nest.[72] Another strategy that males may use against sneakers is to adjust their ejaculate size according to the assessed risk of sperm competition from other males.

Males that were visually confronted with virtual rival males that displayed courting behaviour produced significantly more sperm compared to those being exposed to virtual males performing brood-caring behaviour.[73] Also, the size of the virtual stimuli affected sperm production; males ejaculated more sperm after being presented to larger virtual rivals compared to smaller ones.[75] These results should be viewed in the light of the finding that males switch off their sexual activity during their paternal phase through hormonal changes (see Chapter 8). Thus, it should be adaptive for a male to regard males that perform courting behaviour as bigger threats in sperm competition than males performing parental activities. Not only behaviour but also size itself may affect male sperm production. De-Fraipont and coworkers[71] demonstrated that although sneaking behaviour may be an alternative mating strategy conducted by mostly younger (smaller) males, the stimulus of a bigger male also increased sperm production in males. This may be explained by the fact that bigger males, in most cases, become dominant over smaller ones, making them more successful in combats over mating opportunities. Another explanation may be that the male presented to a stimulus of a bigger male simply gets into the "sneaking mood," because he is the smaller of the two.

For the victims of sneaking, lost paternity as well as the energy wasted on nonrelated offspring should provide strong selection for evolving abilities to detect that sneaking has occurred. However, there is little evidence that such mechanisms exist. The possibility that a male could distinguish his own eggs from those fertilised by others is not much supported in the literature. Indeed, most studies on sticklebacks suggest that males cannot make such discriminations[42,64,76] (but see Jamieson and Colgan[69]).

5.4.3 Stealing Eggs

During a raid, stickleback males may not only steal fertilisations from other males, but they may also steal already fertilised eggs from each other, either by removing clutches or by overtaking other males' nests that contain eggs.[63–65,70,77,78]

Whereas the advantage of sneaking as an alternative mating strategy is easily understood, it is more difficult to explain the benefit of stealing or napping eggs from other males and putting them into their own nest to care for. However, there are some possible explanations for this behaviour. Females may prefer to mate with males that already have eggs in their nests. Thus, a stickleback male may be able to attract more females by supplementing his nest with stolen eggs[9,79] (but see Jamieson and Colgan[69]). That eggs may attract females has been seen in both sticklebacks and other fish,[79–86] and there may be several reasons for this:

1. The female reduces the chance of her eggs being eaten though a simple dilution effect.[87]
2. The chances of egg survival might increase with brood size if males provide more care to larger broods.[88,89]
3. A male with more eggs may be more attractive, because he demonstrates that he is prepared or able to care for the eggs.[38]

4. Many eggs indicate that the male has attracted many females, and a female may use egg number as a cue that allows her to copy the choices of other females, thereby reducing her time and costs spent on searching for high-quality mates.[38,90,91]

Jamieson and Colgan[69] found that males that were the last to complete the construction of their own nests and thus were the last to spawn in their nests were the ones that raided other nests most frequently. They also found that raids were initiated during spawning of neighbouring males, and that sneak spawning then preceded egg theft. Interestingly, in their study males stole eggs they themselves had potentially fertilised.

Two or more of the preceding mechanisms may of course work in concert. Moreover, stolen eggs may serve as an extra food supply, reducing the consequences of male filial cannibalism.

5.4.4 Eating Eggs

Stickleback males sometimes consume their own eggs,[1,2,62,64,65,92–99] and the selective mechanism behind this behaviour is still poorly understood. Rohwer[87] predicted in his model that a fasting guardian male would benefit from cannibalising on his own offspring (filial cannibalism) because it would help him to undertake additional reproductive cycles during the breeding season, thereby increasing rather than decreasing his fitness over the reproductive period. However, it is not always the case that males are fasting during their paternal care phase and that egg consumption mainly is performed through filial cannibalism.[94,95,100] Males may actually reveal their "cannibalistic mood" to females by the intensity of male red nuptial colouration. However, Candolin[98,99] found that this signal was not totally honest over the season and that males with low survival prospects increased their signal strength.

Egg consumption may be very common in a stickleback population. In fact, it can in some populations account for 32% of the diet,[100] and may have consequences for the operational sex ratio (ratio of ready-to-mate males to ready-to-mate females) in parental fishes.[55] Both males and females raid and eat eggs. Rohwer[87] termed the behaviour of eating unrelated offspring *heterocannibalism*. When raiding, females commonly form shoals and consume all eggs in the nests of the males.[94,95,100–105] It is believed that females raid nests and consume eggs because of other reasons than the need for the extra nutrition the eggs may give them, especially as it has been shown that they raid and consume eggs despite high levels of food.[94] Vickery and coworkers[104] proposed a model based on the idea that females create opportunities to spawn under better conditions through nest raiding, i.e., females will use raiding to force a male to rebuild a nest or allow her to spawn in a nest containing no or fewer eggs, which should be an advantage when considering the oxygen supply to her eggs.[106]

Studying two different populations in which one was cannibalistic and the other was not, Foster[96] showed that the intensity of courtship differed between the two. In the noncannibalistic populations males performed more conspicuous courtship behaviour compared to the cannibalistic population. This difference may be related to a difference in water type and visibility (oligotrophic vs. eutrophic) between the two populations. Foster and Baker[107] suggested that filial cannibalism in some populations may also be dependent on morphological traits of the fish, such as body size or a trait correlating with body size. However, whereas they found no correlation between body size and cannibalistic tendencies, they did find that body size affects the effectiveness of cannibalism by females. Also Rohwer[87] proposed that cannibalistic tendencies should be related to body size, suggesting that larger males need more energy than smaller males and should thus cannibalise more on their offspring.

5.5 COURTSHIP AND MATE CHOICE

Courtship behaviour has been thoroughly described earlier[1,46,108] and I do not intend to describe it here in great detail; instead, give an overview. During springtime, both males and females migrate from deeper water to more shallow areas. Here, females remain in their shoals while males settle, take up territories, and build nests. In most populations the male develops red colouration on his cheeks and belly and his eyes become shiny blue. It is quite common for the rest of the body to become greenish but there are some populations in which males develop only a black or white colouration over their body. When a female approaches a male, he usually performs a zigzag dance toward the female and sometimes bites her on different places on the body. He then swims back to the nest in the same zigzag way. If the female is interested, she will follow the male and stay very close by. At this stage the male often repeatedly puts his head into his nest. He sometimes shows a fanning display or swims through or over the nest, possibly imitating the fertilising action. This swimming is usually called sweeping. If the female is interested, she develops a more contrasting pattern on her body, adopts a head-up posture, and begins swimming toward the male. The male usually bites her and repeats his nest-related activities, such as fanning, gluing, and sweeping behaviour. When the female decides to spawn, the male normally bites her on her tail as she creeps though the nest and lays her eggs. When the female has spawned, the male immediately pushes himself through the nest and fertilises the eggs inside, whereupon he makes sure the female leaves the nest area. There are, of course, many variations to this behaviour both between individuals and populations.

During the last decades, female choice has been in focus as one of the mechanisms behind the development of male secondary sexual traits in animals.[109–111] Today, sexual selection is viewed as a strong evolutionary force with widespread consequences for various morphological traits, behaviours, mating systems, and life histories.

Males are most often the sex on which sexual selection has the strongest impact.[112] This is explained by the so-called Bateman gradient,[113] where male reproductive success is only limited by the times he can mate, whereas female fecundity does not increase if she mates with a large number of males (see also Jones et al.[114] for sex-role-reversed species). In species with conventional sex roles (males compete), males may offer females only genes (indirect benefits) or resources that will increase female fecundity or immediate fitness (direct benefits). In the sticklebacks, males offer the females both genes and care of their offspring, and they have conventional sex roles.

5.5.1 Female Choice on Male Bodily and Behavioural Traits

5.5.1.1 Male Colour and Female Mate Choice

Mating colouration has been investigated for over 50 years in the three-spined stickleback and today we can even include UV-lit patterns in the channels of communication between sticklebacks (see Chapter 6). Some of the first studies to investigate whether the males' red colouration could be explained by female preference for red was performed by Leiner[47] and Wunder[115] in the early 1930s, but they failed to conclude that red was important to females. The first study to conclude that red colouration was important for female choice was published in 1937 by Ter Pelkwijk and Tinbergen,[116] who found that females followed dummies to an artificial nest only if the dummies had a red belly. Since then, many studies have confirmed the importance of the red nuptial colouration for female preference.[117–121] Cronly-Dillon and Sharma[122] showed that the spectral sensitivity of the visual system for red was similar in males and females outside the reproductive season. However, during the reproductive season the female's sensitivity for red increases, reaching a higher level than that of the male, which indicates that female preference for red may be a reason for the evolution of male red nuptial colouration.

The extent of the red colouration differs between and within different populations.[123] In some populations, males turn black rather than red in the mating season,[124] and in others, the male does not develop any colouration at all, a form called the *white stickleback*.[48–53,125] However, in most populations males become more or less extensively red, and females show a preference for mating with males having red nuptial colouration.[1,108,120]

The red colour is due to the carotenoid pigments astaxanthin and tunaxanthin/lutein.[126,127] These red pigments cannot be synthesised by the animals themselves, but are received via their food,[128] primarily from small crustaceans.[1] The degree of redness in the males correlates with the amounts of carotenoids in their diet.[129] Carotenoids are also important in physiological processes,[130] and red colouration has been shown to correlate to male condition,[118,129,131] resistance against parasites[132] (see also Chapter 9), courtship effort,[130] nest defence,[133] and mating success.[108,121] Candolin[98] showed that in a situation where food was restricted, large males that completed several breeding cycles during a season also increased their red colouration more than smaller males that completed only a few breeding cycles. This pattern did not repeat itself when food was unlimited, indicating that the red colouration of the male may, in some circumstances, constitute an honest signal to the female of the male's parental ability.

Smith and coworkers[134] suggested that mate preference for red nuptial colouration is an effect of a receiver bias in the perceptual or cognitive system of the three-spined stickleback for the colour of red that may have arisen in the context of foraging. In foraging experiments, the authors found that female and male three-spined sticklebacks preferred red objects outside a mating situation. Bakker[120] showed that there was a positive genetic correlation between male expression of red in their nuptial colouration and female preference for redder males.

Still, whereas female preference for red may be important in many populations of sticklebacks, there may also be variation within and between different populations, and as we shall see next, an apparent preference may be brought about by the experimental conditions.[135,136]

5.5.1.2 Female Choice and Male Dominance

The red nuptial colouration of males works as a signal to not only females, but also males. Dominant males express redder colouration than subordinate males.[137] However, in the three-spined stickleback, laboratory studies on female preference for dominant and colourful males might be misleading, and the outcome may depend on how the experimental setup is designed. In nature, females may escape a dominant male by swimming out of his territory, which is not possible in many experimental setups. If partition screens separate the males from the female, she can inspect the males without them chasing her around, but she will in such a setup have to rely on only one sense, vision. During recent years, more effort has been put into obtaining a better picture of how an animal senses the world around it, and in this respect much new information on sticklebacks has been provided (see Chapter 6). This new information emphasizes that putting animals behind glass walls will deprive them of important cues.

A novel experimental design involves leashing the males with ultrathin lines, so that they are unable to physically interact with each other, whereas females can freely swim between the males and utilize not only distant vision, but also cues such as olfaction, tactile stimuli, and close visual inspection.[136] In such an experiment, it was found that females did not choose the dominant males (which were the redder ones) when the males were leashed, whereas they did "prefer" red dominant males when the males were unleashed.[136] In the same study, females did choose redder males in a classic setup, with the males behind transparent screens and when she was limited to using vision as the only cue for mate choice.

5.5.1.3 Female Choice on Male Paternal Skills

In species such as sticklebacks, in which males provide females with direct benefits such as paternal care, one would expect females to select males on traits that were linked to paternal quality. A study by Bakker and Mundwiler[138] showed that males possessed relatively larger pectoral fins than females in both wild-caught and laboratory-bred fish. This may be related to the fact that males use their pectoral fins when fanning their eggs in their nest. Kunzler and Bakker[140] found that males with relatively larger pectoral fins hatched out offspring of higher quality. In addition, relatively larger pectoral fins in males were correlated to the health status of the males. Thus, males with large pectoral fins were in better physical condition and had more food in their stomachs than those with smaller fins. In addition, poor physical condition and small pectoral fins were often associated with infection by the intestinal parasite *Pomphorhynchus laevis.*[138] As mentioned earlier, male red nuptial colouration may also be an honest signal for parental skills in males.[99]

5.5.2 Female Choices on Male Extra-Bodily Traits

5.5.2.1 The Nest as an Ornament

Females may not only choose males on morphological and behavioural cues, but also on material traits related to his territory or nest. The main function of a nest is to keep the eggs safe in one place in the male's territory: in other words, to prevent the eggs from drifting away or getting eaten by predators. In some populations, females prefer males inhabiting areas with dense vegetation where the nests are better concealed.[140,141] It has been shown that offspring of males that choose to build concealed nests have a higher survival rates and are also more effective in fanning their nests,[140,141] two variables that can be explained by a reduced risk of predation and lower level of male–male interaction in dense vegetation.

Males also use the nest itself as a quality-revealing ornament that gives females information about male health status. It has been shown that males under immunological stress build less "compact" and, thus, less "neat" nests.[142] However, the "neatness" of the nest may not be the only attraction for females. As mentioned, in an experiment where males were given colourful foil sticks and spangles that they could put into their nests (Figure 5.1), males preferred to decorate their nests with red sticks in preference to other colours and shapes, and the females preferred to mate with males having nests that contained sticks and spangles.[28] Early studies investigated which nest material colours were preferred by the males, by giving the males cotton threads.[27] It was also found that males marked their nest entrance with a colour different from that of the rest of the nest.[25,26] However, female preference was not investigated in those studies. It may seem odd that females prefer nests with reduced nest camouflage, and this is in apparent contrast to earlier studies on this species in which females showed a preference for concealed nests.[140,141] However, female preference for ornamented nests may indeed be in line with earlier findings of female preference for more concealed nests. A safe nest may be advertised by the male either by a more intense courtship (as in the study by Candolin and Voigt[141]) or by nest decorations. By displaying a large number of nest decorations males, may show their ability to maintain the nest for some period of time, which could indicate a low density of either egg predators or nest-raiding neighbours around. There may also be other reasons why males decorate their nests. One put forward in 1930 by Wunder[25] was that they mark their nest entrances to make it easier for females to find them, or, as suggested in 1958 by Morris,[27] decoration of the nest may work as an address tag, helping the male to recognize or find his way back to his own nest (for a review on nest-building material and nest building, see Rowland[108]). A male–male competition factor may also be involved, i.e., males show that they are dominant by being able to keep all this material in their nest, and they may also show that they are able to steal material in short supply from neighbouring males, as we have seen them do in aquaria (Östlund-Nilsson and Holmlund, unpublished). The ability to retain the material in their nest may be a signal to females that the male is good at protecting eggs. Finally, as red colouration is attractive to a female in her choice of males, perhaps bringing red objects into the nest lets the male present a "super stimuli" for the females.

5.5.2.2 Mate Choice and Egg Contents in the Nest

When discussing the role of nests in mate choice, it should again be mentioned that the decisions of other females, indicated by the presence of eggs in the nest, also may be an important factor in female choice. The reasons why females should prefer to mate with males whose nests contain eggs not only include copying (which has the advantage of saving her search time and reducing her risk of being predated), but also the fact that laying her eggs next to the eggs of other females could, through dilution, reduce the risk of her eggs being cannibalised upon by the rearing male or by other neighbouring males,[1,2,5,62,65,87,92–95,97–100] or by passing females.[94,95,100–105] Generally, for nest-building teleosts, there is contradictory evidence on whether females prefer nests with eggs or not. Too many eggs in a male's nest could be negative because of oxygen depletion, particularly in species such as sticklebacks where the eggs are laid densely packed together.[108] A reason why females appear to choose males with eggs in their nest may be that these males have more intense courtships or, for other reasons, have been more successful in leading the females into their nests.[67]

REFERENCES

1. Wootton, R.J., *The Biology of the Sticklebacks*, Academic Press, London, 1976.
2. Wootton, R.J., *A Functional Biology of Sticklebacks*, Croom-Helm, London, 1984.
3. Bell, M.A. and Foster, S.A., *The Evolutionary Biology of the Threespine Stickleback*, Oxford University Press, New York, 1994.
4. Warington, R., Observations on the habits of the stickleback (being a continuation of a previous paper), *Ann. Mag. Nat. Hist.*, 16(2), 330, 1855.
5. Van den Assem, J., Territory in the three-spined stickleback (*Gasterosteus aculeatus* L.), an experimental study in intra-specific competition, *Behaviour*, Suppl. 16, 1, 1967.
6. Black, R., Hatching success in the three-spined stickleback (*Gasterosteus aculeatus*) in relation to changes in behaviour during the paternal phase, *Anim. Behav.*, 19, 532, 1971.
7. Stanley, B.V. and Wootton, R.J., Effects of ration and male density on the territoriality and nest-building of male three-spined sticklebacks (*Gasterosteus aculeatus* L.), *Anim. Behav.*, 34, 527, 1986.
8. Ward, G. and FitzGerald, G.J., Effects of sex ratio on male behaviour and reproductive success in a field population of threespine sticklebacks (*Gasterosteus aculeatus*) (Pisces: Gasterosteidae), *J. Zool.*, 215, 597, 1988.
9. Goldschmidt, T. and Bakker, T.C.M., Determinants of reproductive success of male sticklebacks in the field and in the laboratory, *Neth. J. Zool.*, 40, 664, 1990.
10. Candolin, U. and Voigt, H.R., Correlation between male size and territory quality: consequence of male competition or predation susceptibility?, *Oikos*, 95, 225, 2001.

11. Kraak, S., Bakker, T.C.M., and Mundwiler, B., Sexual selection in sticklebacks in the field: correlates of reproductive, mating and paternal success, *Behav. Ecol.,* 10, 696, 1999.
12. Candolin, U. and Voigt, H.R., Size dependent selection an arrival times in sticklebacks: why small males arrive first, *Evolution*, 57, 862, 2003.
13. Bolyard, K.J. and Rowland, W.J., The effects of spatial context and social experience on the territorial aggression of male threespine stickleback, *Behaviour*, 137(7–8), 845, 2000.
14. Kraak, S.B.M., Bakker, T.C.M., and Hocevar, S., Stickleback males, especially large and red ones, are more likely to nest concealed in macrophytes, *Behaviour*, 137(7–8), 907, 2000.
15. FitzGerald, G.J. and Lachance, S., Paternal investment in the blackspotted stickleback *Gasterosteus wheatlandi, Acta Oecologica — Int. J. Ecol.*, 14(1), 17, 1993.
16. Tinbergen, N., Social releasers and the experimental method required for their study, *Wilson Bull.,* 60, 6, 1948.
17. Rowland, W.J., Bolyard, K.J., and Halpern, A.D., The dual effect of stickleback nuptial coloration on rivals: manipulation of a graded signal using video playback, *Anim. Behav.*, 50(1), 267, 1995.
18. Bolyard, K.J. and Rowland, W.J., Context dependent response to red colouration in stickleback, *Anim. Behav.*, 52, 923, 1996.
19. Hentschel, H., The kidney of *Spinachia spinachia* (L.) Flem. (Gasterosteidae, Pisces) 1. Investigations of juvenile sticklebacks: anatomy, circulation and fine structure, *Z. Mikrosk. Anat. Forsch. Leipzig,* 91, 4, 1977.
20. Hentschel, H., The kidney of a teleost, *Spinachia spinachia* II. Histochemical identification of sialic acid-containing glycoprotein and fine structure of mucus secreting cells, *Tissue Cell,* 3, 517, 1979.
21. Jakobsson, S., Borg, B., Haux, C., and Hyllner, S.J., An 11-ketotestosterone induced kidney-secreted protein: the nest building glue from male three-spined stickleback, *Gasterosteus aculeatus Fish. Physiol. Biochem.,* 20, 79, 1999.
22. ones, I., Lindberg, C., Jakobsson, S., Hellqvist, A., Hellman, U., Borg, B., and Olsson, P.E., Molecular cloning and characterization of spiggin: an androgen-regulated extraorganismal adhesive with structural similarities to Von Willebrand factor-related proteins, *J. Biol. Chem.,* 276, 17857, 2001.
23. Hagen, D.W., Isolating mechanisms in threespine sticklebacks (*Gasterosteus*), *J. Fish. Res. Board Can.,* 24, 1637, 1967.
24. Schüts, E., Die wirkung von untergrund und nestmaterial auf das nestbauverhalten des dreistachligen stichlings (*Gasterosteus aculeatus*), *Behaviour*, 72, 242, 1980.
25. Wunder, W., Experimentelle untersuchungen an dreistachligen stichling (*Gasterosteus aculeatus* L.) während der Laichzeit, *Z. Morphol. Ökol. Tiere*, 16, 453, 1930.
26. Leiner, M., Ökologisches von *Gasterosteus aculeatus* L., *Zool. Anz.*, 93, 317, 1931.
27. Morris, D., The reproductive behaviour of the ten-spined stickleback (*Pygosteus pungitius* L.), *Behaviour*, Suppl., 6, 1, 1958.
28. Östlund-Nilsson, S. and Holmlund, M., The artistic stickleback, *Behav. Ecol. Soc.,* 53, 214, 2003.
29. Breder, C.M. and Rosen, D.E., *Modes of Reproduction in Fishes*, Natural History Press, New York, 1966.
30. Blumer, L.S., Male parental care in the bony fishes, *Q. Rev. Biol.*, 54, 149, 1979.
31. Gross, M.R. and Shine, R., Parental care and mode of fertilization in ectothermic vertebrates, *Evolution,* 35, 775, 1981.

32. Gross, M.R. and Sargent, R.C., The evolution of male and female parental care in fishes, *Am. Zool.,* 25, 807, 1985.
33. Clutton-Brock, T.H., *The Evolution of Parental Care*, Princeton University Press, Princeton, NJ, 1991.
34. Trivers, R.L., Parental investment and sexual selection, in *Sexual Selection and the Descent of Man 1871–1971*, Campbell, B., Ed., Heinemann, London, 1972, p. 132.
35. Dawkins, R. and Carlisle, T.R., Parental investment, mate desertion and a fallacy, *Nature*, 262, 131, 1976.
36. Alexander, R.D., The evolution of social behaviour, *Annu. Rev. Ecol. Syst.*, 4, 325, 1974.
37. Williams, G.C., *Sex and Evolution*, Princeton University Press, Princeton, NJ, 1975.
38. Ridley, M., Paternal care, *Anim. Behav.*, 26, 904, 1978.
39. Baylis, J.R., The evolution of parental care in fishes, with reference to Darwin's rule of male sexual selection, *Environ. Biol. Fish.,* 6, 223, 1981.
40. Craig-Bennett, M.A., The reproductive cycle of the three-spined stickleback. *Gasterosteus aculeatus, Linn., Philos. Trans. R. Soc. B*, 219, 197, 1931.
41. Künzler, R. and Bakker, T.C.M. Pectoral fins and paternal quality in sticklebacks, *Proc. R. Soc. Lond. B*, 267, 999, 2000.
42. van Iersel, J.J.A., An analysis of the parental behaviour of the male three-spined stickleback (*Gasterosteus aculeatus* L.), *Behaviour*, Suppl. 3, 1, 1953.
43. Hancock, A., Observations on the nidification of *Gasterosteus aculeatus* and *Gasterosteus Spinachia, Ann. Mag. Nat. Hist.*, 2(10), 241, 1852.
44. Tinbergen, N., *The Study of Instinct*, Oxford University Press, New York and Clarendon, London, 1951.
45. Sevenster, P., A casual analysis of a displacement activity (fanning in *Gasterosteus aculeatus* L.), *Behaviour*, Suppl., 9, 1, 1961.
46. Whoriskey, F.G. and FitzGerald, G.J., Ecology of the threespine stickleback on the breeding grounds, in *The Evolutionary Biology of the Threespine Stickleback*, Bell, M.A. and Foster, S.A., Eds., Oxford University Press, New York, 1994, chap. 7.
47. Leiner, M., Fortsetzung der Ökologischen studien an *Gasterosteus aculeatus, Z. Morphol. Ökol. Tiere.*, 16, 499, 1930.
48. Haglund, T.R., Buth, D.G., and Blouw, D.M., Allozyme variation and the recognition of the white stickleback, *Biochem. Syst. Ecol.*, 18, 559, 1990.
49. Blouw, D.M. and Hagen, D.W., Breeding ecology and evidence of reproductive isolation of a widespread stickleback fish Gasterosteidae in Nova Scotia Canada, *Biol. J. Linn. Soc.,* 39, 195, 1990.
50. Jamieson, I.G., Blouw, D.M., and Colgan, P.W., Parental care as a constraint on male mating success in fishes a comparative study of threespine and white sticklebacks, *Can. J. Zool.,* 70(5), 956, 1992.
51. Mac Donald, J.F., Bekkers, J., Macisaac, S.M., and Blouw, D.M., Intertidal breeding and aerial development of embryos of a stickleback fish (Gasterosteus), *Behaviour,* 132, 1183, 1995.
52. MacDonald, J.F., Macisaac, S.M., Bekkers, J., and Blouw, D.M., Experiments on embryo survivorship, habitat selection, and competitive ability of a stickleback fish (*Gasterosteus*) which nests in the rocky intertidal zone, *Behaviour,* 132, 1207, 1995.
53. Blouw, D.M., Evolution of offspring desertion in stickleback fish, *Ecoscience,* 3, 18, 1996.
54. Jones, J.C. and Reynolds, J.D., Costs of egg ventilation for male common gobies breeding in conditions of low dissolved oxygen, *Anim. Behav.*, 57, 181, 1999.

55. Smith, C. and Wootton, R.J., The effect of brood cannibalism on the operational sex ratio in parental teleost fishes, *Rev. Fish Biol. Fish.*, 5, 372, 1995.
56. Wootton, R.J., Energy allocation in the threespine stickleback, in *The Evolutionary Biology of the Threespine Stickleback*, Bell, M.A. and Foster, S.A., Eds., Oxford University Press, New York, 1994, chap. 5.
57. Chellappa, S., Huntingford, F.A., Strang, R.H.C., and Thompson, R.Y., Annual variation in energy reserves in male three-spined stickleback, *Gasterosteus aculeatus* L., *J. Fish. Biol.*, 35, 275, 1989.
58. Östlund-Nilsson, S., Fifteen-spined stickleback females prefer males with more secretional threads in their nests: an honest condition display by males, *Behav. Ecol. Soc.,* 50, 263, 2001.
59. FitzGerald, G.J., Guderley, H., and Picard, P., Hidden reproductive costs in the threespine stickleback, *Exp. Biol.,* 48, 295, 1989.
60. Chellappa, S. and Huntingford, F.A., Depletion of energy reserves during reproductive aggression in male three-spined stickleback, *Gasterosteus aculeatus* L., *J. Fish. Biol.*, 35, 315, 1989.
61. Smith, C. and Wootton, R.J., Parental energy expenditure of the male three-spined stickleback, *J. Fish. Biol.*, 54, 1132, 1999.
62. FitzGerald, G.J. and van Havre, N., The adaptive significance of cannibalism in sticklebacks (Gasterosteidae: Pisces), *Behav. Ecol. Sociobiol.,* 20, 125, 1987.
63. Mori, S., Factors associated with and fitness effects of nest-raiding in the three-spined stickleback, *Gasterosteus aculeatus*, in a natural situation, *Behaviour,* Suppl. 16, 132(13–14), 1011, 1995.
64. Li, S.K. and Owings, D.H., Sexual selection in the three-spined sticklebacks: II. Nest raiding during the courtship phase, *Behaviour,* 64, 298, 1978.
65. Kynard, B., Breeding behavior of a lacustrine population of threespine stickleback (*Gasterosteus aculeatus* L.), *Behaviour,* 67, 178, 1978.
66. Sargent, R.C., Territory quality, male quality, courtship intrusions, and female nest-choice in the threespine stickleback, *Gasterosteus aculeatus, Anim. Behav.*, 30, 364, 1982.
67. Rico, C., Kuhnlein, U., and FitzGerald, G.J., Male reproductive tactics in the threespine stickleback: an evaluation by DNA fingerprinting, *Mol. Ecol.*, 1, 79, 1992.
68. Jamieson, I.G. and Colgan, P.W., Eggs in the nest of males and their effect on mate choice in the three-spined stickleback, *Anim. Behav.*, 38, 859, 1989.
69. Jamieson, I.G. and Colgan, P.W., Sneak spawning and egg stealing by male threespine sticklebacks, *Can. J. Zool.*, 70, 963, 1992.
70. Goldschmidt, T. and Bakker, T.C.M., Determinants of reproductive success of male sticklebacks in the field and in the laboratory, *Nether. J. Zool.*, 40(4), 664–687, 1990.
71. De-Fraipont, M., Fitzgerald, G.J., and Guderly, H., Age related differences in reproductive tactics in the three-spined stickleback, *Gasterosteus aculeatus*, *Anim. Behav.,* 46, 961, 1993.
72. Le Comber, S., Faulkes, C.G., Formosinho, J., and Smith, C., Response of territorial males to the threat of sneaking in the three-spined stickleback (*Gasterosteus aculeatus*): a field study, *J. Zool.,* 261, 15, 2003.
73. Zbinden, M., Mazzi, D., Kunzler, R., Largiader, C.R., and Bakker, T.C.M., Courting virtual rivals increase ejaculation size in sticklebacks (*Gasterosteus aculeatus*), *Behav. Ecol. Soc.,* 54, 205, 2003.
74. Candolin, U., Effects of algae cover on egg acquisition in male three-spined stickleback, *Behaviour*, 141(11–12), 1389, 2004.

75. Zbinden, M., Largiader, C.R., and Bakker, T.C.M., Body size of virtual rivals affects ejaculate size in sticklebacks, *Behav. Ecol.,* 15, 137, 2004.
76. Östlund Nilsson, S., Does paternity or paternal investment determine the level of paternal care and does female choice explain egg stealing in the fifteen-spined stickleback?, *Behav. Ecol.,* 13(2), 188, 2002.
75. Wootton, R.J., A note on nest-raiding behavior of male sticklebacks, *Can. J. Zool.*, 49, 960, 1971.
76. Jones, A., Östlund-Nilsson S., and Avise J., A microsatellite assessment of sneaked fertilizations and egg thievery in the fifteen-spined stickleback, *Evolution,* 52, 848, 1998.
79. Ridley, M. and Rechten, C., Female sticklebacks prefer to spawn with males whose nests contain eggs, *Behaviour*, 76, 152, 1981.
80. Marconato, A. and Bisazza, A., Males whose nests contain eggs are preferred by female *Cottus gobio* L. (Pisces, Cottidae), *Anim. Behav.*, 34, 1580, 1986.
81. Unger, L.M. and Sargent, R.C., Allopaternal care in the fathead minnow, *Pimephales promelas,* female prefer males with eggs, *Behav. Ecol. Sociobiol.,* 23, 27, 1988.
82. Knapp, R.A. and Sargent, R.C., Egg mimicry as a mating strategy in the fantail darter, *Ethiostoma flabellare,* females prefer males with eggs, *Behav. Ecol. Sociobiol.,* 25, 321, 1989.
83. Sikkel, P.C., Egg presence and developmental stage influence spawning-site choice by female garibaldi, *Anim. Behav.*, 38, 447, 1989.
84. Kraak, S.B.M. and Videler, J.J., Mate choice in *Aidablennius sphynx* (Teleostei, blennidae): females prefer nests containing more eggs, *Behaviour*, 119, 242, 1991.
85. Goldschmidt, T., Bakker, T.C.M., and Feuth-de Bruijn, E., Selective copying in mate choice of sticklebacks, *Anim. Behav.*, 45, 541, 1993.
86. Forsgren, E., Karlsson, A., and Kvarnemo, C., Female sand gobies gain direct benefits by choosing males with eggs in their nests, *Behav. Ecol. Sociobiol.*, 39, 91, 1996.
87. Rohwer, S., Parent cannibalism of offspring and egg raiding as a courtship strategy, *Am. Nat.*, 112, 429, 1978.
88. Coleman, R.M., Gross, M.R., and Sargent, R.C., Parental investment decision rules: a test in bluegill sunfish, *Behav. Ecol. Sociobiol.,* 18, 59, 1985.
89. Sargent, R.C., Paternal care and egg survival both increase with clutch size in the fathead minnow, *Pimephales promelas*, *Behav. Ecol. Sociobiol.,* 23, 33, 1988.
90. Losey, G.S., Stanton, F.G., Telecky, T.M., Tyler, W.A., III, and the Zoology 691 Graduate Seminar Class, Copying others, an evolutionarily stable strategy for mate choice: a model, *Am. Nat.*, 128, 653, 1986.
91. Kraak, S.B.M., Female preference and filial cannibalism in *Aidablennius sphyx* (Teleostei, Blenniidae): a combined field and laboratory study, *Behav. Process,* 36, 85, 1996.
92. Bruijn, E.F. and Sevenster, P., Parental reactions to young in sticklebacks (Gasterosteus aculeatus L.), *Behaviour*, 83, 186, 1983.
93. Salfert, I.G. and Moodie, G.E., Filial egg-cannibalism in the brook stickleback *Culaea inconstans* (Kirkland), *Behaviour*, 93, 82, 1985.
94. Whoriskey, F.G. and FitzGerald, G.J., Sex, cannibalism and sticklebacks, *Behav. Ecol. Sociobiol.*, 18, 15, 1985.
95. Hyatt, K.D. and Ringler, N.H., Role of nest raiding and egg predation in regulating population density of threespine sticklebacks *Gasterosteus aculeatus* in a coastal British Columbia lake, Canada, *Can. J. Fish. Aquat. Ser.,* 46(3), 1989.
96. Foster, S.A., Understanding the evolution of behaviour in the threespine stickleback: the value of geographic variation, *Behaviour*, 132(15–16), 1107, 1995.

97. Sillett, K.B. and Foster, S.A., Ontogenetic niche shifts in two populations of juvenile threespine stickleback, *Gasterosteus aculeatus*, that differ in pelvic spine morphology, *Oikos*, 91(3), 468, 2000.
98. Candolin, U. Changes in expression and honesty of sexual signalling over the reproductive lifetime of sticklebacks, *Proc. R. Soc. Biol. Sci. B,* 267(1460), 2425, 2000.
99. Candolin, U., Increased signalling effort when survival prospects decrease: male-male competition ensures honesty, *Anim. Behav.*, 60(4), 417, 2000.
100. Hyatt, K.D. and Ringler, N.H., Egg cannibalism and the reproductive strategies of threespine sticklebacks Gasterosteus aculeatus in coastal British Columbia Canada lake, *Can. J. Zool.*, 67(8), 2036, 1989.
101. Snyder, R.J., Seasonal variation in the diet of the threespine sticklebacks, *Gasterosteus aculeatus*, in Contra Costa County, California, *Calif. Fish Game*, 70, 167, 1984.
102. Foster, S.A., Diversionary displays of parental sticklebacks: defenses against cannibalistic groups, *Behav. Ecol. Sociobiol.*, 22, 335, 1988.
103. Ridway, M.S. and McPhail, J.D., Raiding shoal size and distraction display in male sticklebacks (*Gasterosteus*), *Can. J. Zool.*, 66, 201, 1988.
104. Vickery, W.L., Whoriskey, F.G., and FitzGerald, G.J., On the evolution of nest raiding and male defensive behaviours in sticklebacks (Pisces: Gasterosteidae), *Behav. Ecol. Sociobiol.*, 22, 185. 1988.
105. Belles-Isles, J.C.D. and FitzGerald, G.J., Female cannibalism and male courtship tactics in threespine sticklebacks, *Behav. Ecol. Sociobiol.,* 26, 363, 1990.
106. Reebs, S.G., Whoriskey, F.G., and FitzGerald, G.J., Diel patterns of fanning activity, egg respiration, and the nocturnal behavior of male three-spined sticklebacks *Gasterosteus aculeatus* L. (*f. trachurus*), *Can. J. Zool.,* 62, 329, 1984.
107. Foster, S.A. and Baker, J.A., Evolutionary interplay between ecology, morphology and reproductive behavior in threespine stickleback, *Gasterosteus aculeatus, Environ. Biol. Fish*, 44(1–3), 1995.
108. Rowland, W., Proximate determinants of stickleback behaviour: an evolutionary perspective, in *The Evolutionary Biology of the Threespine Stickleback*, Bell, M.A. and Foster, S.A., Eds., Oxford University Press, New York, 1994, chap. 11.
109. Halliday, T.R., The study of mate choice, in *Mate Choice*, Bateson, P., Ed., Cambridge University Press, Cambridge, 1983, pp. 3–32.
110. Andersson, M., *Sexual Selection,* Princeton University Press, Princeton, NJ, 1994.
111. Andersson, M. and Iwasa, Y., Sexual selection, *Trends Ecol. Evol.*, 11, 53, 1996.
112. Berglund, A., Bisazza, A., and Pilastro, A., Armaments and ornaments: an evolutionary explanation of traits of dual utility. *Biol. J. Linn. Soc.*, 58, 385–399, 1996.
113. Bateman, A.J., Intrasexual selection in Drosophila, *Heredity*, 2, 349–368, 1948.
114. Jones, A.G.J., Rosenqvist, G., Berglund, A., Arnold, S.J., and Avise, J.C., The Bateman gradient and the cause of sexual selection in a sex-role-reversed pipefish, *Proc. R. Soc. Lond. B*, 267, 677–680, 2000.
115. Wunder, W., Gattenwahlversuche bei Stichlingen und Bitterlingen, *Zool. Anzeig.*, Suppl. 7, 152, 1934.
116. TerPelkwijk, J.J. and Tinbergen, N., Eine reizbiologische analyse einiger Verhaltensweisen von *Gasterosteus aculeatus* L., *Z. Tierpsychol.,* 1, 193, 1937.
117. Semler, D.E., Some aspects of adaptation in a polymorphism for breeding colors in the threespine stickleback (*Gasterosteus aculeatus*), *J. Zool.*, 165, 291, 1971.
118. Milinski, M. and Bakker, T.C.M., Female sticklebacks use male coloration in mate choice and hence avoid parasitized males, *Nature*, 344, 330, 1990.

119. McLennan, D.A. and McPhail, J.D., Experimental investigations of the evolutionary significance of sexually dimorphic nuptial coloration in *Gasterosteus aculeatus* (L.): the relationship between color and female behaviour, *Can. J. Zool.,* 68, 482, 1990.
120. Bakker, T.C., Positive genetic correlation between female preference and preferred male ornament in sticklebacks, *Nature*, 363, 255, 1993.
121. Bakker, T.C.M., and Mundwiler, B., Female mate choice and male red coloration in a natural three-spined stickleback (*Gasterosteus aculeatus*), *Behav. Ecol.,* 5, 74, 1994.
122. Cronly-Dillon, J. and Sharma, S.C., Effect of season and sex on the photopic spectral sensitivity of the three-spined stickleback, *J. Exp. Biol.,* 49, 679, 1968.
123. Mckinnon, J.S., Video mate preferences of female three-spined sticklebacks from populations with divergent male coloration, *Anim. Behav.*, 50, 1645, 1995.
124. Reimchen, T.E., Loss of nuptial colour in three-spined sticklebacks (*Gasterosteus aculeatus*), *Evolution*, 43, 450, 1989.
125. Jamieson, I.G., Blouw, D.M., and Colgan, P.W., Field observation on the reproductive biology of a newly discovered stickleback *Gasterosteus*, *Can. J. Zool.,* 70(5), 1057, 1992.
126. Brush, A.H. and Reisman, H.M., The carotenoid pigments in the three-spined stickleback, *Gasterosteus aculeatus*, *Comp. Biochem. Physiol.*, 14, 121, 1965.
127. Wedekind, C., Meyer, P., Frischknecht, M., Niggli, U.A., and Pfander, H., Different caroteniods and potential information content of red colouration of male three-spined stickleback, *J. Chem. Ecol.,* 24, 787, 1998.
128. Ronnestad, I., Hemre, G.I., Finn, R.N., and Lie, O., Alternate sources and dynamics of vitamin A and its incorporation into the eyes during the early endotrophic and exotrophic larval stages of Atlantic halibut (*Hippoglossus hippoglossus* L.), *Comp. Biochem. Physiol.*, A, 119, 787, 1998.
129. Frischknecht, M., The breeding coloration of male threespined sticklebacks (*Gasterosteus aculeatus*) as an indicator of energy investment vigour, *Evol. Ecol.*, 7, 439, 1993.
130. Olson, V.A. and Owens, I.P.F., Costly sexual signals: are carotenoids rare, risky or required?, *Trends Ecol. Evol.,* 13, 510, 1998.
131. Barber, I., Arnott, S.A., Braithwaite, V.A., Andrew, J., Mullen, W., and Huntingford, F.A., Carotenoid-based sexual coloration and body condition in nesting male sticklebacks, *J. Fish Biol.,* 57, 777, 2000.
132. Barber, I., Nairn, D., and Huntingford, F.A., Nests as ornaments: revealing construction by male sticklebacks, *Behav. Ecol.,* 12, 390, 2001.
133. Mckinnon, J.S., Red coloration and male parental behaviour in the threespine stickleback, *J. Fish. Biol.*, 49, 1030, 1996.
134. Smith, C., Barber, I., Wootton, R.J., and Chittka, L., A receiver bias in the origin of three-spined stickleback mate choice, *Proc. R. Soc. Lond. B,* 271, 949, 2004.
135. Rowland, W.J., The effects of male nuptial coloration on stickleback aggression: a re-examination, *Behaviour,* 80, 118, 1982.
136. Östlund Nilsson, S. and Nilsson, G.E., Free female choice in stickleback: lack of preference for male dominance traits, *Can. J. Zool.,* 78, 1251, 2000.
137. Bakker, T.C.M. and Sevenster, P., Determinants of dominance in male sticklebacks (*Gasterosteus aculeatus* L.), *Behaviour,* 86, 55, 1983.
138. Bakker, T.C.M. and Mundwiler, B., Pectoral fin size in a fish species with paternal care: a condition dependent sexual trait revealing infection status, *Freshwater Biol.,* 41, 543, 1999.
139. Kunzler, R. and Bakker, T.C.M., Pectoral fins and paternal quality in sticklebacks, *Proc. R. Soc. Biol. Sci. B*, 267, 999, 2000.

140. Sargent, R.C. and Gebler, J.B., Effects of nest site concealment on hatching success, reproductive success, and paternal behaviour of the threespine stickleback, *Gasterosteus aculeatus*, *Behav. Ecol. Soc.*, 7, 137, 1980.
141. Candolin, U. and Voigt, H.R., Predator-induced nest site preference: safe nests allow courtship in sticklebacks, *Anim. Behav.*, 56, 1205, 1998.
142. Barber, I., Nairn, D., and Huntingford, F.A., Nests as ornaments: revealing construction by male sticklebacks, *Behav. Ecol.*, 12, 390, 2001.

6 The Umwelt of the Three-Spined Stickleback

Deborah A. McLennan

CONTENTS

6.1 INTRODUCTION

At all times and in all places, the Earth is awash in electromagnetic radiation, sound waves, vibrations, and chemical stimuli. No species can detect and respond to all that sensory information; each has carved out a small portion of the whole to form its own sensory niche. Technology has allowed us to describe these niches for other species even if we cannot truly appreciate or experience them. For example, try to envision being a mormyrid and singing to your mate with electricity or being any fish and touching an object without being in physical contact with it. This is the essence of *umwelt*, the term used by von Uexkull[1] to call our attention to these unique sensory worlds, many parts of which are beyond the perceptual capabilities of mere biologists. In the following pages I shall discuss the umwelt of the three-spined stickleback, with particular reference to how that world interacts with and shapes intraspecific communication during the breeding season. When data are scarce for threespines, I shall fall back on information from other gasterosteids, under the assumption that the conservative nature of (some) evolutionary diversification will allow extrapolation from one close relative to the other. By the end of this chapter, I hope to have shown that sticklebacks are immersed in a unique world of images and scents, and possibly tastes and sounds. These different sensory modalities are employed to a greater or lesser extent throughout their lives, but it is during the breeding season that multimodal and multicomponent signals really shine.

6.2 VISION

6.2.1 The Medium: Transmission Properties

Light transmission is more complicated in aquatic ecosystems than on land. Photons crossing the air–water interface will eventually be absorbed or scattered, either by organic particles suspended in the water or by the water molecules themselves. Absorption of light on its way from the sender to the receiver leads to image degradation because information is lost. Scattering of light leads to image degradation because some information is lost (from the object), and irrelevant information is added to the visual pathway (from the background). This extraneous information decreases contrast between the object and the background: distant objects appear faint and blurred, as if seen through fog (the veiling effect). The upshot of all this absorbing and scattering is that the composition of light in any given environment depends on the distance that light has travelled since entering the water and what is in the water. For example, chlorophylls shift the transmission maximum of light to wavelengths of around 500 to 600 nm (greenish yellow: majority of coastal waters, lowland ponds and rivers). Add tannins and lignins to the picture, and little light penetrates below 3 m, with the transmission maximum pushed to well over 600 nm (reddish brown: some swamps, marshes). Because of this differential transmission, the quality (intensity and wavelength composition) of light may vary dramatically along different lines of sight radiating from the same point.[2]

The three-spined stickleback (or the entities contained within the *G. aculeatus* species complex; see Chapter 1) inhabits a wide range of environments, including tidepools, brackish marshes, eutrophic lakes, and tea-stained ponds. Combining data collected specifically from some of those habitats[3,4] with information gathered for other fishes allows us to envision this range of spectral environments by following an anadromous threespine on her way to the breeding grounds. In the open ocean her world is veiled in a blue-green haze, which shifts gradually toward green as she approaches the shoreline. She moves upstream through the brackish estuary into a small eutrophic lake, in which wavelength composition is shifted to green-yellow. A video camera submerged in this environment records objects surrounded by moderately bright greenish fog (spacelight). Finally, she passes into a deeply tea-stained creek, an environment that favours progressively longer wavelengths depending on the concentration of tannins and lignins. Not only is the spectral quality of light modified, but the overall amount of available light is reduced. Our video camera records objects enclosed within a dim, reddish-brown spacelight. So, in the course of maybe one or two weeks she has passed through bright blue-green to moderate green-yellow to dim red spacelight. Can one "species" possess a wide enough range of photopigments and processing abilities to successfully occupy all of these habitats, or has natural selection, within the plesiomorphic background of the gasterosteid visual system, produced populations fine-tuned to particular spectral environments? I will return to this question in a moment.

6.2.2 The Cue

6.2.2.1 Male Nuptial Colour

Anadromous threespine males from British Columbia develop a temporally fluid but distinct series of nuptial colour mosaics that correspond to particular stages of the breeding cycle.[5] A nest-building male is generally quite inconspicuous, with pale blue eyes, a dull red throat, and medium grey body. During the courtship stage, a pulse of red sweeps across his entire lateral surface, accompanied by a surge in eye colour intensity and a marked reduction in dorsal melanism. As courtship proceeds, he suddenly flushes snowy white upon completion of a creeping through or gluing bout, then performs a very intense, lateral zigzag approximately one quarter to one third of the way above his nest toward the female. Prior to this striking colour change, gluing and creeping through are usually followed by a zigzag dance all the way up to the female, a combined dance and lunge or simply a lunge. Both the white flush and intense zigzag are extremely distinctive. If the female begins to drop, the male turns, swims rapidly to the nest and performs the dorsal roll, nest-show display. Under normal courtship conditions, the snowy flush is a brief, powerful predictor that the male is now "ready to spawn." Although red is still visible during this flush, the overall effect, to my eye at least, is one of covering the red with a film of fine, white gauze. Once the female is in the nest and the male has begun quivering, a wave of grey sweeps forward from his caudal fin. This colour change intensifies during and immediately following fertilisation, but the courtship mosaic returns once the male is ready to spawn again. Parental care is accompanied by a gradual decrease in the distribution of red, as a wave of medium to dark grey sweeps forward from the caudal fin. Once the fry appear outside the nest, the male displays a new mosaic, an intensely red throat and jaw, bright blue eyes, and dark grey lateral, dorsal, and ventral surfaces. During all these changes, the dorsal surface of the fish reflects a metallic blue sheen when viewed from above and, occasionally, from the side.

This complex mosaic of red-orange, blue, and black reflects the structure of the basic vertebrate dermal chromatophore unit:

1. Xanthophores and erythrophores in the outer layer contain carotenoid pigments, complex molecules that selectively absorb short wavelengths and thus appear yellow-red.
2. Iridiophores in the middle layer contain crystalline platelets of guanine, hypoxanthine, or uric acid. Iridiophores reflect light back to the receiver's eyes, the exact wavelength and intensity of which depends on the thickness and spacing of the platelets. In the threespine, these structures are responsible for producing the iridescent blue-greens in the iris and along the dorsal and lateral surfaces.
3. Melanophores in the basal layer contain highly polymerized pigments synthesized from tyrosine that absorb wavelengths across the whole visible spectrum, into ultraviolet.

Dispersion of melanosomes in the pigment cells causes skin to darken (more light is absorbed), whereas aggregation causes lightening. Dendritic processes from melanophores extend up over the xanthophores, so the red-yellow signal produced by the top layer of the chromatophore unit can be concealed or revealed rapidly without movement of carotenoid pigments. Because these three components evolve independently of one another, the potential exists for selection to shape complicated signals using relatively few components. For example, depending on the lighting conditions, increasing the reflectivity of the iridiophore layer can make it easier or more difficult for a receiver to discriminate between differences in the carotenoid intensity of two signals. Those same signals are optimally displayed when underlying melanosomes are aggregated because this increases the amount of light that is reflected back through the chromatophore; dispersing the melanosomes increases photon absorption, damping down the signal (reviewed in Grether et al.[6]). Overall then, nuptial colouration in the threespine is a three-dimensional composite signal that is manifested on horizontal (distribution of colour across the surface of the fish), vertical (interactions among different components of the dermal chromatophore units which form each colour patch), and temporal (changes in the signal over short and long time scales) axes.

Theoretical descriptions aside, what do we really know about the actual composition of the male mosaic? As far as I know there have been no investigations of the iridiophore platelets and melanin pigments in threespines, so the following discussion will focus on carotenoids. Carotenoids have a variety of physiological functions, including the detoxification of free radicals, modulation of the immune system,[7] and protecting developing sperm from damage.[8] A carotenoid-based signal thus has the potential to be an honest indicator of quality because its development requires that pigments be diverted from these other critical functions. Only high-quality individuals should be able to pay this price (reviewed in Alonso-Alvarez et al.[9]).

Animals cannot synthesise carotenoids *de novo*, although many species are capable of modifying carotenoids extracted from their diet. Intuitively, there should be a positive link between pigment display and success at capturing carotenoid-containing prey items, which means that the intensity of an individual's carotenoid-based signal should be a truthful advertisement of foraging abilities.[10] For variability in the signal to be a reliable indicator of foraging "success," however, carotenoids must be a limited resource.[11] Is there evidence for such a limitation in the stickleback's world? Threespines are opportunistic feeders, capable of adjusting to differential prey availability across months and consuming a wide variety of prey species,[12–20] the majority of which are a potential source of important carotenoids such as astaxanthin (Table 6.1). Certainly, not all habitats are equally carotenoid rich,[3] which may explain some interpopulation differences in male hue and intensity (as in guppies[21]), but are pigments ever limiting on a consistent enough basis across a large enough number of populations to produce a reliable link between colour and foraging success on an evolutionary timescale? Answer: we do not know.

Additionally, the pathway from food capture to pigment deposition is a complicated one, involving an individual's ability to acquire (foraging efficiency), extract (vertebrates do not assimilate carotenoids efficiently[21]), modify, transport, and deposit carotenoids. Heritable differences in male colour within[22] and between populations could thus theoretically be caused by variability at any of these steps.[9,23]

So what happens to the carotenoids once they have been extracted from the food? Few researchers have asked this question, but the answers, although sketchy, are intriguing (Table 6.2):

1. Intersexual differences: Within a population, mature males and females store virtually the same pigments but in different relative amounts; females retain hydroxy or epoxy compounds (83.1% of total content) and males retain ketones (86.7% of total content).
2. Ontogeny of colour: Changes in the male's pigment profile occur during maturation, involving both a moderate alteration in the types of carotenoids present and a major shift in the relative proportions of the remaining pigments (see also Brush and Reisman[24]); the percentage concentration of ketones increases. These changes are accompanied by the proliferation of erythrophores in the mouth, and on the cheeks, throat, opercula, abdomen, and lateral surfaces of the maturing male.
3. Male nuptial colouration: Red colouration has been linked to the presence of two β-ketones, astaxanthin and canthaxanthin, in a variety of teleost species.[25] These pigments predominate in the mandibular region of the breeding male threespine (81% of total content) and are the major carotenoids stored in the skin, muscles, and liver (70% of total content).[25] Males thus carry an internal pool of carotenoids but whether that pool can be accessed continually during the breeding season is unknown. Research with guppies, for example, indicates that carotenoids were extracted from storage to fill nonsexual physiological roles before pigments were sequestered for colour patterns.[26]

TABLE 6.1
Carotenoid Distribution in Representative Prey Items of *Gasterosteus aculeatus*

Species[a]	Pigments									
	1	2	3	4	5	6	7	8	9	Others
Branchiopoda										
Artemia salina	x	x	x	x	x	x		x	x	10,12–15
Daphnia magna	x		x	x	x			x	x	14,15
D. longispina	x		x	x	x			x		
Copepoda										
Cyclops kolensis[b]	x		x	x	x			x		11,13
C. strenuous strenuous	x	x	x	x	x		x	x		
Diaptomus bacillifer and D. castor	x		x			x			x	
Hemidiaptomus amblydon	x		x			x			x	
Eudiaptomus amblydon	x	x	x					x	x	
Ostracoda										
Heterocypris incongruens	x		x							
Cyclocypris laevis	x		x		x					
Amphipoda										
Gammarus lacustris	x	x	x		x	x	x	x	x	10,11,14,16–18
G. pulex[c]	x		x		x			x		
Insecta										
Chironomus annularis	x		x		x		x		x	11
Stickleback eggs[d]	x		x					x		15–17, Unidentified

Note: Numbers refer to pigments: 1 = β-carotene; 2 = Isozeaxanthin; 3 = Astaxanthin; 4 = Echininone; 5 = Canthaxanthin; 6 = Violaxanthin; 7 = α-carotene; 8 = Lutein; 9 = γ-carotene; 10 = Isocryptoxanthin; 11 = β-cryptoxanthin; 12 = Phoeniconone; 13 = Crustaxanthin; 14 = Phoenicoxanthin; 15 = Neoth-xanthin; 16 = α-Doradexanthin; 17 = Tunaxanthin; 18 = Mutatochrome.

Sources: Unless otherwise noted references are in [a]McLennan, D.A., M.Sc. thesis, University of British Columbia, Vancouver, 1988; [b]Czeczuga, B. et al., *Folia Biol.*, 48, 77, 2000; [c]Gaillard, M. et al., *Comp. Biochem. Physiol. B*, 139, 2004; [d]Nordeide, J.T. et al., *J. Evol. Biol.*, 19, 431, 2006. All with permission.

A quick perusal of Table 6.2 indicates that the "red" component of the male nuptial signal tends to be based on more than one carotenoid pigment. The variety of different pigments underlying the "signal" may be one of the reasons there is so much variability in its the hue and development within and among populations. A signal made of many pigments with different absorption profiles is more sensitive to spectral fine-tuning by selection than a one-pigment cue[6]; something that is clearly advantageous to fishes inhabiting such diverse spectral regimes. Different pigments may also transmit different bits of information about male quality to potential mates,[27] although how that information is coded is currently open to question. Does a male appear to be "one hue," the final outcome of the interaction between light and a variety of pigments distributed

TABLE 6.2
Percentage Distribution of Carotenoids in Different Populations of *Gasterosteus aculeatus*

Pigments	Freshwater Poland[a]			Baltic Sea[a]	Freshwater Japan[b]	Freshwater Scotland[c]	Freshwater Switzerland[d]
	Juvenile Male	Adult Female	Adult Male	Adult Male	Adult Male	Adult Male	Adult Male
β-Carotene	5.4	7.3	0.0	0.0	0.0		
Total	**5.4**	**7.3**	**0.0**	**0.0**	**0.0**		
α-Cryptoxanthin	0.0	0.0	0.0	0.0	1.5		
Cryptoxanthin	16.6	14.1	1.0	0.0	0.0		
Cynthiaxanthin	0.0	0.0	0.0	0.0	7.2		
Diatoxanthin	0.0	0.0	0.0	0.0	3.3		
Lutein	15.0	39.9	8.4	4.9	20.4	Present	Present
Neothxanthin	0.0	11.2	0.0	0.0	0.0		
Tunaxanthin	0.0	0.0	0.0	0.0	39.3		Present
Zeaxanthin	35.9	0.0	0.0	15.1	12.3		
Total Hydroxy	**67.5**	**65.2**	**9.4**	**20.0**	**84.0**		
Epoxide	0.0	9.6	2.6	41.1	0.0		
Mutatochrome	0.0	8.3	1.3	8.6	0.0		
Total Epoxy	**0.0**	**17.9**	**3.9**	**49.7**	**0.0**		
Astaxanthin	24.1	8.6	60.7	22.8	16.0	Present	Present
Canthaxanthin	0.0	0.0	12.8	0.0	0.0		
Doradexanthin	3.0	1.0	13.2	7.5	0.0		
Total Ketones	**27.1**	**9.6**	**86.7**	**30.3**	**16.0**		

Note: All table entries are percentages.

Sources: From [a]Czeczuga, B., *Hydrobiologia*, 74, 7, 1980; [b]Matsuno, T. and Katsuyama, M., *Bull. Jpn. Soc. Fish.*, 42, 761, 1976; [c]Barber, I. et al., *J. Fish Biol.*, 57, 777, 2000; [d]Wedekind, C. et al., *J. Chem. Ecol.*, 24, 787, 1998. All with permission.

evenly in the signal, or does he appear to be multihued? For example, in anadromous Salmon River fishes, low-intensity males appeared chrome orange (Munsell notation = 2.5 YR), whereas bright males appeared flame scarlet (Munsell notation = 10.0 R) to my eye. Variability in hue occurred between males at the same stage of the breeding cycle, and within an individual male across his breeding cycle. So, male hue hovered at the orange end of the spectrum during the initial stages of nest building and moved toward red as the intensity of the male's signal increased to its peak during courtship.[5] I assumed that these hue changes were caused by absolute changes in carotenoid concentrations, but they might also reflect relative differences in the importance of particular pigments at different times in the breeding cycle. The former explanation, which is a property of pigment biophysics, seems more plausible than the latter,[24] which requires a very complex pattern of pigment deposition and control.

4. Interpopulation differences: Astaxanthin has been detected in all *G. aculeatus* populations analysed to date and its concentration is positively correlated with the intensity of male colour in at least one population.[28] Apart from that, carotenoid content is variable, but not totally random, across populations. For example, males from the Japanese and Polish freshwater populations are radically different from one another both in the type and proportion of stored pigments (share only 2 of 12 carotenoids). The two Polish populations, on the other hand, share 67% of their pigments (approximately 85% of their total carotenoid content) despite the fact that one group is freshwater and the other marine. Interestingly, this observation supports the hypothesis that freshwater threespine populations are the descendants of a marine ancestor.

Although the majority of threespines display the classic blue-red-black mosaic, there are some melanic populations along the Pacific coast of North America, ranging from "entire body black" similar to the brook stickleback to possible hybrids displaying black bodies with red throats.[29–36] There is also a widespread Atlantic Canada "morph" in which males are described as turning an iridescent white dorsally and laterally during the breeding season.[37] The authors' description of "white" sounds similar to the white flush I saw in the Salmon River males. Could "white" in this unusual population represent an extension of that flush, something that has been selected either because it is more visible through the algae in which these fish breed or because it co-opted the female's tendency to drop toward a flushing male and follow him to the nest?

Clearly we need to quantify the extent of intra- and interpopulational variability in cue structure, as well as the mechanisms controlling the distribution of iridiophore platelets and melanin pigments (in addition to carotenoids), before we can begin to decrypt the messages being transmitted in this multicomponent signal. It would, for example, be interesting to document the eye colour of melanic males. Because all the other gasterosteid species have yellow eyes, is "blue/blue-green" a synapomorphy for the entire *G. aculeatus* species group (including melanics) or has the origin of this novel eye colour had a more complicated evolutionary history within that group? Is a message being transmitted by the novel hue, by a change in contrast with the rest of the male nuptial mosaic, both, or neither?

6.2.2.1.1 The Latest Thing: Ultraviolet Radiation

Ultraviolet-A radiation is composed of extremely short wavelengths (320–400 nm) and is easily scattered by water molecules and solutes. Short wavelengths are also high energy and thus can cause damage to biological systems. Given these two constraints, it is not surprising that many biologists assumed that UV radiation would not be an important part of courtship communication in fishes. The recent discovery of UV receptors in the retina of many freshwater fishes including sticklebacks (see next section) has, however, caused us to rethink this assumption.[38] A nuptially coloured male threespine reflects ultraviolet radiation, the quantity and quality of

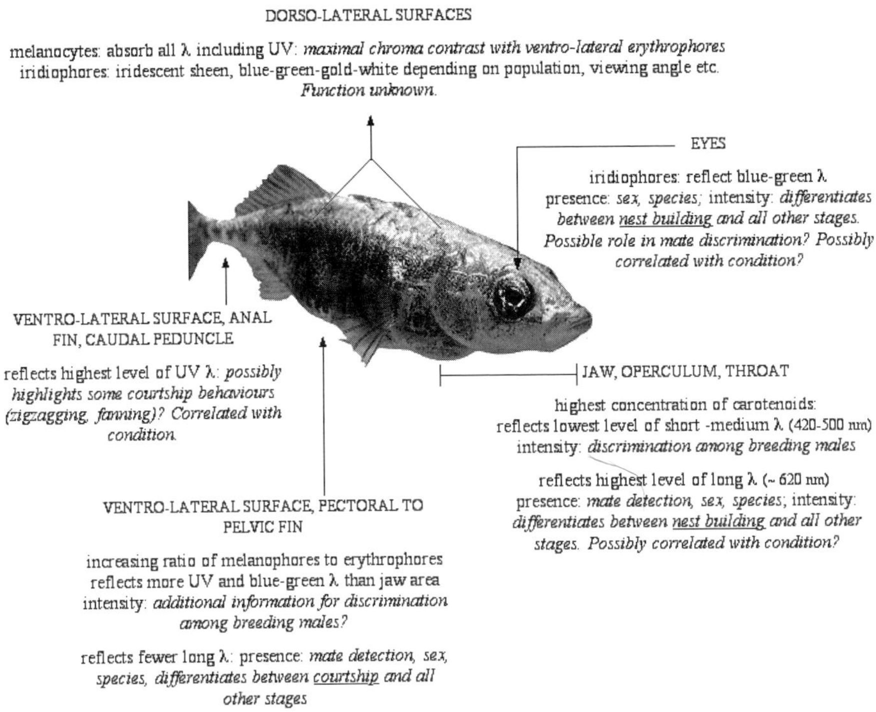

FIGURE 6.1 Summary of the different components of the male nuptial signal and the possible messages encoded therein. (From Oxford Scientific Films, 1976. With permission.)

which depends on which part of his body is being viewed. For example, strength of reflection increased from the dorsal and dorsolateral surfaces << opercular area (red pigment) < pelvic spine, anal fin area (no red). Interestingly, overall body condition was only correlated with peak UV reflectance contrast from the area around the anal fin[39] (Figure 6.1). Could this, to us hidden, component of the nuptial signal be transmitting any information to a prospective mate? Mate choice experiments will be required before we can answer this question. Because UV light is transmitted over such short distances, this "signal" may only come into play when the courting partners are close together.[40] In the threespine system, this might include behaviours that highlight the anal fin/caudal peduncle region such as the zigzag dance and fanning in the midst of courtship.

Four papers have appeared since I wrote the preceding paragraph, all of which indicate that UV light is indeed an important part of the stickleback's umwelt. Nonbreeding threespines learned to locate a hidden food source using information from two landmarks differing only in the presence or absence of reflected UV radiation. To human observers, these two landmarks looked identical; only animals sensitive to UV radiation would perceive them as being different, and thus conveying information about the location of the food reward.[41] Females from different threespine populations prefer males illuminated by a daylight regime that included UV-A radiation (freshwater Scotland[42]; Germany[43]), whereas nonbreeding fish preferred the conspecific shoal that was viewed through a UV-transmitting filter[44] (but see Boulcott et al.[42]). This fourth component of the male nuptial mosaic may function by increasing the brightness of uncoloured (silvery) areas along the posterior ventral surface, which would in turn produce yet another contrast area in the signal overall, between the ventral surface and the weakly reflective red flanks (which in turn contrast with the dark dorsal surface).

6.2.2.2 Female Nuptial Colour

Three-spined stickleback females from populations in Russia, eastern North America, and southern British Columbia become bright gold with an intensified, vertical barring pattern along their dorsal and lateral surfaces during the breeding season.[45] Although there appears to be little difference in the general structure of this signal among different populations, there does seem to be some interpopulational variability in its intensity. For example, Pacific coast females are intensely golden along their entire dorsolateral surface with very light bars, whereas Lake Ontario and New York females have gold concentrated along the lateral plates and the dorsal surface and extremely dark bars (Rowland et al.[46]; unpublished observation). What pigments could be responsible for this colour change? Lutein, which is yellow, is the dominant carotenoid in the epidermis of mature females from freshwater population in Japan (Table 6.2) and it is also one of the major pigments found in potential prey items (Table 6.1). Could lutein be the female nuptial carotenoid, or is it simply stored in epidermal cells for transportation to developing eggs (Table 6.2) or other physiological functions? The gold signal is metallic, indicating that iridiophores play an important role in reflecting wavelengths either in the yellow or the white-silver range, increasing the brightness, and perhaps amplifying the spectral properties of the female's signal.[6] Once again, there are currently no answers to these questions.

Females from the Little Campbell River, British Columbia, possess an unusual trait, the development of red throats during the breeding season.[47] Although the intensity of red is lower in females than males, the colour is conspicuous, particularly during the female head-up courtship display. This system is fascinating because it is so paradoxical. Coupling "red throat" with female courtship should send conflicting messages (I am an intruding territorial male vs. I am a receptive female), which might confuse a prospective suitor. Why do males not react to this flash of red with aggression or with at least reduced courtship? Extrapolating from a study by Nordeide[48] indicates that this might, in fact, be just what they do. Nordeide investigated yet another strange population, one in which many females, similar to all males, have red pelvic spine membranes. When allowed to choose, males courted red-spined females less intensely, implying that they were adversely affected by the conflicting messages. More importantly, the intensity of a female's pelvic spine colour was negatively correlated with her size, age, and the amount of carotenoids in her gonads. Because these egg-bound carotenoids may improve larval survival, males who spawn with redder females will not gain anything by that choice, and may even sacrifice some aspects of offspring fitness.[49] In this particular system then, two aspects of the female morphology, body size and pelvic spine colour, reinforce the message of relative "clutch superiority" (large, dull female = more eggs, higher-quality eggs).

Given the redundancy of the signal components, why is red maintained in this population? One possibility is that the development of red pelvic spines is genetically correlated in threespines; redder fathers produce redder sons and daughters. In this case the daughters must simply "make the best of a bad situation" by sharing the reproductive success of their preferred brothers. Additional compensation for their lack of attractiveness could occur if red plays a role in female–female competition for access to mates. Males did direct some courtship toward the red-spined female in the laboratory, so they might accept her in the field if she were persistent enough. Finally, an individual female might become increasingly dull coloured as she aged (this population contains individuals up to 4 years old), which would indicate that forces other than just genetic correlation were playing a role in trait origin, development, and maintenance.

6.2.3 The Receiver: How Sticklebacks See

The distribution of light-sensitive pigments in the eyes of freshwater fishes tends to correspond to the spectral properties of the environment. For example, diurnal fishes living in shallow, clear waters usually have ultraviolet, blue, green, and orange-red photopigments (e.g., poeciliids and cyprinodonts), whereas crepuscular fishes in green-shifted habitats are sensitive to only green and red (e.g., pike, bowfins).[2] Sticklebacks generally group with other diurnal, shallow water fishes. Female *Gasterosteus aculeatus*, *G. wheatlandi*, *A. quadracus*, and *P. pungitius* from brackish pools and ditches around Long Island, New York, have four visual pigments in their retina, with peak spectral sensitivities of around 360 nm (UVS [ultraviolet-sensitive]: 360–365 nm) and 445 nm (SWS [short-wavelength-sensitive]: 440–450 nm) in single cones, and 530 nm (MWS [medium-wavelength-sensitive]: 519–535 nm), and 605 nm (LWS [long-wavelength-sensitive]: 575–625 nm) in the two halves of the double cone[50] (see also threespines from the Little Campbell River; 598 nm, 530 nm, 438 nm, and 362 nm[51]). Interestingly, all nuptially coloured males in the Long Island population reflect at approximately 630 nm, stimulating the female's LWS double cones. The horizontal spacelight in this habitat is shifted toward green-yellow, so a nuptial male looks bright against a dark background when "viewed" through the offset LWS cones. Increased sensitivity of the female eye to long wavelengths during the breeding season[52] will therefore increase her chances of detecting a male from a distance while she is swimming through the breeding grounds (mate recognition). Oddly though, the LWS pigment is relatively insensitive to variation in male throat colour and is thus not very useful for discriminating among males. At first this result seems counterintuitive. If the photopigment most sensitive to the wavelength of the male signal is not involved in mate discrimination what is? Rowe et al.[50] solved the problem when they discovered that there is substantial intermale variability in reflectance at the short-medium end of the spectrum (400–550 nm). Because carotenoid pigments selectively absorb short wavelengths, the concentration of these wavelengths in reflected light will transmit information about the intensity of a male's signal; males with fewer carotenoids will reflect more short wavelengths and *vice versa* (mate discrimination). Input from the long- and short-wavelength channels can thus efficiently code for subtle differences in male nuptial colour, allowing a female to locate, identify, and choose a mate.[50]

McDonald and Hawryshyn[4] expanded the analysis of stickleback vision from photoreceptors to ganglion cells, the first-order neurons in the visual system. They uncovered a fascinating shift in responsiveness toward increasingly longer wavelengths correlated with the extent of red shift in the ambient background light. So, both ON and OFF ganglia from threespines living in an extremely tea-stained lake had virtually identical long-wavelength peak spectral sensitivities (in the 600–620 nm range), whereas fish inhabiting a eutrophic lake showed peak spectral sensitivities at medium (OFF: 540 nm) and longer (ON: 540–600 nm) wavelengths. In both habitats the OFF response appears to be tuned to the prevailing spacelight,[53] so long wavelengths, which stimulate the ON ganglion cells, will appear brighter against a dark background. The structure of the visual system at this level thus reinforces the contrast of the male's nuptial signal with the spectral qualities of the environment.

What about populations with melanic males? "Black" will always contrast with the prevailing spacelight in freshwater systems, regardless of habitat. It is thus interesting that melanic threespines tend to be restricted primarily to heavily tannin-stained waters.[3,36] The general (but not absolute) absence of red males from these habitats makes sense biologically because a signal reflecting photons matching the surrounding spacelight should appear invisible from a distance. MacDonald et al.[54] tested this scenario by presenting mirror images of males to females from a red-throated population. The images differed only in nuptial colour (red or black) and the spectral composition of the background (blue vs. red). When the two males were presented against a blue background, the females displayed the expected preference for red. When the suitors appeared in red spacelight, however, the females transferred their attention to the more conspicuous black males. So why are the melanic populations restricted mainly to tea-stained waters? After all, *Pungitius pungitius* and *Culaea inconstans* live in freshwater habitats similar to their threespine relatives yet have only melanic males. Clearly there is no prohibition to being a black stickleback male in green-shifted waters. To answer this question we first need to determine whether all populations of melanic threespines are plesiomorphically black (which seems very unlikely), have secondarily lost the red signal, or are a combination of both scenarios. This requires that we have a robust and comprehensive phylogeny for threespine populations, including all of the melanics.

6.2.4 Function of the Cue

6.2.4.1 The Male–Female Dialogue

The role for red in male–male interactions was extensively discussed by Rowland,[55] so let us focus on the role for colour in courtship. In most populations studied to date, a male responds to the presence of a courting female by increasing the saturation of colour in his eyes (blue) and body (red)[5,51] (similar cycling of black body in the Wapato Lake, Washington, population,[56] but different cycling of black body, red throat in a California population,[35] and black body in other Washington populations[57]), producing not only a more intense nuptial signal but also increasing the contrast between the two colours.[50,51] The reliability of the information transmitted by the signal depends, in part, on its continuity between different stages in the breeding cycle. In the Salmon River population, there was a significant positive correlation among all stages for the intensity of red, but not for blue eye colour, which peaks during courtship, then plateaus. The strongest association between the two signal components did, however, occur during courtship,[5] supporting the "increasing contrast" hypothesis.

I have already discussed the changes in the receiver's system that make "red" a more conspicuous signal in many stickleback habitats, and the fact that threespine females are more responsive to red during the breeding season. We must now move from the present to the past and ask whether the male-based signal and the female response to it originated in the same ancestor or, as suggested by the sensory bias hypothesis,[58,59] did the female response evolve before the male trait? Smith et al.[60] investigated this question by recording the number of bites three-spined and nine-spined sticklebacks directed toward different coloured plastic strips. They discovered that males and females of both species showed the same preferential response to red, with orange a very distant second. Given that ninespine males are not red (which eliminates the response being sexually motivated) and that threespines use red as a cue for identifying prey items,[18] it appears that the structure of the gasterosteid visual system may have been, in part, shaped by selection for increased foraging efficiency. In other words, female threespines were predisposed (biased) toward a red-based male signal when that signal originated. This hypothesis can be explored further by adding *Apeltes*, *Culaea*, and *Spinachia* to the database, which will allow us to pinpoint when the bias toward red originated within the Gasterosteidae (unless it is even older, of course).

One interesting aside here: females from a Long Island population either preferred a model of a male with a yellow belly to the red-bellied alternative (low-response individuals) or were attracted to both (high-response individuals). Compared to a green-bellied model, red and yellow provided the highest chroma contrast with the model's grey flank. Additionally, all females found an all-grey or an all-red model equally unattractive. Taken together, these results suggest that it is the appropriate contrast between the belly flank and the dorsal flank surfaces and not red *per se* that attracts females.[61] So, how would a threespine female respond to a golden male? *Gasterosteus wheatlandi* males are golden-green, but their nuptial signal covers their entire body. Presumably, the lack of the appropriate contrast (belly flank vs. dorsal flank) would render them as unattractive as the all-red model. More intriguingly, given that *Pungitius pungitius* and *G. aculeatus* both have chroma contrast (dorsal/lateral vs. ventral/lateral colour) in the male nuptial signal, whereas *G. wheatlandi* and *Culaea inconstans* do not, is the response to chroma contrast plesiomorphic or independently evolved (assuming that *Pungitius* females show the same reaction), or unique to *G. aculeatus* (assuming that *Pungitius* females do not respond to chroma contrast)?

The evolution of spectral sensitivity will be affected by the transmission properties of the environment,[62–64] which, in turn, will impose constraints on the parts of the spectrum that are available for building "nuptial colouration."[4,54] Given enough underlying variability in both the receiver and the sender, sensory drive may eventually produce signal divergence correlated with habitat types and geographical distributions.[63,64] So, for example, females from some of the melanic populations are less sensitive to red than are females from red populations.[65] Once again though, it is difficult to determine whether these females display the plesiomorphic response to red, in which case there was an increase in sensitivity to red in populations with red males, or whether the melanic females have undergone a derived decrease in sensitivity. We really do need a robust phylogeographic analysis of threespine relationships!

"Taking advantage of a bias in the female's sensory system" translates into female preference for a novel male trait at its point of origin. That preference is maintained today in most extant populations of threespines; females tend to prefer more intensely red males,[22,61,66–73] although that preference is not absolute either within or across populations. If "red" transmits "male, territorial, three-spined stickleback," does "intensity of red" convey any information beyond just detectability? There are several possible answers to this question. First, intensity may not signal anything about male quality, but may be selected because of thc indirect benefit to females who produce sexy sons.[22] Alternatively, a choosy female may be getting a mate who is healthier, more dominant, a more effective forager, a better father, more fertile, with higher quality sperm, any one of which will directly influence female fitness.[27–30,74–82]

Unfortunately, or interestingly, depending on your point of view, attempts to find a relationship between the intensity of red and either male reproductive success or some aspect of male quality have proved equivocal. Intensity of red was correlated with (1) lower growth rate but higher resistance to parasitic infection in offspring,[83] (2) resistance to some parasite species but not others,[67,84–86] (3) hatching success in the laboratory when males cared for only one clutch,[87] (4) male condition at some sites[28,67,70,80,88] but not others,[51,55,70,84,89,90] (5) changes in condition over time,[75] and (6) the number of eggs males received in one year,[65] but not several years later at the same and nearby sites.[91] Some of the noise in the system may be introduced by naive assumptions about the interaction between colour development and external variables. For example, many researchers tend to treat a variety of parasite species as equivalent entities when examining the association between the development of secondary sexual characteristics and resistance to parasitism. But parasites are not all "the same." Some species do far more damage to their hosts than others, which may explain the ambiguous relationship between the intensity of red and resistance to parasites.[85,92] This ambiguity may also be caused by differences between testing conditions in the field and the laboratory. Candolin and Voigt[93] showed that territory holders in the field were less heavily infected with *Schistocephalus solidus* plerocercoids, in better condition, and more intensely coloured than infected males. The latter males, however, were capable of boosting their colour signal, building nests and courting females in the laboratory, which gave a somewhat different picture of the parasite's effect on the development of male colour and behaviour. The authors argued that ample food supply and absence of predators in the laboratory decreased the costs of reproduction, benefiting the more heavily parasitised males. This in turn implies that the intensity of sexual selection on male colour will fluctuate across seasons depending on environmental conditions.

Noise in the system may also be introduced by the way we measure "female preference" and "male colour." So, let us begin with female behaviour. Dividing females into high and low responders in mate choice tests indicates that the female "preference" for red increases with increased sexual responsiveness; possibly because females with low sexual motivation respond to the threat component of the signal more strongly than the courtship component.[61,72,94] Sexual motivation is, in turn, tied to the point a female has reached in her ovulatory cycle.[95] Threespines ovulate all of their mature oocytes in one clutch, then have a limited time in which to spawn before those eggs become overripe or berried.[96–98] Not surprisingly, females become less choosy as time from ovulation increases and they still have not spawned.[68,99] Females in good condition continue to assess males, but are increasingly likely to spawn as they approach impending clutch loss, whereas energy-stressed or exhausted females spend less time in assessment and go straight to spawning.[100] In both these cases, an individual may end up spawning with a less preferred suitor, someone who just happened to be in the right place at the right time. These constraints of female biology may be extremely important in the field where females rarely have the luxury of waiting for Mr. Right, but rather must content themselves with picking the best of whichever males happen to be available at that point in time. Consider a simple system in which the most intensely red male has received all the clutches he can handle, so newly ovulated females transfer their attentions to preferred suitor number 2 until he has all the eggs he can handle and so forth. Now add a female about to drop her clutch, who responds to the very next zigzag directed toward her by diving into the nest regardless of male colour. Depending on the size of the breeding population, the intensity of the sexual selection vector driving toward amplification of the signal will be weakened by the vagaries of reality.

Female condition and limited male availability aside, the second variable influencing the female's response to male-based colour is her ability to distinguish between signals. Braithwaite and Barber[101] demonstrated that females are very attracted to the brighter male when the difference between suitors is large, but as that difference decreases so does the strength of the attraction (see also McLennan and McPhail[66] and Bakker and Mundwiler[70]). The authors suggested that females rely on other cues to choose a mate when their suitors are closely matched for colour (and in the case of their experiment, for size as well). Östlund Nilsson and Nilsson[90] drew attention to this possibility when they demonstrated that female threespines from a Swedish population demonstrated a very strong preference for redder males when they were separated from the males by dividers. When, however, the males were tethered so they could not interact with each other but could still court, female preference for red disappeared, prompting the authors to hypothesize that choice was shifted to some other male-based trait once the female was close to both suitors.

Rowland[102] anticipated these studies when he suggested that females might be responding to particular components of the male mosaic at different points of the courtship sequence. Noting that males zigzag more intensely further from their nests,[103,104] he proposed that dancing may initially attract a female's attention and draw her into the territory, at which point her attention shifted from behaviour to the colour/size aspects of her potential mate. In other words, the mate choice process involved a series of decisions from initial approach to nest entry, with the female being capable of rejecting the male at any point of that process.[100] This hierarchical mate choice process is similar to Ibrahim and Huntingford's[18] proposal that threespines make foraging decisions based on the interaction of their response to four prey stimuli: red colour, fast movement, straight shape, and large size.

In an elegant choice test based on computer-generated male images, Kunzler and Bakker[105] presented female threespines with males differing in the intensity of red throat colour, the intensity of courtship, and body size. They confirmed that the females' strongest preference for any single cue was based on intensity of colour, but overall attraction to males increased when courtship was added to red, and large body[106] added to red + courtship. In a perfect corollary, females were confused when presented with males sending conflicting signals (bright red, bad dancer vs. dull red, intense dancer). Females thus have an expectation of male quality based on the positive interaction of at least three visual cues (colour, courtship, and size). Given that the intensity of colour and of behaviour are correlated in at least two populations[68,88] and that body size is a crucial component of dominance interactions,[77,107] including the ability to prevent a neighbour from sneaking fertilisations,[108] it appears that these variables are transmitting redundant and truthful information about male quality in some cases.

The third confounding variable in the evolution of red is the dynamic nature of the male signal itself, which is far more fluid across both short and long time scales than the relatively fixed colour displays of other flamboyantly coloured species such as darters.[109] Such fluidity can wreak havoc in mate choice experiments, or in tests designed to search for correlations between the intensity of colour and a marker of male quality. For example, males in some populations maintain a certain colour level in the presence of rivals, amplify their colour when a female enters the scene, and then drop back to precourtship levels when she leaves (social facilitation sensu; Reisman[110]).[66,111] This fluidity may depend, in part, on male condition. Rush et al.[51] reported that high-condition males in one population at least were capable of rapidly adjusting the long-wavelength reflectance of their signal in response to the presence of a female, reducing their colour intensity more in her absence, and thus presumably not being as conspicuous to passing predators, than did low-condition males. Fluidity may also depend on male motivation. For example, Candolin[81,82,112] discovered that food-deprived (and hence poor-condition) males dramatically increase the intensity of their signal (a curvilinear relationship between intensity and condition; see also Baube[89]) and take more risks when exposed to predators than males in good condition. She attributed this false advertising to a kind of "last hurrah" scenario; males with a low expectation of survival have nothing to lose and everything to gain by investing their remaining resources in one intensive breeding bout. As an aside, it is impossible to determine in the field whether low-condition males are intrinsically inferior, or are simply exhausted after several breeding bouts (or possibly both).[51,82] In other words, a male with limited abilities to change his colour intensity may be either a low-condition, low-quality male, or a low-condition, high-quality male at the end of his parental duties; confounding the search for a relationship between male colour and male quality. From a female's perspective, of course, both males are equally bad choices. If she spawns with a high-quality but exhausted male who cannot successfully complete another parental cycle, then the fact that her offspring would have had their father's "good genes" had they survived, is moot.

Deception introduced by the interaction among male colour, condition, and motivation is reduced when other males are added to the system.[87,111] Although a bright male in poor condition might fool a female, he is unlikely to be so lucky with a rival in good condition; in the latter case the male must both "talk the talk and walk the walk."[113,114] The preceding experiments showed decoupling between the intensity of red and overall male condition, so now it would be interesting to document whether the low-condition males also changed the intensity of their sexual behaviour to match their inflated signal. Presumably this would be a more difficult signal to fake, setting up the dynamic "bright red + poor dancer" that so confused females in Kunzler and Bakker's simulations. If males cannot inflate their courtship, then there are two checks on the system reinforcing honesty in the signal, one due to male–male competition (intrasexual selection) and the other to male–female courtship displays (intersexual selection).

The final complication in the system involves the relationship between nest placement and male colour. In some populations, females prefer males who build close to vegetation,[115] possibly because eggs in that locality are more likely to survive predation and raiding attempts until at least hatching,[32,56,116,117] cover provides protection during vigorous threespine courtship,[118] and males dance more intensely when rivals are not visually present.[119] Males who nest in the open in predator-rich waters are indeed on the horns of a dilemma. If they decrease their colour and courtship intensity, which many do, they trade mates for longevity because females are less attracted to them.[118,120] Males who nest within dense vegetation face a different set of problems. Although they may be able to display their full colour and courtship intensity, they take longer to begin nest construction[121] and are more difficult to find, paying for security with fewer female visits.[122,123] Kraak et al.[124] noted that large, red males were more likely to nest within the (presumably moderate) cover of plants, leaving us with a chicken and egg problem; which came first, the preference for big red males (side effect, female gets a better nest site) or the preference for nest site (side effect, female gets a bigger, redder male)? Nest placement aside, Barber et al.[125] found no correlation between the intensity of male colour and other "nest quality" variables (e.g., compactness, number of threads, weight of sand placed over nest) in the laboratory, whereas Guderley and Guevara[126] reported a positive correlation between colour intensity, nest structure, and glue production in the field. Do these results reflect real interpopulation variability, or are they another example of Candolin and Voigt's suggestion that the laboratory and the field represent two very different environments to fishes?

All of the preceding studies reinforce Rowland's proposal that females make decisions about male suitability at various points in the courtship interaction, from initial attraction to nest entry. If such a hierarchy based on numerous male traits does exist, it is possible that the intensity of selection will vary across the hierarchy and across populations. This, in turn, might produce subtle nonlinear selection on male traits, confounding studies investigating the interaction between female preferences, the development of preferred male traits, and male reproductive success.[127]

Before leaving this section I would like to suggest a possible mechanism for Bakker's[22] demonstration that there is the genetic correlation between the intensity of a male's nuptial signal and the choosiness of his daughters. Photopigments are binary molecules, a vitamin A (carotenoid) derivative and a protein (opsin). Although β-carotene is the major precursor for the biosynthesis of vitamin A (retinol) in mammals, the picture is quite different in the few piscine species studied to date. Astaxanthin is the major precursor to vitamin A and its derivatives in Atlantic salmon,[128] rainbow trout (plus zeaxanthin and canthaxanthin[129,130]), and *Tilapia nilotica* (plus zeaxanthin, canthaxanthin, and tunaxanthin[131]). Given that most of the preceding carotenoids play an important role in nuptial signal development in threespines (Table 6.2), it is possible that the father–daughter association may reflect a correlation between the amount of particular carotenoids available for deposition in chromatophores (sender) and for use in making photopigments in the female's eye (receiver). This relationship may reach all the way to egg formation. Early eye development in the Atlantic halibut is dependent on forms of vitamin A deposited in the yolk by the female.[132] If this mechanism also occurs in threespines, we can ask whether daughters of bright males (red body) deposit more vitamin A derivatives or precursors in their eggs than daughters of dull males, providing yet another connection between the intensity of a father's colour and the quality of his offspring.

6.2.4.2 The Female–Male Dialogue

Female threespines signal courtship readiness by approaching a nesting male and holding in a head-up, concavely curved, position. This position, which resembles lordosis in rodents,[133] emphasises the female's extended, egg-filled abdomen, releasing and reinforcing sexual behaviour in her suitor.[74,75,134,135] For their part, males in most populations direct their courtship preferentially toward larger or plumper (abdomen more distended) females[55,134–138] (but see Rowland and Sevenster[139]). Why should males be choosy? Male threespines invest a substantial amount of energy in territory defence and parental care, and this decreases the likelihood that an individual will be able to renest after completing one breeding cycle. Intuitively this might suggest that every territorial male should court any passing female and strive to acquire as many clutches as possible, but two details of male reproductive biology argue against this. First, males may have limited sperm supplies[140] and thus should allocate sperm wisely, which is exactly what Zbinden et al.[141,142] discovered. Just as males were able to adjust the intensity of their colour signal according to the social situation, they also adjusted ejaculate size and courtship length according to the size and courtship status of a rival, larger ejaculate and shorter courtship if the rival is large, larger ejaculate if the rival is courting. Second, because embryo survival is dependent on a constant supply of oxygenated water via paternal fanning and the detection and early removal of diseased eggs, there will be an upper limit to the number of eggs that any male can effectively care for.[143]

It thus behoves a male to maximise his (probably) one chance to breed, which returns us, full circle, to male preference for larger females. Female fecundity is correlated with size,[96] so choosing to spawn with a larger female means the acquisition of more eggs. Larger females also provide relatively more resources per egg,[144] so a male who chooses such a mate will get both more, and more heavily yolked, eggs than a male who mates randomly. Interestingly, preference for large size does not appear to be a general male trait; dull males do not appear to differentiate among females, whereas attractive males are very choosy.[138,145] This result is intriguing, for it suggests that a male may have an internal perception of his own attractiveness. A male who perceives himself as unattractive increases his chances of spawning by courting every female who comes near him (cf. increased number of gonopodial thrusts by sneaker poeciliid males[146]), whereas a more attractive male can afford to reject mates, accepting only high-quality females. The male's response is not a simple reflexive one. It is moulded by experience. Rowland[133] presented males with epoxy female models in a horizontal and vertical position, then recorded courtship and bites directed toward each model over a 56-min trial period. The males did not differentiate between the two models for the first 4 min of the test, but by the second interval, and throughout the rest of the trial period, the male directed more of his courtship dancing and less of his biting toward the head-up model. Apparently male sticklebacks have a bias toward the head-up posture, which causes a more rapid habituation to the horizontal posture, producing the differential male response. It is also possible that the males were increasingly selective as they became more sexually aroused. Could habituation to particular stimuli, which would decrease the overall range of cue acceptability and give the impression of "enhanced selectivity," be the outcome of increased sexual motivation?[94]

As with every visually based trait discussed so far, positive reaction to a larger female is not universal in the threespine's world, particularly when different ecomorphs come into contact with one another. Limnetic males, which are smaller and more gracile than benthic fishes, retain the ancestral preference for large females when collected from allopatric populations, but display a derived preference for small females (size-assortative mating) when collected from populations sympatric with benthics.[147] The shift in the plesiomorphic preference may have been driven, in part, by the high cost to small males of courting large female benthics who are adept at egg cannibalism.[148,149] Size-assortative mating was also reported for females from freshwater (small body morph) and anadromous populations collected in British Columbia, Iceland, Alaska, Norway, Scotland, and Japan.[150] Presumably, males in these populations also display a preference for similarly sized mates.

In addition to general shape, female sticklebacks contribute information to the male–female dialogue in the form of a nuptial signal. Three-spined sticklebacks respond more to bold over fine chequered patterns,[151] so the intensification of vertical barring seen in the female signal may take advantage of a perceptual or sensory bias in the male's visual system. Information transferred by this signal is a reliable indicator of sex and behavioural motivation. In both three-spined[46] and brook[152] sticklebacks, the appearance of female nuptial colouration is tightly coupled with ovulation and the appearance of female courtship behaviour, and is as distinct from the male signal as possible. Both the redundancy between female colour and female behaviour, and the disparate natures of the male and female signal, interact to reinforce the female's message of courtship readiness. And males do respond to this message, courting nuptially coloured females more intensely than gravid, but uncoloured alternatives (*G. aculeatus*[46]; *C. inconstans*[153]).

6.3 OLFACTION

6.3.1 THE MEDIUM: TRANSMISSION PROPERTIES

Olfactory cues serve as long-distance signals in moving water,[154] allowing a fish to detect the presence of another, be it predator, prey, conspecific, or heterospecific, before that "other" can actually be seen. Olfactory cues are transmitted via a turbulent and unpredictable medium, which intuitively might lead one to question the efficacy of scent in anything other than a general arousal role. Computer simulations, coupled with experimental manipulations, have, however, revealed that variability in concentration fluctuations, pulse and interpulse intervals, temporal filter properties of chemoreceptor cells, and the active sampling movements of the receiver can all combine to extract temporal, intensity, and directional information from an outwardly chaotic odour plume.[154–157]

As with light, the transmission of chemical cues through the channel (water) will be affected by properties of that channel, the most important of which are the flow speed of the water, substrate type, and the size and shape of the waterway.[158,159] Because of turbulence, odour does not diffuse outwards in a simple concentration gradient, but rather forms a complex plume with intermittent packets of scent; the signal is thus patchy in space and time. This means that tropotaxis like that displayed by forked-tongued squamates is unlikely in aquatic systems. Balkovsky and Shraiman[160] proposed that receivers could track complex odour plumes efficiently by employing three simple rules: (1) search actively in lateral motions perpendicular to the flow, (2) once you encounter a packet, assume that it lies somewhere within a cone with the end upstream, and begin searching within the cone in a zigzag manner, and (3) reposition the cone based on the new information and repeat (2). This model, which works well for insects, appears to work moderately well for fishes as well. Freshwater eels tracking an odour relied initially on the direction of the water flow, possibly detected by the lateral line, to orient. They then moved upstream in a relatively straight fashion. As they got closer to the source, they decreased their speed and began swimming perpendicular to the flow whenever they lost contact with the plume.[161]

Fishes living in slow-moving or still waters are presented with a different set of transmission parameters; odours can form a concentration gradient more easily in still waters. So, for example, bullheads living in quiet habitats swim more directly to a food source, making fewer course corrections and swimming more slowly, when placed in a no-flow situation. When faced with current, they swim faster and make more adjustments to their pathway (more like the eels discussed previously). Even with the behavioural changes, the catfish did not locate the cue source as effectively in the flow situation.[162] These results support the hypothesis that an animal's sensory systems and behaviour will closely match the sensory conditions of its environment.[163] Given the number of different environments inhabited by three-spined sticklebacks, it is likely that both the behavioural and physiological mechanisms involved in olfactory detection will show the same variability as that reported for the visual system.

6.3.2 Structure of the Cue

6.3.2.1 Alarm Signals

The Ostariophysi are renowned for possessing specialised epidermal cells that produce and release Schreckstoff (fright substance) when damaged.[164,165] When conspecifics detect the cue, they respond with species-specific antipredator behaviours such as increased shoaling, freezing, or decreased movement. Schreckstoff was initially considered to be a unique synapomorphy for the Ostariophysi, but recent research has uncovered alarm pheromones in a wide variety of nonostariophysan fishes, including salmonids, cottids, gasterosteids, centrarchids, poeciliids, and percids.[166] Given the wide taxonomic diversity of these fish, it seems unlikely that alarm pheromones have arisen only once throughout the history of the ray-finned fishes, or indeed that the pheromone is the same in each group, although there appears to be some overlap. Brown et al.[167] demonstrated that the nitrogen oxide functional group was the prime molecular trigger in the ostariophysan alarm cue. They hypothesized that the trigger could be coupled with various molecules, including hypoxanthine-3-*N*-oxide and pyridine-*N*-oxide, to produce an array of alarm signals within the clade. Nothing is currently known about the structure of the alarm cue in sticklebacks.

6.3.2.2 Social Signals

Although behavioural and physiological studies have demonstrated that pheromones play an important role in the breeding system of many fishes, we actually know very little about the structure of most putative chemical cues. Generalising from the species that have been studied in some detail (e.g., the African catfish, the goldfish, and various salmonids), male-based pheromones usually comprise conjugated and free steroid metabolites and prostaglandins or, more rarely, bile acids. Males themselves are extremely sensitive to F-type prostaglandins, which potentially allows them to distinguish ovulated from unovulated females.[168] Conjugated steroids and prostaglandins are large molecules, making them ideal candidates for pheromones because the number of information-transmitting structural isomers of a compound increases exponentially with molecular size. The level of the stimulatory effect of these substances depends on their structure and their relative concentrations, implying that detailed and species-specific information may be transmitted by a multicomponent olfactory cue (syntactic coding[169]).

6.3.3 The Receiver: How Sticklebacks Smell

As a fish swims, water moves through each external naris into the olfactory cavity, over the olfactory rosette lying on the floor of the cavity and out of the internal naris. The rosette consists of a series of lamellae, some of which contain either ciliated (respond to amino acids) or microvillar (respond to pheromones) receptors on the dendrites of olfactory receptor neurons (ORN). ORN axons join to form the olfactory nerve (cranial nerve I), which runs to the olfactory bulb, where it synapses with second-order neurons. Stimulus processing at this level varies depending on species. For example, although individual neurons are not specific to particular amino acids in zebra fish, the combination of cells firing and the temporal patterning of that response can produce distinctive odour images.[170] In brown trout, distinctive odour images are produced by the reactions of specific neurons organised in clumps located in different regions of the olfactory bulb.[171] Axons from the olfactory bulb neurons form the olfactory tract, which is divided into three distinct bundles in *Carassius* and *Gadus* species, the lateral olfactory tract (LOT), medial olfactory tract (MOT), and the medial bundle of the MOT (mMOT). In *G. aculeatus*, the LOT remains compact until its fibres diverge into the lateral part of the *area dorsalis telencephali*. The MOT, on the other hand, comprises a complex of nerve bundles, terminating in the *area ventralis telencephali*, the lateral part of the *area dorsalis telencephali* and the *nucleus praeopticus*.[172] Although not named as such, it is probable that one of these bundles is homologous with the goldfish mMOT. The LOT mediates feeding behaviour (amino acids) and the mMOT is involved in alarm responses. Processing of pheromones is associated with the MOT,[173] which innervates, among other things, areas of the telencephalon and preoptic area that affect GtH production, which in turn regulates reproductive physiology and behaviour.[174] Research with lake whitefish has recently uncovered PGF-responsive neurons in the tissue connecting the olfactory bulb to the telencephalon. Because these neurons do not respond with typical synchronous oscillatory discharges to the pheromonal cue, they might be responsible for amplifying and processing pheromones transmitted via a dynamic odour plume. Similar synaptic contact between primary olfactory nerves and this area of the brain has been reported for threespines.[175] Laberge and Hara[176] hypothesized that there are two functional olfactory subsystems in the Actinopterygii, paralleling the vomeronasal (cf. the PGF responsive cells) — primary olfactory systems of tetrapods. This discovery is particularly relevant to studies of reproductive behaviour in fishes because the vomeronasal system is responsible for detecting most pheromones in mammals and reptiles.

Three-spined sticklebacks, similar to their *Spinachia* cousins, are monotrematous, that is, each olfactory organ has only a single naris.[177] To date, no one has published a detailed description of the functional morphology of the threespine's nose, so I will rely on research using the sea stickleback, *Spinachia spinachia*.[178] In that species, subdivisions within the olfactory organ allow water to flow in one direction across the epithelium, increasing the effectiveness of chemical detection because "used" and "fresh" waters are not mixed on the way in and out of the naris. A bifurcate accessory ventilation sac, which communicates with its olfactory organ through a slitlike aperture, lies between each olfactory cavity and the buccal membrane. The sac is affected by pressure changes in the buccal cavity during respiration, expanding during inspiration and drawing water in, contracting during expiration and pushing water out of the naris. Opening and closing the mouth suddenly amplifies this bellowslike effect. The fish inhales, then expels water in one intense jet from each naris.

Threespines have a small olfactory epithelium (only two folds in each nasal rosette[179]), which may indicate that olfaction does not play an important role in the gasterosteids' world.[180] Early research supported this proposition. Bilateral cauterization of the nares,[173] frontal or bilateral lesions of the telencephalon, which included sectioning of the olfactory nerves and removal of the *bulbus olfactoris*,[181,182] and mediocaudal lesions of the telencephalon[183] did not prevent males from completing a reproductive cycle; although removing areas of the forebrain did alter the "normal" fluctuation of aggressive, sexual, and parental behaviours over the cycle. All these experiments were performed when the males had completed a nest and were courting. Segaar et al.[184,185] limited neurophysiological manipulations to males that had just initiated nest building and discovered that (1) an intact olfactory nerve (n1) was important for the initiation of nest building and indispensable for the appearance of zigzag dancing; (2) transection of n1, the *ramus pretrematicus* of the first branch of the vagus nerve (X1pre), and the *ramus posttrematicus* of the glossopharyngeal nerve (IXpost) decreased fanning across the entire cycle; and (3) nerve fibres that ramify from IXpost and disperse into the pharynx between the first and second gill arch affect different aspects of the breeding cycle. Sectioning these fibres stimulated fanning (section IXpost distal to the F-anastomosis of IXpost with X1pre), suppressed courtship when the fry hatched (section IXpost at a more proximal level near or just dorsal to where it crosses the first *arteria branchialis*), and stimulated sexual behaviour (section IXpost at an even more proximal level halfway between the crossing of the first and second *arteria branchialis*). Clearly, olfactory stimuli are necessary to the development of a male threespine's breeding cycle, although the exact role for such stimuli varies depending on the nature of the stimulus and the stage a male has reached in the cycle.

6.3.4 Function of the Message

6.3.4.1 Alarm Signals and Predation

Brook sticklebacks respond to alarm pheromones from injured conspecifics and a sympatric prey guild member, the fathead minnow (*Pimephales promelas*), in the laboratory by shoaling and moving downwards when in a group and with a decrease in activity when alone.[186] In the field, they tend to avoid areas marked with either scent, although that avoidance is by no means absolute (conspecific cue: 40.8% caught in marked traps; fathead minnow cue: 37.5% caught in marked traps), and is a far cry from the strong antipredator reaction shown by fathead minnows (conspecific cue: 4% caught in marked traps). Avoidance of the "predation-event" site after the stimulus had been removed lasted up to 8 h for fish present during cue release; naïve individuals moved constantly through the area (only the cyprinid cue was tested[187,188]). Sticklebacks caught in conspecific traps were smaller than expected (3.2–4.3 cm), implying that the development of the response might be mediated by experience.[189] For example, sticklebacks in a pike-free lake (Marshy Creek, Saskatchewan) only responded to the scent of pike if it had been eating sticklebacks. This implies that (1) there is a genetic component in the alarm response and (2) predators somehow become "labelled" with the alarm cues of their prey. Fish from Pike Lake, on the other hand, decreased their activity regardless of whether the pike had been eating conspecifics, fathead minnows, or swordtails. If sticklebacks are responding like fathead minnows, then the Pike Lake individuals have learned to associate the scent of the alarm cue with the scent of the pike, then generalised to just "scent of the predator" independent of that predator's diet.[190,191]

Both Schutz[192] and Hoogland et al.[193] reported that European *G. aculeatus* did not respond to conspecific skin extracts. Threespines from Morice Lake, New Brunswick, however, reacted to cues from their own population, a neighbouring population, and a close relative and prey guild member, *Apeltes quadracus*. Their reactions included an increase in the time spent under cover, near the ground, freezing with spines raised and oriented head down,[194] all of which are common responses shown by solitary individuals to the presence of a predator. Are the North American fishes simply more alarm oriented than their European relatives (evidence of interpopulational and interspecific differences) or are the conflicting results an artefact of methodology (e.g., gasterosteids are far less flamboyant in their reactions than minnows, hunger level may affect the intensity of response, etc.)? For that matter, do gasterosteids display the sexually dimorphic responses to conspecific skin extracts shown by the green swordtail, *Xiphophorus helleri*,[195] or any of the sophisticated learning abilities shown by *Pimephales promelas*, in which an individual's ability to associate the release of heterospecific (*Culaea inconstans*) alarm substance with a predator depends on the complexity of the habitat and heterospecific density?[196,197] Moving from behaviour to physiology, we must ask: What is the structure of the gasterosteid alarm cue? How does it vary between species? Where is it produced?

If the release of alarm substance occurs by damage to the epidermis, why are these cues not released during vigorous activities such as biting, fighting, tail dragging, and dorsal pricking, not to mention abrasive activities like creeping through or struggling into a nest, which may contain sharp stems or small pebbles? In other words, are individuals at risk of freezing, fleeing, or hiding following the inadvertent release of Schreckstoff during spawning? Smith[198] proposed that this problem is overcome in many ostariophysan species by a change in the production of the signal or the reaction of the receiver to that signal during the breeding season, or both. For example, male fathead minnows lose the ability to produce alarm substance during the breeding season but they retain their fright reaction to it, as do females.[199] In fact, males will decrease territorial activity and even go as far as to abandon their nests if exposed to alarm pheromone or the scent of a minnow-fed predator.[200] It is thus possible that males use the intensity of such olfactory cues to adjust their reproductive behaviour and to choose nesting sites that are as predator free as possible. Species breeding in shallow, slow-moving waters, or in confined places under rocks and branches, might be expected to be more strongly affected by the mistaken release of alarm cues during courtship than species breeding in open or fast-moving waters. The clarity of the water may be important as well; alarm behaviour in the laboratory is more intense when the water is turbid rather than clear.[201] Overall then, the relationship between alarm cues and breeding will be affected by variables of the breeding habitat, the presence or absence of abrasive courtship, and changes in cue production and receiver response. None of these possibilities have been explored for any gasterosteid.

6.3.4.2 Social Behaviour: Shoaling

Olfaction has been demonstrated to play a role in many components of the gasterosteid breeding system (Table 6.3). Threespine fry demonstrate a preference for shoaling with full siblings over nonsiblings that decreases as they age (from 18 to 45 d post hatching[202]) until no detectable preference is expressed (61–91 d post hatching[203]). The fry were able to see and smell the stimulus fish, so it is unclear whether visual and olfactory cues are involved in this discriminatory ability.[202] Whatever the basis for the attraction, it apparently has some genetic component (fish raised with nonsiblings still prefer siblings, albeit at a lower level[204]), which is subject to modification as the fish ages. Given that threespine fry from 14 to 43 d old joined groups of same-sized blackspotted stickleback fry with great alacrity in the lab, and are found in mixed shoals in the field, it seems probable that the antipredator benefits of shoaling with similar individuals regardless of genetic relationship far outweigh the costs of swimming in either a small group or alone, searching for siblings. This suggestion is corroborated by experiments demonstrating that wild-caught young adults prefer the scent of familiar to unfamiliar shoal mates after just 2 weeks of confinement together in arbitrarily constructed groups. This attraction extends to unfamiliar fish that have spent 2 weeks under the same environmental conditions (water type and diet[205]). Juvenile threespines also preferred the odour from the heavier of two nonkin shoals, but that preference disappeared when the scent of a pike, which had just dined on sticklebacks, was added to the water.[203] This response may have been an artefact of the lab design. The fish were tested in a fluvarium in which they were confined to a test area downstream of two currents carrying olfactory stimuli. Under these conditions, and with no visual cues to rapidly direct them to a shoal, the juveniles might have moved as far as possible from the source of the alarm cue (and presumed location of the predator), resulting in "no choice" for a particular shoal size. Overall, then, juvenile threespines are capable of extracting information from odour-based cues about kinship (early in development), recent association based on habitat and diet (familiarity), and shoal size. They act on this information by following the general rule, when possible: shoal with familiar fish.

6.3.4.3 Social Behaviour: Reproduction

As discussed previously, the visual components of the stickleback's courtship repertoire include a species-specific mosaic colour pattern, courtship behaviours, and morphological traits such as body size. Together, these signals form a (generally) cohesive multicomponent display. Such displays increase the receiver's ability to detect, discriminate, and remember the sender (when learning is involved in the communicative process). As can be seen from Table 6.3, the stickleback's breeding repertoire is also multimodal; information is transmitted in different sensory modalities (reviewed in Rowe[206]). Olfactory cues contain information about sex and species that is redundant with information encoded in visual signals, increasing the probability that the message will be received.[207] These cues also serve a variety of other functions during the breeding season, most of which have not been explicitly investigated for gasterosteids.

TABLE 6.3
Olfactory-Based Studies in Gasterosteids

Potential Information in Olfactory Cue	Sex, Species, and Age of Test Fish	Results
Offspring recognition	Both sexes, *G. aculeatus* reproductive adults	Males did not recognize their own clutch, females direct fewer bites to their own eggs than to eggs from another female[264]
	G. aculeatus reproductive females	As above, but discrimination disappeared when the clutches were mixed, containing 33 or 20% of the test female's eggs[265]
Kin/shoaling	Both sexes, *G. aculeatus* fry	18–20-d-old fry discriminate sibs from nonsibs (visual and olfactory cues), preference is weakened if sibs are raised with nonsibs[205]
	Both sexes, *G. aculeatus* fry	Preference to shoal with sibs decreases from 18–45 d post hatching[203]
	Both sexes *G. aculeatus* juveniles	No preference to shoal with kin 61–91 d post hatching[204]
	Both sexes, *G. aculeatus* nonreproductive adults (familiar: unrelated but raised together for two weeks)	Prefer familiar (vs. unfamiliar) shoalmates (visual + olfactory; olfactory alone), unfamiliar shoalmates raised under the same (vs. different) conditions (freshwater vs. saline; diet of *Daphnia* vs. bloodworms) (visual + olfactory)[206]
	Both sexes, *G. aculeatus* nonreproductive adults (familiar: related, raised together)	Prefer familiar kin vs. unfamiliar nonkin (visual and olfactory cues), shoals were not significantly different in visual traits (body size, mass, condition), perhaps implicating olfaction[226]
Gender	*G. aculeatus* reproductive females	More time with male than female cue[227]
	C. inconstans reproductive males	Equal time with both cues but only court female scent[210]
Male reproductive state	*G. aculeatus* reproductive females, nonterritorial males	More bites to cue from displaying rather than nondisplaying male[226]
	G. aculeatus reproductive females	Prefer cue from nest building males vs. males without nests[227]
	S. spinachia reproductive females	Prefer cue from male + nest vs. nesting materials alone[267]
Mhc complement	*G. aculeatus* reproductive females	Prefer males whose *Mhc* genotype, in concert with their own genotype, produces offspring with an intermediate number of *Mhc* alleles[228–230,235,236]
Species	*G. aculeatus* and *C. inconstans* reproductive females	Recognize each other's males but prefer conspecific[209]
	C. inconstans reproductive males	Recognize but are not attracted to threespine female, prefer conspecific[210]

Note: Unless specifically stated, the experiments used only olfactory cues.

6.3.4.3.1 *Physiological Functions of the Olfactory Cue*

Male-based olfactory cues act as physiological primers, stimulating ovulation in maturing females and helping to synchronise the courtship dialogue in a variety of fish species. The same is true in reverse; preovulatory cues in the form of gonadal steroids from females function as potent physiological primers, preparing the male for spawning by triggering changes in circulating levels of GtH and increasing milt production. Ovulation is associated with the release of F-type prostaglandins and their metabolites (e.g., cyprinids, catostomids, salmonids, and cobitids) or gonadal steroids (e.g., poeciliids), which release and synchronise male and female courtship.[174] F-type prostaglandins in particular are extremely stable in aqueous solution, indicating that they may be very effective long-term cues. It thus seems plausible that these chemical cues will play a role in synchronising the final stages of male and female sexual maturation and in releasing appropriate sexual behaviours in gasterosteids. The suggestion that the chemical cues are involved in mate synchronization is supported by two observations from my olfactory experiments.[208,209] First, female brook sticklebacks had a longer interspawning interval when they were only exposed to visual stimulation from a courting male ([visual + olfactory]: 2.6 ± 0.5 d vs. [visual] 4.7 ± 1.7 d). Second, the presence of the chemical cue alone stimulated the release of species-specific courtship behaviours; threespine females hovered head up, brook females displayed head-up ground-sink, and a proportion of the brook males actually directed pummels toward the glass wall nearest the cue drip. Control fish did not show any of these behaviours.

6.3.4.3.2 *Alerting Function of the Olfactory Cue*

Chemical cues will generally be perceived before visual ones if the receiver is within the diffusion range of the source. Studies of interactions between temporally displaced signals indicate that the first cue (in this case, chemical) functions to alert the receiver to the presence of the second cue (visual), increasing the probability of its detection and recognition.[206,210] As female threespines approach ovulation, they begin to wander further and further from their foraging areas toward the males' territories. At this stage it seems more probable that the soup of male-based olfactory cues emanating from those territories acts to alert the female to the presence of nesting males rather than transmitting information about a particular male. Receiving this message from a distance increases the female's probability of finding a mate (decreases search time and costs[156,211]), which is not a trivial matter when the males are nesting in or close to vegetation. It seems unlikely that female-based chemical cues act as alerting stimuli to male sticklebacks, because males would only detect the presence of an individual female before she appeared if she were approaching with the current. Because these fishes nest in relatively quiet waters, this information would reach the male in a random fashion, decreasing the overall importance of such a function.

The way in which olfaction (alerting or first stimulus) and vision (second stimulus) interact is not well understood in fishes, but may have something to do with a relationship between the olfactory system and the enigmatic terminal nerve (*nervus terminalis*: NT[212]). The NT does not appear to be chemosensitive, but it does contain cells that are immunoreactive for LHRH (luteinising-hormone-releasing hormone) and NMDA (*N*-methyl-*D*-aspartate), which modulate GtH secretion.[213] Some authors have suggested that the NT cells can be divided into two general classes, one associated with the MOT and forebrain, the other projecting to the optic tecum and retina.[214] In three-spined sticklebacks, nerve fibres from the MOT do indeed synapse with a ganglion of cells (*nucleus olfactoretinalis*) that is part of the terminal nerve complex. The synaptic contacts did not stain equally well in all individuals, supporting the hypothesis that the TN is only sensitive to input from the olfactory system at particular stages of the breeding cycle.[175]

The relationship between the NT and the retina appears to be unique to teleosts,[215] prompting Stell and Walker[214] to hypothesize that the visual NT is part of an "integrating pathway, through which visual and olfactory components of complex reproductive behaviours are regulated and coordinated" (: 94). It is thus possible that the alerting function of olfactory cues is modulated by the MOT/NT's effect on the visual system in the short term, and in the long term coupling of visual and olfactory information during early development. Such intersensory integration, which is a common feature across vertebrate taxa and sensory systems,[207] may have been an important component of the selection arena in which multimodal signals evolved in the gasterosteids.

6.3.4.3.3 Associative Learning and Olfaction

Unlike visual and auditory systems, the olfactory system is a shallow (cortical interpretation is only two neurons away from receptors), low-bandwidth sense that appears to favour global perception rather than precise segmentation.[216] Global perception implies that responses to particular scents are not hard-wired *per se*; rather, the structure of the neurological system permits associative learning, linking scents with a variety of objects including home stream, individual conspecifics, prey items, and predators. For example, pike naïve brook sticklebacks learned to respond with antipredator behaviour to the scent of a pike after only one exposure if that scent was coupled with alarm pheromone from either conspecifics or the fathead minnow.[192] Could there be a role for this type of associative learning in the gasterosteid breeding system? All gasterosteids fan their nests, driving oxygenated and odour-bearing water from the male's immediate vicinity into the nest.[217] An increase in parental care occurred in the ancestor of the nuptially coloured *Gasterosteus* + *Pungitius* + *Culaea* clade; males generally guard their offspring at least until the fry are free swimming. During this period it would be possible for the developing fry to learn to associate the scent of their father and surrounding territorial males with species-specific body shape and colour. Such imprinting would set the developmental stage for the alerting function of olfactory cues when the fry become sexually mature. Depending on the species, retention periods for predator olfactory cue-alarm pheromone-associative learning range from 3 months[218] to 1 year[219] following a single conditioning event. Stickleback fry are intimately associated with their father for at least 3 d post hatching, and with each other in family-based, then larger, groups until the following breeding season, which would allow for prolonged and multiple conditioning events, further reinforcing the learned association between conspecific odour and visual cues.

Sexually mature fish would then already possess a chemical and visual search image or template for prospective mates before returning to the breeding ground. That template however, may be modified by the way olfactory information is processed. In the visual system, the lateral inhibition network involving mitral cells and their decoders acts to increase local contrast, and hence separation, between stimuli. In the olfactory system, on the other hand, lateral inhibition does not appear to aid in discrimination, but rather in generalisation.[217] If a honeybee that has learned to associate scent A with a sugar reward is presented with a mixture of scent A and a similar scent B, it will respond to the mixture as if it is more A-like. Untrained bees react to the mixture as a unique A–B combination.[220] This process, called *blocking*, is one way in which an individual's sensory history can influence its future response to a similar odour. The greater the similarity between the two stimuli, the greater the blocking effect, and hence the stronger the generalisation. Imprinting as a process involves generalisation across an array of novel but similar stimuli (at least in visual cues[221]). So, if stickleback fry imprint on olfactory cues, and that process enhances generalisation across cues because of the way the olfactory system is organised, then it is not surprising that individuals would recognise scents from close relatives (heterospecifics) at a later date (Table 6.3).

6.3.4.3.4 *The Cue and Mate Quality*

The female brook stickleback's response to the male-based cue fluctuates with her ovulatory cycle, changing from attraction at the time of ovulation, to avoidance within 24 h of spawning.[222] This pattern is similar to that documented for changes in the female courtship display across the ovulatory cycle in response to visual cues from a nuptially coloured male (threespine[95]; brook[223]). The mechanisms underlying this cyclical fluctuation are complex, involving the influence of hormones on the female's ability to detect, interpret, and respond to mating signals from potential partners. Sarkar and Subhedar[224] reported that production of β-endorphin and its transfer to the olfactory bulb increased dramatically during the prespawning period in female catfish (*Clarias batrachus*). As maturation approached, β-endorphin-sensitive first-order neurons penetrated further into the olfactory bulb, eventually contacting the second-order neurons just prior to the spawning phase. The authors proposed that this change in β-endorphin activity represented a gearing up by the olfactory system, resulting in increased responsiveness to olfactory cues during the spawning stage of the breeding cycle.

Women rate how a man smells naturally as one of their major criteria for mate choice. Women are also more sensitive to odours and process those odours faster around the time of ovulation. During that time they prefer the scent of symmetrical men, the image of more masculine-looking men, and fantasise more about interacting with men other than their partners, all of which may indicate that they can distinguish higher-quality males during their fertile period or that they are more preoccupied with mate quality during this time.[225] Given that stickleback females are more responsive to male-based cues immediately following ovulation, can they, like human females, extract information about male quality based on those cues? The answer to this question is, possibly. Female threespines can differentiate between males with and without a nest, and between males and females based on odour cues alone.[226,227] But does this discrimination extend to differences among individual males?

To answer this question, we must turn to the realm of immunology. Genes of the major histocompatibility complex (*Mhc*) produce the cell surface receptors that selectively bind foreign antigens, immobilising them for attack by specific T-lymphocytes. This complex contains the most diverse genes found in vertebrates, which intuitively may lead to the conclusion that it is better to be as polymorphic as possible because you can mount an attack against more foreign invaders. The immune system is, however, more intricate. A delicate balance exists between *Mhc* genes and T-cells, too many different *Mhc* alleles have a negative effect on the development of T-cells, producing an immune system in which the capacity to recognize foreign antigens is not matched by the capacity to attack and nullify the invader. Individuals with too many *Mhc* alleles are thus expected to show the same resistance to disease as individuals with too few, so the optimal number of *Mhc* alleles should be somewhere between the two extremes.[228]

Threespines have seven *Mhc* class IIB genes,[229] which could translate into hundreds of possible allelic combinations; however, samples from a German population indicated that the average number of alleles per individual is between five and six.[230] Fish with an intermediate number of alleles were the least heavily infected with parasites in the field,[231,232] and displayed the most resistance to infection (*Glugea anomala*, *Schistocephalus solidus*) in the laboratory.[233] Although individuals with suboptimal *Mhc* diversity can compensate somewhat by altering levels of gene expression (fish with too few alleles upregulate *Mhc* expression and *vice versa*),[234] having an intermediate number of *Mhc* alleles is the best solution, overall.

Does this have anything to do with mate choice? Initial investigations of female threespines' responses to *Mhc* odours were puzzling because the females did not discriminate between males similar to, or different from, themselves.[230] Subsequent research, however, solved the puzzle: A female was attracted to the male whose genotype, in combination with her own, produced offspring with an intermediate number of *Mhc* alleles.[235] In an extremely elegant experiment Milinski et al.[236] manipulated a female's response to male scent by supplementing male-based water with synthetic *Mhc* peptides. As predicted, supplementation increased the attractiveness of males with too few unique *Mhc* alleles, while decreasing the attractiveness of optimal males. This type of interaction does not produce a single "most preferred" male phenotype, but rather a distribution of preferred males mirroring the distribution of *Mhc* alleles in the population of choosing females. This of course begs the question, how does a female "know" her *Mhc* genotype? The most obvious answer to this question is that individuals are familiar with their own scent, but this suggestion needs to be examined experimentally before the obvious can be accepted as the explanation.

Returning to the question of the olfactory cue and male quality, studies have shown that the cue transmits information about sex, species, and reproductive state (territory holder for males, receptive and courting for females; see Table 6.3). This information complements the message sent by visual cues and allows individuals to locate a group of potential mates. So, how does an individual choose among members of that group? Although the olfactory cue transmits information about a male's *Mhc* genotype, that information is useful only relative to the female's genotype. In other words, *Mhc*-based odours may be necessary, but they are not sufficient to find the "best" or highest-quality mate, assuming that quality is a combination of many factors such as health, vigour, dominance status, and experience.

The interaction between male quality and genetic dissimilarity vis-a-vis a given female produces an evolutionary conundrum: What does a female do if the *Mhc*-preferred male is less vigorous than a nonpreferred alternative?[237] For sticklebacks, at least, the answer may be to use many different cues during courtship. There are three general ways in which a female stickleback can benefit from mate choice. She can use (1) one or more components in male scent to locate the appropriate species and sex (decreases search time, increases offspring survival based on genetic compatibility), (2) the intensity/chroma contrast of the nuptial signal to secure a male with good genes (increases offspring fitness) and a male that will be a good father (increases offspring survival), and (3) another component in the olfactory cue (*Mhc*-based) to find a male with the appropriate genotype (increases offspring fitness). Females can integrate these different bits of information by using a decision-making hierarchy[238]; the multimodal extension of Rowland's suggestion that female's make decisions along a hierarchy of visually based cues. So, female threespines may be initially attracted to the sensory soup of breeding threespine males, selectively attend to bright red males within that soup, and then only respond to bright red males with the optimal amount of genetic dissimilarity.

6.4 WHAT SENSORY SYSTEMS ARE LEFT TO STUDY?

There are at least three more sensory systems that may form part of the threespines' umwelt: gustation, sound, and "near touch" (the response to the displacement of water molecules by specialised cells in the lateral line).

6.4.1 Gustation

Gustation clearly plays a role in foraging behaviour, which in turn is intimately connected with fitness. Although diet, rate of food consumption, and predatory behaviour based on visual cues have all been investigated in threespines (e.g., Ibrahim and Huntingford[18,239,240]), until Mussen and Peeke[241] reported that marine threespines were stimulated to display feeding behaviour (substrate biting) by water that had previously held brine shrimp, there were no studies of the relationship between olfaction and prey detection. Mussen and Peeke's results corroborate the hypothesis that in many fishes, olfactory cues, processed by the olfactory bulb-LOT connection, alert the hunter to the presence of prey. To my knowledge, no one has investigated the influence of gustatory cues in the detection of, or preference for, prey items in any gasterosteid. Nor has anyone asked whether taste plays any role in reproduction. For example, in poeciliids, ovulation ruptures follicles releasing oestrogen-based steroids bathing the developing ovules, and these steroids drain through the gonopore. A male samples a female's scent by nibbling around her gonopore. This behaviour, in conjunction with the demonstration that male Mexican mollies, *Poecilia mexicana*, can identify gravid females with their nares blocked,[242] and studies showing taste buds concentrated on the upper side of the head in various poeciliids (*P. mexicana*, *P. velifera*, and *X. helleri*), hints that male recognition of female receptivity may involve gustatory cues in some fishes.

6.4.2 Acoustical

Sound is a more effective long-distance signal in water than light, electricity, or odour, so, not surprisingly, fishes from over 50 families sing to one another in social contexts.[166] Of these fishes, one species in particular may serve as a template for threespine studies. The channel catfish, *Ictalurus punctatus*, produces rasping sounds by moving its pectoral fin. Fin movement drives the dorsal process from the central articulating surface (primary pivot point between the pectoral girdle and the pectoral spine) backwards and forwards across a slight depression in the cleithrum called the *spinal fossa*. The ventrolateral surface of the dorsal process bears numerous fine ridges, which rub across the rough textured surface of the fossa producing a rasping pulse. Stridulations of this type are extremely variable, pulse rate depending on the speed of fin movement and the distance between ridges on the dorsal process, and number and dominant frequency of the pulses depending on how long and how hard the fish drives the dorsal process across the fossa.[243] Channel catfish are sensitive to stridulation pulse frequencies, so the potential exists for the sound to become part of the communication system. Sörensen[244] suggested that the coupling of the sound produced by the movement of the fins and locking of the pectoral spine with the visual effect of spine raising might function to warn predators of the inadvisability of dining on catfish (message: "I am large, you will choke"). Warning signals are more effective when multimodal.[211] Spine raising is employed during social displays, so it is also possible that the stridulation-raised spine couplet is serving a threat role in agonistic encounters (message: "I am large, you will lose") and is reinforcing body size in courtship (message: "I am large, choose me"). Could such a mechanism be operating in *G. aculeatus*, a species with relatively formidable dorsal and pelvic spines? Does sound reinforce the effects of female preference for males with more symmetrical pelvic spines?[245,246] If spine locking communicates information during breeding, is there any compensation for the loss of sound production in populations without a pelvic girdle? We do not know the answers to any of these questions.

6.4.3 Near Touch: The Lateral Line (Mechanoreceptive) System

The threespine's lateral line system consists of free neuromasts arranged in single lines along the mandible, around the eye and the naris, and along the dorsal-lateral surface from the postotic region of the head to the caudal peduncle (in the middle of the lateral plates), and a double line of free neuromasts along the caudal peduncle, above and below the causal keel.[247,248] Threespines from river populations learned to locate a food patch using only water flow cues faster than threespines from ponds, indicating that the lateral line (1) plays a role in at least foraging behaviour and (2) may be yet another sensory system that can be fine-tuned to reflect local environmental conditions.[249] The lateral line system responds to the displacement of water molecules, providing a kind of hydrodynamic imaging, replacing vision with touch.[250,251] Mexican mollies, *Poecilia mexicana*, living in the perpetual deep night at the end of a large cave, have a hyperdeveloped lateral line system. Interestingly, males have a very simplified agonistic repertoire, retaining only tail beating performed when the two combatants are close to one another,[252] and females are capable of differentiating between small and large males in complete darkness, something epigean Mexican mollies cannot do. Threespines are not found in such extreme environments, but some populations do live in the dim light of tea-stained creeks and ponds. Do these populations show heightened responses to chemical cues or changes in their lateral line that would compensate, in the same manner as the Mexican cave mollies, for the decreased availability of visual information?

6.5 UMWELT AND US

The umwelt of *Gasterosteus aculeatus* is defined by the transmission properties of the habitats in which the numerous, variable threespine populations live. Anthropogenic intervention can disrupt the breeding system by operating on any of the relevant transmission parameters. For example, increased turbidity degrades the transmission of visual cues, increasing hybridization between sympatric congeners using visual cues in mate choice.[253] Noise from boats contains low-frequency components, overlapping the transmission and detection frequencies of many fishes. This interference may damage a fish's hearing abilities in the long term, while disrupting communication in the short term.[254,255] The olfactory system is directly exposed to the environment and thus extremely susceptible to damage by dissolved pollutants, including copper, lead, mercury, nickel, zinc, sliver, cadmium, and organophosphates.[256] Relatively minor acidification of holding waters (from neutral to 6.0) eliminates the response to alarm cues by fathead minnows, fish that overlap with brook and three-spined sticklebacks in many areas. Acidification alters the chemical structure of the antipredator trigger, eliminating its biological functionality.[257] Because acidification damages both the receiver (sensory system) and the sender (the cue), its effects can sweep rapidly through the system.

We tend to think of the threespine as being a widespread, almost weedy species, assuming that it will survive the environmental insults we hurl at it. Recent extirpation of the Hadley Lake benthic-limnetic species pair, and the endangered status of the remaining three species pairs[258,259] challenges that belief, reminding us to be more proactive in the preservation of this extroverted, but fragile, fish.

REFERENCES

1. Von Uexkull, J., *Strafzuege durch die Umwelten von Tieren und Menschen*, Springer-Verlag, Berlin, 1934.
2. Lythgoe, J.N., Ed., *Vision in Fishes: Ecological Adaptations*, Plenum Press, New York, 1980, p. 431.
3. Reimchen, T.E., Loss of nuptial colour in threespine sticklebacks (*Gasterosteus aculeatus*), *Evolution*, 43, 450, 1989.

4. McDonald, C.G. and Hawryshyn, C.W., Intraspecific variation of spectral sensitivity in threespine stickleback (*Gasterosteus aculeatus*) from different photic regimes, *J. Comp. Physiol. A,* 176, 255, 1995.
5. McLennan, D.A. and McPhail, J.D., Experimental investigations of the evolutionary significance of sexually dimorphic nuptial coloration in *Gasterosteus aculeatus* (L.): temporal changes in the structure of the male mosaic signal, *Can. J. Zool.*, 67, 1767, 1989.
6. Grether, G.F., Kolluru, G.R., and Nersissian, K., Individual colour patches as multicomponent signals, *Biol. Rev.*, 79, 583, 2004.
7. Lozano, G.A., Carotenoids, immunity, and sexual selection: comparing apples and oranges?, *Am. Nat.*, 158, 200, 2001.
8. Blount, J.D., Møller, A.P., and Houston, D.C., Antioxidants, showy males and sperm quality, *Ecol. Lett.*, 4, 393, 2001.
9. Alonso-Alvarez, C. et al., An experimental test of the dose-dependent effect of carotenoids and immune activation on sexual signals and antioxidant activity, *Am. Nat.*, 164, 651, 2004.
10. Kodric-Brown, A. and Brown, J.H., Truth in advertising: the kinds of traits favored by sexual selection, *Am. Nat.*, 124, 309, 1984.
11. Endler, J.A., Natural selection on color patterns in *Poecilia reticulata*, *Evolution*, 34, 76, 1980.
12. Hynes, H.B.N., The food of fresh-water sticklebacks (*Gasterosteus aculeatus* and *Pygosteus pungitius*), with a review of methods used in studies of the food of fishes, *J. Anim. Ecol.*, 19, 36, 1950.
13. Worgan, J.P. and FitzGerald, G.J., Habitat segregation in a salt marsh among adult sticklebacks (Gasterosteidae), *Environ. Biol. Fish*, 6, 105, 1981.
14. Worgan, J.P. and FitzGerald, G.J., Diel activity and diet of three sympatric sticklebacks in tidal salt marsh pools, *Can. J. Zool.*, 59, 2375, 1981.
15. Ward, G. and FitzGerald, G.J., Fish predation on the macrobenthos of tidal salt-marsh pools, *Can. J. Zool.*, 61, 1358, 1983.
16. Walsh, G. and FitzGerald, G.J., Resource utilization and coexistence of three species of sticklebacks (Gasterosteidae) in tidal salt-marsh pools, *J. Fish Biol.*, 25, 405, 1984.
17. Ukegbu, A.A. and Huntingford, F.A., Age structure and growth in Scottish populations of the three-spined stickleback, *Scott. Nat.*, 101, 19, 1989.
18. Ibrahim, A.A. and Huntingford, F.A., Laboratory and field studies on diet choice in three-spined sticklebacks, *Gasterosteus aculeatus* L., in relation to profitability and visual features of prey, *J. Fish Biol.*, 34, 245, 1988.
19. Pont, D., Crivelli, A.J., and Guillot, F., The impact of three-spined sticklebacks on the zooplankton of a previously fish-free pool, *Freshwater Biol.*, 26, 149, 1991.
20. Sanchez-Gonzales, S., Ruiz-Campos, G., and Contreras-Balderas, S., Feeding ecology and habitat of the threespine stickleback, *Gasterosteus aculeatus microcephalus*, in a remnant population of north-western Baja California, Mexico, *Ecol. Freshwater. Fish*, 10, 191, 2001.
21. Grether, G.F., Hudon, J., and Millie, D.F., Carotenoid limitation of sexual coloration along an environmental gradient in guppies, *Proc. R. Soc. Lond. B*, 266, 1317, 1999.
22. Bakker, T.C.M., Positive genetic correlation between female preference and preferred male ornament in sticklebacks, *Nature*, 363, 255, 1993.
23. Olson, V.A. and Owens, I.P.F., Costly sexual signals: are carotenoids rare, risky or required?, *Trends Ecol. Evol.*, 13, 510, 1998.
24. Brush, A.H. and Reisman, H.M., Carotenoid pigments in three-spined stickleback *Gasterosteus aculeatus*, *Comp. Biochem. Physiol.*, 14, 121, 1965.

25. Czeczuga, B., Carotenoids in fish. XXVI. *Pungitus pungitus* (L.) and *Gasteosteus aculeatus* L. (Gasterosteidae), *Hydrobiologia*, 74, 7, 1980.
26. Hudon, J., Grether, G.F., and Millie, D.F., Marginal differentiation between sexual and general carotenoid pigmentation of guppies (*Poecilia reticulata*) and a possible visual explanation, *Physiol. Biochem. Zool.*, 76, 776, 2003.
27. Wedekind, C. et al., Different carotenoids and potential information content of red coloration of male three-spined stickleback, *J. Chem. Ecol.*, 24, 787, 1998.
28. Barber, I. et al., Carotenoid-based sexual coloration and body condition in nesting male sticklebacks, *J. Fish Biol.*, 57, 777, 2000.
29. McPhail, J.D., Predation and evolution of a stickleback (*Gasterosteus*), *J. Fish. Res. Board Can.*, 26, 3183, 1969.
30. Semler, D.E., Some aspects of adaptation in a polymorphism for breeding colours in threespine stickleback (*Gasterosteus aculeatus*), *J. Zool.*, 165, 291, 1971.
31. Moodie, G.E.E., Morphology, life-history, and ecology of an unusual stickleback (*Gasterosteus aculeatus*) in Queen-Charlotte Islands, Canada, *Can. J. Zool.*, 50, 721, 1972.
32. Moodie, G.E.E., Predation, natural selection and adaptation in an unusual threespine stickleback, *Heredity*, 28, 155, 1972.
33. Bell, M.A., Melanism in a high elevation population of *Gasterosteus aculeatus*, *Copeia*, 1982, 829, 1982.
34. Hagen, D.W. and Moodie, G.E.E., Polymorphism for breeding colors in *Gasterosteus aculeatus*. I. Their genetics and geographic distribution, *Evolution*, 33, 641, 1979.
35. von Hippel, F.A., Black male bellies and red female throats: color changes with breeding status in a threespine stickleback, *Environ. Biol. Fish*, 55, 237, 1999.
36. Scott, R.J., Sensory drive and nuptial colour loss in the three-spined stickleback, *J. Fish Biol.*, 59, 1520, 2001.
37. Jamieson, I.G., Blouw, D.M., and Colgan, P.W., Field observations on the reproductive biology of a newly discovered stickleback (*Gasterosteus*), *Can. J. Zool.*, 70, 1057, 1992.
38. Losey, G.S. et al., The UV visual world of fishes: a review, *J. Fish Biol.*, 54, 921, 1999.
39. Ingolf, P.R., Modarressie, R., and Bakker, T.C.M., Male three-spined sticklebacks reflect in ultraviolet light, *Behaviour*, 141, 1531, 2004.
40. Fuller, R.C., Patterns in male breeding behaviors in the bluefin killifish, *Lucania goodei*: a field study (Cyprinodontiformes: Fundulidae), *Copeia*, 2001, 823, 2001.
41. Boulcott, P. and Braithwaite, V., Ultraviolet light and visual behaviour in the three-spined stickleback, *Gasterosteus aculeatus*, *Physiol. Biochem. Zool.*, 78, 736, 2005.
42. Boulcott, P.D., Walton, K., and Braithwaite, V.A., The role of ultraviolet wavelengths in the mate-choice decisions of female three-spined sticklebacks, *J. Exp. Biol.*, 208, 1453, 2005.
43. Rick, I.P., Modarressie, R., and Bakker, T.C.M., UV wavelengths affect female mate choice in three-spined sticklebacks, *Anim. Behav.*, 71, 307, 2006.
44. Modarressie, R., Rick, I.P., and Bakker, T.C.M., UV matters in shoaling decisions, *Proc. R. Soc. Lond. B*, 273, 849, 2006.
45. McLennan, D.A., Integrating phylogenetic and experimental analyses: the evolution of male and female nuptial coloration in the stickleback fishes (Gasterosteidae), *Syst. Biol.*, 45, 261, 1996.
46. Rowland, W.J., Baube, C.L., and Horan, T.T., Signalling of sexual receptivity by pigmentation pattern in female sticklebacks, *Anim. Behav.*, 42, 243, 1991.
47. McKinnon, J.S. et al., Female red throat coloration in two populations of threespine stickleback, *Behaviour*, 137, 947, 2000.

48. Nordeide, J.T., Do male sticklebacks prefer females with red ornamentation?, *Can. J. Zool.*, 80, 1344, 2002.
49. Nordeide, J.T., Rudolfsen, G., and Egeland, E.S., Ornaments or offspring? Female sticklebacks (*Gasterosteus aculeatus* L.) trade off carotenoids between spines and eggs, *J. Evol. Biol.*, 19, 431, 2006.
50. Rowe, M.P. et al., Optimal mechanisms for finding and selecting mates: how threespine stickleback (*Gasterosteus aculeatus*) should encode male throat colors, *J. Comp. Physiol. A*, 190, 241, 2004.
51. Rush, V.N. et al., Reflectance spectra from free-swimming sticklebacks (*Gasterosteus*): social context and eye-jaw contrast, *Behaviour*, 140, 1003, 2003.
52. Cronly-Dillon, J. and Sharma, S.C., Effect of season and sex on the photopic spectral sensitivity of the three-spined stickleback, *J. Exp. Biol.*, 49, 679, 1968.
53. Beaudet, L., Browman, H.I., and Hawryshyn, C.W., Optic nerve response and retinal structure in rainbow trout of different sizes, *Vision Res.*, 33, 1739, 1993.
54. McDonald, C.G., Reimchen, T.E., and Hawryshyn, C.W., Nuptial colour loss and signal masking in *Gasterosteus*: an analysis using video imaging, *Behaviour*, 132, 963, 1995.
55. Rowland, W.J., Proximate determinants of stickleback behaviour: an evolutionary perspective, in *Evolutionary Biology of the Threespine Stickleback*, Bell, M.A. and Foster, S.A., Eds., Oxford University Press, Oxford, 1994, p. 297.
56. Kynard, B.E., Breeding behavior of a lacustrine population of threespine sticklebacks (*Gasterosteus aculeatus* L.), *Behaviour*, 67, 178, 1978.
57. Scott, R.J. and Foster, S.A., Field data do not support a textbook example of convergent character displacement, *Proc. R. Soc. Lond. B*, 267, 607, 2000.
58. Ryan, M.J., Signals, species, and sexual selection, *Am. Sci.*, 78, 46, 1990.
59. Ryan, M.J., Sexual selection, sensory systems and sensory exploitation, Vol. 7, *Oxford Surveys in Evolutionary Biology*, Oxford University Press, Oxford, 1990, p. 157.
60. Smith, C. et al., A receiver bias in the origin of three-spined stickleback mate choice, *Proc. R. Soc. Lond. B*, 271, 949, 2004.
61. Baube, C.L., Rowland, W.J., and Fowler, J.B., The mechanisms of colour-based mate choice in female threespine sticklebacks: hue, contrast and configurational cues, *Behaviour*, 132, 979, 1995.
62. Endler, J.A. and McLellan, T., The processes of evolution: towards a newer synthesis, *Annu. Rev. Ecol. Syst.*, 19, 395, 1988.
63. Ryan, M.J., Cocroft, R.B., and Wilczynski, W., The role of environmental selection in intraspecific divergence of mate recognition signals in the cricket frog, *Acris crepitans*, *Evolution*, 44, 1869, 1990.
64. Ryan, M.J., Rand, A.S., and Weigt, L.A., Allozyme and advertisement call variation in the túngara frog, *Physalaemus pustulosus*, *Evolution*, 50, 2435, 1996.
65. Boughman, J.W., Divergent sexual selection enhances reproductive isolation in sticklebacks, *Nature*, 411, 944, 2001.
66. McLennan, D.A. and McPhail, J.D., Experimental investigations of the evolutionary significance of sexually dimorphic nuptial colouration in *Gasterosteus aculeatus* (L.): the relationships between male colour and female behaviour, *Can. J. Zool.*, 68, 482, 1990.
67. Milinski, M. and Bakker, T.C.M., Female sticklebacks use male coloration in mate choice and hence avoid parasitized males, *Nature*, 344, 330, 1990.
68. Bakker, T.C.M. and Milinski, M., Sequential female choice and the previous male effect in sticklebacks, *Behav. Ecol. Sociobiol.*, 29, 205, 1991.

69. Bakker, T.C.M. and Milinski, M., The advantages of being red: sexual selection in the stickleback, *Mar. Behav. Physiol.*, 23, 287, 1993.
70. Bakker, T.C.M. and Mundwiler, B., Female mate choice and male red coloration in a natural three-spined stickleback (*Gasterosteus aculeatus*) population, *Behav. Ecol.*, 5, 74, 1994.
71. McKinnon, J.S., Video mate preferences of female three-spined sticklebacks from populations with divergent male coloration, *Anim. Behav.*, 50, 1645, 1995.
72. Rowland, W.J. et al., Video playback experiments on stickleback mate choice: female motivation and attentiveness to male colour cues, *Anim. Behav.*, 49, 1559, 1995.
73. Rowland, W.J., Bolyard, K.J., and Halpern, A.D., The dual effect of stickleback nuptial coloration on rivals: manipulation of a graded signal using video playback, *Anim. Behav.*, 50, 267, 1995.
74. ter Pelkwijk, J.J. and Tinbergen, N., Eine Reizbiologische analyse einiger Verhaltensweisen von *Gasterosteus aculeatus* L., *Z. Tierpsychol.*, 1, 193, 1937.
75. Tinbergen, N., Social releasers and the experimental method for their study. *Wilson Bull.*, 60, 6, 1948.
76. Tinbergen, N., *The Study of Instinct*, Oxford University Press, Oxford, 1951.
77. Rowland, W.J., The effects of body size, aggression and nuptial coloration on competition for territories in male three spine stickleback, *Gasterosteus aculeatus*, *Anim. Behav.*, 37, 282, 1989.
78. Rowland, W.J., The relationships among nuptial coloration, aggression, and courtship of male three-spined sticklebacks, *Gasterosteus aculeatus*, *Can. J. Zool.*, 62, 999, 1984.
79. Bakker, T.C.M., Aggressiveness in sticklebacks (*Gasterosteus aculeatus* L.): a behaviour-genetic study, *Behaviour*, 98, 1, 1986.
80. Frischknecht, M., The breeding coloration of male three-spined sticklebacks (*Gasterosteus aculeatus*) as an indicator of energy investment in vigor, *Evol. Ecol.*, 7, 439, 1993.
81. Candolin, U., Reproduction under predation risk and the trade-off between current and future reproduction in the threespine stickleback, *Proc. R. Soc. Lond. B*, 265, 1171, 1998.
82. Candolin, U., The relationship between signal quality and physical condition: is sexual signalling honest in the three-spined stickleback?, *Anim. Behav.*, 58, 1261, 1999.
83. Barber, I. et al., Indirect fitness consequences of mate choice in sticklebacks: offspring of brighter males grow slowly but resist parasitic infections, *Proc. R. Soc. Lond. B*, 268, 71, 2001.
84. FitzGerald, G.J., Fournier, M., and Morissette, J., Sexual selection in an anadromous population of three-spine sticklebacks — no role for parasites, *Evol. Ecol.*, 8, 348, 1994.
85. Folstad, I. et al., Sexually selected colour in male sticklebacks: a signal of both parasite exposure and parasite resistance, *Oikos*, 69, 511, 1994.
86. Tierney, J.F., Huntingford, F.A., and Crompton, D.W.T., Body condition and reproductive status in sticklebacks exposed to a single wave of *Schistocephalus solidus* infection, *J. Fish Biol.*, 49, 483, 1996.
87. Candolin, U., Male-male competition ensures honest signaling of male parental ability in the three-spined stickleback (*Gasterosteus aculeatus*), *Behav. Ecol. Sociobiol.*, 49, 57, 2000.
88. Cubillos, E.R. and Guderley, H.E., Analysis of the factors related with mate choice and reproductive success in male three-spined sticklebacks, *J. Fish Biol.*, 56, 1201, 2000.

89. Baube, C.L., Manipulations of signalling environment affect male competitive success in three-spined sticklebacks, *Anim. Behav.*, 53, 819, 1997.
90. Östlund Nilsson, S. and Nilsson, G.E., Free choice by female sticklebacks: lack of preference for male dominance traits, *Can. J. Zool.*, 78, 1251, 2000.
91. Kraak, S.B.M., Bakker, T.C.M., and Mundwiler, B., Sexual selection in sticklebacks in the field: correlates of reproductive, mating, and paternal success, *Behav. Ecol.*, 10, 696, 1999.
92. McLennan, D.A. and Brooks, D.R., Parasites and sexual selection: a macroevolutionary perspective, *Q. Rev. Biol.*, 66, 255, 1991.
93. Candolin, U. and Voigt, H.R., No effect of a parasite on reproduction in stickleback males: a laboratory artefact?, *Parasitology*, 122, 457, 2001.
94. Rowland, W.J., Sexual arousal increases mate selectivity in the stickleback, *Behaviour*, 141, 1371, 2004.
95. Wootton, R.J., Changes in courtship behaviour of female three-spined sticklebacks between spawnings, *Anim. Behav.*, 22, 850, 1974.
96. Wootton, R.J., *The Biology of the Sticklebacks*, Academic Press, London; New York, 1976, p. 387.
97. Lam, T.J. et al., Overripe eggs and postovulatory corpora lutea in the threespine stickleback, *Gasterosteus aculeatus* L., form *trachurus*, *Can. J. Zool.*, 56, 2029, 1978.
98. Guderley, H.E., Physiological ecology and evolution of the threespine stickleback, in *Evolutionary Biology of the Threespine Stickleback*, Bell, M.A. and Foster, S.A., Eds., Oxford University Press, Oxford, 1994.
99. Milinski, M. and Bakker, T.C.M., Costs influence sequential mate choice in sticklebacks, *Gasterosteus aculeatus*, *Proc. R. Soc. Lond. B*, 250, 229, 1992.
100. Luttbeg, B. et al., State-dependent mate assessment and mate selection behavior in female threespine sticklebacks (*Gasterosteus aculeatus*, Gasterosteiformes: Gasterosteidae), *Ethology*, 107, 545, 2001.
101. Braithwaite, V.A. and Barber, I., Limitations to colour-based sexual preferences in three-spined sticklebacks (*Gasterosteus aculeatus*), *Behav. Ecol. Sociobiol.*, 47, 413, 2000.
102. Rowland, W.J., Do female stickleback care about male courtship vigour? Manipulation of display tempo using video playback, *Behaviour*, 132, 951, 1995.
103. Sevenster-Bol, A.C.A., On the causation of drive reduction after a consummatory act in *Gasterosteus aculeatus* L., *Arch. Néerl. Zool.*, 15, 175, 1962..
104. Symons, P.E.K., Analysis of spine-raising in the male three-spined stickleback, *Behaviour*, 26, 1, 1966.
105. Kunzler, R. and Bakker, T.C.M., Female preferences for single and combined traits in computer animated stickleback males, *Behav. Ecol.*, 12, 681, 2001.
106. Rowland, W.J., Mate choice and supernormality effect in female sticklebacks (*Gasterosteus aculeatus*), *Behav. Ecol. Sociobiol.*, 24, 433, 1989.
107. Dufresne, F., FitzGerald, G.J., and Lachance, S., Age and size-related differences in reproductive success and reproductive cost in threespine stickleback, *Behav. Ecol.*, 1, 140, 1990.
108. Largiardèr, C.R., Fries, V., and Bakker, T.C.M., Genetic analysis of sneaking and egg-thievery in a natural population of the three-spined stickleback (*Gasterosteus aculeatus* L.), *Heredity*, 86, 459, 2001.
109. Dallas, J., On rapid change of colour in *Gasterosteus*, *Ann. Mag. Nat. Hist.*, 6, 489, 1895.
110. Reisman, H.M., Effects of social stimuli on the secondary sex characters of male three-spined sticklebacks, *Gasterosteus aculeatus*, *Copeia*, 1968, 816, 1968.

111. Candolin, U., Male-male competition facilitates female choice in sticklebacks, *Proc. R. Soc. Lond. B*, 266, 785, 1999.
112. Candolin, U., Increased signalling effort when survival prospects decrease: male-male competition ensures honesty, *Anim. Behav.*, 60, 417, 2000.
113. Rohwer, S., The social significance of avian winter plumage variability, *Evolution*, 29, 593, 1975.
114. Johnstone, R.A. and Norris, K., Badges of status and the cost of aggression, *Behav. Ecol. Sociobiol.*, 32, 127, 1993.
115. Kraak, S.B.M., Bakker, T.C.M., and Mundwiler, B., Correlates of the duration of the egg collecting phase in the three-spined stickleback, *J. Fish Biol.*, 54, 1038, 1999.
116. Sargent, R.C. and Gebler, J.B., Effects of nest concealment on hatching success, reproductive success, and paternal behaviour of the threespine stickleback, *Behav. Ecol. Sociobiol.*, 7, 137, 1980.
117. Mori, S., The breeding system of the three-spined stickleback, *Gasterosteus aculeatus* (forma leiura) with reference to spatial and temporal patterns of nesting activity, *Behaviour*, 126, 97, 1993.
118. Candolin, U. and Voigt, H.R., Predator-induced nest site preference: safe nests allow courtship in sticklebacks, *Anim. Behav.*, 56, 1205, 1998.
119. Dzieweczynski, T.L. and Rowland, W.J., Behind closed doors: use of visual cover by courting male three-spined stickleback, *Gasterosteus aculeatus*, *Anim. Behav.*, 68, 465, 2004.
120. Candolin, U., Predation risk affects courtship and attractiveness of competing threespine stickleback males, *Behav. Ecol. Sociobiol.*, 41, 81, 1997.
121. Candolin, U., Effects of algae cover on egg acquisition in male three-spined sticklebacks, *Behaviour*, 141, 1389, 2004.
122. Candolin, U. and Voigt, H.R., Correlation between male size and territory quality: a consequence of male competition or predation risk?, *Oikos*, 95, 225, 2001.
123. Candolin, U. and Salesto, T., Effects of increased vegetation cover on nesting behavior of sticklebacks (*Gasterosteus aculeatus*), *Behav. Ecol. Sociobiol.*, 59, 689, 2006.
124. Kraak, S.B.M., Bakker, T.C.M., and Hocevar, S., Stickleback males, especially large and red ones, are more likely to nest concealed in macrophytes, *Behaviour*, 137, 907, 2000.
125. Barber, I., Nairn, D., and Huntingford, F.A., Nests as ornaments: revealing construction by male sticklebacks, *Behav. Ecol.*, 12, 390, 2001.
126. Guderley, H. and Guevara, R.C., Physiological correlates of paternal care in male threespine stickleback, *Ecoscience*, 5, 438, 1998.
127. Blais, J., Rico, C., and Bernatchez, L., Nonlinear effects of female mate choice in wild threespine sticklebacks, *Evolution*, 58, 2498–2510, 2004.
128. Christiansen, R., Lie, Ø., and Torrissen, O., Effect of astaxanthin and vitamin A on growth and survival of Atlantic salmon fry, *Salmo salar* L., *Aquat. Fish. Manage.*, 25, 903, 1995.
129. Schiedt, K. et al., Absorption, retention and metabolic transformations of carotenoids in rainbow trout, salmon and chicken, *Pure Appl. Chem.*, 57, 685, 1985.
130. Al-Khalifa, A.S. and Simpson, K.L., Metabolism of astaxanthin the rainbow trout (*Salmo gairdneri*), *Comp. Biochem. Physiol. B*, 91, 563, 1988.
131. Katsuyama, M. and Matsuno, T., Carotenoid and vitamin A, and metabolism of carotenoids, β-carotene, canthaxanthin, astaxanthin, zeaxanthin, lutein and tunaxanthin in tilapia, *Tilapia nilotica*, *Comp. Biochem. Physiol. B*, 90, 131, 1988.

132. Rønnestad, I. et al., Alternate sources and dynamics of vitamin A and its incorporation into the eyes during the early endotrophic and exotrophic larval stages of Atlantic halibut *Hippoglossus hippoglossus* L.), *Comp. Biochem. Physiol. A*, 119, 787, 1998.
133. Rowland, W.J., Habituation and development of response specificity to a sign stimulus: male preference for female courtship posture in stickleback, *Anim. Behav.*, 60, 63, 2000.
134. Jenkins, J.R. and Rowland, W.J., Stimulus-specific and response-specific habituation in courting stickleback: developmental and functional considerations, *Behaviour*, 137, 933, 2000.
135. Rowland, W.J., Mate choice by male sticklebacks, *Gasterosteus aculeatus*, *Anim. Behav.*, 30, 1093, 1982.
136. Rowland, W.J., The ethological basis of mate choice in male threespine sticklebacks, *Gasterosteus aculeatus*, *Anim. Behav.*, 38, 112, 1989.
137. Sargent, R.C., Gross, M.R., and van den Berghe, E.P., Male mate choice in fishes, *Anim. Behav.*, 34, 545, 1986.
138. Bakker, T.C.M. and Rowland, W.J., Male mating preference in sticklebacks: effects of repeated testing and own attractiveness, *Behaviour*, 132, 935, 1995.
139. Rowland, W.J. and Sevenster, P., Sign stimuli in the threespine stickleback (*Gasterosteus aculeatus*): a re-examination and extension of some classic experiments, *Behaviour*, 93, 241, 1985.
140. Zbinden, M., Largiadèr, C.R., and Bakker, T.C.M., Sperm allocation in the three-spined stickleback, *J. Fish Biol.*, 59, 1287, 2001.
141. Zbinden, M. et al., Courting virtual rivals increase ejaculate size in sticklebacks (*Gasterosteus aculeatus*), *Behav. Ecol. Sociobiol.*, 54, 205, 2003.
142. Zbinden, M., Largiadèr, C.R., and Bakker, T.C.M., Body size of virtual rivals affects ejaculate size in sticklebacks, *Behav. Ecol.*, 15, 137, 2004.
143. Perrin, N., Signalling, mating success and paternal investment in sticklebacks (*Gasterosteus aculeatus*): a theoretical model, *Behaviour*, 132, 1037, 1995.
144. Fletcher, D.A. and Wootton, R.J., A hierarchical response to differences in ration size in the reproductive performance of female three-spined sticklebacks, *J. Fish Biol.*, 46, 657, 1995.
145. Kraak, S.B.M. and Bakker, T.C.M., Mutual mate choice in sticklebacks: attractive males choose big females, which lay big eggs, *Anim. Behav.*, 56, 859, 1998.
146. Bisazza, A. and Pilastro, A., Small male mating advantage and reversed sexual dimorphism in poeciliid fishes, *J. Fish Biol.*, 50, 397, 1997.
147. Albert, A.Y.K. and Schluter, D., Reproductive character displacement of male stickleback mate preference: reinforcement or direct selection?, *Evolution*, 58, 1099, 2004.
148. Foster, S.A., Evolution of the reproductive behaviour of threespine stickleback, in *Evolutionary Biology of the Threespine Stickleback*, Bell, M.A. and Foster, S.A., Eds., Oxford University Press, Oxford, 1994, p. 381.
149. Foster, S.A. and Baker, J.A., Evolutionary interplay between ecology, morphology and reproductive behaviour in threespine stickleback, *Gasterosteus aculeatus*, *Environ. Biol. Fish*, 44, 213, 1995.
150. McKinnon, J.S. et al., Evidence for ecology's role in speciation, *Nature*, 429, 294, 2004.
151. Meesters, A., Über die Organisation des Gesichtsfeldes der Fische, *Z. Tierpsychol.*, 4, 84, 1940.
152. McLennan, D.A., Changes in female colour across the ovulatory cycle in the brook stickleback, *Culaea inconstans* (Kirtland), *Can. J. Zool.*, 72, 144, 1994.

153. McLennan, D.A., Male mate choice based upon female nuptial coloration in the brook stickleback, *Culaea inconstans* (Kirtland), *Anim. Behav.*, 50, 213, 1995.
154. Dusenbery, D.B., *Sensory Ecology: How Organisms Acquire and Respond to Information*, W.H. Freeman, New York, 1992.
155. Moore, P. and Atema, J., A model of a temporal filter in chemoreception to extract directional information from a turbulent odour plume, *Biol. Bull.*, 174, 355, 1988.
156. Jumper, G.Y. and Baird, R.C., Location by olfaction: a model and application to the mating problem in the deep-sea hatchetfish *Argyropelecus hemigymnus*, *Am. Nat.*, 138, 141, 1991.
157. Moore, P.A., A model of the role of adaptation and disadaptation in olfactory receptor neurons: implications for the coding of temporal and intensity patterns in odor signals, *Chem. Senses*, 19, 71, 1994.
158. Moore, P.A., Grills, J.L., and Schneider, R.W.S., Habitat-specific signal structure for olfaction: an example from artificial streams, *J. Chem. Ecol.*, 26, 565, 2000.
159. Vickers, N.J., Mechanisms of animal navigation in odor plumes, *Biol. Bull.*, 198, 203–212, 2000.
160. Balkovsky, E. and Shraiman, B.I., Olfactory search at high Reynolds numbers, *Proc. Natl. Acad. Sci. U.S.A.*, 99, 12589, 2002.
161. Carton, A.G. and Montgomery, J.C., Evidence of a rheotactic component in the odour search behaviour of freshwater eels, *J. Fish Biol.*, 62, 501, 2003.
162. Sherman, M.L. and Moore, P.A., Chemical orientation of brown bullheads, *Ameiurus nebulosus*, under different flow conditions, *J. Chem. Ecol.*, 27, 2301, 2001.
163. Endler, J.A., Multiple-trait coevolution and environmental gradients in guppies, *Trends Ecol. Evol.*, 10, 22, 1995.
164. von Frisch, K., Zur psychologie des Fisch-Schwarmes, *Naturwissenschaften*, 26, 601, 1938.
165. Smith, R.J.F., Alarm signals in fishes, *Rev. Fish Biol. Fish.*, 2, 33, 1992.
166. McLennan, D.A., The sexual behaviour of North American freshwater fish, in *Systematics, Historical Ecology and North American Freshwater Fishes*, Mayden, R.L., Ed., Elsevier Science, in press.
167. Brown, G.E. et al., Ostariophysan alarm pheromones: laboratory and field tests of the functional significance of nitrogen oxides, *J. Chem. Ecol.*, 26, 139, 2000.
168. Sorensen, P.W. and Stacey, N.E., Evolution and specialization of fish hormonal pheromones, in *Advances in Chemical Signals in Vertebrates*, Johnston, R.E., Muller-Schwarze, D., and Sorensen, P.W., Eds., Kluwer Academic/Plenum Publishers, New York, 1999, p. 15.
169. Colombo, L. et al., Pheromones in teleost fish, in *Proc. Int. Symp. Reprod. Physiol. Fish*, Richter, C.J.J. and Goos, H.J.Th., Eds., Wageningen, The Netherlands, 1982, p. 84.
170. Friedrich, R.W. and Laurent, G., Dynamic optimization of odour representations by slow temporal patterning of mitral cell activity, *Science*, 291, 889, 2001.
171. Laberge, F. and Hara, T.J., Electrophysiological demonstration of independent olfactory receptor types and associated neuronal responses in the trout olfactory bulb, *Comp. Biochem. Physiol. A*, 137, 397, 2004.
172. Nieuwenhuys, R., The structure of the telencephalon of the teleost *Gasterosteus aculeatus*. Kon. Akad. Wetensch. (Amsterdam), *Proc. Ser. C*, 62, 341, 1959.
173. Weltzien, F. et al., Does the lateral bundle of the medial olfactory tract mediate reproductive behaviour in male Crucian carp?, *Chem. Senses*, 28, 293, 2003.
174. Laberge, F. and Hara, T.J., Neurobiology of fish olfaction: a review, *Brain Res. Rev.*, 36, 46, 2001.

175. Honkanen, T. and Ekström, P., An immunocytochemical study of the olfactory projections in the three-spined stickleback, *Gasterosteus aculeatus*, *J. Comp. Neurobiol.*, 292, 65, 1990.
176. Laberge, F. and Hara, T.J., Non-oscillatory discharges of an F-prostaglandin responsive neuron population in the olfactory bulb-telencephalon transition area in lake whitefish, *Neuroscience*, 116, 1089, 2003.
177. Solger, B., Notiz über die Nebenhöhle des Geruchsorgans von *Gasterosteus aculeatus* L., *Z. Wiss. Zool.*, 57, 186, 1894.
178. Theissen, B., Functional morphology of the olfactory organ in *Spinachia spinachia* (L.), (Teleostei, Gasterosteidae), *Acta Zool.*, 63, 247, 1982.
179. Teichmann, H., Vergleichende Untersuchungen an der Nase der Fische, *Z. Morphol. Oekol. Tiere*, 43, 171, 1959.
180. Bardach, J.E. and Todd, J.H., Chemical communication in fish, in *Communication by Chemical Signals*, Johnston, J.W., Jr., Moulton, D.G., and Turk, A., Eds., Appelton-Century-Crofts, New York, 1970, p. 205.
181. Segaar, J., Ethophysiological experiments with male *Gasterosteus aculeatus*, in *Structure and Function of the Cerebral Cortex*, Tower, D.B. and Schadé, J.P., Eds., Elsevier, Amsterdam, 1960, p. 301.
182. Segaar, J., Telencephalon and behaviour in *Gasterosteus aculeatus* males, *Behaviour*, 18, 256, 1961.
183. Segaar, J. and Nieuwenhuys, R., New etho-physiological experiments with male *Gasterosteus aculeatus*, with anatomical comment, *Anim. Behav.*, 11, 331, 1963.
184. Segaar, J. et al., Influence of chemical receptivity on reproductive behaviour of the male three-spined stickleback (*Gasterosteus aculeatus* L.): an ethological analysis of cranial nerve functions regarding nest fanning activity and zigzag dance, *Behaviour*, 86, 100, 1983.
185. Segaar, J. and de Bruin, J.P.C., Behavioural consequences of chemoreception during reproductive cycles of the three-spined stickleback (*Gasterosteus aculeatus* L.), *Behaviour*, 93, 139, 1985.
186. Mathis, A. and Smith, R.J.F., Intraspecific and cross-superorder responses to chemical alarm signals by brook stickleback, *Ecology*, 74, 2395, 1993.
187. Wisenden, B.D. et al., The role of experience in risk assessment: avoidance of areas chemically labelled with fathead minnow alarm pheromone by conspecifics and heterospecifics, *Ecoscience*, 2, 116, 1995.
188. Wisenden, B.D., Chivers, D.P., and Smith, R.J.F., Risk-sensitive habitat use by brook stickleback (*Culaea inconstans*) in areas associated with minnow alarm pheromone, *J. Chem. Ecol.*, 20, 2975, 1994.
189. Chivers, D.P. and Smith, R.J.F., Intra-and interspecific avoidance of areas marked with skin extract from brook sticklebacks (*Culaea inconstans*) in a natural habitat, *J. Chem. Ecol.*, 20, 1517, 1994.
190. Gelowitz, C.M., Mathis, A., and Smith, R.J.F., Chemosensory recognition of northern pike (*Esox lucius*) by brook sticklebacks (*Culaea inconstans*): population differences and the influence of predator diet, *Behaviour*, 127, 105, 1993.
191. Chivers, D.P., Brown, G.E., and Smith, R.J.F., Acquired recognition of chemical stimuli from pike, *Esox lucius*, by brook sticklebacks, *Culaea inconstans* (Osteichthyes, Gasterosteidae), *Ethology*, 99, 234, 1995.
192. Schutz, F., Vergleichende Untersuchungen über die Schreckreaktion bei Fischen und deren Verbreitung, *Z. Verg. Physiol.*, 38, 84, 1956.

193. Hoogland, R., Morris, D., and Tinbergen, N., The spines of sticklebacks (*Gasterosteus* and *Pygosteus*) as a means of defence against predators (*Perca* and *Esox*), *Behaviour*, 10, 205, 1956.
194. Brown, G.E. and Godin, J.-G., Anti-predator responses to conspecific and heterospecific skin extracts by threespine sticklebacks: alarm pheromones revisited, *Behaviour*, 134, 1123, 1997.
195. Mirza, R.S., Scott, J.J., and Chivers, D.P., Differential responses of male and female red swordtails to chemical alarm cues, *J. Fish Biol.*, 59, 716, 2001.
196. Pollock, M.S. and Chivers, D.P., The effects of density on the learned recognition of heterospecific alarm cues, *Ethology*, 110, 341, 2004.
197. Pollock, M.S. et al., Fathead minnows, *Pimephales promelas*, learn to recognise chemical alarm cues of introduced brook stickleback, *Culaea inconstans*, *Environ. Biol. Fish.*, 66, 313, 2003.
198. Smith, R.J.F., Testosterone eliminates alarm substance in male fathead minnows, *Can. J. Zool.*, 51, 875, 1973.
199. Smith, R.J.F., Male fathead minnows (*Pimephales promelas* Rafinesque) retain their fright reaction to alarm substance during the breeding season, *Can. J. Zool.*, 54, 2230, 1976.
200. Jones, H.M. and Paszkowski, C.A., Effects of exposure to predatory cues on territorial behaviour of male fathead minnows, *Environ. Biol. Fish.*, 49, 97, 1997.
201. Hartman, E.J. and Abrahams, M.V., Sensory compensation and the detection of predators: the interaction between chemical and visual information, *Proc. R. Soc. Lond. B*, 267, 571, 2000.
202. FitzGerald, G.J. and Morissette, J., Kin recognition and choice of shoal mates by threespine sticklebacks, *Ethol. Ecol. Evol.*, 4, 273, 1992.
203. Steck, N., Wedekind, C., and Milinski, M., No sibling odour preference in juvenile three-spined sticklebacks, *Behav. Ecol.*, 10, 493–497, 1999.
204. Van Havre, N. and Fitzgerald, G.J., Shoaling and kin recognition in the threespine stickleback (*Gasterosteus aculeatus* L.), *Biol. Behav.*, 13, 190, 1988.
205. Ward, A.J.W., Hart, P.J.B., and Krause, J., The effects of habitat- and diet-based cues on association preferences in three-spined sticklebacks, *Behav. Ecol.*, 15, 925, 2004.
206. Rowe, C., Receiver psychology and the evolution of multicomponent signals, *Anim. Behav.*, 58, 921, 1999.
207. Shannon, C.E. and Weaver, W., *The Mathematical Theory of Communication*, University of Illinois Press, Champaign, IL, 1949.
208. McLennan, D.A., The importance of olfactory signals in the gasterosteid mating system: sticklebacks go multimodal, *Biol. J. Linn. Soc.*, 80, 555, 2003.
209. McLennan, D.A., Male brook sticklebacks' (*Culaea inconstans*) response to olfactory cues, *Behaviour*, 141, 1411, 2004.
210. Rowe, C. and Guilford, T., The evolution of multimodal warning displays, *Evol. Ecol.*, 13, 655, 1999.
211. Real, L., Search theory and mate choice. I. Models of single-sex discrimination, *Am. Nat.*, 136, 376, 1990.
212. Fujita, I. et al., The olfactory system, not the terminal nerve, functions as the primary chemosensory pathway mediating responses to sex pheromones in male goldfish, *Brain Behav. Evol.*, 38, 313, 1991.
213. Flynn, K.M., Schreibman, M.P., and Magliulo-Cepriano, L., Developmental changes in NMDA receptor expression in the platyfish brain, *Brain Res.*, 771, 142, 1997.
214. Stell, W.K. and Walker, S.E.B., Functional-anatomical studies on the terminal nerve projection to the retina of bony fishes, *Ann. N.Y. Acad. Sci.*, 519, 80, 1987.

215. von Bartheld, C.S., Central connections of the terminal nerve in ray-finned fishes, *Ann. N. Y. Acad. Sci.*, 519, 392, 1987.
216. Laurent, G., A systems perspective on early olfactory coding, *Science*, 286, 723, 1999.
217. van Iersel, J.J.A., An analysis of the parental behaviour of the male three-spined stickleback (*Gasterosteus aculeatus* L.), *Behaviour*, Suppl. 3, 1, 1953.
218. Csányi, V., Csizmadia, G., and Miklosi, A., Long-term memory and recognition of another species in the paradise fish, *Anim. Behav.*, 37, 908, 1989.
219. Chivers, D.P. and Smith, R.J.F., The role of experience and chemical alarm signalling in predator recognition by fathead minnows (*Pimephales promelas*), *J. Fish. Biol.*, 44, 273, 1994.
220. Hosler, J.S. and Smith, B.H., Blocking and the detection of odour components in blends, *J. Exp. Biol.*, 203, 2797, 2000.
221. Bolhuis, J.J. and Horn, G., Generalisation of learned preferences in filial imprinting, *Anim. Behav.*, 44, 185, 1992.
222. McLennan, D.A., Changes in response to olfactory cues across the ovulatory cycle in brook sticklebacks, *Culaea inconstans*, *Anim. Behav.*, 69, 181, 2005.
223. McLennan, D.A., Changes in female breeding behaviour across the ovulatory cycle in the brook stickleback, *Culaea inconstans* (Kirtland), *Behaviour*, 126, 191, 1993.
224. Sarkar, S. and Subhedar, N., Seasonal changes in endorphin-like immunoreactivity in the olfactory system of the female catfish, *Clarias batrachus* (Linn.), *Gen. Comp. Endocrinol.*, 123, 127, 2001.
225. Gangestad, S.W., Bennett, K.L., and Thornhill, R., A latent variable model of developmental instability in relation to men's sexual behaviour, *Proc. R. Soc. Lond. B*, 268, 1677, 2001.
226. Waas, J.R. and Colgan, P.W., Chemical cues associated with visually elaborate aggressive displays of three-spine sticklebacks, *J. Chem. Ecol.*, 18, 2277, 1992.
227. Haberli, M.A. and Aeschlimann, P.B., Male traits influence odour-based mate choice in the three-spined stickleback, *J. Fish Biol.*, 64, 702, 2004.
228. Milinski, M., The function of mate choice in sticklebacks: optimizing Mhc genetics, *J. Fish Biol.*, 63, 1, 2003.
229. Reusch, T.B.H., Schaschl, H., and Wegner, K.M., Recent duplication and inter-locus gene conversion in major histocompatibility class II genes in a teleost, the three-spined stickleback, *Immunogenetics*, 56, 427, 2004.
230. Reusch, T.B.H. et al., Female sticklebacks count alleles in a strategy of sexual selection explaining *Mhc* polymorphism, *Nature*, 414, 300–302, 2001.
231. Wegner, K.M. et al., Parasite selection for immunogenetic optimality, *Science*, 301, 1343, 2003.
232. Wegner, K.M., Reusch, T.B.H., and Kalbe, M., Multiple parasites are driving major histocompatibility complex polymorphism in the wild, *J. Evol. Biol.*, 16, 224, 2003.
233. Kurtz, J. et al., Major histocompatibility complex diversity influences parasite resistance and innate immunity in sticklebacks, *Proc. R. Soc. Lond. B*, 271, 197, 2004.
234. Wegner, K.M. et al., Genetic variation in MHC class II sequence polymorphism in three-spined sticklebacks, *Mol. Ecol.*, 15, 1153, 2006.
235. Aeschlimann, P.B. et al., Female sticklebacks *Gasterosteus aculeatus* use self-reference to optimise *Mhc* allele number during mate selection, *Behav. Ecol. Sociobiol.*, 54, 119, 2003.
236. Milinski, M. et al., Mate choice decisions of stickleback females predictably modified by MHC peptide ligands, *Proc. Natl. Acad. Sci.*, 102, 4414, 2005.
237. Mays, H.L. and Hill, G.E., Choosing mates: good genes versus genes that are a good fit, *Trends Ecol. Evol.*, 19, 554, 2004.

238. Candolin, U., The use of multiple cues in mate choice, *Biol. Rev.*, 78, 575, 2003.
239. Ibrahim, A.A. and Huntingford, F.A., Laboratory and field studies of the effect of predation risk on foraging in three-spined sticklebacks (*Gasterosteus aculeatus* L.), *Behaviour*, 109, 46, 1989.
240. Ibrahim, A.A. and Huntingford, F.A., The role of visual cues in prey selection in three-spined sticklebacks (*Gasterosteus aculeatus*), *Ethology*, 81, 265, 1989.
241. Mussen, T.D. and Peeke, H.V.S., Nocturnal feeding in the marine threespine stickleback (*Gasterosteus aculeatus* L.): modulation by chemical stimulation, *Behaviour*, 138, 857, 2001.
242. Zeiske, E., Prädispositionen bei *Mollienesia sphenops* (Pisces, Poeciliidae) für einen Übergang zum Leben in subterranean Gewässern, *Z. Verg. Physiol.*, 58, 190, 1968.
243. Fine, M.L. et al., Pectoral spine locking and sound production in the channel catfish *Ictalurus punctatus*, *Copeia*, 1997, 777, 1997.
244. Sörensen, W., Are the extrinsic muscles of the air-bladder in some Siluroiidae and the elastic spring apparatus of others subordinate to the voluntary production of sounds? What is, according to our present knowledge, the function of the Weberian ossicles?, *J. Anat. Physiol.*, 29, 205, 1895.
245. Mazzi, D., Kunzler, R., and Bakker, T.C.M., Female preference for symmetry in computer-animated three-spined sticklebacks, *Gasterosteus aculeatus*, *Behav. Ecol. Sociobiol.*, 54, 156, 2003.
246. Mazzi, D. et al., Inbreeding affects female preference for symmetry in computer-animated sticklebacks, *Behav. Gen.*, 34, 417, 2004.
247. Honkanen, T., Comparative study of the lateral-line system of the three-spined stickleback (*Gasterosteus aculeatus*) and the nine-spined stickleback (*Pungitius pungitius*), *Acta Zool.*, 74, 331, 1993.
248. Grote, W., Vogt, C., and Hofer, B., *Die Süsswasserfische von Mittel Europ*, Dr. Schlüter and Mass, Halle, 1909.
249. Braithwaite, V.A. and Girvan, J.R., Use of water flow direction to provide spatial information in a small-scale orientation task, *J. Fish Biol.*, 63, 74, 2003.
250. Körner, K.E. et al., The role of experience in mating preferences of the unisexual Amazon molly, *Behaviour*, 136, 257, 1999.
251. Plath, M. et al., Sexual selection in darkness? Female mating preferences in surface- and cave-dwelling Atlantic mollies, *Poecilia mexicana* (Poeciliidae, Teleostei), *Behav. Ecol. Sociobiol.*, 55, 596, 2004.
252. Parzefall, J., Gagelmann, U., and Schartl, M., Aggressive behaviour and optomotor response in different populations of Poecilia *mexicana* (Pisces, Poeciliidae), *Mem. Biospeol.*, 24, 63–69, 1997.
253. Seehausen, O., van Alphen, J.J.M., and Witte, F., Cichlid fish diversity threatened by eutrophication that curbs sexual selection, *Science*, 277, 1808, 1997.
254. Scholik, A.R. and Yan, H.Y., Effects of boat engine noise on the auditory sensitivity of the fathead minnow, *Pimephales promelas*, *Environ. Biol. Fish.*, 63, 203, 2002.
255. Scholik, A.R. and Yan, H.Y., Effects of noise on the auditory sensitivity of the bluegill sunfish, *Lepomis macrochirus*, *Comp. Biochem. Physiol. A*, 133, 43, 2002.
256. Jones, J.C. and Reynolds, J.D., Effects of pollution on reproductive behaviour of fishes, *Rev. Fish Biol. Fish.*, 7, 463, 1997.
257. Brown, G.E. et al., The effects of reduced pH on chemical alarm signaling in Ostariophysan fishes, *Can. J. Fish. Aquat. Sci.*, 59, 1331, 2002.
258. Foster, S.A., Baker, J.A., and Bell, M.A., The case for conserving threespine stickleback populations: protecting an adaptive radiation, *Fisheries*, 28, 10, 2003.

259. Wood, P.M., Will Canadian policies protect British Columbia's endangered pairs of sympatric sticklebacks?, *Fisheries*, 28, 19, 2003.
260. McLennan, D.A., Phylogenetic analysis sof behavioral evolution: a case study using gasteroid fishes, M.Sc. thesis, University of British Columbia, Vancouver, 1988.
261. Czeczuga, B., Kozlowska, M., and Czeczuga-Semeniuk, E., Adaptive role of carotenoids and carotenoproteins in *Cyclops kolensis* Lilljeborg (Crustacea: Copepda) specimens to extremely eutrophic conditions, *Folia Biol.*, 48, 77, 2000.
262. Gaillard, M. et al., Carotenoids of two freshwater amphipod species (*Gammarus pulex* and *G. roeseli*) and their common acanthocephalan parasite *Polymorphus minutus*, *Comp. Biochem. Physiol. B*, 139, 2004.
263. Matsuno, T. and Katsuyama, M., Comparative biochemical studies of carotenoids in fishes — XI. Carotenoids of two species of flying fish, mackerel pike, killifish, three-spined stickleback and Chinese eight-spined stickleback, *Bull. Jpn. Soc. Fish.*, 42, 761, 1976.
264. FitzGerald, G.J. and van Havre, N., The adaptive significance of cannibalism in sticklebacks (Gasterosteidae: Pisces), *Behav. Ecol. Sociobiol.*, 20, 125, 1987.
265. Smith, R.S. and Whoriskey, F.G., Jr., Multiple clutches: female threespine sticklebacks lose the ability to recognise their own eggs, *Anim. Behav.*, 36, 1838, 1988.
266. Frommen, J.G. and Bakker, T.C.M., Adult three-spined sticklebacks prefer to shoal with familiar kin, *Behaviour*, 141, 1401, 2004.
267. Östlund, S., Female 15-spined sticklebacks detect males with empty nests by non-visual cues, *J. Fish Biol.*, 47, 1106, 1995.

7 Reproductive Physiology of Sticklebacks

Bertil Borg

CONTENTS

7.1 INTRODUCTION

Sticklebacks are very successful fishes, being able to live and reproduce in the most diverse aquatic environments. This is probably because of the male's nest, which protects the eggs against many predators. The nest is glued together with threads of spiggin, a protein formed in the kidney under androgen stimulation. This protein production is not only of vital importance for the stickleback but also offers a unique model for the study of androgen actions in fish. The ability of sticklebacks to spawn in water of all salinities is another unusual trait contributing to their success. This ability is probably dependent on many adaptations, of which stimulatory effects of ovarian fluid on sperm in hostile environments is one. Traits like these indicate reproductive physiology as a key factor in the evolution of these interesting fishes. This chapter attempts to cover developments in stickleback reproductive physiology from 1991 onward because much of the older literature is already covered by reviews.[1,2] Some aspects of reproductive physiology, such as hormonal control of reproductive behaviour, endocrine disruption, and effects of parasites on reproduction are covered in Chapters 8, 9, and 10. Most studies on stickleback reproductive physiology have been done on the three-spined stickleback, *Gasterosteus aculeatus*, and in this chapter "stickleback" will always refer to this species, unless otherwise specified. Apart from spiggin production and other secondary sexual characters and the biology of sperm and spermatogenesis, the control of seasonal reproduction has been extensively studied in the stickleback. The stickleback is convenient to study in the laboratory and responds in a clear-cut way to light treatments. Thus, the mechanisms in photoperiodic control of reproduction in many respects are better understood in the stickleback than in any other fish.

7.2 SPERMATOGENESIS AND SPERMATOZOA

In this chapter, processes in sperm production and control of sperm motility are described. In particular, the importance of ovarian fluid for sperm viability, and thus fertilisation, in different environments is analysed.

Several studies on seasonal cycles in three-spined, nine-spined, *Pungitius pungitius*, and brook stickleback, *Culaea inconstans*, have shown that spermatogenesis is quiescent in the breeding season, when spermatozoa and a small number of spermatogonia are the only germ cells present.[3,4] Spermatogenesis is inactive during the breeding season also in many other fishes in temperate regions. In many cases, such as salmonids, there is a prespawning spermatogenesis that is completed when breeding starts. Postspawning spermatogenesis occurs in, for example, European perch, *Perca fluviatilis*, pike, *Esox lucius*, and sticklebacks. In sticklebacks, spermatogenesis starts at the end of breeding in summer and is completed in late autumn or early winter several months before the onset of spawning. Control of spermatogenesis is described in Section 7.5.1.

Electron micrographs of stickleback testes showed that the spermatids display a granular pattern of chromatin condensation and the mature spermatozoa a more complete, homogenous condensation.[5] Mitochondria are not in direct contact with the flagellum but are located in a cytoplasmic ring surrounding the flagellar base. As usual in teleosts, the sperm nucleus is rounded, and there is no acrosome.[5] In many other animals, the lytic enzyme–containing acrosome is necessary for the spermatozon's ability to penetrate the egg membrane and reach the oocyte. In contrast, the sperm of teleosts lack an acrosome. They are able to enter through a small opening in the egg membrane called the *micropyle*.

7.2.1 Sperm Proteins

During spermiogenesis, that is, the transformation of spermatids into spermatozoa, which is the final step of spermatogenesis, the nucleus condensates, and the nuclear histone proteins are replaced by protamines. These are small, extremely basic proteins with a very high arginine content, which were first found in sperm from fishes but were later found in many other animals.[6] Compared to the other fish taxa, a large variability in sperm nuclear basic proteins (SNBPs) from mature testes in different stickleback species (*G. aculeatus, G. wheatlandi*, and *P. pungitius*) and in their close relative the tubesnout (*Aulorhynchus flavidus*), as studied by polyacrylamide gel electrophoresis (PAGE), has been observed.[5] Amino acid composition analysis showed a high arginine content in the stickleback SNBPs (about 40%) and also a presence of histidine and lysine. This is in agreement with typical protamines, but the stickleback SNBPs moved slowly in the system compared to these.[5] SNBPs from testes in different stickleback species and in the tubesnout have also been sequenced.[7] Molecular masses calculated from these sequences agreed with mass spectrometry determinations of purified proteins. A bootstrap protein parsimony tree on these sequences and sequences from other fishes was consistent with the generally accepted phylogeny.[7]

Three-spined sticklebacks were collected monthly between October and May, and their SNBPs were analysed.[5] Especially in October, there was, apart from the protamine-like SNBP, also slower-moving proteins (histones) present. Slower-moving proteins disappeared as the breeding season approached, suggesting that as far as nuclear proteins are concerned, sperm maturation is not completed until late spring.[5] Sticklebacks with dark testes were regarded as more mature than those with white testes. Gonadal tissue from the latter had largely lost their histone content and had a decreased extent of protamine phosporylation. The phosphorylated protamines could be distinguished from the nonphosphorylated ones in acetic acid-urea PAGE on account of a slightly increased molecular weight. Phosphorylation was confirmed after alkaline phosphatase digestion.[7]

7.2.2 Sperm Motility

Most teleost species stay in either freshwater or seawater throughout their life. Some species are able to live in both environments for at least part of their life cycle, and many of these undertake regular migrations between the two environments. The activation and survival of gametes pose diametric problems in the hypotonic freshwater and the hypertonic seawater, and the ability of the three-spined stickleback to spawn successfully both in freshwater and in full-strength seawater is very unusual.

To find out how this is possible, stickleback sperm motility has been studied in different media using computer-assisted sperm analysis (CASA).[8] Sperm from sticklebacks of freshwater, brackish water, and marine origin were tested in water with 0, 5.5, and 30 ppm salinity. In all cases, by far the longest period of motility was recorded in the 5.5-ppm medium (mean life span 116–483 min). In freshwater, sperm from freshwater and brackish sticklebacks moved on average 0.5 min, whereas sperm from seawater sticklebacks was immotile. Only sperm from seawater sticklebacks moved in seawater (76 min). Measures of straight line velocity (VSL), curved line velocity (VCL), and motile sperm percentage yielded a similar picture.[8] The short life span of spermatozoa in freshwater is in contrast to observations that it takes at least 15 min for all eggs in the stickleback nest to be fertilised.[9] However, the stickleback eggs are not spawned in open water, but inside a nest where they are expelled together with a viscous fluid from the ovarian lumen. Addition of 25% ovarian fluid increased mean sperm longevity in freshwater to 245 min. This treatment also further increased motility in brackish water and had similar effects on the other studied parameters. It was suggested that the stimulatory effect of ovarian fluid on sperm motility is one of the traits that enabled the originally marine sticklebacks to successfully invade freshwater. In the entirely marine 15-spined stickleback, *Spinachia spinachia*, on the other hand, sperm moved longest in seawater and not at all in freshwater, and sperm motility parameters were not influenced by ovarian fluid.[10]

Osmolarity and ionic levels in the stickleback ovarian fluid were found to increase with the salinity of fish habitat and were 246 mOsmol, 150 m*M* Na^+, 136 m*M* Cl, 4 m*M* K^+, and 2 m*M* Ca^{2+} in brackish water.[11] Based on this composition, artificial ionic ovarian fluids were made, and sperm motility parameters (see previous paragraph) were tested, using CASA in dilution series of natural and artificial ovarian fluids.[11] Both types of ovarian fluid were highly stimulatory — in agreement with the earlier study — and although the effectiveness increased with concentration, even the lowest concentrations tested (0.75% of natural fluid and 3.1% of artificial fluid) were effective. There was no difference in effectiveness in natural and ovarian fluid on any parameter, indicating the ovarian fluid ionic content alone must be responsible for the stimulatory effect. Water samples were taken from stickleback nests after spawning in freshwater. An analysis of the ionic content showed Na concentration corresponding to 13, 11, and 3.2% of that in ovarian fluid after 0, 5, and 15 min respectively.[11] These levels are sufficient to stimulate sperm motility to an extent where it can be important for fertilisation in freshwater. Stimulatory effects of ovarian fluid on sperm have also been found in several other teleosts, but it is so far only in the stickleback that it has been shown to be of biological importance. Although the ionic content alone can explain the positive effect of ovarian fluid on sperm motility, the macromolecules may also play a role by increasing ionic retention in the nest. The ionic content was higher in nests 5 or 15 min after the addition of natural, compared to ultrafiltered, ovarian fluid.[11]

It has also been observed that stickleback sperm that had ceased to move in freshwater could be reactivated with saline water, and it was suggested that sperm deposition in the nest prior to spawning could be a tactic used by a resident male to increase paternity in a sneaking situation.[12] Sticklebacks often sneak, that is, other males than the nest owner can try to "steal" fertilisations. However, it has not been shown that prespawning sperm deposition takes place, and earlier experiments in which males were prevented from entering the nest after the female had spawned suggest that no or very little sperm is released prior to egg deposition.[13] However, it should be pointed out that this was not in a competitive situation.

Sperm were reported to be more motile in 1-year-old than in 2-year-old stickleback males kept in brackish water, which has been related to different reproductive tactics.[14] However, in water with a salinity of 1.5%, the motility only lasted 5–7 min.[14] This is dramatically shorter than what has been observed in brackish water in the preceding investigations as well as others.[8,11,15] A possible source of error in the study of 1- and 2-year-old males was that sperm was collected by "gently squeezing the abdomen."[14] Other researchers, including myself, have not found this possible, but have instead taken sperm directly from dissected testes. I find it likely that sperm collected by squeezing has been contaminated with urine, which can damage the sperm.

7.3 GONADAL HORMONES

The role of androgens in stimulating reproductive behaviour and secondary sexual characters is better known in sticklebacks than in perhaps any other fish. This section deals mainly with androgen levels and their natural changes.

11-ketotestosterone (11KT) is generally regarded as the most important androgen in teleost fishes.[16] In the stickleback male, seasonal levels of 11KT peak in the breeding season and drop to their lowest levels in the postbreeding period when spermatogenesis takes place.[17] It perhaps appears obvious that 11KT should peak in the breeding period, when reproductive behaviour and secondary sexual characters are best developed; but in many other fishes, 11KT and testosterone (T) levels peak in the prespawning rather than in the spawning period.[16]

Circulating androgen levels in the stickleback not only changed seasonally, but also over the nesting cycle. 11KT levels are much higher (300–400 ng/ml) in males with nests without eggs that display courtship behaviour than in the parental phase, when the nest contains eggs, and when parental behaviour rises, and courtship behaviour disappears.[18–20] In the late parental phase, 11KT levels often drop to about 10 ng/ml.[18–20] The decline of both courtship behaviour and 11KT levels occurs sooner when the nest contains large than small numbers of eggs.[18] It should be pointed out that although 11KT levels in late parental males are dramatically lower than in sexual phase males, they are still higher than 11KT levels outside the breeding season.[17] T-levels also declined over the nesting cycle.[20] A decline in androgen levels from the sexual to the parental phase has also been observed in several birds and in a number of other fishes where the male takes care of the offspring, which is common among teleosts.[18] However, the decline in androgen levels in the parental phase is particularly pronounced in the stickleback.

Ovaries are female endocrine organs, with sex steroids as major products. Sex steroid levels have been studied in a few pooled plasma-samples from breeding female sticklebacks captured in the field.[21] T, 11β-hydroxytestosterone (OHT), 11β-hydroxyandrostenedione (OHA), 11-ketoandrostenedione (11KA), and 11KT were measured using RIA. The levels of OHT and OHA in females were similar to those found in males sampled on the same occasion. Females had lower levels of 11KA and, in particular, of 11KT than the males. On the other hand, female levels of T were much higher than in the males, and of the studied androgens T was present at the highest level in females. This sexual pattern in levels of 11KT and T is in agreement with that found in most studied teleosts.[16] Plasma levels of oestradiol (E2) were low (1–2 ng/ml) in breeding females, and 17,20β-dihydroxy-4 pregenen-3-one (17,20P) was not detected.[21] Seasonal patterns in androgens have been studied in females kept under a simulated yearly cycle in captivity. T levels were at their highest in the breeding season, whereas 11KT levels were always low in females.[22]

7.4 SECONDARY SEXUAL CHARACTERS

Secondary sexual characters are here defined as characters apart from the primary sexual characters, i.e., gonads, sexual ducts, and genitalia, which are different in the sexes. The kidney hypertrophy in the stickleback is perhaps the most studied and best-known secondary sexual characteristic in any fish.

7.4.1 KIDNEY HYPERTROPHY

In the breeding season, the kidney of the male stickleback increases dramatically in size through hypertrophy and produces a "glue" that is used in the building of the nest. In the female, which does not take part in nest building, kidney hypertrophy does not normally occur, though it can be induced by androgen administration (see the following pages). The glue is stored in the urinary bladder before being applied as threads to the nest in a characteristic movement (gluing).

The composition of stickleback glue has been studied by collecting threads of glue from stickleback nests and urinary bladder contents from males in breeding condition and separating the proteins by means of sodium dodecyl sulphate-polyacrylamide gel electrophoresis (SDS-PAGE).[23] One major protein with an apparent molecular mass of 203 kDa dominated in both types of samples.[23] This protein was named *spiggin* (*spigg* means "stickleback" in Swedish). Periodic acid Schiff-staining demonstrated that spiggin was a glycoprotein, and following deglycosylation, the apparent molecular mass decreased to about 200 kDa. The amino acid composition of spiggin was also determined from both types of samples; a notable trait was the high content (about 8%) of cysteine, suggesting the presence of numerous stabilising disulphide bridges.[23] These results are in contrast to another study where also the protein in stickleback nest threads separated using SDS-PAGE, and where instead of one major band at 203 kDa, many bands from 7.6 to 80 kDa were found, perhaps due to degradation.[24] Carbohydrate tests on the glue protein demonstrated the presence of pentose and hexose, but not of ketose and hexuronic acid.[24]

Partial spiggin sequences obtained using SDS-PAGE purified and trypsin digested urinary-bladder spiggin, have been used to design oligonucleotides, making it possible to clone and sequence spiggin mRNAs extracted from mature male stickleback kidney cDNA libraries.[25] By means of Northern blot, three spiggin mRNAs — a predominating subunit-α (4.2 kb), subunit-β (2.2 kb), and subunit-γ (1.6 kb) were observed in kidneys from mature males. Spiggin mRNA was not found in female kidneys. Sequences of homologous parts of the different subunits were identical, indicating they are derived by alternative splicing from the same locus. The sequence data were deposited in the GenBank database under the following accession numbers: AF323732 (subunit-α), AF323733 (β), and AF323734 (γ). A 910-amino acid protein was deduced for subunit-α. The location of six peptides sequenced from purified spiggin in the deduced subunit-α sequence confirmed the identity of the spiggin mRNAs. Spiggin was overall hydrophobic and was found to be a novel protein with structural similarities to von Willebrand Factor (vWF, a component in clotting of blood)–related proteins.[25] The deduced proteins were found to contain a 19-residue signal peptide and *N*- and *O*-glycosylation sites and motifs with many similarities to D-domains within vWF. In addition, there were also vicinal cysteine motifs and regions with high cysteine content bisecting the D-domains.[25]

The predicted molecular weight for α-spiggin was 103 kDa and less for β- and γ-spiggin. These weights are all considerably lower than the apparent molecular weight of 203 kDa (200 kDa after deglycosylation) based on SDS-PAGE. Antiserum raised against a part of the spiggin protein recognised SDS-PAGE bands with apparent molecular weights of 130 and 51 kDa as well as a faint band of 90 kDa in kidneys, and a single band of 203 kDa from urinary bladder content, which was absent in the kidney. This indicates that spiggin undergoes multimerisation in the urinary bladder. In vWF and in the related protein, mucin D-domains and disulphide elements are required for multimerisation, and it is likely that this is also the case for spiggin.[25] Further, the multimerisation of vWF is autocatalysed and, depending on vicinal cysteine residues, this is also possibly the case for spiggin. Spiggin mRNA, as determined using slot blot analysis, was observed in kidneys of mature males, but not in other tested tissues, such as liver, testis, brain, and muscle.[25] In a later study, two types of spiggin were cloned from kidneys of mature male sticklebacks.[26] Apart from the type described earlier (Spiggin-I), there was also another type, Spiggin-II, with a deduced open reading frame encoding for 639 amino acids (616 amino acids were similarly deduced for Spiggin-I). There was an amino acid similarity of 80% between Spiggin-I and Spiggin-II.[26]

Kidney hypertrophy in the stickleback has long been known to be androgen dependent; it disappears after castration and can be induced in both sexes by androgen treatment.[16] In a dose–response study on the effectiveness of different androgens on kidney hypertrophy, castrated male sticklebacks were injected daily with androgens at doses ranging (apart from controls) from 0.008 to 25 μg/g body weight/d.[27] The effectiveness was calculated from the amount needed to obtain half of the maximal stimulation of the kidney epithelium height (KEH). 11KT was the most effective androgen tested. Other nonreduced 11-androgens were somewhat less effective, whereas 5α-reduced androgens were still less, and T was surprisingly ineffective. The relative effectivenesses if 11KT is rated as 100 were: OHA: 67, 11KA: 42, 5α-androstane-3,11,17-trione: 4.4, 5α-dihydrotestosterone (DHT): 2.6, 17β-hydroxy-5α-androstane-3,11-dione: 1.6, T: 0.3, androstenedione (A4): <1, and 5β-androstane-3,11,17-trione: <1.[27] The high effectiveness of 11KT compared to other androgens and the observation that 11KT is found at higher levels in the plasma than other androgens in breeding stickleback males indicates that 11KT is the most important circulating androgen in stimulating the kidney.[17,27] Studies on spiggin induction confirm the high effectiveness of 11KT: Spiggin was absent in the urinary bladder of untreated castrated males, but present in sham-operated breeding males and in castrated fish treated with Silastic capsules containing 11KT.[23] Spiggin is so far the only androgen-induced protein known in fishes. Spiggin mRNA could be induced in kidneys of females by implants containing 11KA, which converts to 11KT extratesticularily.[17,28] Implants containing cortisol, E2, T, progesterone, or DHT were ineffective in this study.[25] Spiggin mRNA induction in female kidneys by a high 11KA treatment was detectable after 27 h, but not after 10 h using slot-blot analysis.[25]

Spiggin can also be induced by androgens administered via the water. Methyltestosterone (MT) at doses of 0.01, 0.10, 1, and 10 μg/l for 3 weeks increased KEH and spiggin levels in females and males from 0.10 and 1 μg/l, respectively.[29] Control males had higher KEH and spiggin content than females, and spiggin levels were highly correlated with KEH.[29] DHT administrated similarly at doses of 1, 2, 3, 4, or 5 μg/l to female sticklebacks significantly increased spiggin from 3 μg/l after 3 weeks and from 2 μg/l after 5 weeks.[29] A dose of 20 ng/ml oestradiol administrated similarly had no effect on spiggin in females.[29] Both Spiggin-I and Spiggin-II could be induced in female stickleback kidneys by administration of both MT and DHT.[26] That DHT is effective in some studies and not in others may be related to different methods of administration, leading to different levels in the fish.

Kidney somatic index (KSI, kidney weight or body weight × 100), spiggin mRNA expression, and frequencies of gluing behaviour were lower in stickleback males in the parental phase compared to the nesting, sexual phase.[20] This could be due to the much lower 11KT levels in the parental phase.[18–20] The high effectiveness of 11KT is consistent with studies on secondary sexual characters in other fishes, though the picture is most complete in the stickleback.[16]

Steroids are often metabolised in the target organs into other compounds, which may be the effective forms on the receptor level. A theory has been put forward that 11-androgens, in analogy with the reduction of T to 5α-dihydrotestosterone (5α-DHT) in mammals, obtain full effectiveness first after 5α-reduction.[30] To study this possibility, stickleback kidneys were incubated with tritiated A4 or 11KT, and the metabolites were separated by means of thin layer chromatography.[31] A4 was converted to a large extent into several other compounds including 5α-reduced ones. 11KT was converted to a far smaller extent and no 5α-reduced 11KT was found.[31] The absence of 5α-reduction of 11KT in the kidney and the relatively low effectiveness of 5α-reduced 11-androgens in stimulating the kidney (see previous discussion) do not support the theory that 11-androgens need to be 5α-reduced to become fully effective.[27,30]

7.4.2 Kidney Androgen Receptors

To exert biological actions, hormones need first to bind to specific receptors. After being activated by binding of hormones, intracellular steroid receptors bind to response elements in the DNA and initiate transcription of specific mRNAs in the nucleus. The mRNAs leave the nucleus and are translated into specific proteins in the cytoplasm. Although 11KT or other 11-androgens are more effective than other androgens in stimulating male secondary sexual characters in teleosts, androgen receptors (AR) binding 11KT selectively have not been found. On the other hand, teleost ARs and AR-binding sites have been found to bind selectively to T or DHT, and cell systems transfected with teleost AR are activated by 11KT to a similar extent as by other androgens.[32] Absence of 11KT-specific receptors can, at least in the case of the stickleback kidney, hardly be explained by T or 5α-reduced 11KTs being the physiologically relevant hormones.

In an early study, no cytosolic or nuclear binding of either T or 11KT was found in stickleback kidneys.[33] A specific binding of 11KT, but not of T, on the other hand, was found in kidney tissue pieces (but not in other tested organs) and in the membrane fraction of kidneys.[33] The specific 11KT binding in the tissue displayed an ED50 value (50% of displaceable binding) of 28 n*M*, which is consistent with a receptor function. Not only 11KT, but also 5α-reduced 11KT and 5αDHT, and to a lesser extent T and progesterone, could displace radiolabelled 11KT.[33] The lack of specific binding in the cytosolic and nuclear fraction and its presence in the membrane fraction have led to the suggestion that 11KT exerts its effects via receptors other than the intracellular steroid hormone receptors.[33] This is, however, in contrast with spiggin induction in female sticklebacks exposed to 1 μg/l MT or 5 μg/l DHT via the ambient water, the former being the most effective.[34] These androgenic effects were suppressed by the addition of the anti-androgen flutamide (500 μg/l), suggesting that the androgens acted via an androgen receptor of the classic mammalian type.[34]

Using RNA from mature male stickleback kidneys and oligonucleotide probes based on conserved regions in AR from other teleosts, a stickleback AR has been cloned and sequenced.[32] A single AR gene with two splicing variants, ARβ1 (Gene Bank Accession no. AY24706) and ARβ2 (AY24707) was found.[32] Sequence comparison analysis defined the stickleback AR as an ARβ isotype.[32] The closest overall similarity was found with AR from the red seabream (76.1%), which, similar to the stickleback, belongs to the Percomorph Acanthopterygians. The amino acids thought to be involved in direct ligand interactions with DHT in human AR were conserved in the stickleback. Using radiolabelled 5α-DHT as a ligand, high-affinity (K_d = 18.7 n*M* in cytoplasm), saturatable androgen-specific binding sites were found in the cytosolic and nuclear fractions, but not in membranes, from stickleback kidneys.[32] Ligand competition assays using kidney cytosolic and radiolabelled 5α-DHT as a ligand showed the highest binding affinity for DHT (EC_{50} = 1.31 n*M*), whereas both 11KT and T had much lower affinities (43 and 69 n*M*, respectively); E2 had a competitive binding affinity of 85 n*M*, whereas the tested progestins did not compete with DHT.[32] Also, in ARβ2 containing extracts from reticulocytes into which mRNA for AR had been introduced, DHT was found to have the highest binding affinity (EC_{50} 0.67 n*M*), followed by 11KT and T (2.2 and 3.0 n*M*, respectively).[32] The high binding affinity of DHT to the stickleback AR is paradoxical, considering that the biological activity of 11KT is much higher. This difficulty may be partly explained by the fact that a lower proportion of circulating 11KT than of DHT and T is probably protein bound and that only unbound androgen may enter cells and activate the intracellular receptors. However, it was also found that both human HepG2 cells and zebra fish ZFL cells transfected with ARβ2 receptors showed a higher AR-activated luciferase response induced by 11KT (10- and 12-fold, respectively) than by DHT (4.1- and 4.5-fold, respectively).[32] Thus, the stickleback has an AR specifically activated by 11KT, so far the only one known in any animal.

Both cytosolic androgen-receptor-binding levels (reflecting AR protein levels) and AR mRNA levels were suppressed in kidneys from female and male-castrated sticklebacks treated with implants containing E2, but not influenced by androgens.[32]

7.4.3 Other Secondary Sexual Characters

The hypertrophied male kidney is not the only secondary sexual character in the stickleback. This chapter deals with the red breeding colour and pectoral fins and muscles in male sticklebacks, traits that are very important in sexual selection and paternal care, respectively.

The ventral side of the male three-spined stickleback turns red in the breeding season. This red nuptial colour depends largely on the presence of red carotenoid pigments in the erythrophores. The reddishness of the fish not only depends on the amount of pigment present, but also on whether it is concentrated or dispersed, which can change rapidly. Carotenoids in unsaponified and saponified extracts of skin from nesting sticklebacks have been studied by means of high pressure liquid chromatography (HPLC).[35] They were identified by means of retention time, UV spectrometry, and mass spectrometry (saponified samples only). Astaxanthin, lutein, and tunaxanthin were identified. Contents of astaxanthin and lutein + tunaxanthin were not correlated with each other. More astaxanthin was found in the redder males and more lutein + tunaxanthin in the more yellowish males.[35]

In the parental phase, the male stickleback ventilates his eggs by fanning, pushing a water current through the nest by beats of the pectoral fins. It has been reported that male sticklebacks had larger pectoral fin area for a given body length than females in the breeding season.[36] Furthermore, longer pectoral fin rays were found in male than in female sticklebacks.[37] Male sticklebacks compensated for removal of part of their pectoral fins by increasing fin beat frequency.[38] Male brook sticklebacks caught in the breeding season had 13.5% larger pectoral fin area than females.[39] In another study, on the other hand, no consistent differences either in pectoral fin area or in fin ray length between male and female sticklebacks and also no effect of castration and androgen treatment on these traits were found.[40] However, males did have relatively larger pectoral muscles than females.[40] Relative weights of pectoral muscles diminished with castration, but was not increased by androgen treatments.[40]

7.5 BRAIN–PITUITARY–GONADAL AXIS

Vertebrate gonads are controlled by pituitary gonadotropic hormones (GTHs); these are in turn controlled by hypothalamic factors. Together, this forms the brain–pituitary gonadal axis, which is under feedback control. This section describes these traits in the stickleback. Because sticklebacks, unlike most other fishes, can be easily castrated, they have been subject to many experimental studies on feedback mechanisms. The unexplained, but well-established, negative relationship between androgens and spermatogenesis in the stickleback is in contrast to most other vertebrates.

7.5.1 Gonadotropic Hormones

Teleost fishes have two gonadotropic hormones (GTHs): luteinising hormone (LH, GTH II) and follicle-stimulating hormone (FSH, GTH I), which consist of two subunits, a common α-subunit and a unique, but related β-subunit.[41,42] Unlike in mammals, the two GTHs are present in different pituitary cells in teleost fishes.[43] In the older literature, based on light microscopy and electron microscopy, most authors, including myself, have been of the opinion that sticklebacks possess only one type of GTH cell, although it has also been proposed that they had two.[44]

FSH-β (EMBL accession number AJ534871) and LH-β (AJ534969) from stickleback pituitary glands have been cloned and sequenced.[45] The cDNA of FSH-β encoded a protein of 122 amino acids and the LH-β cDNA, a protein of 148 amino acids.[45] Compared to other GTH-β subunits sequenced so far, the stickleback GTH-β displayed the highest similarity to those from other percomorph actinopterygians, with 46 to 55% amino acid similarity for the FSH-β and 75 to 77% for the LH-β subunits.[45] The cloned cDNAs were used for constructing probes that could be used in measuring the levels of FSH-β and LH-β mRNA or expression.[45]

Expression of both FSH-β and LH-β was found to be considerably higher in nesting males than in postbreeding males.[46] Under the former condition, 11KT levels are high and androgen-dependent traits such as secondary sexual characters and reproductive behaviour are well developed, whereas spermatogenesis is quiescent. In the postbreeding period, on the other hand, androgen levels are low and androgen-dependent characters absent, whereas spermatogenesis is active. In breeding stickleback males, a positive correlation was found between plasma levels of 11KT and expression of both FSH-β and LH-β.[46]

The seasonal cycle of FSH-β and LH-β expression in female and male sticklebacks kept under seminatural conditions and sampled over a yearly cycle from November to October has also been studied.[22] Both GTHs underwent marked seasonal changes.[22] From relatively low expressions in November–December, male FSH-β rose and peaked in January in both sexes, after a marked decline following May; levels were very low from June onward, but elevated to some extent again in autumn.[22] Male LH-β also started to rise in January, but peaked in May, later than for FSH-β. Similarly as for FSH-β, LH-β expression dropped during May–June and was very low for the rest of the period. LH-β expression was highest at peak breeding in May, when KSI and 11KT plasma levels were also the highest, consistent with GTH-stimulating androgen secretion. However, in the period when spermatogenesis commenced from June, the expression of both GTHs was very low.[22] Also, under different photothermal regimes in winter (see later), the expression of GTHs is consistent with androgen production being stimulated by GTHs, whereas spermatogenesis is also then most active when GTH expression is lowest.[45] The positive relationships between androgens and GTHs are largely consistent with results from other fishes, as is the earlier peak in FSH compared to LH.[47] The negative relationship between spermatogenesis on the one hand and androgens and GTH on the other are not in agreement with results from most other fishes, where available data usually support the view that spermatogenesis is stimulated by androgens. The control of spermatogenesis in fishes has been most extensively studied in the Japanese eel, *Anguilla japonica*.[48] In this species, GTH stimulates the synthesis of androgens, particularly 11KT, which in turn stimulates spermatogenesis, an effect that can also be seen *in vitro*. The negative correlations between androgens and GTH on the one hand and spermatogenesis on the other observed in a number of studies are, however, consistent with each other and by the inhibitory effect that androgen administration exerts at the end of breeding.[3,45,46]

The expression of FSH-β and LH-β also underwent marked seasonal changes in female sticklebacks.[22] FSH-β peaked in January. LH-β expression, which followed a similar pattern as the seasonal changes in ovarian weights, peaked later in the spawning period in May. In postbreeding period in late summer, the expression of both GTHs was low.[22] Although seasonal changes in GTHs in females display different patterns in various studied teleosts, the stickleback pattern does not appear very unusual.

7.5.2 Gonadotropic Hormone Releasing Hormone

In vertebrates, gonadotropin releasing hormone (GnRH) from the brain stimulates the release of gonadotropic hormones. A number of different GnRHs, all with ten amino acids, have been found in different species.[49] By means of HPLC combined with RIA, two forms of GnRH were found in stickleback brains, one of them probably being identical to salmon GnRH (sGnRH), whereas the other one was not identical to any previously identified type.[50] In the first study on the distribution of GnRH in the stickleback brain, immunocytochemistry with an antiserum against mammalian GnRH on paraffin-sections was used.[51] Immunoreactive cells were observed in periventricular areas in the nucleus preopticus magnocellularis, n. dorsomedialis thalami, n. ventromedialis thalami, n. periventricularis posterior, and in the periventricular dorsomedian tegmentum.[51] In a later study, the distribution of GnRH in the stickleback back was studied using four different antisera, including two against sGnRH.[50] GnRH immunoreactive (ir) cells were observed in the periventricular areas where they had also been observed previously.[50,51] However, in addition to that, and more prominently, immunoreactive cells were also found in the nucleus olfactoretinalis, n. anterioris periventricularis, and in the nucleus lateralis tuberis.[50] In teleost fishes, the pituitary is controlled by direct innervation rather than by release of hypothalamic neurohormones in the median eminence as in tetrapods. An extensive distribution of GnRH-ir fibres was found in the proximal pars distalis of the pituitary, where the GTH cells are located, but not in other pituitary areas.[50] All in all, the results of the two immunocytochemical studies were in marked contrast to each other. The later study is technically superior to the older, because it was carried out on cryosections (which are likely to preserve antigenicities better than paraffin sections) and also used a wider variety of antisera, although these largely gave the same picture.[50,51] The presence and distribution of GnRH in the stickleback brain appears not much different from that found in many other fishes.

Receptor-like binding of ^{125}I-labelled D-Arg6-Pro9-salmonGnRH-NEt (sGnRHa) has been studied in pituitaries of male and female sticklebacks under different physiological conditions.[52] In the natural seasonal cycle, binding capacity was highest in the breeding season and nondetectable in late winter–early spring. Postbreeding fish had a lower binding capacity than breeding fish. When sticklebacks caught in winter were exposed to different combinations of photoperiod and temperature (LD 16:8, 20°C; LD 16:8, 4°C; LD 8:16, 20°C; and LD 8:16, 4°C), no effect of photoperiod on binding at high temperature was found, whereas binding was non-detectable in all low-temperature groups.[52]

In numerous fishes, GnRH administration has been shown to stimulate GTH release both *in vivo* and *in vitro*, ovulation, and sexual maturation. In many cases, the effects are increased with the addition of dopamine antagonists, which diminish inhibitory dopamine effects on GTH release. Neither GnRH nor dopamine blockers have been reported to stimulate stickleback reproduction; on the contrary, a number of experiments with negative results have been performed (Eva Andersson, Efthimia Antonopoulou, and Borg, unpublished), and the reason for this lack of effects is not known.

7.5.3 Feedback in the Breeding Season

In fishes, as in other vertebrates, the gonads are stimulated by GTHs, which in turn are under the control of the brain, particularly the hypothalamus. The brain–pituitary–gonad axis is under feedback control. The secretion of GTHs is influenced by gonadal hormones acting on the brain and pituitary level. In mammals, negative feedback is best known, and is the only type present in males. A major function of negative feedback is homeostatic, keeping circulating gonadal hormones at the appropriate levels. If a mammal is hemicastrated, i.e., one of the gonads is removed, negative feedback ensures a compensatory increase in steroidogenesis in the remaining gonad, so that circulating steroid levels rapidly return to normal.[53] To find out if this was also the case in fishes, breeding stickleback males have been hemicastrated or sham-operated.[54] The experiment was terminated after 12 d, plasma was sampled, and levels of androgens were measured using RIA. 11KT and T levels in the hemicastrated males were half of those found in the controls, indicating that no compensatory steroidogenesis had taken place.[54] In a similar experiment on mature Atlantic salmon, *Salmo salar*, and parr males, plasma levels of 11KT and T in hemicastrated males were 63 and 75% of those found in sham-operated males, e.g., a partial compensation.[54] Together with the stickleback results, this suggests that there is no close homeostatic control of androgen levels by negative feedback in teleost males.

Positive feedback effects are well known to occur on the brain–pituitary–gonadal axis in fishes, the most studied case being the stimulatory effect of oestrogens and aromatisable androgens (see later text) on LH in salmonids.[55] In the breeding season, LH-β mRNA levels were considerably lower in castrated than in sham-operated stickleback males, indicating a positive feedback in this phase, whereas there was no effect of castration on FSH-β mRNA levels.[46]

7.6 CONTROL OF THE SEASONAL REPRODUCTIVE CYCLE

To ensure that offspring are produced during the time of the year when survival is optimal, mechanisms by which external and internal cues time reproduction have evolved. The control of the stickleback reproductive cycle has been the subject of a large number of studies. Of these, the single most important is the monumental article by Baggerman in 1957.[56] Since then, there have been a large number of contributions in this field by her and other authors. Many aspects of the control of seasonal reproduction are now better known in the stickleback than in any other fish.

7.6.1 Endogenous Cyclicity

Many biological events display endogenous rhythmicity. Endogenous rhythms are circannual if the period is close to a year. Circannual rhythms in reproduction has been found in many animals kept under constant photoperiod and temperature.[57] They have also been studied in sticklebacks.[56,58] In an early study, nest building was recorded in males kept under constant LD 16:8 and 20°C.[56] The fish showed a spontaneous cessation of breeding also under this photoperiod, which at other seasons is highly stimulatory. The mean duration of the breeding period was 91 d, followed by a mean nonbreeding period of 108 d, i.e., a free-running period of 199 d. The second breeding period lasted shorter than the first, only 47 d. Most sticklebacks survived less than 1 year after the first onset of breeding, but in the few animals that bred a third time, the second free-running cycle had almost exactly the same length of the first. In males exposed to LD 8:16 and 20°C, the breeding period lasted shorter than under long photoperiod (46 as compared to 91 d), and no fish displayed a second reproductive period.[56] In a subsequent investigation, endogenous breeding cycles in male sticklebacks were studied using similar methods, except for the choice of photoperiods and a lower temperature (15°C), which was used to obtain a longer survival time.[58] In animals displaying circannuality, the median lengths of the first breeding period under LD 16:8 were 49 d in both the experiments carried out, the first nonbreeding periods 119 and 217 d, the first free-running cycles 165 and 294 d, and the second free-running cycles similar to the first.[58] In this respect, the results of the two studies were in general agreement and suggested that the endogenous period of the circannual cycle was considerably shorter than 1 year.[56,58] Under constant light, most sticklebacks displayed an extended (median 147 d) first breeding period.[58] This is consistent with the hypothesis that stickleback reproduction is controlled by a circannual cycle in photosensitivity.[59] If such a cycle does control stickleback breeding, it should further be expected that under a shorter photoperiod the first breeding period would be shorter, the total breeding cycles longer, and the proportion of time spent in breeding condition lower than under a longer photoperiod. However, no significant differences in these respects were observed between males kept under constant LD 12:12 and LD 16:8, suggesting that a circannual cycle in photosensitivity is not the whole explanation for the endogenous control of breeding in sticklebacks.[58] Following the first cycle, increasing numbers of males, particularly under LD 12:12, displayed irregular cycles with shortened nonbreeding periods suggesting a desynchronisation of internal oscillators.[58] Surprisingly, a few males kept under LD 12:12 continued their first nesting period for very extensive periods of time (294–511 d), whereas no first nesting period under LD 16:8 exceeded 91 d.[58] It is tempting to compare this with starlings, where a longer photoperiod is needed to stimulate the onset of postreproductive refractoriness than to stimulate the onset of the reproductive period.[60] The rainbow trout, *Oncorhynchus mykiss,* is the only other fish where circannual reproductive cycles in a strict sense have been studied extensively.[61] Unlike the stickleback, the rainbow trout usually displays a consistent circannual spawning cycle, with a period close to a year.

7.6.2 Retinal and Extraretinal Photoreceptors

In mammals, the presence of photoreceptors is established only in the retina. Other vertebrates, however, also have extraretinal photoreceptors. The best known site is the pineal organ, which has been shown to contain photoreceptor cells in many fishes, including the stickleback.[62] There are, however, also other sites of extraretinal photoreception in the brain and elsewhere.[63] It has been demonstrated in many vertebrates that extraretinal photoreceptors are able to mediate photoperiodic effects on reproduction in the absence of eyes. This is also the case in the stickleback, in which both intact and blinded males matured under LD 16:8 but did not mature under LD 8:16, in winter.[64] It has been demonstrated in birds that not only can extraretinal photoreception mediate photoperiodic effects, but also that the eyes are unable to do so if the skull is covered.[65] This possibility has also been tested in the stickleback. The skulls (but not the eyes) of stickleback males were covered with opaque or transparent plastic foil, and the fish were exposed to long photoperiod at varying light intensities in winter.[66] KEH increased with higher light intensities. There was no significant difference in KEH between fish with opaque and transparent covers at the highest intensity and in darkness. At low light intensities, however, the fish with transparent covers had higher KEH than those with opaque covers.[66] This demonstrates that the eyes are less effective than the extraretinal photoreception in stimulating maturation in sticklebacks.[66] Whether the absence of a cover effect at higher intensities is due to the fact that eyes can mediate the effect of the higher intensity or due to the fact that sufficient light is able to enter the skull from the sides and from below cannot be determined. Cecilia Bornestaf and Borg (unpublished) also observed that a lower proportion of female sticklebacks with opaque covers on their heads than with transparent covers matured fully (strippable eggs) under long photoperiod in winter.

7.6.3 Melatonin

In almost all studied vertebrates, including fishes, circulating levels of the indolamine melatonin are higher in the night than during the day. This has also been found to be the case in the stickleback.[67] In mammals, melatonin is known to mediate photoperiodic effects on reproduction, both in long-day breeders where it is inhibitory (e.g., in Djungarian hamster, *Phodopus sungorus*, and in short-day breeders where it is stimulatory (e.g., in sheep, *Ovis aries*).[68,69] Not much is known about the role of melatonin in the control of reproduction in fishes. The pineal organ appears to be the main source of circulating melatonin in fishes as in mammals, because pinealectomy diminishes the melatonin level or rhythm.[70] However, pinealectomy also interrupts an innervation of extensive brain areas that presumably mediates photic information received in the pineal organ. For this reason, and because pineal hormones other than melatonin cannot be excluded, the effects of pinealectomy on reproduction found in several fishes do not necessarily reveal a functional role of melatonin. Inhibitory effects of melatonin injections on reproduction have been found in a number of long-day breeding fishes, including the stickleback.[71] However, this should be regarded with caution because injections give a high, transient peak in melatonin levels, quite unlike the natural nighttime rise.

A more physiological approach has been to administer melatonin via the ambient water (20 or 80 μg/l or solvent alone) for 16 h/d to sticklebacks kept under LD 16:8 or under constant light.[67] Control males were also kept under LD 8:16 in the LD 16:8 experiment. The 20 μg/l treatment led to a daily pattern in plasma melatonin roughly similar to those (455 pg/ml) found at midnight under exposure to short photoperiod, whereas the 80 μg/l dose led to plasma levels clearly above those found naturally.[67] Although short photoperiod (LD 8:16) clearly suppressed maturation, both doses of melatonin failed to prevent long photoperiod stimulation of KEH.[67] Also, females kept under LD 16:8 or constant light have been treated with melatonin as above in winter–early spring.[72] Both 20 and 80 μg/l raised plasma melatonin levels, in this case, the levels found after the higher treatment were more similar to the natural nighttime levels. Plasma melatonin levels in fish sampled 3 to 5 h after being transferred to melatonin-free water were similar to natural daytime levels.[72] The development of full maturation, i.e., the presence of ovulated eggs or spawnings, was studied daily. At the end of the experiments (after a little more than 2 months), maturation had occurred in most (75–100% in the different groups) females kept under LD 16:8 or constant light, but not in controls kept under LD 8:16 (0–3%).[72] Melatonin treatment did not influence the proportion of females maturing, nor was the pace of maturation influenced, except that under LD 24:0 the higher dose slowed maturation marginally.[72] GSI increased under LD 16:8 and LD 24:0, but was not influenced by melatonin treatment.[72] Oocyte maturation, as studied histologically, was not influenced by melatonin treatment. LD 8:16, on the other hand, suppressed GSI and oocyte maturity below the status found in initial controls.[72] The results suggest that melatonin plays no major role in the short photoperiod inhibition of stickleback maturation.[67,72] Actually, there is little evidence that melatonin is involved in control of seasonal reproduction in nonmammalian vertebrates.[70] The pineal organ may influence reproduction in other ways, for example, by the extensive innervation of the brain from the pineal organ, which occurs in many fishes, including the stickleback.[73]

Melatonin levels have been measured in extracts of pooled brains from sticklebacks sampled 2 to 3 h after sunset from two Polish rivers in different seasons.[74] At both localities, brain melatonin levels were higher in spring and autumn than in winter and summer. These authors suggest that the high melatonin levels observed, especially in March and October, set a time frame for the time of spawning and suggest an inhibitory role for melatonin in stickleback reproduction.[74] Seasonal correlations alone are, however, hardly sufficient for this causal conclusion, especially because the natural photoperiods in spring do not suppress stickleback breeding.[56]

7.6.4 Aromatase

Some androgens such as T can be aromatised to oestrogens, whereas others, such as 11KT, cannot. Aromatase is present in the brain of vertebrates, and the aromatase activity is particularly high in the brains of teleosts, including the stickleback.[75] The biological role of the high aromatase activity in teleosts is, however, not known.

To see if aromatase plays a role in photoperiodic control of reproduction, stickleback males kept under LD 16:8 or 8:16 in winter were implanted with Silastic capsules filled with the steroidal aromatase inhibitor 1,4,6-androstatriene-3,17-dione (ATD) or the nonsteroidal aromatase inhibitor Fadrozol (CGS16949).[76] Nonsteroidal inhibitors are less likely than steroidal ones to interfere with other steroid-hormone functions and, therefore, more reliable. In fish implanted with empty capsules, the kidney hypertrophied under long, but not under short, photoperiod, whereas most fish displayed active spermatogenesis under short, but not under long, photoperiod. Both ATD and Fadrozol decreased the proportion of fish with active spermatogenesis. The aromatase inhibitors did not influence the kidney under LD 16:8, whereas ATD, and even more so Fadrozol, had strong stimulatory effects under LD 8:16.[76] In a further study, the effects of aromatase inhibitors on kidney development were confirmed.[77] In addition, the expressions of FSH-β and LH-β were higher under LD 16:8 than under LD 8:16. The expressions were not influenced by aromatase inhibitors under LD 16:8, whereas under LD 8:16 FSH-β expression was increased by Fadrozol, and LH-β expression was increased by both ATD and Fadrozol.[77]

The results given earlier suggest that aromatisation is important for the inhibitory effect of short photoperiod on maturation in the stickleback. An aromatase effect on fish reproduction has also been observed in young Atlantic salmon, where implants of aromatase inhibitor capsules increase the proportion of males undergoing maturation as parr.[78] It is not known how and on what level the aromatase effects on stickleback and salmon maturation are exerted. In some cases, feedback effects of gonadal hormones on fish GTH secretion in fishes have been found to be aromatase dependent.[55] The feedback effects known in the stickleback, however, appear not to be aromatase dependent (see the following section).

7.6.5 Gonadotropic Hormones and Photoperiods

Long photoperiod in combination with high temperature stimulates androgen-dependent characters and GTH expression, whereas when male sticklebacks were exposed to photoperiods of LD 8:16 or 16:8 combined with temperatures of 7 or 20°C in winter, the LD 8:16 20°C group was the only one showing active spermatogenesis.[45] The LD 16:8 20°C fish displayed the most advanced maturation (highest KEH), and the LD 8:16 20°C fish the lowest KEH.[45] Levels of FSH-β and LH-β mRNA were also significantly lower in the LD 8:16 20°C fish than in any other group. Levels of LH-β mRNA in the LD 16:8 20°C group were also higher than in the LD 16:8 7°C group.[45] Also, in another study, both the expression of FSH-β and LH-β was higher in male sticklebacks kept under LD 16:8 than under LD 8:16 combined with high temperature in winter.[77]

Photoperiod can alter the sensitivity of steroid feedbacks on GTH secretion in mammals. For instance, lower doses of T are needed to suppress circulating LH and FSH levels in castrated golden hamster under a nonstimulatory short photoperiod, than under a stimulatory long photoperiod.[79] Photoperiodic effects on feedback mechanisms have later been extensively studied in sheep, where feedback effects are less negative under short photoperiods, which in this species stimulates reproduction.[80]

Changes in feedback may also be part of the mechanisms by which photoperiod affects reproduction in fishes. To study this possibility, stickleback males were exposed to different treatments under LD 8:16 and 16:8 at 20°C in winter.[77] The fish were castrated and implanted with empty capsules or capsules containing T, or 11KA, or sham-operated. After about 40 d under these conditions, the fish were sampled. In agreement with previous studies, KSI was higher in sham-operated fish under the stimulatory LD 16:8 than under the nonstimulatory 8:16.[77] KSI was low under both photoperiods in castrated males and increased after treatment with 11KA, but not with T. Both the expression of FSH-β and of LH-β was higher in sham-operated males kept under LD 16:8 than under LD 8:16.[77] Under LD 16:8, LH-β expression declined in castrated males, whereas under LD 8:16 it was already low in the controls. Compared to castrated controls, LH-β expression was stimulated by both 11KA, and more effectively by T, under both photoperiods. Photoperiod did not affect LH-β expression in the different castration treatments. Thus, LH-β expression was under similar positive feedback control under both photoperiods.[77] The effects of treatments on FSH-β expression were, however, completely different under the two light regimes. Under the long photoperiod, castration lowered FSH-β expression and the androgen treatments raised it, indicating a positive feedback. In contrast, castration under short photoperiod increased FSH-β expression and androgen treatment decreased it.[77] Both the positive and negative feedback effects on GTH expression were exerted by both the aromatisable androgen T and by the nonaromatisable androgen 11KA, indicating that aromatisation is not of critical importance in this respect.[77] It appears likely that the onset of breeding is controlled by FSH, rather than by LH, because FSH-β expression peaks earlier in the season than LH-β expression.[22] A negative feedback under the nonstimulatory short photoperiod could inhibit reproduction, and the positive feedback under a stimulatory long photoperiod could accelerate reproduction to full maturation. This might also explain why sexual maturation in the stickleback is of an all-or-nothing type. The proportion of sticklebacks maturing changes gradually with progressing season and increasing photoperiod, but each individual either matures or does not.[59]

REFERENCES

1. Wootton, R.J., *The Biology of Sticklebacks,* Academic Press, London, 1976.
2. Guderley, H.E., Physiological ecology and evolution of the threespine stickleback, in *The Evolutionary Biology of the Threespine Stickleback,* Bell, M.A. and Foster, S.A., Eds., Oxford University Press, Oxford, 1996, p. 85.

3. Andersson, E., Mayer, I., and Borg, B., Inhibitory effect of 11-ketoandrostenedione and androstenedione on spermatogenesis in the three-spined stickleback, *Gasterosteus aculeatus*, *J. Fish Biol.*, 33, 835, 1988.
4. Sokolowska, E. and Kulczykowska, E., Annual reproductive cycle in two free living populations of three-spined stickleback (*Gasterosteus aculeatus* L.): patterns of ovarian and testicular development, *Oceanologia*, 48, 103, 2006.
5. Lemke, M.J., Chiva, M., Coyle, B., Kulak, D., MacKay, M., and Kasinsky, H.E., Variability of sperm nucelar basic proteins in the three-spined stickleback and related species of Gasterosteidei, *Comp. Biochem. Physiol. B,* 122, 339, 1999.
6. Urich, K., *Comparative Animal Biochemistry,* Springer-Verlag, Berlin, Heidelberg, 1994.
7. Gimenez-Bonafé, P., Laszczak, M., Kasinsky, H.E., Lemke, M.J., Lewis, J.D., Iskander, M., He, T., Ikonomou, M.G., White, F.M., Hunt, D.F., Chiva, M., and Ausio, J., Characterization and evolutionary relevance of the sperm nuclear basic proteins from stickleback fish, *Mol. Reprod. Dev.*, 57, 185, 2000.
8. Elofsson, H., McAllister, B.G., Kime, D.E., Mayer, I., and Borg, B., Long lasting stickleback sperm: is ovarian fluid a key to success in freshwater?, *J. Fish. Biol.,* 63, 240, 2003.
9. Zbinden, M., Sperm Allocation and Sperm Competition in the Three-Spined Stickleback (*Gasterosteus aculeatus*), Doctoral thesis, Rheinischen Friedrich-Willhelms-Universität, Bonn, Germany, 2002.
10. Elofsson, H., van Look, K., Borg, B., and Mayer, I., Influence of salinity and ovarian fluid on sperm motility in the fifteen-spined stickleback, *J. Fish Biol.,* 63, 1429, 2003.
11. Elofsson, H., Van Look, K.J.W., Sundell, K., Sundh, H., and Borg, B., Stickleback sperm saved by salt in ovarian fluid, *J. Exp. Biol.,* accepted, 2006.
12. Le Comber S.C., Faulkes C.G., Van Look K.J.W., Holt W.V., and Smith C., Recovery of sperm activity after osmotic shock in the three-spined stickleback: implications for pre-oviposition ejaculation, *Behaviour*, 141, 1555, 2004.
13. Sevenster-Bol, A.C.A., On the causation of drive reduction after a consummatory act (in *Gasterosteus aculeatus* L.), *Arch. Néerl. Zool.*, 15, 175, 1962.
14. de Fraipont, M., FitzGerald, G.J., and Guderley, H., Age-related differences in reproductive tactics in the three-spined stickleback, *Gasterosteus aculeatus*, *Anim. Behav.*, 46, 961, 1993.
15. Ziuganov, V.V. and Khlebovich, V.V., Analysis of mechanisms which determine differences in the reaction of spermatozoa to water salinity in the three-spined stickleback marine and fresh-water forms, *Ontogenez,* 10, 506, in Russian, 1979.
16. Borg, B., Androgens in teleost fishes, *Comp. Biochem. Physiol.,* 109C, 219, 1994.
17. Mayer, I., Borg, B., and Schulz, R., Seasonal changes in and effect of castration/androgen replacement on the plasma levels of five androgens in the male three-spined stickleback, *Gasterosteus aculeatus* L., *Gen. Comp. Endocrinol.*, 79, 23, 1990.
18. Páll, M., Mayer, I., and Borg, B., Androgen and behavior in the male three-spined stickleback, *Gasterosteus aculeatus* I. — development of 11-ketotestosterone levels during the nesting cycle, *Horm. Behav.*, 41, 377, 2002.
19. Páll, M., Mayer, I., and Borg, B., Androgen and behavior in the male three-spined stickleback, *Gasterosteus aculeatus* II. — castration and 11-ketoandrostenedione effects on courtship and parental care during the nesting cycle, *Horm. Behav.*, 42, 337, 2002.
20. Páll, M.K., Hellqvist, A., Schmitz, M., Olsson, P.-E., Mayer, I., and Borg, B., Changes in reproductive physiology and behaviour over the nesting cycle in male three-spined sticklebacks, *Gasterosteus aculeatus*, *J. Fish Biol.*, 66, 1400, 2005.

21. Borg, B. and Mayer, I., Androgens and behaviour in the three-spined stickleback, *Behaviour*, 132, 1025, 1995.
22. Hellqvist, A., Schmitz, M., Mayer, I., and Borg, B., Seasonal changes in expression of LH-β and FSH-β in male and female three-spined stickleback, *Gasterosteus aculeatus, Gen. Comp. Endocrinol.*, 145, 263, 2006.
23. Jakobsson, S., Borg, B., Haux, C., and Hyllner, S.J., An 11-ketotestosterone induced kidney-secreted protein: the nest building glue from male three-spined stickleback, *Gasterosteus aculeatus*, *Fish Physiol. Biochem.*, 20, 79, 1999.
24. Maeda, M. and Sugimori, D., Adhesive substance of *Gasterosteus aculeatus aculeatus, Chem. Lett.,* 733, 1998.
25. Jones, I., Lindberg, C., Jakobsson, S., Hellqvist, Hellman, U., Borg, B., and Olsson, P.-E., Spiggin: an androgen regulated extraorganismal adhesive with structural similarities to von Willebrand factor-related protein. *J. Biol. Chem.*, 276, 17857, 2001.
26. Kawasaki, F., Katsiadaki, I., Scott, A.P., Matsubara, T., Osatomi, K., Soyano, K., Hara, A., Arizono, K., and Nagae, M., Molecular cloning of two types of spiggin cDNA in three-spined stickleback, *Gasterosteus aculeatus, Fish Physiol. Biochem.,* 28, 425 (abstract), 2003.
27. Borg, B., Antonopoulou, E., Andersson, E., Carlberg, T., and Mayer, I., Effectiveness of several androgens in stimulating kidney hypertrophy, a secondary sexual character, in castrated male three-spined stickleback, *Gasterosteus aculeatus*, *Can. J. Zool.*, 71, 2327, 1993.
28. Mayer, I., Borg, B., and Schulz, R., Conversion of 11-ketoandrostenedione to 11-ketotestosterone by blood cells of six fish species, *Gen. Comp. Endocrinol.*, 77, 70, 1990.
29. Katsiadaki, I., Scott, A.P., Hurst, M.R., Matthiessen, P., and Mayer, I., Detection of environmental androgens: a novel method based on enzyme-linked immunosorbent assay of spiggin, the stickleback *Gasterosteus aculeatus* glue protein, *Environ. Toxicol. Chem.*, 9, 1946, 2002.
30. Zentel, H.J., Experimentelle Untersuchungen zur Geschlechts-differenzierung beim Guppy *Poecilia reticulata* (Peters), Doctoral thesis, Johann-Gutenberg Universität, Mainz, Germany, 1988.
31. Borg, B., Mayer, I., Lambert, J.G.D., Granneman, J., and Schulz, R., Metabolism of androstenedione and 11-ketotestosterone in the kidney of the three-spined stickleback, *Gasterosteus aculeatus*, *Gen. Comp. Endocrinol.*, 86, 248, 1992.
32. Olsson, P.-E., Berg, A. H., von Hofsten, J., Grahn, B., Hellqvist, A., Larsson, A., Karlsson, J., Modig, C., Borg, B., and Thomas, P., Molecular cloning and characterization of a nuclear androgen receptor activated by 11-ketotestosterone, *Reprod. Biol. Endocrinol.*, 3, 37, 2005.
33. Jakobsson, S., Mayer, I., Schulz, R.W., Blankenstein, M.A., and Borg, B., Specific binding of 11-ketotestosterone in an androgen target organ, the kidney of the male three-spined stickleback, *Gasterosteus aculeatus*, *Fish Physiol. Biochem.*, 15, 459, 1996.
34. Katsiadaki, I., Scott, A.P., and Mayer, I., The potential of the three-spined stickleback (*Gasterosteus aculeatus* L.) as a combined biomarker for oestrogens and androgens in European waters, *Mar. Environ. Res.*, 54, 725, 2002.
35. Wedekind, C., Meyer, P., Frischknecht, M., Niggli, U.A., and Pfander, H., Different carotenoids and potential information content of red coloration of male three-spined stickleback, *J. Chem. Ecol.*, 24, 787, 1998.
36. Brønseth, T. and Folstad, I., The effect of parasites on courtship dance in threespine sticklebacks: more than meets the eye?, *Can. J. Zool.*, 75, 589, 1997.

37. Bakker, T.C.M. and Mundwiler, B., Pectoral fin size in a fish species with paternal care: a condition-dependent, sexual trait revealing infection status, *Freshwater Biol.*, 41, 543, 1999.
38. Künzler, R. and Bakker, T.C.M., Pectoral fins and paternal quality in sticklebacks, *Proc. R. Soc. Lond. B.*, 267, 999, 2000.
39. Hechter, R.P., Moodie, P.F., and Moodie, G.E.E., Pectoral fin asymmetry, dimorphism and fecundity in the brook stickleback, *Culaea inconstans*, *Behaviour*, 137, 999, 2000.
40. Hoffmann, E. and Borg, B., Sex differences in pectoral muscles but not in pectoral fins in the three-spined stickleback *Gasterosteus aculeatus*, *J. Fish Biol.*, 68, 1451, 2006.
41. Suzuki, K., Kawauchi, H., and Nagahama, Y., Isolation and characterization of two distinct gonadotropins from chum salmon pituitary glands, *Gen. Comp. Endocrinol.*, 71, 92, 1988.
42. Quérat, B., Sellouk, A., and Salmon, C., Phylogenetic analysis of the vertebrate glycoprotein hormone family including new sequences of sturgeon (*Acipenser baeri*. beta subunits of the two gonadotropins and the thyroid-stimulating hormone, *Biol. Reprod.*, 63, 222, 2000.
43. Natio, N., Suzuki, K., Nozaki, M., Swanson, P., Kawauchi, H., and Nakai, Y., Ultrastructural characteristics of two distinct gonadotropes (GTH I-cells and GTH II-cells. in the pituitary of rainbow trout *Oncorhynchus mykiss, Fish Physiol. Biochem.*, 11, 241, 1993.
44. Slijkhuis, H., Ultrastructural evidence for two types of gonadotropic cells in the pituitary gland of the male three-spined stickleback, *Gasterosteus aculeatus*, *Gen. Comp. Endocrinol.*, 36, 639, 1978.
45. Hellqvist, A., Bornestaf, C., Borg, B., and Schmitz, M., Cloning and sequencing of the FSH-β and LH β-subunit in the three-spined stickleback, *Gasterosteus aculeatus*, and effects of photoperiod and temperature on LH-β and FSH-β mRNA expression, *Gen. Comp. Endocrinol.*, 135, 167, 2004.
46. Hellqvist, A., Schmitz, M., Lindberg, C., Olsson, P.-E., and Borg, B., LH-β and FSH-β mRNA expression in nesting and post-breeding three-spined stickleback, *Gasterosteus aculeatus*, and effects of castration on expression of LH-β, FSH-β and spiggin mRNA, *Fish Physiol. Biochem.*, 25, 311–317, 2001.
47. Swanson, P., Salmon gonadotropins: reconciling old and new ideas, in *Proceedings of the 4th International Symposium on Reproductive Physiology of Fish,* University of East Anglia, Norwich, U.K., Scott, A.P., Sumpter, J.P., Kime, D.E., and Rolfe, M.S., Eds., Fish Symp. 91, Sheffield, 1991, p. 2.
48. Miura, T. and Miura C., Japanese eel: a model for analysis of spermatogenesis, *Zool. Sci.*, 18, 1055, 2001.
49. Adams, B.A., Vickers, E.D., Warby, C., Park, M., Fischer, W.H., Craig, A.G., Rivier, J.E., and Sherwood, N.M., Three forms of gonadotropin-releasing hormone, including a novel form, in a basal salmonid, *Coregonus clupeaformis, Biol. Reprod.*, 67, 232, 2002.
50. Andersson, E., Bogerd, J., Borg, B., Sharp, P.J., Sherwood, N.M., and Goos, H.J.Th., Characterization and localization of gonadotropin-releasing hormone in the brain and pituitary of the three-spined stickleback, *Gasterosteus aculeatus*, *Cell Tissue Res.*, 279, 485, 1995.
51. Borg, B., Goos, H.J.Th., and Terlou, M., LHRH-immunoreactive cells in the brain of the three-spined stickleback, *Gasterosteus aculeatus* L. (Gasterosteidae), *Cell Tissue Res.*, 226, 695, 1982.

52. Andersson, E., Borg, B., and Goos, H.J.Th., Temperature, but not photoperiod, influences gonadotropin-releasing hormone binding in the pituitary of the three-spined stickleback, *Gasterosteus aculeatus*, *Gen. Comp. Endocrinol.*, 88, 111, 1992.
53. Lindgren, S., Damber, J.-E., and Carsten, H., Compensatory testosterone secretion in unilaterally orchidectomized rats, *Life Sci.,* 18, 1203, 1976.
54. Hellqvist, A., Mayer, I., and Borg, B., Effects of hemi-castration on plasma steroid levels in two teleost fishes; the three-spined stickleback, *Gasterosteus aculeatus*, and the Atlantic salmon, *Salmo salar*, *Fish Physiol. Biochem.,* 26, 107, 2002.
55. Crim, L.W., Peter, R.E., and Billard, R., Onset of gonadotropic hormone accumulation in immature trout pituitary gland in response to estrogen or aromatizable androgen steroid hormone, *Gen. Comp. Endocrinol.,* 44, 374, 1981.
56. Baggerman, B., An experimental study on the timing of breeding and migration in the three-spined stickleback (*Gasterosteus aculeatus* L.), *Arch. Néerl. Zool.*, 12, 105, 1957.
57. Turek, F.W. and van Cauter, E., Rhythms in reproduction, in Knobil, E. and Neill, J.D. Eds., *The Physiology of Reproduction*, Vol. 2, 2nd ed., Raven Press, New York, 1994, p. 487.
58. Bornestaf, C. and Borg, B., Endogenous breeding cycles in male threespine sticklebacks, *Gasterosteus aculeatus*, *Behaviour,* 137, 921, 2000.
59. Baggerman, B., The role of biological rhythms in the photoperiod regulation of seasonal breeding in the stickleback *Gasterosteus aculeatus* L., *Neth. J. Zool.*, 35, 14, 1985.
60. Falk, H. and Gwinner, E., Timing of photorefractoriness in the European starling — significance of photoperiod early and late in the reproductive cycle, *Biol. Reprod.,* 39, 1004, 1988.
61. Duston, J. and Bromage, N., Photoperiodic mechanisms and rhythms of reproduction in the female rainbow trout, *Fish Physiol. Biochem.,* 2, 35, 1986.
62. van Veen, Th., Ekström, P., Borg, B., and Møller, M., The pineal complex of the three-spined stickleback, *Gasterosteus aculeatus* L., A light-, electron microscopic and fluorescence histochemical investigation, *Cell Tissue Res.,* 209, 11, 1980.
63. Foster, R.G., Grace, M.S., Provencio, I., Degrip, W.J., and Garcia-Fernandez. J.M., Identification of vertebrate deep brain photoreceptors, *Neurosci. Biobehav. Rev.,* 18, 541, 1994.
64. Borg, B., Extraretinal photoreception involved in photoperiodic effects on reproduction in male three-spined sticklebacks, *Gasterosteus aculeatus*, *Gen. Comp. Endocrinol.,* 47, 84, 1982.
65. Menaker, M., Roberts, R., Elliott, J., and Underwood, H., Extraretinal light perception in the sparrow. III. The eyes do not participate in photoperiodic photoreception, *Proc. Natl. Acad. Sci. U.S.A.,* 67, 320, 1970.
66. Bornestaf, C. and Borg, B., Extraretinal photoreception is more important than retinal photoreception for sexual maturation in the three-spined stickleback (*Gasterosteus aculeatus*), in Bornestaf, C., *Mechanisms in the Photoperiodic Control of Reproduction in the Three-Spined Stickleback Gasterosteus aculeatus*, Doctoral thesis, Stockholm University, 2000, chap. 4.
67. Mayer, I., Bornestaf, B., Wetterberg, L., and Borg, R., Melatonin does not prevent long photoperiod stimulation of secondary sexual characters in the male three-spined stickleback *Gasterosteus aculeatus*, *Gen. Comp. Endocrinol.*, 108, 386, 1997.
68. Carter, D.S. and Goldman, B.D., Antigonadal effects of timed melatonin infusion in pinealectomized male Djungarian hamsters (*Phodopus sungorus sungorus*): duration is the critical parameter, *Endocrinology,* 113, 1261, 1983.

69. Bittman, E.L., Demsey, R.J., and Karsch, F.J., Pineal melatonin secretion drives the reproductive response to daylength in the ewe, *Endocrinology,* 113, 2276, 1983.
70. Mayer, I., Bornestaf, B., and Borg, R., Melatonin in non-mammalian vertebrates: physiological role in reproduction?, *Comp. Biochem. Physiol.,* 118A, 515, 1997.
71. Borg, B. and Ekström, P., Gonadal effects of melatonin in the three-spined stickleback, *Gasterosteus aculeatus* L., during different seasons and photoperiods, *Reprod. Nutr. Dev.*, 21, 919, 1981.
72. Bornestaf, C., Mayer, I., and Borg, B., Melatonin and maturation pace in female three-spined stickleback, *Gasterosteus aculeatus*, *Gen. Comp. Endocrinol.*, 122, 341, 2001.
73. Ekström, P., Central nervous connections of the pineal organ and retina in the teleost *Gasterosteus aculeatus* L., *J. Comp. Neurol.*, 226, 321, 1984.
74. Sokolowska, E., Kalamarz, H., and Kulczykowska, E., Seasonal changes in brain melatonin concentration in the three-spined stickleback (*Gasterosteus aculeatus*): towards an endocrine calendar, *Comp. Biochem. Physiol. A*, 139, 365, 2004.
75. Borg, B., Timmers, R.J.M., and Lambert, J.G.D., Aromatase activity in the brain of the three-spined stickleback, *Gasterosteus aculeatus*. I. Distribution and effects of season and photoperiod, *Exp. Biol.*, 47, 63, 1987.
76. Bornestaf, C., Antonopoulou, E., Mayer, I., and Borg, B., Effects of aromatase inhibitors on reproduction in male three-spined sticklebacks, *Gasterosteus aculeatus*, exposed to long and short photoperiods, *Fish Physiol. Biochem.*, 16, 419, 1997.
77. Hellqvist, A., Schmitz, M., and Borg, B., Effects of photoperiod on the feedback mechanisms on the brain-pituitary-gonadal axis in the three-spined stickleback, *Gasterosteus aculeatus,* in Hellqvist, A., *The Brain-Pituitary-Gonadal Axis and Gonadotropic Hormones in the Three-Spined Stickleback Gasterosteus aculeatus*, Doctoral thesis, Stockholm University, 2003, chap. 5.
78. Antonopoulou, E., Mayer, I., Berglund, I., and Borg, B., Effects of aromatase inhibitors on sexual maturation in Atlantic salmon, *Salmo salar*, male parr, *Fish Physiol. Biochem.,* 14, 15, 1995.
79. Turek, F., The interaction of photoperiod and testosterone in regulating serum gonadotropin levels in castrated male hamsters, *Endocrinology*, 101, 1210, 1977.
80. Rosa, H.J.D. and Bryant, M.J., Seasonality of reproduction in sheep (review), *Small Ruminant Res.,* 48, 155, 2003.

8 Hormonal Control of Reproductive Behaviour in the Stickleback

Ian Mayer and Miklós Páll

CONTENTS

8.1 INTRODUCTION

Teleost fishes are the largest and most diverse vertebrate group, with over 25,000 described species. Fish display an almost unique variety of reproductive strategies and behaviours, and for this reason have attracted the attention of researchers encompassing most disciplines of reproductive biology, including reproductive endocrinology and behaviour. The regulation of reproductive behaviour in fishes is complex, and involves the interaction of a number of physiological as well as environmental and social factors. Although considerable advances have been made in understanding the hormonal control of reproductive behaviour in fish,[1,2] these earlier works are now outdated and significant gaps in our knowledge still exist. One of the species that has contributed greatly to our better understanding of hormone-behavioural relationships in fish is the three-spined stickleback, *Gasterosteus aculeatus*. This small, ubiquitous fish, with its remarkable array of reproductive behaviours has long attracted the attention of behavioural ecologists, dating back to the classical studies of Tinbergen.[3] More recently, the stickleback has become an important model species for studying the hormonal control of reproductive behaviour.[4,5]

There are a number of reasons why the stickleback has become such an important model species in this field of research. First and foremost, the stickleback displays a variety of distinctive reproductive behaviours, which are most pronounced in the males. Breeding males display a variety of strikingly different behaviours over the duration of its nesting cycle, including parental behaviour. In common with other parental-caring species, each nesting cycle of the male stickleback can be temporally divided into an initial sexual phase followed by a parental phase. The sexual phase is characterised by behaviours including territoriality, nest building, and courtship, whereas the dominant behaviour shown during the parental phase is nest fanning, during which the male fans oxygenated water over the eggs using its pectoral fins. Other advantages of the stickleback include its readiness to breed and to display all aspects of male reproductive behaviour in captivity. Further, the suitability of the stickleback for studying hormone-behaviour relationships is enhanced by the fact that males can be easily gonadectomised without testicular remnants regenerating, allowing for classical gonadectomy-hormone replacements studies. Another, often overlooked advantage of the stickleback is that it is a very stress-tolerant species. Individuals will resume feeding and display apparently normal behaviour within minutes after handling.

Thus, its ability to be easily gonadectomised, its readiness to breed in captivity, together with its propensity to display a variety of distinct male reproductive behaviours makes the stickleback an ideal model species to study hormone-behavioural relationships in teleost fishes.

8.2 REPRODUCTIVE BEHAVIOUR IN THE STICKLEBACK

The overriding reason why the three-spined stickleback has commonly been chosen as a species to study hormone-behaviour relationships is that the males display a unique repertoire of reproductive behaviours, including those associated with parental care. During the stickleback's breeding season, which occurs between late spring and early summer, males will go through as many as three nesting cycles, during which the males care for the developing eggs. As mentioned previously, the nesting cycle can be divided into two consecutive phases, an initial sexual phase, followed by a parental phase.

Sexual phase. Mature males will establish a breeding territory that they will guard aggressively against other male sticklebacks, as well as other small fish that intrude into their territory. Soon after establishing its territory, a male will start to build a nest, using for this purpose a glue (spiggin) secreted by the kidney. It should be noted here that in the ethological literature a nest-building phase is in itself often recognised as a separate phase. On completion of its nest the male will actively court gravid females. This well-documented courtship behaviour involves the male swimming toward the female in a series of rapid side-to-side jumps known as the "zigzag dance," after which he immediately swims back to his nest. If the female is receptive (fully gravid), she will follow the male back to the nest and deposit her eggs. The male immediately follows the female through the nest, fertilises the eggs, and then chases the female away.

Parental phase. During the parental phase, which occurs immediately following spawning, the male protects and cares for his nest during egg development. The most prominent behaviour shown by the male during this secondary phase is "fanning," which involves the male positioning himself obliquely head-down in front of the nest and actively fanning oxygenated water over the eggs using his pectoral fins. The male will defend the nest and its eggs vigorously until hatching, which takes approximately 1 week, depending on the water temperature. Besides this parental fanning, a male will also perform fanning over his nest, even before a female has spawned her eggs. This preparental fanning has been termed *displacement* or *irrelevant fanning*. Although the function of displacement fanning is still not clearly understood, it has been suggested that males perform this fanning behaviour to attract females. Displacement fanning has been shown to increase male mating success in a number of fish showing parental care. For example, the mating success of male Florida flagfish, *Jordanella floridae*, has been shown to be positively correlated to the level of premating fanning,[6] while female 15-spined sticklebacks, *Spinachia spinachia*, were found to prefer males that fanned their nests for shorter, more frequent bouts.[7] In the three-spined stickleback, although displacement fanning has yet to be positively correlated to mating success, an early study found that males increased displacement fanning when sexually ripe females entered their territory.[8]

Although the broad repertoire of reproductive behaviours displayed by males has been the primary reason that the three-spined stickleback has endeared itself to behavioural ecologists, it is for the same reason that the stickleback has become a preeminent species in the study of hormone-behaviour relationships in fish. In this respect, the stickleback offers another major advantage in that a number of its reproductive behaviours are readily quantifiable, notably courtship and fanning behaviour. In many fish, the expression of reproductive behaviour is rather ambiguous, and its evaluation can be somewhat arbitrary, and very much dependent upon the individual observer. In contrast, the stickleback displays a number of very distinct unambiguous behaviours, most of which, such as courtship and fanning behaviour, can be accurately quantified in terms of the number of behavioural acts (zigzags, fanning bouts) performed over a fixed time interval.

The reproductive behaviour displayed by female sticklebacks is far less pronounced than that shown by the males, which tends to be a general rule among fishes. The most noticeable female behaviour is the so-called "head-up" posture displayed by sexually ripe females, in the presence of a courting male. The display of this behaviour signals to the male that the female is ready to lay her eggs (ovulated) in the nest.

8.3 GONADAL STEROIDS AND REPRODUCTIVE BEHAVIOUR

In early studies, it was found that castration resulted in the abolition or reduction of all types of reproductive behaviour in male sticklebacks, and that these behaviours could be stimulated in both intact and castrated males following androgen administration.[9–13] These early gonadectomy/hormone-replacement studies clearly demonstrated that the gonadal steroids played a major role in the control of male reproductive behaviour in the stickleback. However, some caution should be applied to these earlier studies as the steroid mostly used was the potent synthetic androgen methyl testosterone (MT), whose effects may not be physiological, especially at the high doses generally used in these earlier studies. Further early evidence supporting a functional role for gonadal steroids in the control of reproductive behaviour in the stickleback came from a study that investigated the effect of an androgen inhibitor on male behaviour. In this study, male sticklebacks given intraperitoneal injections of the androgen inhibitor cyproterone acetate showed reduced aggressive and courtship behaviour, as well as reduced displacement fanning.[14]

In a more recent study, Borg[15] studied the effect of implanting sticklebacks that had been castrated when in breeding condition, but before nest-building, with the natural androgens testosterone (T), androstenedione, 5α-dihydrotestosterone, or 11-ketoandrostenedione (11kA). Again, reproductive behaviour was abolished by castration, and androgen replacement restored to varying degrees all types of reproductive behaviour, including nest building, courtship, and fanning. Of the four androgens tested, 11kA was the most effective in restoring these behaviours. This study further confirmed a functional role for the gonadal steroids in the control of male reproductive behaviour in the stickleback. However, as in other fish, it was expected that the testes of the stickleback would synthesise a number of different steroids, and that steroidogenic activity would vary over the seasonal reproductive cycle. To determine more precisely the major gonadal steroids produced by the testes, Borg et al.[16] studied the steroid metabolism in the testes of both breeding and nonbreeding male sticklebacks by incubating testicular tissue with the androgen precursor ^{3}H-pregnenolone. The results showed that the level of steroid metabolism was considerably higher in the testes of breeding compared to nonbreeding males, and that the major steroid produced by the testes of breeding sticklebacks was 11kA. However, in common with most teleosts, the major androgen measured in the blood of breeding male sticklebacks was found to be the 11-ketotestoserone (11kT), and 11kT was the only measured androgen to show a pronounced peak in plasma level during the breeding season[17] (Figure 8.1). This apparent paradox was resolved following the demonstration that 11kA could be readily converted to 11kT *in vitro* by blood cells of this species.[18] This was further confirmed by castration and implantation experiments in sticklebacks, that showed that while castration reduced plasma steroid levels, castrated males given 11kA implants showed elevated levels of plasma 11kT but not 11kA.[17]

The 11-ketoandrogens, in addition to being most potent in stimulating reproductive behaviour, have also been shown to be the most effective androgens in stimulating secondary sexual characters (e.g., kidney hypertrophy) in male sticklebacks.[19] These observations are in line with the general consensus that the 11-ketoandrogens (in particular, 11kT) are physiologically the most important androgens in male teleosts, including the stickleback.[20]

In most studied teleosts, steroid synthesis by the testes of breeding males switches from androgen to progestin production during spermiation,[21] often at a time

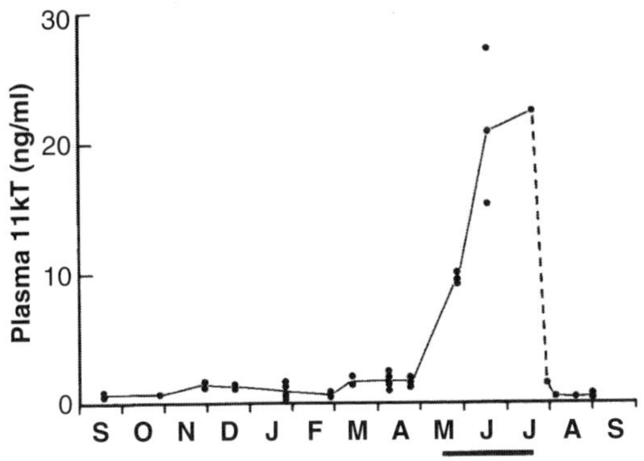

FIGURE 8.1 Plasma levels of 11kT in male three-spined sticklebacks sampled at monthly intervals in the field (except for the two points separated by the broken line, which were males sampled in captivity). Each point represents the value from a pooled plasma sample (10–15 males). The breeding season is indicated by the solid bar. (After Mayer, I. et al., *Can. J. Zool.* 68, 1360, 1990.)

corresponding to the period of most intense male sexual behaviour. For example, in salmonids, the progestin 17,20-dihydroxy-4-pregnen-3-one (17,20-P) and not 11kT is the dominant plasma steroid in actively spawning males.[22,23] Although 17,20-P has been implicated in the control of spawning behaviour in male salmonids,[24] conclusive evidence for a functional role for this progestin in the control of reproductive behaviour in fish is still lacking. A possible role for progestins in stickleback behaviour is presently unknown as plasma levels of 17,20-P are constantly low or nondetectable in breeding males.[17]

8.4 ANDROGEN–BEHAVIOUR RELATIONSHIPS OVER THE NESTING CYCLE

Detailed studies on the temporal changes in hormone levels over the breeding cycle in teleost fish, especially relating to reproductive behaviour, are still relatively few. However, a number of studies have reported androgen changes over the nesting cycle in male fish showing paternal care, including the bluegill sunfish[25] (*Lepomis macrochirus*), garibaldi[26] (*Hypsypops rubicundus*), plainfin midshipman[27] (*Porichthys notatus*), spiny damselfish[28] (*Acanthochromis polyacanthus*), and most recently the stickleback[29] (see Table 8.1 for levels). In all these studies, it was found that plasma 11kT levels were consistently higher in males in the initial sexual phase (nesting and courtship) compared to the later parental phase. Further, higher 11kT levels were reported in nonbrooding compared to brooding males of two species of pipefish (*Syngnathus acus* and *S. typhle*), a species showing sex role reversal.[30] In the stickleback, although these temporal changes in androgen levels are most pronounced for 11kT (plasma 11kT levels being 400 and 12 ng/ml during the sexual and parental phases, respectively[29]), plasma levels of T have been shown to decline similarly during the parental phase. For example, Páll et al.[31] reported that plasma T levels declined from 70 ng/ml in sexual nonspawned males to 14 and 8 ng/ml in

TABLE 8.1
Changes in Plasma 11-Ketotestosterone Levels between the Sexual and Parental Phases of the Nesting Cycle in Five Species of Parental-Caring Fish

Species	Plasma 11kT levels (ng/ml)		Reference
	Sexual Phase	Parental Phase	
Bluegill sunfish (*Lepomis macrochirus*)	56.0	5.0	Kindler et al.[25]
Garibaldi (*Hypsypops rubicundus*)	22.0	7.0	Sikkel[26]
Plainfin midshipman (*Porichthys notatus*)	12.0	0.8	Knapp et al.[27]
Spiny damselfish (*Acanthochromis polyacanthus*)	0.85	0.50	Pankhurst et al.[28]
Three-spined stickleback (*Gasterosteus aculeatus*)	About 400	12.0	Páll et al.[29]

Source: After Mayer et al. 2005.

5- and 8-d paternal males, respectively. Except for being a precursor for other steroids, the physiological role of T is still uncertain in the stickleback, as the 11-androgens are certainly the more important androgens physiologically. For example, T is far less effective in restoring reproductive behaviours in castrated males compared to 11kA,[15] as well as being far less effective in stimulating kidney hypertrophy, a pronounced male secondary sexual character, compared to 11kT.[19]

At face value, these results appear to support the concept of the *challenge hypothesis* proposed by Wingfield and colleagues.[32,33] In this hypothesis, based primarily on observations in birds, Wingfield postulates the existence of three levels of circulating androgens: a low constitutive level, a somewhat higher breeding level sufficient to maintain the expression of reproductive traits, and a physiological maximum level that is attained following social stimulation provided by male–male aggression or by interactions with receptive females. In breeding sticklebacks, the sexual phase of the male's nesting cycle is characterised by more frequent and intense male–male interactions, as well as courtship with receptive females, social interactions known to elevate androgen levels in fish (reviewed by Oliveira[34]). It could be postulated that the high androgen levels (both 11kT and T) measured in male sticklebacks during the sexual phase of their nesting cycle represents the physiological maximum, and is primarily the consequence of the more intense and frequent social interactions, including courtship, that occur during this period. During the subsequent parental phase, when social interactions decrease, plasma androgen levels decline to the lower breeding level. This prediction is also supported by the observation that, in both birds and fish, androgen levels are higher during the period of territory establishment compared to when a territory is already established.[35,36] However, under experimental conditions when nesting male sticklebacks are purposely excluded from male–male interactions,[29] they still show very high androgen levels. This observation appears to detract somewhat from the challenge hypothesis concept.

In teleost fishes as in all vertebrates, the secretion of gonadal steroids is under the control of two gonadotropic hormones, follicle stimulating hormone (FSH) and luteinising hormone (LH), both of which have been shown to stimulate androgen production in fish testes.[37] The marked decline in plasma androgen levels (both T and 11kT) during the parental phase in breeding male sticklebacks is probably ultimately under the control of the gonadotropins. This is supported by the observation by Páll et al.[31] that m-RNA expression of the β-subunits of both FSH and LH, which presumably correlates closely to the secretion and release of these hormones, declines during the parental phase in nesting males.

The intriguing question arises, why do plasma androgen levels, especially 11kT, dramatically decline as paternal males enter the parental phase, and how does this relate to mechanisms controlling reproductive behaviour in these fishes? To investigate this question further, Páll et al.[29] studied in greater detail the temporal changes in plasma 11kT levels in male sticklebacks over a complete nesting cycle, and further, how these levels correlated to changes in reproductive behaviour. Males sampled during the sexual phase, characterised by males that had built a nest and exhibited vigorous courtship behaviour toward a receptive female, had plasma 11kT levels of about 400 ng/ml. These very high 11kT levels rapidly declined to levels of about 10 ng/ml in males that had spawned with one or more females (Figure 8.2). Conversely, plasma 11kT remained high in those nest-building males that were denied access to females to spawn with. In the same study, a strong positive correlation was found between 11kT levels and the expression of male courtship behaviour

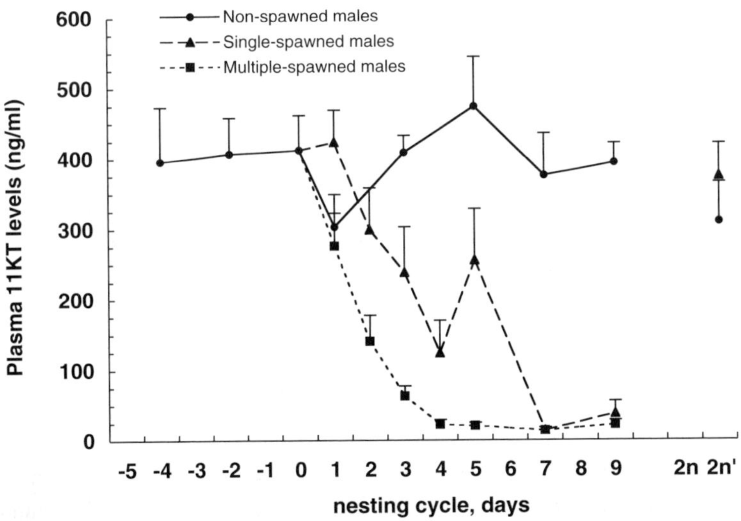

FIGURE 8.2 Temporal changes in plasma 11kT levels in nonspawned, and single- or multiple-spawned males sampled over a nesting cycle. Values represent mean ± SEM, n = 8–12, 2n = start of second nesting cycle. (After Páll, M. et al., *Horm. Behav.* 41, 377, 2002.)

(zigzag dance). During the sexual phase, corresponding to the period of elevated plasma 11kT levels, males showed high levels of courtship behaviour. In contrast, following spawning, courtship behaviour declined sharply concomitant with the decline in plasma 11kT levels. This suggests a functional relationship between 11kT levels and the expression of courtship behaviour in the stickleback.

The results of Páll et al.[38] brought out another point of contention concerning the role androgens play in the control of those behaviours shown during the initial sexual phase (territoriality, nest building, and courtship) in male sticklebacks. Both Smith and Hoar,[13] and later Borg[15] demonstrated that the castration of mature males prior to their entering their nesting cycle resulted in a complete abolition of these behaviours. In contrast, Páll et al.[38] demonstrated that courtship behaviour still continued in males that were castrated after they had entered their nesting cycle. In this study, courtship behaviour (frequency of zigzags) was measured in nesting nonspawned males that had been either castrated, castrated + 11kA implants, or sham-operated. It was observed that the level of courtship behaviour in the castrated males remained as high as in the other two groups for 3 d post castration, after which it started to decline. These results suggest that although the androgens are of primary importance in activating mechanisms controlling the expression of sexual behaviour, they appear to play a more permissive role once these behaviours have been initiated, certainly pertaining to parental behaviour. This is supported by an earlier study by Baggerman,[12] who also demonstrated that castration of males after spawning did not diminish subsequent parental fanning behaviour.

The sharp decline in plasma androgens levels (both 11kT and T) during the parental phase suggests that androgens *per se* do not play an all-important role in the control of parental behaviour in the stickleback, as in other species showing paternal care. This is supported by the results of Páll et al.[38] who studied the effects of castration and castration combined with androgen administration on fanning behaviour in spawned male sticklebacks. In this study, the levels of fanning behaviour in both castrated males and castrated males given 11kA (converted to 11kT) implants did not deviate from those shown by sham-operated males (Figure 8.3). This suggests that once a male has entered its nesting cycle, parental behaviour such as fanning occurs independent of the gonadal steroids, as illustrated by the fact that neither castration (removal of circulating androgens) nor androgen administration (elevated 11kT levels) affected the level of expression of this parental behaviour.

8.5 COST–BENEFIT TRADE-OFFS ASSOCIATED WITH HIGH ANDROGEN LEVELS

Trade-offs are fitness costs that result from the interaction between a beneficial change in one trait and a detrimental change in another.[39] In the stickleback, central to this concept is the apparent trade-offs associated with the very high androgen levels (primarily 11kT) that occur during the sexual phase of the male's nesting cycle. During this breeding period, there is a dynamic balance between the costs and benefits of the various androgen-dependent mechanisms, and that at any one

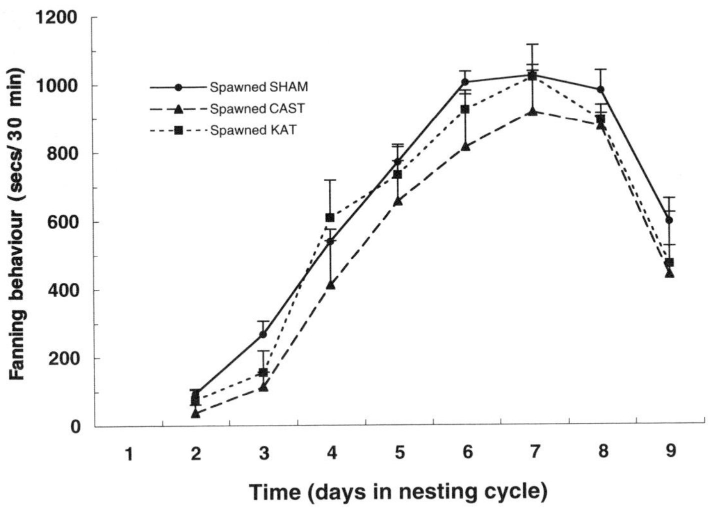

FIGURE 8.3 Temporal changes in fanning behaviour (seconds/30 min) over the nesting cycle in spawned male sticklebacks that had been either sham-operated, castrated, or castrated + 11kA treated. Values represent mean ± SEM, n = 8–12. (After Páll, M. et al., *Horm. Behav.* 41, 337, 2002.)

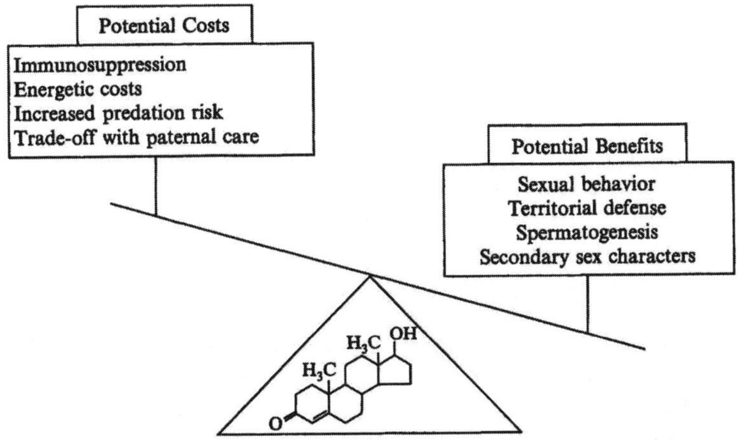

FIGURE 8.4 The selection of androgen-dependent mechanisms depends on the relationship between potential benefits vs. potential costs. For the mechanism to be adaptive, the former must outweigh the latter. (After Oliveira, R.F., *Adv. Stud. Behav.* 34, 165, 2004.)

point of time only those mechanisms where the associated benefits outweigh the associated costs will be selected for[34] (Figure 8.4).

As illustrated in Figure 8.4, there are a number of obvious benefits associated with the high androgen levels observed during the sexual phase of the stickleback's nesting cycle. In male teleosts, the androgens, notably 11kT, are of primary importance in mediating spermatogenesis, and if expressed, secondary sexual characters and reproductive behaviour.[20] In one respect the stickleback is not typical, as males show a postnuptial pattern of spermatogenesis, with spermatogenesis being completed many months before the fish enter their breeding period.[40] However, the stickleback is typical in that the males express very pronounced androgen-dependent secondary sexual characters such as breeding colours and kidney hypertrophy (production of the glue protein spiggin). Nest building, during which the male builds his nest using large amounts of spiggin threads, is an extremely active behaviour during the sexual phase. Kidney hypertrophy and spiggin-mRNA expression are positively correlated with high 11kT levels in nesting males during this period.[31] Another benefit of high androgen levels during the sexual phase is associated with their controlling role in mediating those male reproductive behaviours expressed during this reproductive period, including territoriality and associated aggressive behaviour, nest building, and, later, courtship behaviour. As mentioned previously, the sexual phase of the male's nesting cycle is characterised by more frequent and intense male–male interactions. In this context, increased androgen levels may be beneficial in stimulating the motivational systems controlling male aggressive behaviour.[34] Thus, it is likely that the very high androgen levels measured in male sticklebacks during the sexual phase could serve multiple physiological roles, stimulating both the development of secondary sexual characters as well as sexual behaviour.

The sharp decline in androgen levels during the parental phase of the nesting cycle could be explained by a number of reasons, none of which are mutually exclusive. A number of potential costs associated with high androgen levels have been identified in vertebrates, including fish (Figure 8.4), such as the suppression of immune function, increased energy consumption, increased predation risk, higher incidence of injuries from agonistic interactions, and the trade-off between parental and sexual effort (see reviews by Oliveira[34] and Wingfield et al.[41,42]). In teleosts, as in other vertebrates, a number of studies have demonstrated that high androgen levels can result in immunosupression, both by reducing leucocyte number and antibody production.[43,44] This has led to a number of hypotheses being put forward to explain the relationship between androgens and immunocompetence in vertebrates, generally based on the principle that there is a trade-off between high androgen levels and immune response (reviewed by Oliveira[34]).

The *immunocompetence handicap hypothesis*[45] predicts a trade-off between androgens and immunocompetence, stating that although high androgen levels enhance the expression of secondary sexual ornaments (providing an honest signal of male quality, and thus increasing mating success), they simultaneously suppress immune function. It should be noted here that in the case of the stickleback, kidney hypertrophy is not a more or less honest signal, but has a real function in nest building and, thus, protection of eggs. More recently, Braude et al.[46] proposed the *immunoredistribution hypothesis*, which states that elevated androgen levels promote a reversible relocation of leucocytes to tissues where they are temporarily needed. Alternatively, Wingfield et al.[33] proposed that a major cost of high androgen levels in species showing paternal care was that these androgen levels had the effect of inhibiting the development of paternal behaviour. Thus, the decrease in androgen levels during the parental phase facilitated the expression of paternal care by reducing androgen-driven behaviours such as male–male aggression and courtship. It is worth noting that in the stickleback, plasma 11kT levels again increase to very high levels (> 300 ng/ml) in males entering the sexual phase of their second nesting cycle (as shown in Figure 8.2), at a time when the expression of secondary sexual characters again become of primary importance.[38]

Although all the previously mentioned hypotheses could be applied to the three-spined stickleback, in view of the extremely high plasma androgen levels displayed by males during the sexual phase of their nesting cycle, it is logical to give most credence to the immunocompetence handicap hypothesis as a means of explaining the sharp decline in androgen levels during the subsequent parental phase.

Another possible explanation for the sharp decline in androgen levels during the parental phase is maternal effects. During spawning, a maternal cue, possibly pheromonal, could be given by either the eggs or the viscous ovarian fluid. This is supported by the observation that both courtship behaviour and plasma 11kT levels decline more rapidly in stickleback males that had spawned with multiple females compared to those spawning with only a single female.[29] Finally, it cannot be excluded that the decline in plasma androgens levels is a consequence of the physiological changes related to spermiation and sperm release, which occur during spawning. It is now well established in teleosts that at spawning, steroidogenesis in the testes switches from androgen production to producing the maturation-inducing steroid (MIS, usually a C_{21} steroid). To date, the MIS has not yet been identified in the stickleback.

8.6 PROLACTIN AND BEHAVIOUR

The studies of Páll et al.[29,38] clearly suggested that parental behaviour in male sticklebacks is also under the control of factors other than the gonadal steroids. One of the internal factors that could be responsible for the changes in reproductive behaviour over the nesting cycle and, more specifically, that could play a functional role in the control of paternal behaviour, is prolactin (PRL). This multifunctional peptide hormone has been shown to play an important role in the control of parental behaviour in a wide range of vertebrates, including fish, birds, and mammals (see reviews by Binart et al.[47] and Schradin and Anzenberger[48]). Further support for a controlling role of PRL in the expression of paternal care comes from the fact that an inverse relationship between plasma PRL and androgen levels is often observed during periods of male parental care in both birds and mammals (reviewed by Ziegler et al.[49]). Taken together, these observations in birds and mammals suggest that PRL exerts a negative effect on androgen production and on the expression of androgen-dependent sexual behaviours.

To date, investigations on the functional role of PRL in fish have mainly centred on species displaying parental care, including the stickleback. In an early study, Slijkhuis et al.[50] evaluated pituitary PRL cell activity in breeding male sticklebacks by quantitative electron microscopy and found that cell activity was greater during the parental compared to the sexual phase. These results are in general agreement with a recent study by Tacon et al.,[51] who showed that both plasma and pituitary levels of two PRL isoforms increased during the period when maternal behaviour was expressed in the mouth brooding cichlid, *Oreochromis niloticus*. Unfortunately, due to the present lack of a homologous PRL assay together with the difficulties associated with its small size, it is not yet feasible to measure plasma PRL levels in the stickleback. However, a number of studies have looked at the effect of PRL administration in the stickleback. It was reported that male sticklebacks injected with ovine PRL[52] or implanted with homologous pituitary PRL-lobes[53] during their sexual phase displayed increased fanning of their empty nests. However, the stimulatory effect of PRL on fanning appears to be dose dependent, as in the former study it was observed that at higher doses, PRL administration resulted in an inhibition of fanning.[52]

In addition to the stickleback, PRL administration has been shown to stimulate fanning in a number of other teleost species displaying parental care, including the ocellated wrasse, *Symphodus ocellatus*,[54] and the blue discus fish, *Symphysodon aequifasciatus*, and angelfish, *Pterophyllum scalare*.[55,56] A functional role of PRL in the control of parental behaviour in fish is also supported by the results of Kindler et al.,[57] who showed that parental male bluegills, *Lepomis macrochirus*, injected with the PRL-release inhibitor bromocriptine displayed significantly less defence of their broods compared to control males.

The results of these early studies demonstrating a stimulatory effect of PRL on fanning behaviour is supported by the results of a recent study by Páll et al.[58] In this study, the effects of intraperitoneal injections of either ovine or coho salmon PRL on courtship (zigzag) behaviour and preparental fanning were studied in nesting male sticklebacks. Males injected with both types of PRL displayed a significant increase in fanning behaviour, whereas males injected with salmon PRL (at all tested doses) but not with ovine PRL showed a significant decrease in zigzag behaviour (Figure 8.5). Taken together, these results suggest that the increase in PRL activity previously observed during the parental phase is involved in the control of both the decrease in courtship (zigzag) behaviour and the increase in fanning behaviour that is observed over the nesting cycle as the eggs develop. Thus, similar to what is observed in birds and mammals, the pattern of androgen decrease observed during the parental phase in the stickleback may be due to the physiological constraints imposed by the rise in PRL levels associated with the expression of paternal behaviours.

8.7 OTHER FACTORS

The neuropeptide arginine vasotocin (AVT) has been implicated in the control of reproductive behaviour in a variety of nonmammalian vertebrates.[59,60] A number of recent studies have demonstrated a role for AVT in the control of both aggressive and courtship behaviour in male fish,[61,62] as well as in the expression of both female courtship behaviour and female nuptial colouration in the peacock blenny.[63]

Although less well studied, the neuropeptide isotocin has also been found to be involved in the expression of fish behaviour.[64] It is likely that the gonadal steroids play a permissive role in mediating the actions of these neuropeptides. To date, a

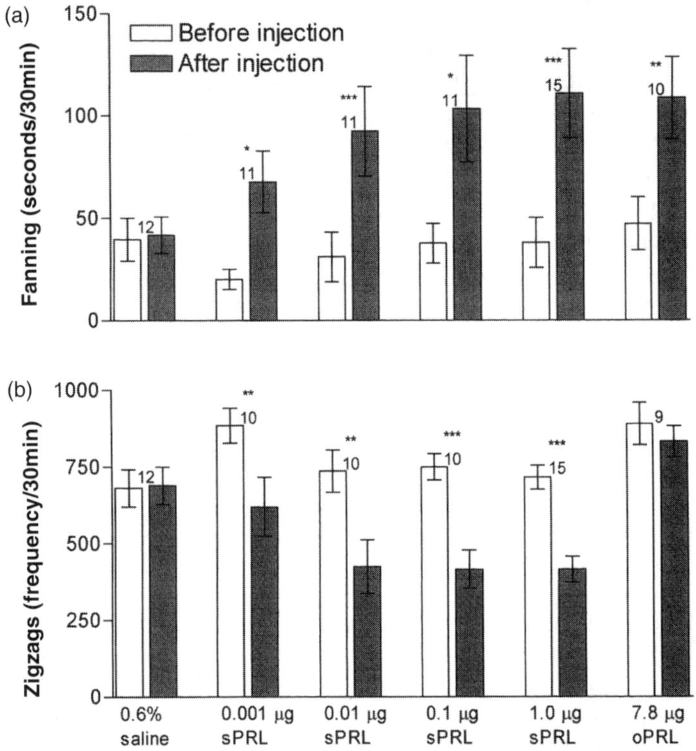

FIGURE 8.5 Effect of injections of coho salmon prolactin (sPRL) or ovine prolactin (oPRL) on (A) fanning and (B) zigzag behaviour in male sticklebacks. Significant differences between before and after injections are indicated as: * = $p < .05$; ** = $p < .01$; *** = $p < .001$; N values are shown above columns. (After Páll, M., et al., *Gasterosteus aculeatus*, *Behaviour* 141, 1511, 2004.)

role for these neuropeptides in the control of reproductive behaviour has not been investigated in the stickleback.

In common with other studied teleosts, a paucity of information exists on the hormonal control of reproductive behaviour in female sticklebacks. The obvious candidate for hormones having a controlling role in the expression of female reproductive behaviour would be the oestrogens, notably 17β-oestradiol (see Stacey[65] for a review). However, the only fish species that has been closely studied in regard to these steroids is the internally fertilised guppy, *Poecilia reticulata*.[66] Some of the primary hormones that appear to be actively involved in the stimulation of female behavioural responses in fish are the prostaglandins, especially prostaglandin-F_2 (PGF_2) (see Stacey[2] for a review). For example, Kobayashi and Stacey[67] demonstrated that PGF_2 induced spawning behaviour in female goldfish, *Carassius carassius*. Again, a functional role for prostaglandins in the control of female reproductive behaviour has not yet been studied in the stickleback.

8.8 BEHAVIOUR AND OTHER STICKLEBACK SPECIES

In view of the fact that reproductive behaviour has been so intensively studied in the three-spined stickleback, it is somewhat surprising that the other six or seven stickleback species (family Gasterosteidae) have been so poorly studied. The nine-spined stickleback (*Pungitius pungitius*) is widespread, and commonly coexists in the same habitats as the three-spined stickleback. Considering the fact that this species displays a similar repertoire of reproductive behaviours as its three-spined cousin, a striking difference being that breeding males turn completely black, it is surprising that it has been studied so little.

The 15-spined stickleback *Spinachia spinachia*, is regarded as a primitive member of the Gasterosteridae[68–70] and is unique to European waters. This large species (120 mm or more in length) inhabits marine or brackish waters, and during the breeding season is found occupying shallow eelgrass meadows or *Fucus* belts. In contrast to most other sticklebacks, the *Spinachia* males do not develop any pronounced breeding colours, and both sexes retain a dark, often mottled cryptic colouration. The breeding males build a large pronounced nest generally high up in the macro algae. During the sexual phase when the males build their nests, social interactions such as male–male aggression are not as frequent or intense compared to the three-spined stickleback. However, *Spinachia* males display pronounced courtship behaviour, involving intense body shakes directed toward the female, and also display a similar repertoire of paternal behaviours, including fanning behaviour.[7,71] In a recent study, *Spinachia* males were sampled directly from the field during their summer breeding period. Males were divided into three breeding categories: (1) stray males — males with no nest; (2) nest-building males — territorial males actively building a nest but with no eggs; and (3) parental males — territorial males having one or more batches of eggs in the nest. Plasma levels of both 11kT and T in all three categories are shown in Figure 8.6 (Páll and Mayer, unpublished data). These results are interesting in two ways. First, plasma levels of both androgens are very much lower in *Spinachia* compared to three-spined stickleback males during the sexual (nest-building) phase, and second, there are no significant differences in plasma levels of both 11kT and T among the three breeding categories. This implies that, in marked contrast to the three-spined stickleback as well as other parental-caring teleosts, there is no difference in plasma androgen levels between the sexual and parental phases of the nesting cycle in *Spinachia*.

8.9 ENDOCRINE DISRUPTION OF REPRODUCTIVE BEHAVIOUR

An increasing number of studies have linked the appearance of reproductive and developmental disorders in a variety of wildlife species to exposure to environmental contaminants that are capable of eliciting responses typically induced by sex steroids (see Tyler et al.[72] for a review). Owing to their associated steroid-like activities, these compounds have been termed *endocrine-disrupting chemicals* (EDCs). These reproductive disorders are particularly prominent in aquatic wildlife, and probably reflects the fact that most man-made contaminants will eventually end up in the

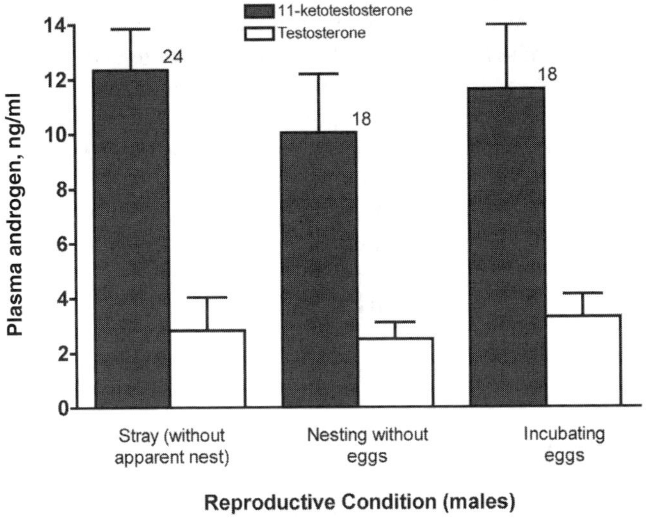

FIGURE 8.6 Plasma levels of 11kT and T in male 15-spined sticklebacks at different stages of their breeding cycle. Males were sampled directly from the field during their summer breeding period. Mean (±SEM) values are shown; *N* values are shown above columns. (Páll and Mayer, unpublished data.)

aquatic environment, which in effect represents a natural sink for these pollutants. As a consequence of this, aquatic species, especially teleost fish, are increasingly being used in the development of biomarkers of endocrine disruption. Indeed, the three-spined stickleback is now attracting much attention as a potential biomarker of both environmental oestrogens and antiandrogens.[73–75]

Although the use of fish models to investigate endocrine disruption is generally based on a number of readily quantifiable reproductive traits such as vitellogenin induction (identification of xenoestrogens), a number of studies have proposed the possibility of employing reproductive behaviour as a valid biomarker of endocrine disruption.[76,77] The test species should meet a number of important criteria. They should be small, easy to maintain and breed in captivity, have a short life cycle, and, most importantly, one or both sexes should display pronounced and easily quantifiable reproductive behaviours. The stickleback certainly meets most of these criteria, and males display a variety of pronounced reproductive behaviours, a number of which, including female courtship and nest-fanning behaviour, are readily quantifiable.[29,38] Further, to date, the stickleback is the only species of fish in which all aspects of reproductive behaviour have been clearly shown to be under the control of the gonadal steroids.[4] For these reasons, the stickleback probably represents the fish species having the best potential for using reproductive behaviour as a biomarker to investigate the possible adverse effects of environmental oestrogens and androgens, and, in turn, reproductive success. Recently, it has been demonstrated that exposure to environmentally relevant levels of the potent synthetic oestrogen, ethinyl estradiol, resulted in decreased aggressive behaviour,[78] as well as increased risk-taking behaviour in male sticklebacks.[79] In another study, it was demonstrated that intraperitoneal injections of 17β-oestradiol (2.0 μg/g body wt.) disrupted some aspects of male reproductive behaviour.[80] Although no differences were observed in nest-building and courtship behaviour, exposed males built their nests significantly later than control males.

The fact that behavioural observations are relatively time consuming, together with the inherent difficulty in quantifying behavioural responses in most species, detract from the use of this reproductive trait as a valid biomarker of endocrine disruption. The general consensus appears to be that although behavioural observations are unlikely to compete with other more rapid and quantifiable assays, they may complement them as a second-tier testing method. A final point of issue, which has been largely overlooked in teleost fishes, is how observed alterations in reproductive behaviour affects reproductive success.

8.10 SUMMARY

Early studies in the stickleback involving gonadectomy and androgen replacement clearly demonstrated a functional role of the androgens in the control of reproductive behaviour in this species. In this respect, it appears that 11kT is physiologically the most important androgen in male sticklebacks, in that it is the most effective tested androgen in stimulating both reproductive behaviour[15] and the development of secondary sexual characters.[19] Further, it is the only measured steroid whose seasonal plasma levels correlate closely with the development of reproductive behaviour.[17] However, a number of intriguing questions arise concerning the role of 11kT. First, why are plasma 11kT levels so high (about 400 ng/ml) in male sticklebacks during the sexual phase of the nesting cycle, and second, why if indeed 11kT is so important, does this androgen (as with T) decline to low levels (about 12 ng/ml) during the parental phase when the male still actively displays reproductive behaviour (fanning)?

A possible reason for the high 11kT levels during the sexual phase is that this androgen is vital in stimulating the pronounced secondary sexual characters (notably kidney hypertrophy and spiggin production) displayed by the male stickleback during nest building. However, nesting males of the 15-spined stickleback *Spinachia spinachia*, which show a similar degree of kidney hypertrophy as nesting three-spined sticklebacks, have considerably lower plasma 11kT levels (about 10–12 ng/ml). This suggests that the unusually high 11kT levels shown by male three-spined sticklebacks is not a function of spiggin production alone, and possibly reflects more the pronounced differences in sexual behaviour shown between the two species.

Although a role for androgens in the control of sexual behaviour (territoriality, male–male aggression, nest-building, and courtship) in male sticklebacks is unequivocal, Baggerman[12] and later Páll et al.[29] have clearly demonstrated that parental behaviour (fanning) occurs independently of the androgens. It is tempting to speculate that the androgens are necessary in activating those mechanisms controlling the initiation of reproductive behaviour, in addition to stimulating the expression of secondary sexual characters. Once males have entered their nesting cycle, androgens appear to play a more permissive role in the control of reproductive behaviours. This is supported by the observation that courtship behaviour still continues in males that had been castrated only after they had completed their nest, that is, had entered their nesting cycle.

Following spawning, present evidence indicates that factors other than the gonadal hormones are more important in the control of parental behaviour in the stickleback. Further studies are needed to investigate the role of other hormones or neuropeptides in the control of reproductive behaviour, especially paternal behaviour, in the stickleback. In this respect prolactin (PRL), a peptide hormone known to be important in mediating parental behaviour in birds and mammals, seems to be a likely candidate. Indeed, a number of studies have demonstrated a correlation between PRL levels and parental behaviour in fishes, including the stickleback where PRL administration has been shown to suppress courtship behaviour and stimulate fanning behaviour in nesting male sticklebacks.[58]

In conclusion, the control of reproductive behaviour in teleost fishes is complex and probably involves interaction among a number of regulatory mechanisms. Although studies on the stickleback have clearly indicated a major role for the androgens, notably 11kT, in the control of male reproductive behaviour, major knowledge gaps still exist in understanding all the components of the regulatory system. In this respect, the stickleback will continue to be a model species in this field of research.

REFERENCES

1. Liley, N.R. and Stacey, N.E., Hormones, pheromones, and reproductive behavior in fish, *Fish Physiol.* 9B, 1, 1983.
2. Stacey, N.E., Role of hormones and pheromones in fish reproductive behavior, in *Psychobiology of Reproduction: An Evolutionary Perspective,* Crews, E.D., Ed., Prentice-Hall, Englewood Cliffs, NJ, 1987, pp. 28–69.
3. Tinbergen, N., *The Study of Instinct*, Oxford University Press, Oxford, 1951.
4. Borg, B. and Mayer, I., Androgens and behaviour in the three-spined stickleback, *Behaviour* 132, 1025, 1995.
5. Mayer, I., Borg, B., and Páll, M., Hormonal control of male reproductive behaviour in fishes: a stickleback perspective, *Behaviour* 141, 1499, 2005.
6. St. Mary, Noureddine, C.G., and Lindström, K., Environmental effects on male reproductive success and parental care in the Florida flagfish, *Jordanella floridae*, *Ethology* 107, 1035, 2001.
7. Östlund, S. and Ahnesjö, I., Female fifteen-spined sticklebacks prefer better fathers, *Anim. Behav.* 56, 1177, 1998.
8. Van Iersel, J.J.A., An analysis of the parental behaviour of the male three-spined stickleback (*Gasterosteus aculeatus*), *Behaviour* Suppl. III, 1, 1953.
9. Hoar, W.S., Hormones and the reproductive behaviour of the male three-spined stickleback (*Gasterosteus aculeatus*), *Anim. Behav.* 10, 247, 1962.
10. Hoar, W.S., Reproductive behavior of fish, *Gen. Comp. Endocrinol.* Suppl. 1, 206, 1962.

11. Wai, E.H. and Hoar, W.S., The secondary sex characters and reproductive behaviour of gonadectomized sticklebacks treated with methyltestosterone, *Can. J. Zool.* 41, 611, 1963.
12. Baggerman, B., On the endocrine control of reproductive behaviour in the male three-spined stickleback (*Gasterosteus aculeatus* L.), *Symp. Soc. Exp. Biol.* 20, 427, 1966.
13. Smith, R.J.F. and Hoar, W.S., The effect of prolactin and testosterone on the parental behavior of the male stickleback, *Gasterosteus aculeatus*, *Anim. Behav.* 15, 342, 1967.
14. Rouse, E.F., Coppenger, C.J., and Barnes, P.R., The effect of androgen inhibitor on behavior and testicular morphology in the stickleback, *Gasterosteus aculeatus*, *Horm. Behav.* 9, 8, 1977.
15. Borg, B., Stimulation of reproductive behaviour by aromatizable and non-aromatizable androgens in the male three-spined stickleback, *Gasterosteus aculeatus* L., in *Proc. V. Congr. Eur. Ichthyol.*, Stockholm, 1987, pp. 269–271.
16. Borg, B., Schoonen, W.G.E.J., and Lambert, J.G.D., Steroid metabolism in the testes of the breeding and nonbreeding three-spined sticklebacks, *Gasterosteus aculeatus*, *Gen. Comp. Endocrinol.* 73, 40, 1989.
17. Mayer, I., Borg, B., and Schulz, R., Seasonal changes in and the effect of castration/androgen-replacement on the plasma levels of five androgens in the male three-spined stickleback, *Gasterosteus aculeatus* L., *Gen. Comp. Endocrinol.* 79, 23, 1990.
18. Mayer, I., Borg, B., and Schulz, R., Conversion of 11-ketoandrostenedione to 11-ketotestosterone by the blood cells of six fish species, *Gen. Comp. Endocrinol.* 77, 70, 1990.
19. Borg, B. et al., Effectiveness of several androgens in stimulating kidney hypertrophy, a secondary sexual character, in castrated male three-spined sticklebacks, *Gasterosteus aculeatus*, *Can. J. Zool.* 71, 2327, 1993.
20. Borg, B., Androgens in teleost fishes, *Comp. Biochem. Physiol.* 109C, 219, 1994.
21. Miura, T. et al., The role of hormones in the acquisition of sperm motility in salmonid fish, *J. Exp. Zool.* 261, 359, 1992.
22. Mayer, I. et al., Seasonal endocrine changes in Baltic salmon, *Salmo salar*, immature parr and mature male parr. I. Plasma levels of five androgens, 17α-hydroxy-20-dihydroprogesterone, and 17-estradiol, *Can. J. Zool.* 68, 1360, 1990.
23. Olsén, H. K. et al., Spawning behaviour and sex hormone levels in adult and precocious brown trout (*Salmo trutta* L.) males and the effect of anosmia, *Chemoecology* 8, 9, 1988.
24. Mayer, I., Liley, N.R., and Borg, B., Stimulation of spawning behavior in castrated rainbow trout (*Oncorhynchus mykiss*) by 17α,20-dihydroxy-4-pregnen-3-one, but not by 11-ketoandrostenedione, *Horm. Behav.* 28, 181, 1994.
25. Kindler, P.M. et al., Serum 11-ketotestosterone and testosterone concentrations associated with reproduction in male bluegill (*Lepomis macrochirus*: Centrarchidae), *Gen. Comp. Endocrinol.* 75, 446, 1989.
26. Sikkel, P.C., Changes in plasma androgen levels associated with changes in male reproductive behavior in a brood cycling marine fish, *Gen. Comp. Endocrinol.* 89, 229, 1993.
27. Knapp, R., Wingfield, J.C., and Bass, A.H., Steroid hormones and parental care in the plainfin midshipman fish (*Porichthys notatus*), *Horm. Behav.* 35, 81, 1999.
28. Pankhurst, N.W., Hilder, P.I., and Pankhurst, P.M., Reproductive condition and behavior in relation to plasma levels of gonadal steroids in the spiny damselfish *Acanthochromis polyacanthus*, *Gen. Comp. Endocrinol.* 75, 446, 1999.

29. Páll, M., Mayer, I., and Borg, B., Androgen and behavior in male three-spined stickleback, *Gasterosteus aculeatus* I. Developement of 11-ketotestosterone levels during the nesting cycle, *Horm. Behav.* 41, 377, 2002.
30. Mayer, I. et al., Plasma levels of sex steroids in three species of pipefish (Syngnathidae), *Can. J. Zool.* 71, 1903, 1993.
31. Páll, M. et al., Changes in reproductive physiology over the nesting cycle in male three-spined sticklebacks, *J. Fish Biol.* 66, 1400, 2005.
32. Wingfield, J.C. et al., Testosterone and aggression in birds: tests of the challenge hypothesis, *Am. Sci.* 75, 602, 1987.
33. Wingfield, J.C. et al., The "challenge hypothesis": theoretical implications for patterns of testosterone secretion, mating systems, and breeding strategies, *Am. Nat.* 136, 829, 1990.
34. Oliveira, R.F., Social modulation of androgens in vertebrates: mechanisms and function, *Adv. Stud. Behav.* 34, 165, 2004.
35. Wingfield, J.C. et al., Towards an ecological basis of hormone-behavior interactions in reproduction in birds, in *Reproduction in Context*, Wallen, K. and Schneider, J.J., Eds., MIT Press Cambridge, MA, 2000, pp. 85–128.
36. Oliveira, R.F. et al., Social modulation of androgens in male teleost fish, *Comp. Biochem. Physiol.* 132B, 203, 2002.
37. Planas, J.V. and Swanson, P., Maturation-associated changes in response of the salmon testis to the steroidogenic actions of gonadotropins (GTH-I and GTH-II) *in vitro*, *Biol. Reprod.* 52, 697, 1995.
38. Páll, M., Mayer, I., and Borg, B., Androgen and behavior in male three-spined stickleback, *Gasterosteus aculeatus* II. Castration and 11-ketotandrostenedione effects on courtship and parental behavior during the nesting cycle, *Horm. Behav.* 42, 337, 2002.
39. Stearns, S.C., Trade-offs in life-history evolution, *Funct. Ecol.* 3, 259, 1989.
40. Andersson, E., Mayer, I., and Borg, B., Inhibitory effect of 11-ketoandrostenedione and androstenedione on spermatogenesis in the three-spined stickleback, *Gasterosterus aculeatus* L., *J. Fish Biol.* 33, 835, 1988.
41. Wingfield, J.C., Jacobs, J., and Hillgarth, N., Ecological constraints and the evolution of hormone-behavior interrelationships, *Ann. N.Y. Acad. Sci.* 807, 22, 1997.
42. Wingfield, J.C., Lynn, S.E., and Soma, K.K., Avoiding the "costs" of testosterone: ecological bases of hormone-behavior interactions, *Brain Behav. Evol.* 57, 239, 2001.
43. Slater, C.H. and Schreck, C.B., Testosterone alters the immune response of Chinook salmon (*Oncorhynchus tshawytcha*), *Gen. Comp. Endocrinol.* 89, 291, 1993.
44. Slater, C.H. and Schreck, C.B., Physiological levels of testosterone kill salmonid leucocytes *in vitro*, *Gen. Comp. Endocrinol.* 106, 113, 1997.
45. Folstad, I. and Karter, A., Parasites, bright males, and the immunocompetence handicap, *Am. Nat.* 139, 603, 1992.
46. Braude, S., Tang-Martinez, Z., and Taylor, G.T., Stress, testosterone, and the immunoredistrobution hypothesis, *Behav. Ecol.* 10, 345, 1999.
47. Binart, N. et al., Prolactin overview, in *Encyclopedia of Reproduction*, Knobil, E. and Neill, J.D., Eds., Academic Press, New York, 1999, pp. 31–38.
48. Schradin, C. and Anzenberger, G., Prolactin, the hormone of paternity, *News Physiol. Sci.* 14, 223, 1999.
49. Ziegler, T.E., Hormones associated with non-mammalian infant care: a review of mammalian and avian studies, *Folia Primatol.* 71, 6, 2000.

50. Slijkhuis, H. et al., Parental fanning behavior and prolactin cell activity in the male three-spined stickleback *Gasterosteus aculeatus* L., *Gen. Comp. Endocrinol.* 54, 297, 1984.
51. Tacon, P. et al., Effect of egg deprivation on sex steroids, gonadotropin, prolactin, and growth hormone profiles during the reproductive cycle of the mouthbrooding cichlid fish *Oreochromis niloticus, Gen. Comp. Endocrinol.* 117, 54, 2000.
52. Molenda, E. and Fiedler, K., Die Wirkung von Prolaktin auf das Verhalten von Stichlings- (*Gasterosteus aculeatus* L.), *Z. Tierpsychol.* 28, 463, 1971.
53. de Ruiter, A.J.H. et al., The effect of prolactin on fanning behavior in the male three-spined stickleback, *Gasterosteus aculeatus* L., *Gen. Comp. Endocrinol.* 64, 273, 1986.
54. Fiedler, K., Die Wirkung von Prolactin auf das Verhalten des Lippfisches *Crenilabrus ocellatus* (Forskål), *Zool. Jahrb. Abt. Allg. Zool. Physiol. Tiere.* 69, 609, 1962.
55. Blüm, V. and Fiedler, K., Hormonal control of reproductive behaviour in some cichlid fish, *Gen. Comp. Endocrinol.* 5, 186, 1965.
56. Blüm, V., Die Rolle des Prolaktins der Cichlidenbrutpflege, *Fortschr. Zool.* 22, 310, 1974.
57. Kindler, P.M. et al., Hormonal regulation of parental care behaviour in nesting male bluegills: do the effects of bromocriptine suggest a role for prolactin? *Physiol. Zool.* 64, 310, 1991.
58. Páll, M., Liljander, M., and Borg, B., Prolactin diminishes courtship and stimulates fanning in nesting male three spined sticklebacks, *Gasterosteus aculeatus*, *Behaviour* 141, 1511, 2004.
59. Moore, F.L., Evolutionary precedents for behavioral actions of oxytocin and vasopressin, *Ann. N.Y. Acad. Sci.* 652, 156, 1992.
60. Goodson, J.L. and Bass, A.H., Social behavior functions and related anatomical characteristics of vasotocin/vasopressin systems in vertebrates, *Brain Res. Rev.* 35, 246, 2001.
61. Salek, S.J., Sullivan, C.V., and Godwin, J., Arginine vasotocin effects courtship behavior in male white perch (*Morone Americana*). *Behav. Brain Res.* 133, 177, 2002.
62. Semsar, K., Kandel, F.L.M., and Godwin, J., Manipulations of the AVT system shift social status and related courtship and aggressive behavior in the bluehead wrasse, *Horm. Behav.* 40, 21, 2001.
63. Carneiro, L.A. et al., The effect of arginine vasotocin on courtship behaviour in a blenniid fish with alternative reproductive tactics, *Fish Physiol. Biochem.* 28, 241, 2003.
64. Goodson, J.L., Evans, A.K., and Bass, A.H., Putative isotocin distributions in sonic fish: relation to vasotocin and vocal-acoustic circuitry, *J. Comp. Neurol.* 462, 1, 2003.
65. Stacey, N.E., Hormonal regulation of female reproductive behavior in fish, *Am. Zool.* 21, 305, 1981.
66. Liley, N.R., The effects of estrogens and other steroids on the sexual behavior of the female guppy, *Poecilia reticulata*, *Gen. Comp. Endocrinol.* Suppl. 3, 542, 1972.
67. Kobayashi, M., and Stacey, N., Prostaglandin-induced female spawning behavior in goldfish (*Carassius auratus*), *Horm. Behav.* 27, 38, 1993.
68. Wootton, R.J., *The Biology of the Stickleback*, Academic Press, London, 1976, 387 pp.
69. McLennan, D.A., Phylogenetic relationships in the Gasterosteidae: an updated tree based on behavioral characters with a discussion of homoplasy, *Copeia* 2, 318, 1993.
70. Keivany, Y. and Nelson, J.S., Phylogenetic relationships of sticklebacks (Gasterosteidae), with emphasis on ninespine sticklebacks (*Pungitius spp.*), *Behaviour* 141, 1485, 2005.

71. Östlund-Nilsson, S., Are nest characters of importance when choosing a male in the fifteen-spined stickleback (*Spinachia spinachia*)? *Behav. Ecol. Sociobiol.* 48, 229, 2000.
72. Tyler, C.R., Jobling, S., and Sumpter, J.P., Endocrine disruption in wildlife: a critical review of the evidence, *Crit. Rev. Toxicol.* 28, 319, 1998.
73. Katsiadaki, I., The use of the stickleback as a sentinel and model species in ecotoxicology, in *Biology of the Three-Spined Stickleback*, Östlund-Nilsson, S., Mayer, I., and Huntingford, F., Eds., Taylor & Francis, Boca Raton, FL, 2006, Chapter 10 in this volume.
74. Katsiadaki, I., Scott, A.P., and Mayer, I., The potential of the three-spined stickleback (*Gasterosteus aculeatus* L.) as a combined biomarker for oestrogens and androgens in European waters, *Mar. Environ. Res.* 54, 725, 2002.
75. Katsiadaki, I. et al., Detection of environmental androgens: a novel method based on enzyme-linked immunosorbent assay of spiggin, the stickleback (*Gasterosteus aculeatus*) glue protein, *Environ. Toxicol. Chem.* 21, 1946, 2002.
76. Jones, C.J. and Reynolds, J.D., Effects of pollution on reproductive behaviour in fishes, *Rev. Fish Biol. Fish.* 7, 463, 1997.
77. Kobayashi, M., Sexual behavior as a parameter for environmental disrupter assessment, *4th International Symposium on Fish Endocrinology*, July 31–August 3, 2000, Seattle, Washington (abstract).
78. Bell, A.M., Effects of an endocrine disrupter on courtship and aggressive behaviour of male three-spined stickleback, *Anim. Behav.* 62, 775, 2001.
79. Bell, A.M., An endocrine disrupter increases growth and risky behavior in threespined stickleback (*Gasterosteus aculeatus*). *Horm. Behav.* 45, 108, 2004.
80. Wibe, Å.E., Rosenqvist, G., and Jenssen, B.M., Disruption of male reproductive behavior in threespine stickleback *Gasterosteus aculeatus* exposed to 17β-estradiol, *Environ. Res. Sect.* A 90, 136, 2002.

9 Host–Parasite Interactions of the Three-Spined Stickleback

Iain Barber

CONTENTS

This chapter provides an examination of the parasite taxa that regularly infect three-spined sticklebacks in natural environments and their impacts on host biology, with the aim of developing an understanding of their role as agents of natural and sexual selection in host populations. The taxonomic and life cycle diversity of stickleback parasites is introduced, and patterns of infection in stickleback populations are examined. The various behavioural mechanisms that sticklebacks can use to avoid parasites or otherwise reduce their own risk of infection or that of their offspring are reviewed. The biology of infected fish is discussed, and sections are included that address the impact of infection on host morphology, physiology, growth, and sexual development. Many stickleback parasites also affect the behaviour of their hosts, and behavioural changes resulting from parasite infections can have potentially significant consequences for the ecology of host individuals and the evolution of populations. In some cases, parasites may actually benefit from the changes in behaviour they induce in their hosts. The impact of parasites on the behaviour of host sticklebacks has been extensively tested, and the various changes that have been recorded and their likely physiological basis and consequences for host ecology are discussed in detail.

9.1 INTRODUCTION

Biologists are increasingly aware of the importance of parasites as agents of selection in ecology and evolution, and many characteristics of animals are now attributed directly or indirectly to the selection pressures that infections impose on host populations. Of course, parasites have considerable potential to impact the survival of hosts through infection-induced mortality. This may be particularly true in the case of newly acquired infections introduced to host populations that were previously unexposed and so have not had time to adapt through coevolutionary processes with the parasite (e.g., *Gyrodactylus salaris* in Norwegian Atlantic salmon *Salmo salar* populations[1]). However, sublethal effects of infections may also create strong selective pressures. In recent years, parasites, and the threat of infection they pose, have been implicated in the evolution of mate choice, ornamental characteristics, and even for the evolution of sex itself, as well as giving rise to complex immune responses and antiparasite behaviours.

Many key studies on the ecological and evolutionary aspects of host–parasite interactions have been undertaken using the three-spined stickleback *Gasterosteus aculeatus* L. as a host, and sticklebacks have become important model host organisms in the study of host–parasite interactions. There are a number of reasons why the three-spined stickleback is such an attractive model host. First, for a fish of noneconomic importance, there is an unsurpassed wealth of background information on its ecology, behaviour, and evolutionary biology.[2–4] This is coupled with the species' utility as a laboratory model, its geographically wide distribution and availability, and the existence of a number of experimentally amenable parasite infections. The recent completion of the stickleback genome sequence[5] will undoubtedly mean that the stickleback will remain a key model host in parasitology throughout and beyond the postgenomic revolution currently sweeping biology.

With this in mind, the aim of the present chapter is to summarise the current state of knowledge regarding the effects of parasites on the biology and ecology of individual sticklebacks and their populations, and to highlight potentially fruitful areas of research. In this chapter there is no attempt to provide an in-depth coverage of infection-associated pathology, any detailed immunology, or any practical advice on how to diagnose or treat infectious diseases in sticklebacks; for such information the reader is referred to general, comprehensive texts by Iwana and Nakanishi[6] (immunology), Roberts[7] (pathology), and Noga[8] (diagnosis and treatment). Rather, in this chapter, the focus is on studies that examine the importance of parasites as agents of natural and sexual selection in sticklebacks, to understand the role played by parasites in the natural ecology and evolution of the fish.

9.1.1 Sticklebacks as Hosts for Parasites

In common with other predatory fishes, the three-spined stickleback serves as host to a large number of parasitic taxa.[2] Some of these parasites, or at least specific development stages of these parasites, are stickleback "specialists" that rarely, if ever, infect other fish species. However, the widespread distribution of the three-spined stickleback, and its adaptation to almost every aquatic habitat type throughout its geographic distribution mean that a wide variety of generalist parasites, with nonspecific fish host requirements, are encountered and acquired. Furthermore, their omnivorous diet[2,9] exposes sticklebacks to a wide range of food-borne infections. This, combined with the role of sticklebacks as prey of a wide range of aquatic and terrestrial predators, suggests they are likely to provide ideal intermediate hosts for a range of trophically transmitted parasites. It is therefore not surprising that in two recent studies of brackish water fish communities in the Baltic Sea, sticklebacks were found to be the most heavily parasitised fish species present, both in terms of parasite numbers and parasite diversity.[10,11]

9.1.2 Taxonomic Diversity of Stickleback Parasites

The total number of parasitic taxa infecting the three-spined stickleback worldwide is difficult to estimate. There are a number of difficulties in ascertaining the total number of stickleback parasites, or those of any host taxon. First, the definition of what constitutes a parasite is open to question; liberal definitions of fish parasites may include a taxonomic spread from viruses to vertebrates (lampreys), though fish parasitologists more commonly restrict their studies largely to protozoan and metazoan parasites, and this is the convention used in this chapter. Second, there is often insufficient knowledge regarding the divergence of parasites of sticklebacks between geographically separated host populations; to what extent, for example, do *Schistocephalus solidus* cestodes — common and ecologically important parasites of sticklebacks — from North America and Europe differ, and do they qualify as separate species or subspecies?[12] In general, there is insufficient knowledge regarding speciation in parasite taxa to answer these questions (see Zietra and Lumme[13] for a recent example). A related problem is in determining whether species that infect one host taxon (such as the three-spined stickleback) are taxonomically distinct from those infecting other host taxa; hence, separate species of *Schistocephalus* have been claimed, infecting different host stickleback species (*S. solidus* in *G. aculeatus* and *S. pungitii* in the nine-spined stickleback *Pungitius pungitius*). Detailed accounts of the parasite fauna of the three-spined stickleback have been provided by Hoffman[14] (North America, freshwater), Margolis and Arthur[15] (Canada), Kennedy[16] (U.K., freshwater), and Wootton[2] (global), with numerous other lists focusing on specific parasite groups (typically helminths, often with crustaceans) of selected populations or geographic regions. The parasite species list presented at the end of this chapter (Table 9.2) uses the data presented in these various publications to update the initial list provided by Wootton,[2] yet it is still unlikely to be complete. A global stickleback parasite biodiversity survey, undertaken by a worldwide network of fish parasitologists, is presently underway and is likely to provide a more comprehensive list of stickleback parasites (details available at http://www.nrel.colostate.edu/projects/iboy/whatandwhere.html#stickleback). The aim of the survey is to create a database that will allow increased understanding of the biodiversity, food web structure, and ecosystem stress of stickleback communities, and to help scientists use parasites as rapid bioindicators of ecosystem health. The results of this project will undoubtedly provide an immensely valuable resource for stickleback researchers and others interested in host–parasite evolution.

9.1.3 Life Cycle Diversity of Stickleback Parasites

Perhaps the most remarkable aspect of fish parasites is the bewildering variety of life cycles that they exhibit. The details of parasite life cycles may sometimes be complex and are often unresolved, but they are not trivial. Without an understanding of the transmission route, the existence and identity of other hosts in the life cycle, and the particular developmental stage of the parasite harboured by the stickleback, it becomes very difficult to make predictions about the likely effect of infection on host biology, the ecological consequences of infection-associated modifications, the potential for rapid colonisation following introduction into a field (or laboratory) population, or to interpret any such changes in terms of their evolutionary history.

Two broad life cycle types can be identified. Some parasites are capable of living their whole life cycle on or in a single host stickleback (homoxenous [or monoxenous] parasites), whereas others need to be transmitted between a number of hosts to complete their life cycle (heteroxenous parasites). In the case of heteroxenous parasites, those hosts may be of the same or different species. For those heteroxenous parasites that need to utilise different host species, sticklebacks may serve either as a definitive host (i.e., harbouring the reproductively active stage) or as an intermediate host (harbouring a nonreproductive developmental stage). Parasites may either be transmitted directly between individual sticklebacks (for example, through physical contact between individuals, or through the movement of motile parasite stages) or be transmitted indirectly (for example, through the food web [trophic transmission]). Transmission strategies can therefore vary from those of ectoparasitic monogeneans (such as *Gyrodactylus* spp.), which are passed directly from one host stickleback to another through close contact, to those of diplostomatid trematodes, which typically utilise aquatic snails and then sticklebacks as intermediate hosts, before achieving sexual maturity in the intestine of a piscivorous bird. A detailed knowledge of any parasite's life cycle and its route of transmission to host fish are also prerequisite for their potential use as bioindicators of fish diet and habitat use.[17,18]

9.1.4 Parasitological Terminology: A Primer for the Gasterosteologist

As with all disciplines, quantitative parasitology has its own terminology, and a brief primer is provided here for the nonparasitologist (a more complete glossary of terms is provided by Bush et al.[19]). The extent of parasitism can be quantified either at the level of the individual or the population. At the level of the individual, infections can be quantified in a number of ways. For macroparasites, the *intensity* of infection, defined as the number of parasites of a particular species harboured by an individual host, is frequently reported. For microparasites, which are generally capable of rapid asexual propagation within a single host, direct counts of the numbers of individual parasites are generally impractical, and estimates of infection level may be achieved using standard microbiological techniques.[20] In other cases, where each invasion event leads to the development of a visible cyst or colony, these may be counted to provide an estimate of relative infection level (e.g., Pélabon et al.[21]; Ward et al.[22]). An important concept at the host population level is that of the *prevalence* of the infection, defined as the proportion of potential hosts in a population that is infected with a particular parasite species. Another important concept frequently used to characterise populations and to compare statistically between them is the average population intensity of infection (either mean or median intensity, usually with some indication of variation). Frequently, both average (mean or median) intensity and prevalence values are given separately for males and females, or for different host size, age, or classes. Finally, for large-bodied parasites such as *Schistocephalus solidus* (Figure 9.1) — where the total mass of parasite tissue may be expected to be at least as relevant to the biological outcomes of infection as the number of parasites infecting a host — the concept of *parasite index* (PI, expressed as the proportion of the infected fish's weight that is contributed by the parasite) has been introduced.[23]

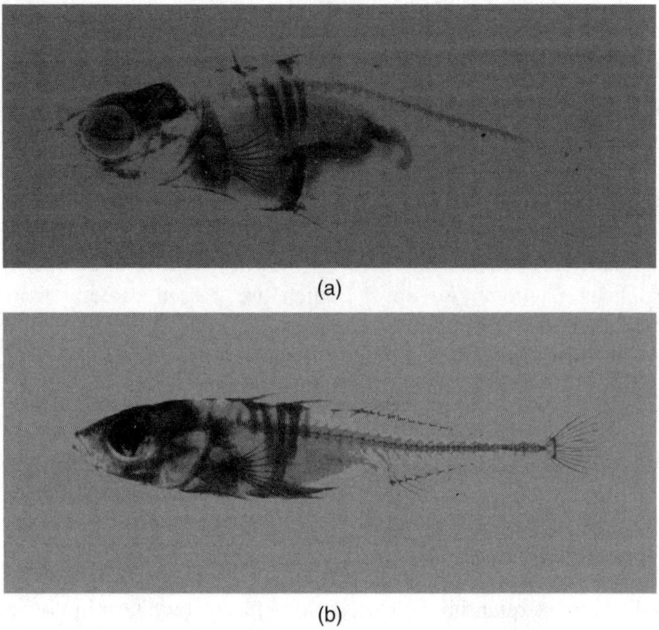

FIGURE 9.1 *Schistocephalus solidus* infections in three-spined stickleback *Gasterosteus aculeatus* (a) cleared and Alizarin-stained parasitised specimen from Llyn Frongoch, Wales, U.K., showing the disruption to pelvic girdle associated with *S. solidus* infection, compared to (b) a noninfected specimen from the same population.

9.2 PATTERNS OF INFECTION IN THREE-SPINED STICKLEBACKS

Not all sticklebacks are equally infected with parasites, and considerable variation in the infection levels and parasite fauna of individuals and populations is observed. The factors that lead to this variation are of considerable interest to ecologists and evolutionary biologists. Here, some of the major factors that have been identified as important in determining the parasite load of sticklebacks are reviewed.

9.2.1 Seasonal Variation in Infection Level

Many fish parasites exhibit seasonal patterns of infection, which are generally related to seasonal shifts in the abundance of intermediate hosts, or to changes in abiotic factors such as temperature (see reviews by Chubb[24–27]). In long-lived species, these seasonal trends can be relatively easily identified, but the typically short (often annual) lifespan of the three-spined stickleback throughout much of its range makes separating true seasonal variation in parasite acquisition from factors related to host ontogeny rather difficult.

For parasites that can be "aged" by examining their size, or their developmental stage within a host, it is possible to gain information about the temporal dynamics of infection, and they can be used to identify periods of peak invasion in regular sampling programmes. There is evidence to suggest that, in some populations at least, *S. solidus* infections are acquired by the new stickleback cohort in a single wave during late summer.[28] This time period coincides with the peak availability of copepods (the first intermediate host) and the period when the fish attain a size when copepods are likely to provide a significant part of the diet. It is during this time that the largest proportion of small (<0.01 g) plerocercoids, indicative of newly acquired infections, are observed.[29] Similarly, *Diplostomum gasterostei* infections have been shown to increase dramatically between April and August,[29] and Chappell[30] demonstrated that the proportion of small (newly acquired) *Diplostomum* metacercariae rose from 14 to 92% during the period May–August. Temperature-dependent release of infective cercariae from trematode-infected snails (e.g., Lyholt and Buchmann[31]) is likely to be at least partly responsible for these observations; however, other factors, including the movements and habitat selection of sticklebacks, and temperature-dependent changes in their immune responses, cannot be ruled out without detailed experimentation. Pennycuick[29] examined the infection of three-spined sticklebacks in Priddy Pool, Somerset, U.K., and showed that levels of infection with *S. solidus*, *Diplostomum gasterostei*, and *Echinorhynchus clavula* increased during the summer (June–September), though variation between years exceeded that observed between seasons.

Distinguishing between environmental factors that may generate temporal changes in levels of infection may be particularly difficult in populations where changes in temperature correlate with other environmental changes. For example, studies of stickleback parasites in a Turkish stream population showed that levels of infection with the monogenean *Gyrodactylus armatus*, trichodinids *Trichodina domerguei*, and *T. tenidus* increased during February–May, during which time temperature increases occurred concurrently with a switch in the salinity of the stream from brackish to freshwater.[32]

9.2.2 Infections in Males and Females

Discrepancies in the infection levels of males and females, both in terms of the prevalence or intensity of infection with a single parasite, or in the specific parasites harboured, may arise for a variety of reasons. Behavioural and morphological characteristics of males and females, including differences in diet, body size, habitat use, home range size, or social contact, may lead to differential exposure to parasites.[33–35] Furthermore, there may be intrinsic variation between the sexes in their capacity to resist infective parasites; males may be more likely than females to harbour parasites because androgens (in fish, primarily 11-ketotestosterone) are known to have immunosuppressive effects.[36] However, in a meta-analysis of parasitological surveys of vertebrate taxa, Poulin[33] demonstrated that, although among mammals and birds males typically exhibited higher levels of infection than females, among the fishes, gender had little effect on parasite load.

A number of field studies have examined stickleback infections in relation to sex, though little consensus has emerged. Chappell[37] examined sticklebacks from a pond in Yorkshire, U.K., over a 12-month period and identified no sex-level variation in eight parasite species. Similarly, Pennycuick[38] found no significant host sex-related differences in the prevalence, nor the intensity, of infection with *S. solidus*, *D. gasterostei*, or *E. clavula*. Among *S. solidus*–infected fish, however, both PI and the mean weight of individual parasites were significantly higher in females than in males, though the differences were only slight. Conversely, analysis of infection data on four macrohelminth species (*S. solidus*, *Cyathocephalus truncatus*, *Bunodera* sp., and *Eustrongylides* spp.) in a large number (about 20,000) of stickleback collected over a remarkable 15-year sampling period from Boulton Lake, British Columbia, allowed Reimchen and Nosil[39] to identify sex differences in the prevalence of *S. solidus* (F>M), *C. truncatus* (M>F), and *Bunodera* sp. (M>F). In agreement with Pennycuick,[38] they also found females to be more likely to harbour large (>15 mm contracted length) *S. solidus* plerocercoids than males. Combining data from all infections, the proportion of males harbouring at least one of the parasites (26.4%) was slightly, but significantly, higher than that of females (22.5%), and infected males were also more likely to harbour multiple species infections than infected females. Breaking down infection levels by developmental stage and year, however, revealed that effects were not absolutely consistent across all body size and year classes. Reimchen and Nosil[39] suggest an ecological explanation for the sex differences in parasite infection; analysis of stomach contents revealed an increased frequency of pelagic prey (which harbour infective stages of *S. solidus*) in females than males, which exhibited an excess of benthic prey (the hosts of *C. truncatus* and *Bunodera* sp. infections).

Other studies have identified male biases in infection load. Arnold et al.[40] used molecular (AFLP-based) sexing techniques (described by Griffiths et al.[41]) to determine the prevalence of *Glugea anomala* infections in males and females over a 12-month period for a 0+ cohort in a Scottish pond and found adult males to have a higher prevalence of dermal *G. anomala* cysts than adult females. At the end of the breeding season, there was a disproportionate increase in *G. anomala* parasitism in adult males, possibly suggesting a breakdown in resistance to infection among reproducing or postreproductive males. In a study of 605 breeding adults from the St. Lawrence River, Quebec, Blais et al.[42] found males to harbour significantly higher intensities of skin-encysted trematode metacercariae. Conversely, in a Turkish stream population, Özer[32] demonstrated statistically significant differences between the sexes for two trichodinid species, with female fish harbouring significantly higher levels of infection than males.

In general, although sex difference in prevalence and intensity of infections are similar between males and female sticklebacks, slight differences, most often with males harbouring heavier infections, are sometimes identified, but usually only in studies with large sample sizes that allow correspondingly high statistical power. There also appears to be a likely effect of body size, linked to gender, on the growth of individual parasites (such as *S. solidus*) that normally achieve a large size relative to their host, with females sustaining larger individual parasites or heavier total burdens. Experimental infection studies would significantly improve our understanding of sex-based differences in parasite infections, particularly regarding the mechanisms that generate any observed differences.

9.2.3 Effect of Host Age and Body Size

The potential role of body size as a correlate of infection level has already been briefly addressed. Recent studies show that there may be relationships between the size of large-bodied parasites, such as *S. solidus*, and host body size,[43] suggesting that the space available for the parasite to grow may constrain parasite size in some cases. Furthermore, because many parasites are accumulated throughout the life span of stickleback hosts, infection levels consequently often exhibit positive relationships with age and body size. Pennycuick[38] quantified infections with *S. solidus*, *D. gasterostei*, and *E. clavula* among a large-bodied, long-lived population in Priddy pool, Somerset, and demonstrated increases in the prevalence and the mean intensity of infection for all infections up to a host body size of about 60 mm; beyond 60 mm *S. solidus* and *E. clavula* infection intensities tended to fall, possibly suggesting selective mortality of the most heavily infected fish. Among three-spined stickleback caught in the Gullmarsfjord, Sweden, the number of individual *Cryptocotyle lingua* metacercariae, which form easily counted pigmented cysts on the skin surface of host sticklebacks in marine populations, correlated closely with host stickleback body length[44] (Figure 9.2a).

9.2.4 Benthic-Limnetic Species Pairs

Fish species that have evolved distinct morphotypes within the same body of water, such as Arctic char *Salvelinus alpinus*,[45,46] present an ideal opportunity to test hypotheses regarding the evolution of host–parasite relationships. Where the separation of morphotypes is based on trophic specialisation, we may expect morph-specific parasite communities that reflect the diet or habitat use by the hosts, and these are generally identified in parasitological investigation. For example, benthic morphs of Arctic char typically harbour parasite communities dominated by acanthocephalans (the infective stages of which are harboured by benthic prey), and pelagic morphs harbour cestode-dominated communities.[47,48] Indeed, such parasitological evidence may even be used as evidence to support ecological separation of putative morphs. The existence of well-documented benthic and limnetic "species pairs" of three-spined stickleback in British Columbia[49] and Alaska,[50] which probably came to coexist through successive waves of invasion of the lakes, would appear to present an ideal opportunity to test host–parasite relationships. Recent studies at Paxton and Priest lakes, British Columbia, suggest that there are significant differences in the level of parasite infection in limnetics and benthics (Andrew MacColl, personal communication). In a detailed survey of the parasite fauna of sticklebacks at the two sites, benthics were found to exhibit a higher prevalence of skin-encysted and internal trematode metacercariae, mussel glochidia, *Crepidostomum* sp., and *Gyrodactylus alexanderi*, whereas the mean intensity of *G. alexanderi*, *Thersitina* sp., *S. solidus*, *Bunodera* sp., *Neoechinorhynchus* sp., *Proteocephalus* sp., and *Diplostomum scudderi* was higher in limnetics. Furthermore, despite being the

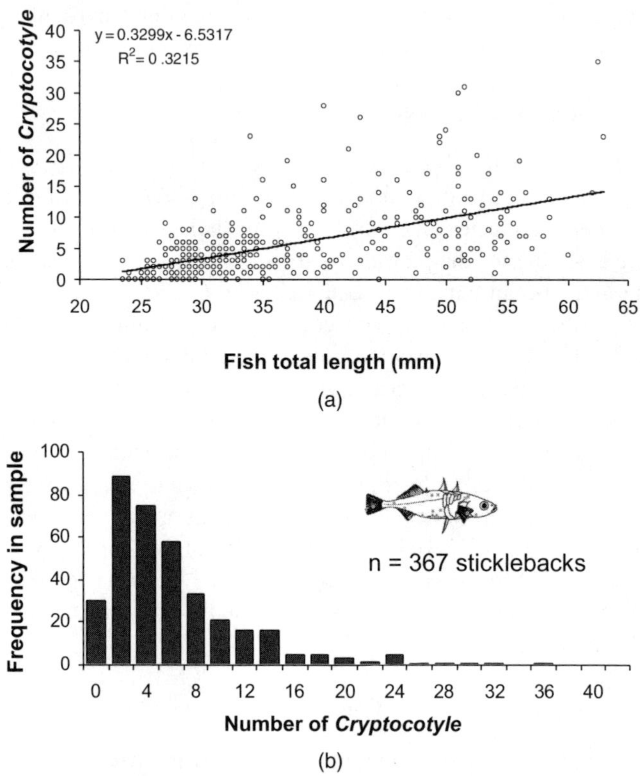

FIGURE 9.2 *Cryptocotyle lingua* infections in three-spined stickleback *Gasterosteus aculeatus* from the Gullmarsfjord, Sweden: (a) Frequency distribution showing typical overdispersion among the host population; (b) relationship between fish body size and *C. lingua* infection intensity. (Data from Barber, I., *Oikos* 101, 331, 2003. With permission.)

smaller bodied of the two "species," limnetics harboured a higher average number of internal helminths than benthics (18 vs. 10 individual worms, including metacercariae, per fish). Further studies on the parasites of benthic-limnetic species pairs are likely to be of considerable value in developing our understanding of the evolutionary basis of host–parasite interactions.

9.2.5 Marine, Freshwater, Brackish, and Anadromous Populations

Because directly transmitted ectoparasites often have limited salinity tolerances[51,52] and indirectly transmitted parasites often have specific host requirements that cannot be equally met in all aquatic habitats, there is typically little overlap between the parasites harboured by marine and freshwater fish. As may be expected, the parasite fauna characterising marine and freshwater populations of three-spined sticklebacks, therefore, typically differ substantially (see Table 9.2). However, few studies have examined geographically proximate marine and freshwater populations, to separate geographic from local environmental factors. An interesting exception to this is a study by Marcogliese,[53] examining nine-spined stickleback *Pungitius pungitius*, which compared the parasite fauna of fish from a brackish water pond (26%) on Sable Island (Nova Scotia) with those of fish in two adjacent freshwater ponds. Although many of the parasites were found in both salinity environments, one notable parasite, *Schistocephalus pungitii* (a putative sister species to *S. solidus*), was only present in the sticklebacks in the freshwater ponds, suggesting that host requirements may not have been fulfilled at the higher-salinity site. The parasite fauna of anadromous populations of three-spined stickleback have not been well studied, but it is likely the endoparasite fauna will reflect both marine and freshwater sources of infection, whereas ectoparasites are likely to reflect the recent salinity regime encountered. Certainly, *S. solidus* is known to occur in seaward-migrating three-spined stickleback (S.A. Arnott, personal communication).

In freshwaters, there may be substantial differences between the parasite fauna of sticklebacks from rivers and lakes. Wegner et al.[54] surveyed eight natural stickleback populations, from four lakes, three rivers, and an estuary, and quantified infections with 15 different macroparasites. The authors showed that parasite diversity varied across populations, with the fish from rivers consistently exhibiting the lowest parasite diversity. The increased diversity of parasites in lacustrine populations may suggest that sticklebacks invading freshwaters face increased pressure from infections. Parasites therefore have the potential to play an important role as selective agents in the ecological divergence of three-spined sticklebacks.[55]

9.2.6 Individual Variation in Parasite Load

Clearly, differences in infection level may be evident between the sexes and between different populations inhabiting different environmental types. However, even among individuals of the same size and sex from the same population, substantial differences in the levels of infection typically remain. Many parasites exhibit negative binomial ("overdispersed") distributions among host stickleback populations, with most individuals harbouring low levels of infection and a small number harbouring much higher infection levels.[38,55] Figure 9.2b shows a characteristically overdispersed distribution of *Cryptocotyle lingua* among a population of sticklebacks from the Gullmarsfjord, Sweden; in this sample just over 20% of host fish harboured more than 50% of all the parasites found among the sample.

The factors responsible for variation in the parasite loads that individuals harbour, even when matched for gender and age within a population, are of particular interest to ecologists and parasitologists. As with gender differences in infection, variation may be due to differences in exposure or susceptibility to infection following exposure. In the case of infection with *S. solidus*, experimental studies have shown that susceptibility to infection following a controlled challenge appears to be related to paternal quality, indicated by sexual ornamentation,[56] whereas the rate at which the parasite grows once established apparently reflects the major histocompatibility complex (*Mhc*) diversity of the host[57] (see Section 9.3.5). Long-term studies by Reimchen and coworkers show that, among adults during the warmest time of the year, variation in the level of infection with a range of helminth parasites covaries with the symmetry of the host pelvic girdle, with asymmetric phenotypes having a disproportionately high prevalence and intensity of infection.[58] One explanation is that some underlying aspect of genetic "quality" is revealed both by the degree of symmetry and the level of parasite resistance, which then show a phenotypic covariance.[39] Alternatively, because dietary differences between symmetric and asymmetric phenotypes have also been documented, with asymmetric phenotypes taking a more benthic diet,[59] it is also possible that differences in exposure generate the observed infection patterns.[60] Unravelling the factors that lead to individual level variation in parasite load will rely on combining controlled laboratory experiments with insights from carefully designed field studies.

9.2.7 Geographical Variation

Parasite community richness can vary across geographical scales.[61,62] Given their wide geographical distribution, both marine and freshwater forms of the three-spined stickleback would appear to offer exceptional model host systems for studying the parasite species richness of a single host taxon across a range of latitudes; however, although parasites have been investigated throughout the distribution of the host, no systematic analysis of geographical patterns in stickleback parasite has yet been undertaken. One major problem is that survey techniques, the range of parasites studied, and species identification vary between published studies. The implementation of standardised techniques, and the data collected in the current stickleback parasite biodiversity programme,[63] should facilitate such analyses in the future.

9.2.8 Environmental Stress and Parasitism

Lafferty and Kuris[64] highlight three major ways in which environmental stressors, such as chemical, biological, or thermal pollution, habitat alteration, or species introductions, might impact on the relationships between native aquatic hosts and their parasites. First, stressors may alter infection levels by changing host susceptibility, by directly impacting the survival of free-living stages, or by altering the abundance of intermediate hosts. Second, habitat alterations, such as changed flow regimes, can affect intermediate host populations, with knock-on consequences for the abundance of attendant parasites. Finally, introduced species may introduce new parasites to susceptible native populations, or may gain an advantage if infections from their native range are suppressed.

Laboratory studies have provided support for the first mechanism in stickleback populations and, given the utility of sticklebacks as a model for the examination of host–parasite interactions in both the laboratory and field, tests of the second and third mechanisms should be possible. Axelsson and Norrgren[65] held marine three-spined sticklebacks under flow-through conditions for $5^1/_2$ months under a range of dilutions of effluent from a Swedish pulp mill and demonstrated the intensity of infection with two skin-dwelling ciliates (of the genera *Trichodina* and *Apiosoma*) to be significantly higher in the exposed fish populations, particularly in those exposed to the higher effluent doses (effluent diluted 400×) (see also Lehtinen[66]). Because these parasites are directly transmitted between fish and are capable of reproducing once established on a fish host, the results suggest that the capacity of effluent-treated sticklebacks to control infections was impaired. Toxicity studies have also demonstrated that sticklebacks infected with physiologically demanding *S. solidus* infections succumb more quickly than noninfected fish under chronic exposure to toxic pollutants such as cadmium,[67] suggesting that multiple stressors can significantly impact the health and survival of stickleback populations.

9.2.9 Community Ecology of Stickleback Parasites

Sticklebacks rarely harbour a single parasite species in isolation; it is far more common to find that individual hosts harbour a suite of infections. The study of parasite community ecology attempts to understand the processes that account for the structure and dynamics of parasite assemblages within hosts.[68] The factors governing the development of parasite communities within fish hosts are likely to be complex, and include the spatial distribution of the host, its age, diet, and its inter- and intraspecific relationships, including its trophic position. There may also be direct interactions between parasites that occur in close spatial proximity, and this can lead to the competitive exclusion of some species.[69]

A limited number of studies have focused on the parasite communities of sticklebacks. A survey designed to look for six target helminth parasites among 31 potential host fish species from the Baltic Sea identified sticklebacks as harbouring the highest diversity of helminths, four of the six target species: *Triaenophorus nodulosus*, *Diphyllobothrium dendriticum*, *D. ditremum*, and *S. solidus*.[10] In a later investigation of the parasite communities of littoral invertebrates and small fish in a region of the Baltic Sea experiencing extreme eutrophication, Zander et al.[11] identified three-spined sticklebacks as the most heavily infected of 17 species examined (including 3 other fish: common goby *Pomatoschistus microps*, nine-spined stickleback, and juvenile flounder *Pleuronectes flesus*). Among the sample of sticklebacks collected (n = 516), positive co-occurrences between the two *Diphyllobothrium* species, between *S. solidus* and each of the two *Diphyllobothrium* species, and between *T. nodulosus* and *D. dendriticum* were identified. A significant three-way association between the two *Diphyllobothrium* species and *S. solidus* (all of which are acquired by eating infected copepods, and require a bird as the final host) was also found. One explanation for these results is that dietary variation, perhaps linked to habitat use, may expose certain individuals to infections with trophically transmitted parasites, which are then acquired as a suite of infections. Alternatively, acquiring one infection by chance may lead to other infections being acquired as a result of behavioural changes associated with the primary parasite, or its potential immunosuppressive effect. The fact that sticklebacks potentially harbour such a wide range of parasites, both taxonomically and with respect to life cycle types, make the fish extremely useful as a bioindicator, because the presence or absence of particular suite of parasite may indicate the presence or relative abundance of other hosts in the life cycles.

9.3 AVOIDING INFECTIONS: BEHAVIOURAL AND IMMUNOLOGICAL RESISTANCE TO PARASITES

Animals are capable of using behavioural mechanisms to limit their exposure to infective parasites, and "behavioural resistance" can therefore be described as the first line of defence in countering parasite invasion,[70] reducing the demands placed on structural and immunological defences.[71] However, as there are likely to be costs associated with behavioural alterations that serve to limit exposure to parasites (e.g., by avoiding potentially rich foraging habitats that also pose a risk of acquiring infections), behavioural resistance mechanisms are only likely to evolve if the fitness benefits of avoiding infections outweigh the fitness costs of performing the behaviour.[72,73] Hart[70] and Lozano[74] have outlined the mechanisms by which the threat held out by the presence of infective parasites may influence the behaviour of actual or potential hosts. These include the avoidance of free-living parasite stages, of other individuals carrying contagious infections, or of prey harbouring infective parasites, as well as parasite-reducing behaviours, such as visiting heterospecific cleaners or scraping the body against the substrate to remove ectoparasites.

9.3.1 Habitat Selection

Working on tide-pool populations of three-spined stickleback on Isle Verte, Quebec, Poulin and FitzGerald[75] demonstrated that the risk of acquiring infections with the ectoparasitic louse *Argulus canadensis* was significantly increased in pools with lower water levels, presumably because these forced fish into contact with the substratum where the parasites could launch an attack. In a series of laboratory studies, Poulin and FitzGerald[76] further demonstrated that sticklebacks can detect the presence of the ectoparasite *Argulus canadensis* and alter normal habitat preferences for specific water depths and vegetation cover to minimise their occupancy of "risky" habitat when lice are present. Karvonen et al.[77] recently demonstrated that individual rainbow trout *Oncorhynchus mykiss* can limit their contact with infective *Diplostomum spathaceum* cercariae by avoiding the infection source (lymnaeid snails); restricting the habitat selection opportunities of trout also increased infection levels. As sticklebacks are susceptible to a range of diplostomatid trematodes, including *D. spathaceum*, similar avoidance behaviours may be expected.

9.3.2 Diet Selection

Conversely, there is little evidence that sticklebacks modify their diet to avoid trophically acquired infections. Wedekind and Milinski[78] examined the effects of *S. solidus* infection on the behaviour and susceptibility to predation of copepods, the first intermediate hosts of the parasite. Copepods harbouring infective procercoids were identifiable, exhibiting altered swimming behaviour and reduced escape behaviours compared to noninfected copepods. However, sticklebacks did not select against them in their diet; instead, sticklebacks actually fed on them preferentially. Furthermore, when given the choice between copepods (which potentially harbour infections) and size-matched *Daphnia* (which do not), sticklebacks from naturally parasitised populations did not prefer *Daphnia* prey. Similar studies by Urdal et al.[79]

also failed to identify a preference for noninfected over infected copepods, despite similar infection-associated behavioural modifications in the copepods. Sticklebacks may even preferentially attack invertebrate prey that harbour infective parasites over noninfected prey (e.g., gammarids infected with *Pomphorhynchus laevis*[80,81]).

In considering the likelihood of host taxa evolving mechanisms of avoidance of prey harbouring infective parasites, a number of factors need to be taken into account. First, it must be possible to detect infected prey and second, there must be a significant fitness cost of ingesting the parasite; importantly, the cost of acquiring an infection must outweigh the potential fitness benefit of foraging on infected prey items that may be more visible, easier to catch, or both.[82] If the probability of acquiring an infection after ingesting is low, or if the cost of harbouring an infection is minimal, then taking advantage of more easily caught, infected prey may be energetically worthwhile.

9.3.3 Partner Choice

Choice of partner, both in mating and social contexts, is also likely to influence an individual stickleback's exposure to directly transmitted, contagious parasites. Dugatkin et al.[83] tested the shoal choice of noninfected sticklebacks given a choice of stimulus shoals consisting of size-matched conspecifics, either noninfected or infected with *Argulus canadensis*, and found clear preferences for nonparasitised shoals. Because *Argulus* can reduce growth and increase mortality of infected sticklebacks,[84] the ability to recognise and avoid parasitised shoals may be an important mechanism of behavioural resistance.

However, although shoaling with *Argulus*-infected conspecifics is avoided, the formation of groups can provide effective protection against infection by this parasite, which can also "attack" fish from the substrate (much as a sit-and-wait predator may do). Fish in larger groups, therefore, potentially gain dilution benefits (*sensu* Hamilton[85]) from group membership.[86] In support of this, shoal size has been shown to reduce the average per capita rate of parasite attacks from *Argulus* parasites,[87] and in the laboratory sticklebacks formed larger shoals in the presence of the parasite, suggesting that shoaling may be an effective antiparasite mechanism against this infection.[88] Yet, although forming larger groups may provide dilution benefits against parasites that attack like predators, being a member of a larger group potentially increases the level of exposure to contagious parasites.[89,90] Consequently, although Poulin[91] demonstrated a negative (though nonsignificant) relationship between shoal size and the abundance of *Argulus* infections among 14 complete shoals of sticklebacks collected from tide pools on Isle Verte (Quebec), he also identified significant positive relationships between shoal size and both the intensity and abundance of ectoparasitic copepods *Thersitina gasterostei*, which are transmitted directly between fish via short-lived planktonic larval stages. The consequences of increased shoal size for infection risk are clearly complex, and depend largely on the infection strategies of the parasites involved.

Infected fish may be rejected as shoalmates even if the parasites they harbour cannot be transmitted directly to conspecifics; for example, *S. solidus*–infected sticklebacks are avoided both by noninfected and by other *S. solidus*–infected conspecifics.[92,93] Although reasons for this avoidance are speculative, it may reflect the possibility that infected fish make poor-quality shoalmates.

9.3.4 Mate Choice

A further way in which sticklebacks can limit their contact with contagious parasites is through their mate choice decisions. In a landmark study, Milinski and Bakker[94] demonstrated that the orange-red carotenoid-based ornamentation of the male three-spined stickleback was used by females in mate choice decisions, and because male ornamentation and the level of infection with *Ichthyophthirius multifiliis* (whitespot disease) covaried negatively, females exhibiting ornament-based mate choice avoided the most heavily infected males. Females potentially gain both direct benefits (by avoiding a male, or a nest site, that harbours a contagious infection), and indirect benefits (possibly by selecting parasite resistance genes for their offspring) by avoiding parasitised males.[95] To determine whether females choosing brightly ornamented males as mates might potentially select parasite resistance for their offspring, Barber et al.[56] fertilised split-clutches of eggs from individual females with sperm from bright- and dull-coloured males to generate groups of maternal half-siblings, which were then raised in the laboratory and exposed to controlled challenges with *S. solidus*. Juveniles sired by brightly coloured males were significantly less likely to develop infections following challenge than were half-siblings that shared the same mother, but had been sired by a dull male. These results strongly suggest that females choosing the most brightly coloured males gain indirect, parasite resistance benefits through ornament-based mate choice, as well as direct parasite avoidance benefits.

9.3.5 The Mhc, Mate Choice, and Parasite Resistance

In recent years, there has been an explosion of research into the role of immune genes, and in particular genes of the major histocompatibility complex (*Mhc*) in determining the susceptibility of animals to parasite infection, and sticklebacks have played a key role in these investigations. Class I and class II genes of the *Mhc* encode cell surface molecules that are capable of binding and presenting short peptides to host T cells, and as such are an important component of the immune response to invading pathogens, including parasites. There is now substantial evidence, from field studies and controlled laboratory experiments, that an individual stickleback's allelic diversity is related to its level of infection with parasites.[96,97] Wegner et al.[98] surveyed eight natural stickleback populations, from rivers, lakes, and estuaries and undertook substantial parasite analyses of the fish they caught, quantifying infections with 15 different macroparasites, while typing each individual stickleback's *Mhc* Class IIb allelic diversity (i.e., the number of different alleles, which can potentially range from 1 to 12). The authors showed that parasite diversity and *Mhc* diversity varied substantially across populations, with fish from rivers having both the fewest parasites and the lowest *Mhc* diversity. Within populations, however, fish harbouring intermediate allelic diversity (i.e., six different alleles) were those that harboured the fewest parasite species and the lowest numbers of individuals per parasite species. Further evidence that intermediate, rather than maximal or minimal, allelic diversity is beneficial for parasite resistance is provided by Kurtz et al.[57] Following exposure to two important and ecologically relevant parasites (*Glugea anomala* and *Schistocephalus solidus*), fish with too many or too few alleles suffered more from infection than those with intermediate *Mhc* allelic diversity. In the case of *S. solidus*, fish with intermediate allelic diversity harboured the slowest growing plerocercoids. Fish with the lowest *Mhc* diversity also exhibited the highest amount of innate immune activity, measured both as respiratory burst and the proportion of granulocytes in the blood, suggesting that fish with suboptimal *Mhc* diversity exhibit the strongest activity of the innate immune system after challenge by parasites.

Mhc alleles can have both direct and indirect effects on the odour of animals[99] and can potentially be used as selection criteria in mate choice. Studies on both mice[100] and humans[101] have shown that *Mhc*-based odour cues can be used by animals to select mates with complementary *Mhc* composition, potentially maximising *Mhc* heterozygosity among offspring. In sticklebacks, Reusch et al.[102] examined the potential for odour-based *Mhc* preferences, and demonstrated that females could discriminate between males based on their odour alone (apparently due to differences in peptide ligands[103]), and that males with the highest *Mhc* allele diversity were preferred as mates. Subsequent studies by Aeschlimann et al.[104] demonstrated that when females that varied in *Mhc* allelic diversity (some females had high levels of diversity and others had fewer *Mhc* alleles) were given a choice of male odours, their preferences were for males that offered a complementary level of *Mhc* allelic diversity. Females with many alleles therefore selected males with few alleles and *vice versa*. Any offspring produced would therefore tend to have intermediate *Mhc* diversity, which provides maximal protection against parasites. The authors suggest that, because the *Mhc* diversity "chosen" for offspring through female mate choice is close to the most frequent diversity found naturally in individual fish, then mate choice can explain population level patterns of heterozygosity.[104]

The reasons why intermediate numbers of alleles, rather than the highest possible diversity, appear to be beneficial to sticklebacks in terms of minimising parasite impacts is not particularly clear. One suggestion is that whereas having too few different *Mhc* alleles leaves an individual with insufficient diversity to recognise a large number of parasite antigens, having too many *Mhc* alleles may lead to self-immunoreactivity. Intermediate level of *Mhc* diversity, apparently sought by females for their offspring and also the most commonly observed in natural populations,[54] may therefore reflect an optimal balance between parasite protection and self-tolerance.[96] Interestingly, despite the protection apparently offered by intermediate numbers of alleles, and the strong preferences of females for mates with complementary numbers of alleles, stickleback in natural populations exhibit considerable variation in allele number (from 2 to 9),[54] and the population genetic mechanisms that could maintain this observed variation are currently under debate.[105,106]

9.4 MORPHOLOGICAL AND PHYSIOLOGICAL EFFECTS OF INFECTION

Parasite infections may impact on the morphology and physiology of host fish as a result of a number of mechanisms.[7,107] Ectoparasites may be externally visible, whereas endoparasites can cause detectable morphological changes in hosts if they are large-bodied, or alter normal patterns of host pigmentation. Damage to skin or

FIGURE 9.3 A leech *Piscicola geometra* infecting a three-spined stickleback *Gasterosteus aculeatus*. (Photograph courtesy of Joanne Cable.)

gills, as a consequence of invasion or invasive connections (e.g., by trematode metacercariae or ectoparasites such as *Gyrodactylus* or leeches; Figure 9.3) can alter ionic balance and osmoregulatory capacity, whereas metabolically demanding parasites can impose considerable energetic burdens on host fish, with consequences for growth, condition, and reproduction. Parasites can also impair the performance of sensory or locomotory tissues damaged during their invasion, migration, or growth phases, and biochemical or neurochemical disruption can occur as a consequence of the build-up of waste products or targeted parasite secretions. Many of these physiological effects of parasites can have behavioural consequences (see Section 9.5). In sticklebacks, most studies of the physiological impacts of infections have focused on a small number of model systems, and because of their considerable metabolic demand, *S. solidus* infections have been most frequently investigated. Table 9.1 summarises studies that have examined aspects of stickleback biology that are impacted by *S. solidus* infection.

9.4.1 Changes in Morphology Associated with Infection

Some of the most severe morphological changes associated with infection are exhibited by fish infected with *S. solidus* plerocercoids (Figure 9.1). Heavily infected fish exhibit a grossly distended body cavity,[23,108] which, although superficially resembling the distension caused by developing ovaries in females, characteristically bulges to the anterior as well as the posterior of the pelvis. In Alaska, a small number of stickleback populations additionally exhibit a white ("demalanised") phenotype when harbouring exceptionally large *S. solidus* plerocercoid burdens[109,110] (Figure 9.4), though whether this is a consequence of a unique host or parasite genotype is not known. Heavy infections with large encysted nematode larvae (Figure 9.5) can also cause abdominal swellings. Infections with the microsporidian *Glugea anomala* are characterised by conspicuous xenomae: creamy-white pustules reaching several

TABLE 9.1
Studies Examining Traits Associated with *Schistocephalus solidus* Infections in Sticklebacks

Aspect of Stickleback Biology Studied	Lab/ Field	Infection-Associated Trait	Natural/ Experimental Infections	Reference
Antipredator behaviour	Lab	Fish with infective plerocercoids showed reduced antipredator behaviours	Natural	177
	Lab	Infected fish showed reduced antipredator response	Natural	151
	Lab	Infected fish showed increased risk taking under threat of predation	Natural	153
	Lab	Infected fish fed closer to a predator	Natural	152
	Lab	Infected fish begin foraging sooner following disturbance by a predator	Natural	164
	Field	Infected fish showed reduced antipredator behaviour	Natural	110
Competitive ability	Lab	Infected fish were poor competitors for food under certain conditions	Natural	161
Foraging behaviour	Lab	Infected fish ingested fewer prey and exhibited increased handling time	Natural	127
	Lab	Infected fish selected a higher proportion of smaller prey items	Natural	155
	Lab	Infected fish fed selectively on larger prey items	Natural	156
	Field	Infected fish had reduced stomach fullness and fed on smaller prey	Natural	157
	Field	Seasonal differences detected in stomach fullness and diet composition of infected and noninfected fish	Natural	118
Shoaling behaviour	Lab	Infected fish show reduced shoaling tendency when satiated	Natural	159
Habitat use	Field	Infected fish found further from cover in autumn	Natural	167
Male sexual display	Field + Lab	Infected males less likely to court on the wild, but not in the lab	Natural	174
	Field	Infected fish developed less red nuptial colouration	Natural	173
Oxygen requirements	Lab	Infected fish consumed more oxygen during swimming	Natural	162
	Lab	Infected fish spent more time breathing at surface under hypoxia and had higher lethal pO_2	Natural	163
	Lab	Infected fish began breathing at surface earlier following decreased pO_2	Natural	158
Growth rate	Lab	Infected fish grew more quickly than noninfected fish	Experimental	126

TABLE 9.1 (CONTINUED)
Studies Examining Traits Associated with *Schistocephalus solidus* Infections in Sticklebacks

Aspect of Stickleback Biology Studied	Lab/ Field	Infection-Associated Trait	Natural/ Experimental Infections	Reference
	Lab	No effect of infection on growth under threat of predation	Experimental	154
Body condition	Field	Infected fish had reduced body condition in the spring and autumn	Natural	28
Gonadogenesis	Field	Infected females show reduced clutch development	Natural	133
	Field	Infected females less likely to be gravid	Natural	132
Brain neurochemistry	Lab	Altered levels of brain monoamine activity in infected fish	Natural	180
Morphology	Field	Infected fish had less symmetric lateral plate counts	Natural	59
	Field	Adults with asymmetric pelvis had increased incidence of infection; pattern reversed in 0+ fish	Natural	58
	Field	Infected fish were demelanised	Natural	109
	Field	Infected fish were demelanised	Natural	110

FIGURE 9.4 The white, demelanised phenotype characteristic of heavy infections in some Alaskan stickleback populations. (Photograph courtesy of Michael A. Bell.)

millimetres in diameter on the skin surface, under the opercula, and in the intestine[111,112] (Figure 9.6). In marine or brackish water habitats, a number of trematode metacercariae encyst under the skin and in the fins, where the host tissue response results in the melanisation and formation of small (<1 mm diameter)

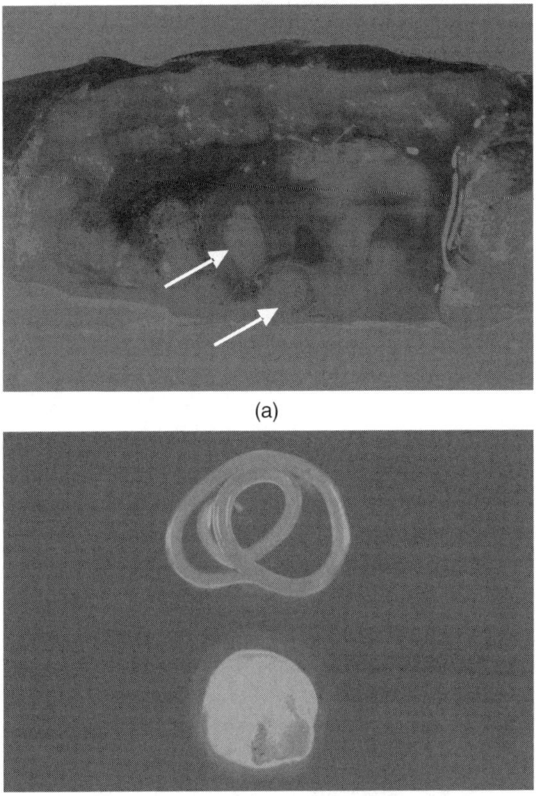

(a)

(b)

FIGURE 9.5 Nematode infection of three-spined stickleback *Gasterosteus aculeatus* from Inverleith pond, Scotland, U.K.: (a) showing the flattened coiled larvae in cysts attached to connective tissue in the body cavity, (b) showing a larva removed from its cyst.

localised black spots. An ectoparasitic dinoflagellate *Haidadinium ichthyophilum* has an unusual effect on the morphology of host sticklebacks in an acid bog on the Queen Charlotte Islands, British Columbia. Fish infected with the dinoflagellate exhibit epithelial hyperplasia, producing layers of host cells that enclose the parasites and giving the appearance of a thick "gelatinous" coating[113] (Figure 9.7). Parasites such as these, which alter host morphology, potentially have implications for stickleback behaviour and ecology, and this is examined in Section 9.5. Other parasite infections can affect secondary sexual traits of male sticklebacks, and the important effects these can have for mate choice and sexual selection are discussed in Section 9.5.4.

9.4.2 Impacts on Sensory Physiology

Sticklebacks in freshwaters are frequently infected with metacercariae of diplostomatid trematodes, particularly *Diplostomum spathaceum*, *D. gasterostei*, and *Tylodelphys clavata*, which occupy the lens, retinal tissue, and humour of the eye,

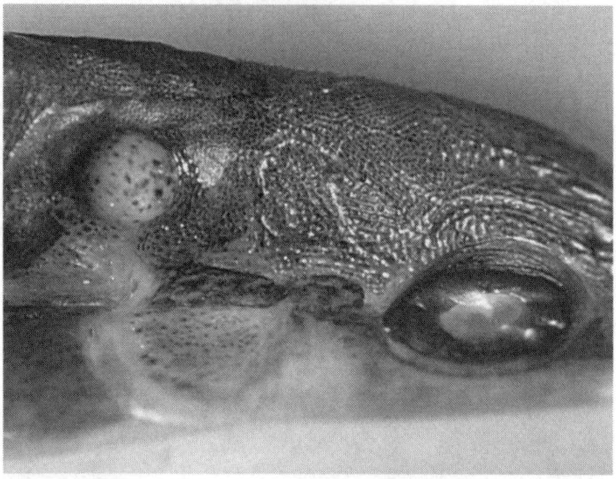

(a)

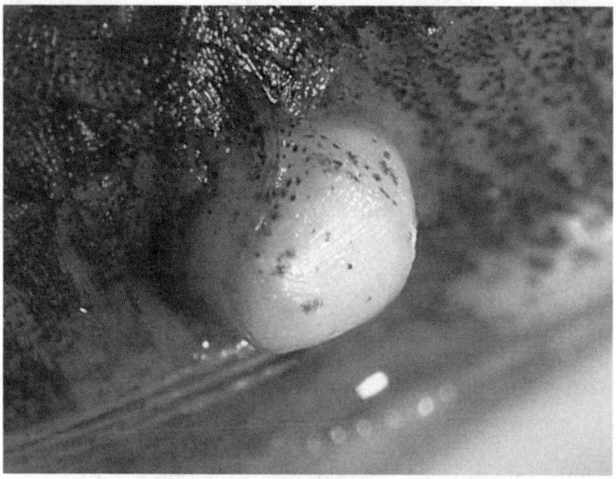

(b)

FIGURE 9.6 *Glugea anomala* infection in three-spined stickleback *Gasterosteus aculeatus* from Shediac, New Brunswick: (a) Xenomae present on the dorsal surface and under the operculum of an infected fish; (b) close-up of a xenoma showing host pigment cells in the xenomae wall. (Photographs courtesy of Monica Lanteigne.)

respectively. Heavy infections of *D. spathaceum* are associated with damage to the lens,[114] and are associated with reduced visual acuity and altered foraging behaviour in stickleback hosts[115] (see Section 9.5.2).

9.4.3 Impacts on Host Energetics

Parasites are only capable of achieving growth and development by utilising host-derived energy resources, and so energetic effects of infection are among the most commonly recorded. Parasites can impact host energy budgets in a variety of ways, including the direct utilisation of host nutrients to sustain their growth and development, by inducing energetically demanding immune responses, by impairing nutrient acquisition, or by reducing the efficiency of locomotion. Predictably,

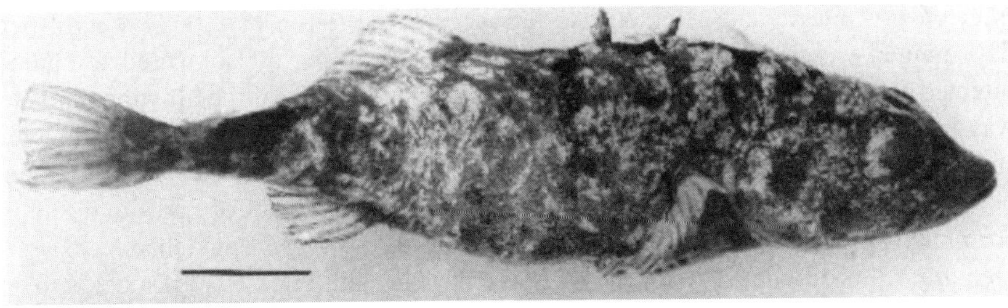

FIGURE 9.7 A three-spined stickleback *Gasterosteus aculeatus* from an acid bog on the Queen Charlotte Islands, British Columbia, infected with the parasitic dinoflagellate *Haidadinium ichthyophilum*. The white patches on the skin represent areas of infection where the parasite has been encapsulated by hyperplastic host epithelial cells, giving the characteristic "gelatinous" appearance. (Photograph courtesy of Thomas Reimchen.)

parasites such as *S. solidus*, which grow to a large size within the stickleback host, have been most closely examined in this regard.

9.4.3.1 Body Condition

In exceptional cases, wild-caught sticklebacks infected with *S. solidus* can harbour a mass of parasite tissue that exceeds that of the host fish,[23,29,116] with infections in which PI exceeds 30% being common in parasitised populations. Evidence from laboratory and field studies shows that during the parasite growth phase *S. solidus* plerocercoids impose a considerable energetic drain on stickleback hosts, and this has significant consequences for host energy stores and condition. Infected fish collected from wild populations typically have lower somatic body condition and liver energy,[23,28,117–119] as do experimentally infected fish held under a fixed ration of 8% body weight per day over the parasite growth phase.[120] In the latter study, a reduction in the mass of the perivisceral fat body in experimentally infected sticklebacks was also identified. Although equivalent data are not available for sticklebacks infected with *S. solidus*, the amino acid composition of roach *Rutilus rutilus* infected with the closely related pseudophyllidean *Ligula intestinalis* are consistent with those of starved fish,[121] confirming that infection with plerocercoids has a significant impact on the nutritional condition of fish hosts. Probably as a result of their reduced energetic condition and increased metabolic demand, naturally infected sticklebacks have been shown to succumb under starvation conditions before noninfected conspecifics in laboratory studies,[122,123] and this is likely to lead to disproportionate mortality during periods of food shortage. Other parasites can also be associated with reduced body condition. Both sexually mature males harbouring *Pomphoryhchus laevis* and females infected with high levels of skin-encysted trematodes exhibit reduced body condition in natural populations.[42,124,125]

9.4.3.2 Growth

Under conditions of limited food availability, growth is predictably negatively impacted by *S. solidus* infection compared to noninfected fish. Experimentally infected juvenile female sticklebacks fed a restricted ration (8% body weight per day) gained weight at a reduced rate over a 16-week period compared to sham-infected females of the same initial body size, once the mass of the growing parasite was taken into account.[120] However, under particularly high rations, experimental studies have shown that infected sticklebacks can grow faster and attain a larger body size than noninfected conspecifics, even when the weight of parasite tissue is accounted for. Arnott et al.[126] demonstrated that three-spined sticklebacks, experimentally infected with *S. solidus* and provided with unlimited food resources, outgrew sham-infected controls, suggesting that infections increased feeding motivation and total food intake. Although untested, one possible mechanism underlying this "gigantism" in infected sticklebacks could be the increased time that infected fish typically spend foraging, which when food is effectively unlimited could lead to the observed rapid growth. Detailed studies of the growth of infected and noninfected sticklebacks under natural conditions in the wild have yet to be undertaken.

When attempting to sustain the growth of a rapidly developing plerocercoid, *S. solidus*-infected sticklebacks face the further problem that the growing plerocercoid reduces the space available for stomach expansion, limiting meal size.[127,128] Considering that infected fish are typically poor competitors for food (see Section 9.5.2), it is unlikely that infected fish are capable of increasing their ration under conditions of food competition, and this has profound implications for a fish's ability to regulate its growth following a period of food shortage. Fish under nutritional stress, such as occurs after a period of food deprivation, normally exhibit hyperphagia on refeeding, to maximise their food intake.[129] This typically leads to a period of increased ("compensatory" or "catch up") growth, allowing the growth trajectory of previously starved fish to converge with the expected size-at-age.[129] However, recent studies suggest that *S. solidus*-infected sticklebacks may be incapable of undertaking compensatory growth when food becomes available after food deprivation, as a result of their inability to become sufficiently hyperphagic, even in the absence of competition (Wright et al., submitted ms). Parasites that impact a host's capacity to elevate food intake therefore have the potential to interfere with mechanisms of growth regulation in natural populations.

9.4.3.3 Sexual Maturation

Parasites can impact host sexual maturation through two main types of mechanism: directly, by impacting the hormonal control of gonadogenesis or indirectly, by reducing the energy available for reproduction. The tapeworm *Ligula intestinalis*, which infects cyprinids, achieves host castration by interfering with the pituitary–gonadal axis of fish hosts,[130] with plerocercoids restricting gonadotropin releasing hormone (GnRH) secretion.[131] *S. solidus*, on the other hand, appears to mediate its effects on sexual maturation through largely energetic mechanisms. The reduced body condition of infected fish has consequences for sexual maturation, particularly in females, because condition is linked closely to gonadogenesis in sticklebacks,[2,119] and *S. solidus* can have particularly severe consequences for sexual development.[132] However, the extent to which *S. solidus*–infected female sticklebacks are capable of maturing sexually appears to depend on both the level of infection and on the population under investigation. Whereas *S. solidus* infections can have a devastating and almost complete effect on sexual maturation in infected females in some populations,[28,132] in others, a high proportion of infected females may become sexually mature and produce clutches despite high levels of infection.[133] To what extent this variation reflects difference in the virulence of the parasite, the genotype of the host, or the result of a coevolutionary process is not known. Among infected fish, the degree of female sexual maturation is often related to PI (the proportion of host body weight contributed by the parasite), with more heavily infected fish being less likely to mature.[23,28,134]

S. solidus infections can also impact the size of female gametes. In nine three-spined stickleback populations in lakes of the Matanuska-Susitna Valley and the Kenai Peninsula of Alaska, mean ovum mass of the infected was significantly reduced, by between 8 and 32%, compared to uninfected females. Furthermore, in two populations studied in detail, a significant, negative relationship between mean ovum mass and *S. solidus* PI was identified.[135] As egg size has important consequences for offspring performance and survival in fish,[136] these results suggest that even in populations in which infected females can mature and produce clutches, heavily infected females may be unlikely to contribute significantly to future generations.

Few studies have focused on the effects of parasite other than *S. solidus*. An exception is that of FitzGerald et al.,[125] who quantified the intensity of infection with "blackspot"-causing trematode metacercariae among a breeding population of anadromous sticklebacks and identified a small but statistically significant negative impact on female clutch size.

9.5 BEHAVIOURAL EFFECTS OF INFECTIONS

Parasites not only influence the behaviour of potential hosts by the threat posed by their presence in natural environments (Section 9.3), but they can also alter the behaviour of host fish after infection, with significant implications for host biology and ecology.[137–139] The proximate and ultimate control of infection-associated host's behavioural change is subject to considerable debate, with three types of possible explanations. First, behavioural modifications may be host adaptations to infection, serving to counter the debilitating effects of the parasite. Second, the behavioural changes may benefit the parasites, for example, if they enhance the probability of successful trophic transmission from the fish host to subsequent hosts in the life cycle by altering interactions with predators.[140,141] Third, it is possible that behavioural changes arise as unavoidable side effects of infection that benefit neither parasite nor host, but simply exist because of the debilitating nature of the infection. Distinguishing between the various explanations remains a considerable challenge to biologists.[142–144] The three-spined stickleback has proved a popular model species for the investigation of infection-associated behavioural traits, and a range of behaviours has been investigated.

9.5.1 Swimming Behaviour

Sticklebacks swim using both the pectoral fins and the tail,[2] so parasites that affect either the body musculature or the fins are likely to have an impact on the kinematics of swimming. Parasites that impact the energy reserves or stamina of hosts also have the potential to interfere with swimming behaviour and activity patterns.[107] Furthermore, parasites that affect body shape (for example, those that grow to a large size and distend the body cavity, or large ectoparasites that affect the surface profile of the fish) potentially interfere with the otherwise streamlined shape of the host and are likely to alter the efficiency of swimming.[145] To date, no studies have addressed the effects of parasites on the mechanics of swimming in sticklebacks, but recent analyses of the normal swimming dynamics and kinematics of noninfected stickleback, such as those provided by Bergstrom,[146] Walker,[147] and Blake et al.[148] considerably facilitate their examination.

9.5.2 Foraging Behaviour

Parasites are known to alter the foraging behaviour of three-spined stickleback hosts through their effects on sensory systems. As discussed in Section 9.4.3, infections with *Diplostomum spathaceum* metacercariae, which are located in the lens tissues of the eye, are associated with reduced visual acuity in the three-spined stickleback, with fish harbouring small numbers of metacercariae initiating approaches toward motile *Daphnia* prey at significantly reduced distances compared to noninfected fish.[115] This loss of acuity, and subsequent inefficiency in foraging, is common in other host species and results in infected fish spending a greater proportion of their time foraging.[149] The majority of studies examining interactions between parasite infection and host foraging behaviour in sticklebacks, however, have focused on the energetic effects of infections and their impacts on the trade-off between foraging and antipredator behaviour.

For parasites that make significant energetic demands on their hosts, such as rapidly growing *S. solidus* plerocercoids, host sticklebacks are expected to have an increased nutrient requirement, and therefore, a need to maximise their food intake rate. Although increasing the size of individual meals is probably not possible for fish harbouring heavy infections (see Section 9.4.3.2), they may be able to partially alleviate the combined effects of infection by altering their foraging behaviour in other ways.[150] For example, infected sticklebacks can take advantage of reduced competition for "risky" food. A number of experimental studies have shown that when faced with an opportunity to forage in the vicinity of a model predator, sticklebacks harbouring established *S. solidus* infections take advantage of the availability of food significantly more than noninfected fish, suggesting that they may be able to mediate the demands imposed by the growing parasite by increased risk taking[151–153] (but see Aeschlimann et al.[154]). Food selection strategies of infected fish may also be altered to focus on prey types for which there is less competition. Both Milinski[155] and Ranta[156] documented shifts in the prey size preferences of infected fish when in competition for food with uninfected sticklebacks, suggesting that infected fish might attempt to maximise their individual food intake by selecting uncontested food items. Field data presented by Tierney[118] and Bergersen[157] also suggest that the prey sizes and types selected by infected fish in wild populations differ from those selected by noninfected fish in natural populations. Infected fish may also modify their time budget to increase the proportion of time spent foraging.[158] In the case of sticklebacks infected with *S. solidus*, the increase in foraging time appears to be made at the expense of aspects of antipredator behaviours, including shoaling[159] and sheltering.[160] Such changes in foraging behaviour and time budget are likely to have significant consequences for the susceptibility to predation of infected sticklebacks in natural populations.

Parasites can also alter the foraging success of stickleback hosts under competition from conspecifics, but their ability to compete for food appears to be dependent on the type of competitive environment experienced. In competition with uninfected conspecifics, infected fish compete better for prey items that are presented sequentially than if they are presented simultaneously.[161] Whereas in the first situation poor competitors may be able to mediate their disability by spending more time waiting for food, in the second situation success depends on efficient prey handling, which is impaired in infected fish.[127]

9.5.3 Habitat Use

Fish may change their habitat preferences as a consequence of the infections they acquire, though separating cause and effect in field studies is extremely difficult. A number of studies have demonstrated differences in the spatial distribution of *S. solidus*–infected sticklebacks compared to fish not harbouring infections. Lester[162] showed that the increased oxygen consumption of sticklebacks harbouring *S. solidus* plerocercoids led to an observed preference for shallower water habitats that offered higher oxygen tension. Both three-spined and nine-spined sticklebacks infected with *S. solidus* exhibit an increased frequency of aquatic surface respiration when exposed to experimentally lowered oxygen levels,[163,164] and infected nine-spined sticklebacks were found to have a higher lethal oxygen level.[163] Preferences for high partial pressures of oxygen are likely to lead sticklebacks to the surface layers of waters, and to shallow vegetated margins of lakes, both of which carry significant risks of increased predation pressure from aerial predators.[165,166] *S. solidus*–infected sticklebacks in a Norwegian lake exhibited altered habitat use, compared to noninfected fish, with infected fish being found at greater distances from aquatic vegetation,[167] resulting in introduced Atlantic salmon *Salmo salar* feeding disproportionately on infected sticklebacks.

9.5.4 Effects on Reproductive Behaviour and Sexual Ornamentation

In the run up to the breeding season, male three-spined sticklebacks compete intensely for nesting territories and for nest-building materials, construct a nest, and perform vigorous courtship displays to attract a female to spawn.[2] In addition, males develop carotenoid-based orange-red nuptial colouration[168] and the pectoral fins enlarge to facilitate nest fanning. Debilitating parasite infections therefore have considerable potential to impact a male's success in both intra- and intersexual competition, and it is likely that many of the traits on which females select males as mates have evolved precisely because they provide reliable information about male's level of parasite infection.[169–171] *Ichthyophthirius multifiliis* infections are known to significantly impact the brightness of a male's sexual ornamentation,[94] whereas infections with both *Pomphorhychus laevis* and *S. solidus* are associated with altered pectoral fin size in breeding males, which are potentially amplified during courtship.[124,172] Folstad et al.[173] undertook an analysis of sexually selected colour and parasitism in male three-spined sticklebacks from a Norwegian lake, and found that whereas two infections (with *Diphyllobothrium diremum* and *D. dendriticum*) were found to correlate positively with redness intensity, another (the more pathogenic *S. solidus*) correlated negatively. As all three parasites are acquired by fish feeding on carotenoid-rich copepods, the authors argue that the observed intensity of redness may confer reliable information to females about the exposure of the host to parasites and its current infection status.

Processes of intense intrasexual competition potentially limit the variability among males available to females in natural populations. Candolin and Voigt[174] present field data to show that, in natural populations, *S. solidus*–infected males are rarely found to be in possession of a nest, were less brightly ornamented than uninfected fish, and were more likely to be found in shoals. However, when the same authors housed infected males individually in the laboratory and presented them with nesting materials, most infected males built nests and courted females, suggesting that intrasexual rivalry over limited nest sites in the field was responsible for the reduced conventional breeding activity of infected fish. Interestingly, the capacity of *S. solidus*–infected males to construct nests in the absence of competition is not universal; infected males from Llyn Frongoch, an upland lake in Wales, U.K., failed to build nests despite individual housing and favourable husbandry[175] (see also Arme and Owen[23]). These results suggest that *S. solidus*–infected males may be unlikely to obtain spawnings through conventional breeding, though the extent to which they engage in alternative strategies such as sneaking[176] remains unknown.

9.5.5 Effects on Antipredator Behaviour

For parasites that utilise sticklebacks as intermediate hosts and are transmitted trophically to higher level predators, impairing the antipredator behaviour of the stickleback host is likely to have important consequences for transmission. Consequently, a number of studies have focused on the antipredator behaviour of parasitised sticklebacks, and again there is a predictable bias toward studies of *S. solidus*. As the definitive host of *S. solidus* is a piscivorous bird, these studies have typically examined the escape responses, shelter use, or recovery from disturbance following simulated attacks by model avian predators. *S. solidus*–infected sticklebacks typically exhibit a reduced propensity to react to a striking model heron,[110,151,153,160,177] as do roach infected with the closely related pseudophyllidean cestode *Ligula intestinalis*.[178] Both Tierney et al.[177] and Barber et al.[160] have demonstrated a switch in the antipredator responses of infected sticklebacks, with those harbouring plerocercoids infective to the definitive host (>50 mg) returning more quickly to a food source after being disturbed, and using shelters less frequently in the immediate period following an "attack." Earlier studies by Milinski,[152] Giles,[158] and Godin and Sproul[153] also demonstrated that *S. solidus*–infected fish take greater risks by approaching potential predators in return for increased foraging opportunities.

What are the likely physiological mechanisms that lead to such dramatic changes in the antipredator behaviour of *S. solidus*–infected sticklebacks? Many of the observed behavioural changes may be explained simply in terms of the increased nutritional requirements of infected fish (see Section 9.4.3.1), and it is perhaps suggestive that many of the behavioural changes of *S. solidus*–infected sticklebacks closely match those of growth-hormone-implanted salmonids.[179] However, research by Øverli et al.[180] provides tantalising evidence that behaviour may be altered through neuromanipulation by the parasites. Such mechanisms of host behaviour manipulation are well documented in parasitised invertebrates[181] and mammals.[182] Øverli et al.[180] investigated the effects of *S. solidus* infection status on the concentrations of neuromodulators, including norepinephrine (NE) and serotonin, in the brains of three-spined sticklebacks. Infected fish had elevated concentrations of NE and serotonin in the telencephalons, but the clearest effect of infection was on the serotonin metabolite, 5-hydroxy-indoleacetic acid (5-HIAA), which increased in the brainstem of infected sticklebacks. As a consequence, the ratio of 5-HIAA:serotonin — an indicator of stress in fish — was elevated in both the hypothalamus and brainstem regions of infected sticklebacks. One explanation for these patterns is that the parasite induces chronic stress, but it is also possible that host behavioural change in this system is mediated through neurochemical manipulation. Further research is required to separate the potential direct and indirect mechanisms of infection-induced behavioural change in *S. solidus*–infected sticklebacks.

If parasites can reduce host antipredator behaviour to facilitate transmission, then it may also be possible for directly transmitted parasites, or indirectly transmitted parasites that have yet to attain infectivity, to increase host antipredator behaviour to maximise their own fitness. Although a definitive study has yet to be undertaken, there is some evidence that this may occur; Milinski[152] demonstrated that sticklebacks infected with *Glugea anomala* approached a potential predator less closely than noninfected fish, whereas Tierney et al.[177] showed that fish infected with preinfective (i.e., <0.05 g[183]) *S. solidus* plerocercoids exhibited less activity immediately following a simulated avian attack. Further research, preferably using experimentally induced infections and tracking the behaviour of individuals, could be used to clarify these and other aspects of parasite-associated changes in host antipredator behaviour.

9.6 FITNESS CONSEQUENCES OF INFECTIONS IN NATURAL POPULATIONS

The results of studies examining the impacts of parasite infections on a wide variety of fitness correlates, including growth, survival, and aspects of reproductive development and behaviour, suggest that infections are likely to have a significant impact on survival and reproductive success of host sticklebacks in natural populations.

TABLE 9.2
Parasites of the Three-Spined Stickleback

Taxonomic Group		Species	Parasite Survey	Infection Site	Recent Key Reference
Plantae					
	Dinophyceae	*Haidadinium ichthyophilum*		Skin	187
Fungi					
		Saprolegnia parasitica	a	Nostrils	
		Saprolegnia sp.	a	Skin	
Mesomycetozoea					
		Dermocystidium gasterostei	a c	Skin	188
Protozoa					
	Sarcomastigophora				
		Cryptobia branchialis	a c	Gills	
		Costia (= *Ichthyobodo*) *necatrix*	a b	Gills	
		Hexamita salmonis	a b	Intestine	
	Apicomplexa				
		Eimeria gasterostei	a b c	Liver	
		Eimeria sp.	c		
		Eimeria (= *Goussia*) *zarnowskii*		Intestine	189
		Goussia aculeati			190
	Microspora				
		Glugea anomala	a b c	Connective tissue, skin, intestine	
		Thelohania baueri	a	Eggs	
		Nosema anomala	c		
	Myxozoa				
		Myxidium gasterostei	a b c	Gall bladder	

			a	b	c	d	e	f	g	h	Site	Ref.
		Sphaerospora elegans	a	b	c						Kidney, urinary bladder	191
		Myxobilatus gasterostei	a	b	c						Kidney, urinary bladder	192
		Myxobilatus medius	a		c						Kidney	
		Myxobilatus sp.		b								
		Henneguya sp.	a									
		Henneguya pungitii		b								
		Ceratomyxa sp.	a	b							Intestine	
Ciliophora												
		Trichophrya intermedia	a		c						Gills	
		Trichodina sp.								h		
		Trichodina gracilis	a		c						Gills, skin	
		Trichodina domerguei	a	b	c		e				Skin, fins	32
		Trichodina domerguei f. latispina	a									
		Trichodina reticulata	a									
		Trichodina tenuidens	a	b			e				Gills	32
		Apiosoma sp.	a	b						h		
		Epistylis livoffi (?)	a	b	c						Gills, skin	
		Glossatella amoebas	a		c							
		Ichthyophthirius multifiliis	a	b							Skin	
		Hemiophrys (= Amphileptus) branchiarum	a		c						Gills, skin	
Platyhelminthes												
	Monogenea											
		Gyrodactylus sp.			c		e			h	Skin, fins, gills	
		Gyrodactylus aculeati			c							
		Gyrodactylus alexanderi	a	b							Skin	193
		Gyrodactylus arcuatus	a		c						Gills, fins	194
		Gyrodactylus avalonia	a	b							Fins	
		Gyrodactylus branchicus										13
		Gyrodactylus bychowsk(y)i	a		c						Gills	
		Gyrodactylus canadensis	a	b							Gills	

TABLE 9.2 (CONTINUED)
Parasites of the Three-Spined Stickleback

Taxonomic Group	Species	Parasite Survey	Infection Site	Recent Key Reference
	Gyrodactylus elegans	a c	Gills, fins	
	Gyrodactylus gasterostei			195
	Gyrodactylus lairdi	a b	Fins	
	Gyrodactylus memorialis	a b	Fins	
	Gyrodactylus pungitii	a		
	Gyrodactylus rarus	a c	Fins	
	Gyrodactylus sp.	c e	Skin, fins, gills	
	Gyrodactylus terranovae	a	Fins	
	Gyrodactyoidea gen. sp.	b d		
	Dactylogyrus sp.	b		
	Diplozoon paradoxum	c		
Digenea				
Echinostomata	*Echinochasmus* sp. (?)	h	Gills	55
Strigeata	*Apatemon gracilis*	a b	Intestine	
	Cotylurus pileatus	a c	Swimbladder	
	Cyathocotyle prussica	h	All tissues	55
	Derogenes varicus	a b	Alimentary canal	
	Diplostomum gasterostei	a	Eye (retina)	
	Diplostomum scudderi	a b	Eye (choroid)	
	Diplostomum spathaceum	a b c h	Eye (lens)	
	Holostephanus lukei	c		
	Lecithaster gibbosus	a b c	Intestine	
	Lecithaster salmonis	a	Intestine	

	Magnibursatus caudofilamentosa						f		
	Neascus sp.			c					
	Posthodiplostomum cuticola	a		c				Skin, fins	
	Posthodiplostomum minimum	a	b	c					
	Tetracotyle sp.		b						
	Tylodelphys clavata	a						Eye (humour)	
Plagiorchata	*Bunodera (= Allobunodera) mediovitellata*	a	b					Intestine	
	Bunodera luciopercae	a						Intestine	
	Bunoderina eucaliae		b	c					
	Crepidostomum cooperi	a	b						
	Crepidostomum farionis	a	b					Intestine	
	Nanophyetus salmincola	a		c					
	Peracreadium gasterostei			c					
	Phyllodistomum spp.								196
	Phyllodistomum folium	a						Urinary bladder	
	Podocotyle atomon	a	b	c		e	f	Intestine	
	Podocotyle reflexa	a		c				Intestine	
	Podocotyle simplex		b						
	Podocotyle sinusacca		b						
	Podocotyle staffordi		b						
Opisthorchiata	*Apophallus brevis*		b						
	Brachyphallus crenatus	a		c		e	f	Stomach	
	Cryptocotyle concavum						f		
	Cryptocotyle lingua						f		
	Metorchis bilis						f		
Cestoda									
	Cestoda gen. sp.		b						
Spathebothriidea	*Cyathocephalus truncatus*	a	b	c				Intestine	197
Pseudophyllidea	Bothriocephalidae gen. sp.		b						
	Bothriocephalus claviceps	a		c				Intestine	

TABLE 9.2 (CONTINUED)
Parasites of the Three-Spined Stickleback

Taxonomic Group		Species	Parasite Survey	Infection Site	Recent Key Reference
		Bothriocephalus scorpi(i)	a b c	Intestine	
		Bothriocephalus sp.	b		
		Diphyllobothrium dendriticum	a c d g	Body cavity, viscera	
		Diphyllobothrium ditremum	g		
		Diphyllobothrium norvegicum	a c	Stomach, intestine, liver	
		Diphyllobothrium sp. plerocercoid	b d e	Liver (encysted)	
		Eubothrium crassum	a	Intestine	
		Schistocephalus solidus	a b c d f g	Body cavity	
		Schistocephalus sp. plerocercoid	b c	Body cavity	
		Triaenophorus nodulosus	a g	Liver	
	Proteocephalidea	*Proteocephalus cernuae*	a c	Intestine	
		Proteocephalus filicollis	a b c d h	Intestine	
		Proteocephalus pugetensis	a b c		
		Proteocephalus sp.	b d		
	Tetrabothriidea	*Phyllobothrium* sp. plerocerocoid	b		
	Cyclopyllidea	*Valipora campylancristrota*	h		
Aschelminthes					
	Nematoda	*Pseudoterranova decipiens*			198
	Anisakidae	*Anisakis* sp. larvae	b		
		Anisakis simplex			199
		Contracaecum aduncum	e	Intestine	
		Contracaecum aduncum (larvae)	e	Body cavity	
		Contracaecum sp.	b f h	Body cavity	

		Raphidascaris acus	a		c			Liver, body cavity, intestine, gonads	
		Raphidascaris cristata	a						
	Cystidicolidae	*Ascarophis morrhuae*		b					
		Cystidicola farionis	a		c			Swim bladder	
	Camallanidae	*Camallanus lucustris (= lacustris)*	a		c		h	Intestine	
		Camallanus truncatus	a		c			Intestine	
	Spiruridae	*Paracuaria adunca*							198
		Cosmocephalus obvelatus							198
	Anguillicolidae	*Anguillicola crassus*					h	Swim bladder	
	Dioctophymatidae	*Eustrongylides* sp. larvae	a	b	c			Musculature, body cavity	
	Philometridae	*Philonema agubernaculum*		b					
	Acanthocephala								
	Echinorhynchidae	*Acanthocephlaus anguillae*				d			
		Acanthocephlaus clavula				d	h	Intestine	
		Acanthocephlaus lucii	a		c		h	Intestine	
		Echinorhynchus truttae				d			
		Metechinorhynchus lateralis		b					
		Metechinorhynchus salmonis	a		c			Intestine	
		Pseud(o)echinorhynchus clavula	a		c			Intestine	
	Neoechinorhynchidae	*Neoechinorhynchus rutili*	a	b	c			Intestine	
	Pomphorhynchidae	*Pomphorhynchus laevis*	a		c	d		Intestine	
		Pomphorhynchus proteus			c				
	Polymorphidae	*Corynosoma semerme*	a		c				
		Corynosoma strumosum	a		c			Coelom, musculature, internal organs	
Annelida									
	Hirudinea								
		Piscicola geometra	a		c			External surface	
Arthropoda									
	Crustacea								
	Copepoda	*Bomolochus cuneatus*		b					

TABLE 9.2 (CONTINUED)
Parasites of the Three-Spined Stickleback

Taxonomic Group	Species	Parasite Survey	Infection Site	Recent Key Reference
	Caligus clemensi	a b	Fins	
	Caligus lacustris	a c		
	Copepoda gen. sp.	b		
	Ergasilus auritus	a b c	Gills, fins	
	Ergasilus sp.	c		
	Ergasilus turgidus	a b c		
	Lernaea cyprinacea	a	Skin	
	Lernaea esocina	a c	Skin, gills	
	Salmincola sp.	a c		
	Thersitina gasterostei	a b c e	Operculum, gills	11
Branchiura	*Argulus alosae*	c		
	Argulus canadensis	a		
	Argulus foliaceus	a c	Skin, gills	
	Argulus stizostethii	b		
Mollusca				
	Unionidae gen. sp. (glochidia larvae)	a c h	Gills	
	Anodonta sp. (glochidia larvae)	b	Gills	

Note: a = Table in Wootton[2]; b = Canada (Margolis and Arthur[15]); c = U.S. (Hoffman[14]); d = Ireland (Holland and Kennedy[200]); e = Loch Etive, Scotland (Dartnall and Walkey[201]); f = Orther Bight, Baltic Sea (Zander et al.[11]); g = Bothnian Bay, Finland (Andersen and Valtonen[10]); h = Schleswig-Holstein, Germany (Kalbe et al.[55]); i = Sable Island, Nova Scotia (Marcogliese[197]).

However, it is not until recently that researchers have attempted to quantify the actual survival, demographic fitness, and consequences of infections for hosts in the wild.[184,185]

Kraak et al.[186] examined correlates of reproductive success in two stickleback populations in Switzerland that differed in their ecology and found that in the Wohlensee population, which exhibited high prevalence of infection with *Pomphorhynchus laevis*, reproductive success in terms of the number of eggs collected in nests was reduced among infected males. In a detailed study of the factors associated with mating success in a natural population, Blais et al.[42] collected all nesting males, along with their nests and the embryos contained within them, from a population of anadromous sticklebacks in a section of the St. Lawrence River, Quebec, before seining the area to collect nonnesting males. They then used microsatellite genotyping to quantify the success of each male in attracting females to spawn and analysed the male characteristics (including the intensity of infection with skin-encysted trematode metacercariae) that were associated with success in nesting and gaining spawnings. Although no significant differences in the infection levels of nest holding and nonnest holding males were found, among nest-holding males, those with nonzero mating success (i.e., with at least one clutch present in the nest) had significantly lower infection intensities than territorial males with zero mating success. This suggested that females discriminated between courting males at least in part on their parasite load. Further studies are now required that combine laboratory and field studies to improve our understanding of how, and to what extent, documented effects of parasites can impact host fitness in natural populations and to identify the mechanisms responsible for observed fitness effects of infection.

REFERENCES

1. Heggberget, T.G. et al., Interactions between wild and cultured Atlantic salmon — a review of the Norwegian experience, *Fish. Res.* 18, 123, 1993.
2. Wootton, R.J., *The Biology of the Sticklebacks*, Academic Press, London, 1976.
3. Wootton, R.J., *A Functional Biology of the Sticklebacks*, California University Press, Berkeley, CA, 1984.
4. Bell, M.A. and Foster, S.A., *The Evolutionary Biology of the Threespine Stickleback*, Oxford University Press, Oxford, 1994.
5. Kingsley, D., Sequencing the genome of the threespine stickleback *Gasterosteus aculeatus*, available on the Internet at http://www.genome.gov/Pages/Research/Sequencing/SeqProposals/SticklebackSEQ.pdf, 2003.
6. Iwana, G. and Nakanishi, T., Eds., *The Fish Immune System: Organism, Pathogen, and Environment*, Academic Press, San Diego, CA, 1996.
7. Roberts, R.J., *Fish Pathology,* 3rd ed., W.B. Saunders, London, 2001.
8. Noga, E.J., *Fish Disease: Diagnosis and Treatment*, University of Iowa Press, Iowa, 2000.
9. Hart, P.J.B. and Gill, A., Evolution of foraging behaviour in the threespine stickleback, in *The Evolutionary Biology of the Threespine Stickleback*, Bell, MA. and Foster, S.A., Eds., Oxford University Press, Oxford, 1994, p. 207.
10. Andersen, K.I. and Valtonen, E.T., Segregation and cooccurrence of larval cestodes in freshwater fishes in the Bothnian Bay, Finland, *Parasitology* 104, 161, 1992.

11. Zander, C.D. et al., Parasite populations and communities from the shallow littoral of the Orther Bight (Fehmarn, SW Baltic Sea), *Parasitol. Res.* 88, 734, 2002.
12. Chubb, J.C. et al., Characterization of the external features of *Schistocephalus solidus* (Muller, 1776) (Cestoda) from different geographical regions and an assessment of the status of the Baltic ringed seal *Phoca hispida botnica* (Gmelin) as a definitive host, *Syst. Parasitol.* 32, 113, 1995.
13. Zietra, M.S. and Lumme, J., The crossroads of molecular, typological and biological species concepts: two new species of *Gyrodactylus* Nordmann, 1832 (Monogenea: Gyrodactylidae), *Syst. Parasitol.* 55, 39, 2003.
14. Hoffman, G.L., *Parasites of North American Freshwater Fishes*, University of California Press, Berkeley, 1967.
15. Margolis, L. and Arthur, J.R., Synopsis of the Parasites of Fishes of Canada, Department of Fisheries and Oceans, Ottawa, 1979.
16. Kennedy, C.R., A checklist of British and Irish freshwater fish parasites with notes on their distribution, *J. Fish Biol.* 6, 613, 1974.
17. MacKenzie, K. et al., Parasites as indicators of water quality and the potential use of helminth transmission in marine pollution studies, *Adv. Parasitol.* 35, 85, 1995.
18. Marcogliese, D.J. and Cone, D.K., Food webs: a plea for parasites, *Trends Ecol. Evol.* 12, 320, 1997.
19. Bush, A.O. et al., *Parasitism: The Diversity and Ecology of Animal Parasites*, Cambridge University Press, Cambridge, 2001.
20. Austin, B. and Austin, D.A., *Bacterial Fish Pathogens: Diseases of Farmed and Wild Fish,* 2nd ed., Springer-Verlag, Berlin, 1993.
21. Pélabon, C. et al., Do microsporidian parasites affect courtship in male two-spotted gobies?, *Mar. Biol.* 148, 189, 2005.
22. Ward, A.J.W. et al., Shoaling behaviour of sticklebacks infected with the microsporidian parasite, *Glugea anomala, Environ. Biol. Fish,* 72, 155, 2005.
23. Arme, C. and Owen, R.W., Infections of the three-spined stickleback, *Gasterosteus aculeatus* L., with the plerocercoid larvae of *Schistocephalus solidus* with special reference to pathological effects, *Parasitology* 57, 301, 1967.
24. Chubb, J.C., Seasonal occurrence of helminths in freshwater fishes, Part I. monogenea, *Adv. Parasitol.* 15, 133, 1977.
25. Chubb J.C., Seasonal occurrence of helminths in freshwater fishes, Part II. trematoda, *Adv. Parasitol.* 17, 141, 1979.
26. Chubb, J.C., Seasonal occurrence of helminths in freshwater fishes, Part III. larval cestoda and nematoda, *Adv. Parasitol.* 18, 1, 1980.
27. Chubb, J.C., Seasonal occurrence of helminths in freshwater fishes, Part IV. adult cestoda, nematoda and acanthocephala, *Adv. Parasitol.* 20, 1, 1982.
28. Tierney, J.F., Huntingford, F.A., and Crompton, D.W.T., Body condition and reproductive status in sticklebacks exposed to a single wave of *Schistocephalus solidus* infection, *J. Fish Biol.* 49, 483, 1996.
29. Pennycuick, L., Seasonal variations in parasite infections in a population of 3-spined sticklebacks, *Gasterosteus aculeatus* L., *Parasitology* 63, 373, 1971.
30. Chappell, L.H., Parasites of 3-spined stickleback *Gasterosteus aculeatus* L from a Yorkshire pond: 1. Seasonal variation of parasite fauna, *J. Fish Biol.* 1, 137, 1969.
31. Lyholt, H.C.K. and Buchmann, K., *Diplostomum spathaceum*: effects of temperature and light on cercarial shedding and infection of rainbow trout, *Dis. Aquat. Organisms* 25, 169, 1996.

32. Özer, A., The occurrence of *Trichodina domerguei* Wallengren, 1897 and *Trichodina tenuidens* Faure-Fremiet, 1944 (Peritrichia) on three-spined stickleback, *Gasterosteus aculeatus* L., 1758 found in a brackish and freshwater environment, *Acta Protozool.* 42, 41, 2003.
33. Poulin, R., Sexual inequalities in helminth infections: a cost of being a male?, *Am. Nat.* 147, 287, 1996.
34. Moore, S.L. and Wilson, K., Parasites as a viability cost of sexual selection in natural populations of mammals, *Science* 297, 2015, 2002.
35. Brei, B. and Fish, D., Comment on "parasites as a viability cost of sexual selection in natural populations of mammals," *Science* 300, 55a, 2003.
36. Folstad, I. and Karter, A.J., Parasites, bright males, and the immunocompetence handicap, *Am. Nat.* 139, 603, 1992.
37. Chappell, L.H., Parasites of 3-spined stickleback *Gasterosteus aculeatus* L from a Yorkshire pond: 2. Variation of parasite fauna with sex and size of fish, *J. Fish Biol.* 1, 339, 1969.
38. Pennycuick, L., Differences in parasite infections in 3-spined sticklebacks (*Gasterosteus aculeatus* L) of different sex, age and size, *Parasitology* 63, 407, 1971.
39. Reimchen, T.E. and Nosil, P., Ecological causes of sex-biased parasitism in threespine stickleback, *Biol. J. Linn. Soc.* 73, 51, 2001.
40. Arnold, K.E. et al., Sex-specific survival and parasitism in three-spined sticklebacks: seasonal patterns revealed by molecular analysis, *J. Fish Biol.* 63, 1046, 2003.
41. Griffiths, R. et al., DNA sex identification in the three-spined stickleback, *J. Fish Biol.* 57, 1331, 2000.
42. Blais, J., Rico, C., and Bernatchez, L., Nonlinear effects of female mate choice in wild threespine sticklebacks, *Evolution* 58, 2498, 2004.
43. Barber, I., Parasites grow larger in faster growing fish hosts, *Int. J. Parasitol.* 37, 137, 2005.
44. Barber, I., Parasites and size assortative schooling in three-spined sticklebacks, *Oikos* 101, 331, 2003.
45. Malmquist, H.J. et al., Diet differences in polymorphic Arctic charr in Thingvallavatn, Iceland, *J. Anim. Ecol.* 61, 21, 1992.
46. Jonsson, B. and Jonsson, N. Polymorphism and speciation in Arctic charr, *J. Fish Biol.* 58, 605, 2001.
47. Dörücü, M. et al., How fish-helminth associations arise: an example from Arctic Charr in Loch Rannoch, *J. Fish Biol.* 47, 1038, 1995.
48. Knudsen, R., Kristoffersen, R., and Amundsen, P.A., Parasite communities in two sympatric morphs of Arctic charr, *Salvelinus alpinus* (L.), in northern Norway, *Can. J. Zool.* 75, 2003, 1997.
49. McPhail, J.D., Speciation and the evolution of reproductive isolation in sticklebacks (*Gasterosteus*) of south-western British Columbia, in *The Evolutionary Biology of the Threespine Stickleback*, Bell, M.A. and Foster, S.A., Eds., Oxford University Press, Oxford, 1994, p. 399.
50. Cresko, W.A. and Baker, J.A., Two morphotypes of lacustrine threespine stickleback, *Gasterosteus aculeatus*, in Benka Lake, Alaska, *Environ. Biol. Fish.* 45, 343, 1996.
51. Dartnall, H.J., Salinity tolerance of trichodinids (Protozoa) parasitic on 3-spined stickleback *Gasterosteus aculeatus, J. Zool.* 172, 207, 1974.
52. Soleng, A. and Bakke, T.A., Salinity tolerance of *Gyrodactylus salaris* (Platyhelminthes, Monogenea): laboratory studies, *Can. J. Fish. Aquat. Sci.* 54, 1837, 1997.

53. Marcogliese, D.J., Comparison of parasites of mummichogs and sticklebacks from brackish and fresh-water ponds on Sable Island, Nova-Scotia, *Am. Midl. Nat.* 133, 333, 1995.
54. Wegner, K.M., Reusch, T.B.H., and Kalbe, M., Multiple parasites are driving major histocompatibility complex polymorphism in the wild, *J. Evol. Biol.* 16, 224, 2003.
55. Kalbe, M., Wegner, K.M., and Reusch, T.B.H., Dispersion patterns of parasites in 0+ year three-spined sticklebacks: a cross population comparison, *J. Fish Biol.* 60, 1529, 2002.
56. Barber, I. et al., Indirect fitness consequences of mate choice in sticklebacks: offspring of brighter males grow slowly but resist parasitic infections, *Proc. R. Soc. Lond. B* 268, 71, 2001.
57. Kurtz, J. et al., Major histocompatibility complex diversity influences parasite resistance and innate immunity in sticklebacks, *Proc. R. Soc. Lond. B* 271, 197, 2004.
58. Reimchen, T.E., Parasitism of asymmetrical pelvic phenotypes in stickleback, *Can. J. Zool.* 75, 2084, 1997.
59. Reimchen, T.E. and Nosil, P., Dietary differences between phenotypes with symmetrical and asymmetrical pelvis in the stickleback *Gasterosteus aculeatus*, *Can. J. Zool.* 79, 533, 2001.
60. Reimchen, T.E. and Nosil, P., Lateral plate asymmetry, diet and parasitism in threespine stickleback, *J. Evol. Biol.* 14, 632, 2001.
61. Rohde, K. *Ecology of Marine Parasites*, 2nd ed., CAB International, Wallingford, 1993.
62. Rohde, K. and Heap, M., Latitudinal differences in species and community richness and in community structure of metazoan endo- and ectoparasites of marine teleost fish, *Int. J. Parasitol.* 28, 461, 1998.
63. Marcogliese, D.J. *Survey of Stickleback Parasites*, available on the web at http://www.nrel.colostate.edu/projects/iboy/whatandwhere#stickleback, 2002.
64. Lafferty, K. D. and Kuris, A. M., How environmental stress affects the impacts of parasites, *Limnol. Oceanogr.* 44, 925, 1999.
65. Axelsson, B. and Norrgren, L., Parasite frequency and liver anomalies in 3-spined stickleback, *Gasterosteus aculeatus* (L.), after long-term exposure to pulp-mill effluents in marine mesocosms, *Arch. Environ. Contam. Toxicol.* 21, 505, 1991.
66. Lehtinen, K.J., Survival, growth and disease of 3-spined stickleback, *Gasterosteus aculeatus* L. broods exposed to bleached kraft mill effluents (Bkme) in mesocosms, *Ann. Zool. Fenn.* 26, 133, 1989.
67. Pascoe, D. and Cram, P., Effect of parasitism on toxicity of cadmium to 3-spined stickleback, *Gasterosteus aculeatus* L., *J. Fish Biol.* 10, 467, 1977.
68. Sousa, W.P., Patterns and processes in communities of helminth parasites, *Trends Ecol. Evol.* 9, 52, 1994.
69. Chappell, L.H., Competitive exclusion between 2 intestinal parasites of 3 spined stickleback, *Gasterosteus aculeatus* L., *J. Parasitol.* 55, 775, 1969.
70. Hart, B., Behavioural adaptations to pathogens and parasites: five strategies, *Neuro. Biobehav. Rev.* 14, 273, 1990.
71. Hart, B., Behavioural defense against parasites: interaction with parasite invasiveness, *Parasitology* 109 (Suppl.), S139, 1994.
72. Minchella, D.J., Host life-history variation in response to parasitism, *Parasitology* 90, 205, 1985.
73. Keymer, A.D. and Read, A.F., Behavioural ecology: the impact of parasitism, in *Parasite-Host Associations: Coexistance or Conflict?*, Toft, C.A., Aeschliman, A., and Bolis, L., Eds., Oxford University Press, Oxford, 1991, p. 37.

74. Lozano, G.A., Optimal foraging theory — a possible role for parasites, *Oikos* 60, 391, 1991.
75. Poulin, R. and FitzGerald, G.J., Water temperature, vertical distribution, and risk of ectoparasitism in juvenile sticklebacks, *Can. J. Zool.* 66, 2002, 1988.
76. Poulin, R. and FitzGerald, G.J., Risk of parasitism and microhabitat selection in juvenile sticklebacks, *Can. J. Zool.* 67, 14, 1989.
77. Karvonen, A., Seppala, O., and Valtonen, E.T., Parasite resistance and avoidance behaviour in preventing eye fluke infections in fish, *Parasitology* 129, 159, 2004.
78. Wedekind, C. and Milinski, M., Do three-spined sticklebacks avoid consuming copepods, the first intermediate host of *Schistocephalus solidus*?, An experimental analysis of behavioural resistance, *Parasitology* 112, 371, 1996.
79. Urdal, K., Tierney, J.F., and Jakobsen, P.J., The tapeworm *Schistocephalus solidus* alters the activity and response, but not the predation susceptibility of infected copepods, *J. Parasitol.* 81, 330, 1995.
80. Bakker, T.C.M., Mazzi, D., and Zala, S., Parasite-induced changes in behavior and color make *Gammarus pulex* more prone to fish predation, *Ecology* 78, 1098, 1997.
81. Mazzi, D. and Bakker, T.C.M., A predator's dilemma: prey choice and parasite susceptibility in three-spined sticklebacks, *Parasitology* 126, 339, 2003.
82. Lafferty, K.D., Foraging on prey that are modified by parasites, *Am. Nat.* 140, 854, 1992.
83. Dugatkin, L.A., Fitzgerald, G.J., and Lavoie, J., Juvenile 3-spined sticklebacks avoid parasitized conspecifics, *Environ. Biol. Fish.* 39, 215, 1994.
84. Poulin, R. and FitzGerald, G.J., The potential of parasitism in the structuring of a salt marsh stickleback community, *Can. J. Zool.* 65, 2793, 1987.
85. Hamilton, W.D., Geometry for the selfish herd, *J. Theor. Biol.* 31, 295, 1971.
86. Mooring M.S. and Hart B.L., Animal grouping for protection from parasites: selfish herd and encounter dilution effects, *Behaviour* 123, 173, 1992.
87. Poulin, R. and FitzGerald, G.J. Shoaling as an anti-ectoparasitic mechanism in juvenile sticklebacks (*Gasterosteus* spp.), *Behav. Ecol. Sociobiol.* 24, 251, 1989.
88. Puttnam, C., Shoaling sticklebacks give parasites the slip, *New Scientist* 123, 27, 1989.
89. Moore, J., Simberloff, D., and Freehling, M., Relationships between bobwhite quail social-group size and intestinal helminth parasitism, *Am. Nat.* 131, 22, 1988.
90. Côté, I.M. and Poulin, R., Parasitism and group size in social animals: a meta-analysis, *Behav. Ecol. Sociobiol.* 95, 159, 1995.
91. Poulin, R., Parasitism and shoal size in juvenile sticklebacks: Conflicting selection pressures from different ectoparasites?, *Ethology* 105, 959, 1999.
92. Barber, I., Downey, L.C., and Braithwaite, V.A., Parasitism, oddity and the mechanism of shoal choice, *J. Fish Biol.* 52, 1365, 1998.
93. Krause, J. and Godin, J.-G.J., Influence of parasitism on shoal choice in the banded killifish (*Fundulus diaphanus*, Teleostei, Cyprinodontidae), *Ethology* 102, 40, 1996.
94. Milinski, M. and Bakker, T.C.M., Female sticklebacks use male coloration in mate choice and hence avoid parasitized males, *Nature* 344, 330, 1990.
95. Read, A.F., Parasites and the evolution of host sexual behaviour, in *Parasitism and Host Behaviour*, Barnard, C.J. and Behnke, J.M., Eds., Taylor and Francis, London, 1990, p. 95.
96. Milinski, M., The function of mate choice in sticklebacks: optimizing MHC genetics, *J. Fish Biol.* 63, 1, 2003.
97. Kurtz, J., Sex, parasites and resistance — an evolutionary approach, *Zoology* 106, 327, 2003.

98. Wegner, K.M., et al., Parasite selection for immunogenetic optimality, *Science* 301, 1343, 2003.
99. Penn, D. and Potts, W., How do major histocompatibility complex genes influence odor and mating preferences?, *Adv. Immunol.* 69, 411, 1998.
100. Potts, W.K., Manning, C.J., Wakeland, E.K., Mating patterns in seminatural populations of mice influenced by MHC genotype, *Nature* 352, 619, 1991.
101. Wedekind, C. et al., MHC-dependent mate preferences in humans, *Proc. R. Soc. Lond. B.* 260, 245, 1995.
102. Reusch, T.B.H. et al., Female sticklebacks count alleles in a strategy of sexual selection explaining MHC polymorphism, *Nature* 414, 300, 2001.
103. Milinski, M. et al. Mate choice decisions of stickleback females predictably modified by MHC peptide ligands, *Proc. Natl. Acad. Sci.* 102, 4414, 2005.
104. Aeschlimann, P.B. et al., Female sticklebacks *Gasterosteus aculeatus* use self-reference to optimize MHC allele number during mate selection, *Behav. Ecol. Sociobiol.* 54, 119, 2003.
105. Hedrick, P.W., Comment on "parasite selection for immunogenetic optimality," *Science* 303, 957, 2004.
106. Wegner, K.M. et al., Response to comment on "parasite selection for immunogenetic optimality," *Science* 303, U2, 2004.
107. Barber, I. and Wright, H.A., Effects of parasites on fish behaviour: interactions with host physiology, in *Behaviour and Physiology of Fish (Fish Physiology, Vol. 24)*, Sloman, K., Balshine, S., and Wilson, R.W., Eds., Academic Press, Amsterdam, 2006, p. 504.
108. Barber, I., A non-invasive morphometric technique for estimating cestode plerocercoid burden in small freshwater fish, *J. Fish Biol.* 51, 654, 1997.
109. LoBue, C.P. and Bell, M.A., Phenotypic manipulation by the cestode parasite *Schistocephalus solidus* of its intermediate host, *Gasterosteus aculeatus*, the threespine stickleback, *Am. Nat.* 142, 725, 1993.
110. Ness, J.H. and Foster, S.A., Parasite-associated phenotype modifications in threespine stickleback, *Oikos* 85, 127, 1999.
111. Dykova, I. and Lom, J., Tissue reaction of the three-spined stickleback *Gasterosteus aculeatus* L to infection with *Glugea anomala* (Moniez, 1887), *J. Fish Dis.* 1, 83, 1978.
112. Dezfuli, B.S. et al., Immunohistochemistry, histopathology and ultrastructure *of Gasterosteus aculeatus* tissues infected with *Glugea anomala*, *Dis. Aquat. Organisms* 58, 193, 2004.
113. Reimchen, T.E. and Buckland-Nicks, J., A novel association between an endemic stickleback and a parasitic dinoflagellate. 1. seasonal cycle and host response, *Can. J. Zool.* 68, 667, 1990.
114. Chappell, L.H., Hardie, L.J., and Secombes, C.J., Diplostomiasis: the disease and host-parasite interactions, in *Parasitic Diseases of Fish*, A.W. Pike and J.W. Lewis, Eds., Samara Publishing, Dyfed, 1994, p. 59.
115. Owen, S.F., Barber, I., and Hart, P.J.B., Low level infection by eyefluke, *Diplostomum* spp. affects the vision of three-spined sticklebacks, *Gasterosteus aculeatus*, *J. Fish Biol.* 42, 803, 1993.
116. Arme, C. and Owen, R.W., Massive infections of 3-spined stickleback with *Schistocephalus solidus* (Cestoda: Pseudophyllidea) and *Glugea anomala* (Sporozoa: Microsporidia), *Parasitology* 54, 10, 1964.
117. Pennycuick, L., Quantitative effects of three species of parasites on a population of three-spined sticklebacks (*Gasterosteus aculeatus* L.), *J. Zool.* 165, 143, 1971.

118. Tierney, J.F., Effects of *Schistocephalus solidus* (Cestoda) on the food intake and diet of the three-spined stickleback, *Gasterosteus aculeatus*, *J. Fish Biol.* 44, 731, 1994.
119. Bagamian, K.H., Heins, D.C., and Baker, J.A., Body condition and reproductive capacity of three-spined stickleback infected with the *cestode Schistocephalus solidus*, *J. Fish Biol.* 64, 1568, 2004.
120. Barber, I. and Svensson, P.A., Effects of experimental *Schistocephalus solidus* infections on growth, morphology and sexual development of female three-spined sticklebacks, *Gasterosteus aculeatus*, *Parasitology* 126, 359, 2003.
121. Soutter, A.M., Walkey, M., and Arme, C., Amino acids in the plerocercoid of *Ligula intestinalis* (Cestoda, Pseudophyllidea) and its fish host, *Rutilus rutilus*, *Z. Parasitkde* 63, 151, 1980.
122. Walkey, M. and Meakins, R.H., An attempt to balance the energy budget of a host-parasite system, *J. Fish Biol.* 2, 361, 1970.
123. Pascoe, D. and Mattey, D., Dietary stress in parasitised and non-parasitised *Gasterosteus aculeatus* L., *Z. Parasitkde* 51, 179, 1977.
124. Bakker, T.C.M. and Mundwiler, B., Pectoral fin size in a fish species with paternal care: a condition-dependent sexual trait revealing infection status, *Freshwater Biol.* 41, 543, 1999.
125. Fitzgerald, G. J., Fournier, M., and Morrissette, J., Sexual selection in an anadromous population of threespine sticklebacks — no role for parasites, *Evol. Ecol.* 8, 348, 1994.
126. Arnott, S.A., Barber, I., and Huntingford, F.A., Parasite-associated growth enhancement in a fish-cestode system, *Proc. R. Soc. Lond. B* 267, 657, 2000.
127. Cunningham, E.J., Tierney, J.F., and Huntingford, F.A., Effects of the cestode *Schistocephalus solidus* on food intake and foraging decisions in the three-spined stickleback *Gasterosteus aculeatus*, *Ethology* 79, 65, 1994.
128. Wright, H.A., Wootton, R.J., and Barber, I., The effect of *Schistocephalus solidus* infection on meal size of three-spined stickleback, *J. Fish Biol.* 68, 801, 2006.
129. Ali, M., Nicieza, A., and Wootton, R.J., Compensatory growth in fishes: a response to growth depression, *Fish Fisheries* 4, 147, 2003.
130. Arme, C., Effects of the plerocercoid larvae of a pseudophyllidean cestode, *Ligula intestinalis*, on the pituitary gland and gonads of its host, *Biol. Bull.* 134, 15, 1968.
131. Arme, C., *Ligula intestinalis*: Interactions with the pituitary-gonadal axis of its fish host, *J. Helminthol.* 71, 83, 1997.
132. McPhail, J.D. and Peacock, S.D., Some effects of the cestode (*Schistocephalus solidus*) on reproduction in the threespine stickleback (*Gasterosteus aculeatus*): evolutionary aspects of a host-parasite interaction, *Can. J. Zool.* 61, 901, 1983.
133. Heins, D.C., Singer, S.S., and Baker, J.A., Virulence of the cestode *Schistocephalus solidus* and reproduction in infected threespine stickleback, *Gasterosteus aculeatus*, *Can. J. Zool.* 77, 1967, 1999.
134. Heins, D.C., Ulinski, B., Johnson, J., and Baker, J.A., Effect of the cestode macroparasite *Schistocephalus pungitii* on the reproductive success of ninespine stickleback, *Pungitius pungitius*, *Can. J. Zool.* 82, 1731, 2004.
135. Heins, D.C. and Baker, J.A. Reduction of egg size in natural populations of threespine stickleback infected with a cestode macroparasite, *J. Parasitol.* 89, 1, 2003.
136. Einum, S., Atlantic salmon growth in strongly food-limited environments: effects of egg size and paternal phenotype?, *Environ. Biol. Fish.* 67, 263, 2003.
137. Barnard C.J. and Behnke J.M., *Parasitism and Host Behaviour*, Taylor and Francis, London, 1990.
138. Barber, I., Hoare, D., and Krause, J., The effects of parasites on fish behaviour: an evolutionary perspective and review, *Rev. Fish Biol. Fish.* 10, 131, 2000.

139. Moore, J., *Parasites and the behaviour of animals*, Oxford University Press, Oxford, 2002.
140. Holmes, J.C. and Bethel, W.M., Modification of intermediate host behaviour by parasites, in *Behavioural Aspects of Parasite Transmission*, Canning, E.U. and Wright, C.A., Eds., Academic Press, London, 1972, p. 123.
141. Dawkins, R., *The Extended Phenotype*, Oxford University Press, Oxford, 1982.
142. Poulin, R., The evolution of parasite manipulation of host behaviour: a theoretical analysis, *Parasitology* 109, S109, 1994.
143. Poulin, R., Adaptive changes in the behaviour of parasitised animals: a critical review, *Int. J. Parasitol.* 25, 1371, 1995.
144. Poulin, R., *Evolutionary Ecology of Parasites: From Individuals to Communities*, Chapman and Hall, London, 1998.
145. Östlund-Nilsson, S. et al., Parasitic isopod *Anilocra apogonae*, a drag for the cardinal fish *Cheilodipterus quinquelineatus*, *Mar. Ecol. Prog. Ser.* 287, 209, 2005.
146. Bergstrom, C.A., Fast-start swimming performance and reduction in lateral plate number in threespine stickleback, *Can. J. Zool.* 80, 207, 2002.
147. Walker, J.A., Dynamics of pectoral fin rowing in a fish with an extreme rowing stroke: the threespine stickleback (*Gasterosteus aculeatus*), *J. Exp. Biol.* 207, 1925, 2004.
148. Blake, R.W. et al., Comparison of the prolonged swimming performances of closely related, morphologically distinct three-spined sticklebacks *Gasterosteus* spp., *J. Fish Biol.* 67, 834, 2005.
149. Crowden, A.E. and Broom, D.M., Effects of the eyefluke, *Dipostomum spathaceum*, on the behaviour of dace (*Leuciscus leuciscus*), *Anim. Behav.* 28, 287, 1980.
150. Milinski, M., Parasites and host decision-making, in *Parasitism and Host Behaviour*, Barnard, C.J. and Behnke, J.M., Eds., Taylor and Francis, London, 1990, p. 95.
151. Giles, N., Behavioural effects of the parasite *Schistocephalus solidus* (Cestoda) on an intermediate host, the three-spined stickleback, *Gasterosteus aculeatus*, *Anim. Behav.* 31, 1192, 1983.
152. Milinski, M., Risk of predation of parasitised sticklebacks (*Gasterosteus aculeatus* L.) under competition for food, *Behaviour* 93, 203, 1985.
153. Godin, J.-G.J. and Sproul, C.D., Risk taking in parasitised sticklebacks under threat of predation: effects of energetic need and food availability, *Can. J. Zool.* 66, 2360, 1988.
154. Aeschlimann, P., Haberli, M., and Milinski, M., Threat-sensitive feeding strategy of immature sticklebacks (*Gasterosteus aculeatus*) in response to recent experimental infection with the cestode *Schistocephalus solidus*, *Behav. Ecol. Sociobiol.* 49, 1, 2000.
155. Milinski, M., Parasites determine a predator's optimal feeding strategy, *Behav. Ecol. Sociobiol.* 15, 35, 1984.
156. Ranta, E., *Schistocephalus* infestation improves prey-size selection by three-spined sticklebacks, *Gasterosteus aculeatus, J. Fish Biol.* 46, 156, 1995.
157. Bergersen, R., Sticklebacks from Greenland, *J. Fish Biol.* 48, 799, 1996.
158. Giles, N., Predation risk and reduced foraging activity in fish: experiments with parasitised and non-parasitised three-spined sticklebacks, *Gasterosteus aculeatus*, *J. Fish Biol.* 31, 37, 1987.
159. Barber, I. Huntingford, F.A., and Crompton, D.W.T., The effect of hunger and cestode parasitism on the shoaling decisions of small freshwater fish, *J. Fish Biol.* 47, 524, 1995.

160. Barber, I., Walker, P., and Svensson, P.A., Behavioural responses to simulated avian predation in female three spined sticklebacks: The effect of experimental *Schistocephalus solidus* infections, *Behaviour* 141, 1425, 2004.
161. Barber, I. and Ruxton, G.D., Temporal prey distribution affects the competitive ability of parasitised sticklebacks, *Anim. Behav.* 58, 1477, 1998.
162. Lester, R.J.G., The influence of *Schistocephalus* plerocercoids on the respiration of *Gasterosteus* and a possible resulting effect on the behavior of the fish, *Can. J. Zool.* 49, 361, 1971.
163. Smith, R.S. and Kramer, D.C., Effects of a cestode (*Schistocephalus* sp.) on the response of ninespine sticklebacks (*Pungitius pungitius*) to aquatic hypoxia, *Can. J. Zool.* 65, 1862, 1987.
164. Giles, N., A comparison of the behavioural responses of parasitised and non-parasitised three-spined sticklebacks, *Gasterosteus aculeatus* L., to progressive hypoxia, *J. Fish Biol.* 30, 631, 1987.
165. Kramer, D.L., Manley, D., and Bourgeois, R., The effect of respiratory mode and oxygen concentration on the risk of aerial predation in fishes, *Can. J. Zool.* 61, 653, 1983.
166. Adams, C.E., Brown, D.W., and Keay, L., Elevated predation pressure associated with inshore migrations of fish in a large lake, Loch Lomond, Scotland, *Hydrobiologia* 290, 135, 1994.
167. Jakobsen, P.J. Johnsen, G.H., and Larsson, P., Effects of predation risk and parasitism on the feeding ecology, habitat use, and abundance of lacustrine threespine stickleback (*Gasterosteus aculeatus*), *Can. J. Fish. Aquat. Sci.* 45, 426, 1988.
168. Barber, I. et al., Carotenoid-based sexual coloration and body condition in nesting male sticklebacks, *J. Fish Biol.* 57, 777, 2000.
169. Zahavi, A., Mate selection — selection for a handicap, *J. Theor. Biol.* 53, 205, 1975.
170. Andersson, M., *Sexual Selection*, Princeton University Press, Princeton, NJ, 1994.
171. Hamilton, W.D. and Zuk, M., Heritable true fitness and bright birds: a role for parasites?, *Science* 218, 384, 1982.
172. Bronseth, T. and Folstad, I., The effect of parasites on courtship dance in threespine sticklebacks: more than meets the eye?, *Can. J. Zool.* 75, 589, 1997.
173. Folstad, I. et al., Sexually selected color in male sticklebacks — a signal of both parasite exposure and parasite resistance, *Oikos* 69, 511, 1994.
174. Candolin, U. and Voigt, H.R., No effect of a parasite on reproduction in stickleback males: a laboratory artefact?, *Parasitology* 122, 457, 2001.
175. Rushbrook, B.J. and Barber, I., Nesting, courtship and kidney development in *Schistocephalus*-infected male three-spined sticklebacks from an upland lake, *J. Fish Biol.*, 69, 870, 2006.
176. Le Comber, S.C. et al., Response of territorial males to the threat of sneaking in the three-spined stickleback (*Gasterosteus aculeatus*): a field study, *J. Zool.* 261, 15, 2003.
177. Tierney, J.F. Huntingford, F.A., and Crompton, D.W.T., The relationship between infectivity of *Schistocephalus solidus* (Cestoda) and anti-predator behaviour of its intermediate host, the three-spined stickleback, *Gasterosteus aculeatus, Anim. Behav.* 46, 603, 1993.
178. Loot, G. et al., Experimental demonstration of a behavioural modification in a cyprinid fish, *Rutilus rutilus* (L.), induced by a parasite, *Ligula intestinalis* (L.), *Can. J. Zool.* 80, 738, 2002.
179. Jönsson, E. Johnsson, J.I., and Thrandur Bjornsson, B., Growth hormone increases predation exposure of rainbow trout, *Proc. R. Soc. Lond. B.* 263, 647, 1996.

180. Adamo, S.A., Modulating the modulators: parasites, neuromodulators and host behavioral change, *Brain Behav. Evol.* 60, 370, 2002.
181. Øverli, Ø. et al., Effects of *Schistocephalus solidus* infection on brain monoaminergic activity in female three-spined sticklebacks, *Gasterosteus aculeatus*, *Proc. R. Soc. Lond. Ser. B.* 268, 1411, 2001.
182. Kavaliers, M., Colwell, D.D., and Choleris, E., Parasites and behavior: an ethopharmacological analysis and biomedical implications, *Neurosci. Biobehav. Rev.* 23, 1037, 1999.
183. Tierney, J.F. and Crompton, D.W.T., Infectivity of plerocercoids of *Schistocephalus solidus* (Cestoda: Ligulidae) and fecundity of the adults in an experimental definitive host, *Gallus gallus*, *J. Parasitol.* 78, 1049, 1992.
184. Lafferty, K.D. and Morris, A.K., Altered behavior of parasitized killifish increases susceptibility to predation by bird final hosts, *Ecology* 77, 1390, 1996.
185. Finley, R.J. and Forrester, G.E., Impact of ectoparasites on the demography of a small reef fish, *Mar. Ecol. Prog. Ser.* 248, 305, 2003.
186. Kraak, S.B.M., Bakker, T.C.M., and Mundwiler, B., Sexual selection in sticklebacks in the field: correlates of reproductive, mating, and paternal success, *Behav. Ecol.* 10, 696, 1999.
187. Buckland-Nicks, J., Reimchen, T.E., and Garbary, D.J., *Haidadinium ichthyophilum* gen.nov. et sp.nov. (Phytodiniales, Dinophyceae), a freshwater ectoparasite on stickleback (*Gasterosteus aculeatus*) from the Queen Charlotte Islands, Canada, *Can. J. Bot.* 75, 1936, 1997.
188. Elkan, E., *Dermocystidium gasterostei* n. sp., a parasite of *Gasterosteus aculeatus* L. and *Gasterosteus pungitius* L., *Nature* 196, 958, 1962.
189. Jastrzebski, M. and Komorowski, Z., Light and electron microscopic studies on *Goussia zarnowskii* (Jastrzebski, 1982) — an intestinal coccidium parasitizing the 3-spined stickleback, *Gasterosteus aculeatus* (L.), *J. Fish Dis.* 13, 1, 1990.
190. Jastrzebski, M., Ultrastructural study on the development of *Goussia aculeati*, a coccidium parasitizing the 3-spined stickleback *Gasterosteus aculeatus*, *Dis. Aquat. Organisms* 6, 45, 1989.
191. Feist, S.W., Chilmonczyk, S., and Pike, A.W., Structure and development of *Sphaerospora elegans* Thelohan 1892 (Myxozoa, Myxosporea) in the sticklebacks *Gasterosteus aculeatus* L. and *Pungitius pungitius* L. (Gasterosteidae), *Eur. J. Protistol.* 27, 269, 1991.
192. Lom, J., Pike, A.W., and Feist, S.W., Myxosporean vegetative stages in the choroidal *rete mirabile* of *Gasterosteus aculeatus* infected with *Myxobilatus gasterostei* and *Spaerospora elegans*, *Dis. Aquat. Organisms* 11, 67, 1991.
193. Lester, R.S.G. and Adams, J.R., *Gyrodactylus alexanderi*: reproduction, mortality, and effect on its host *Gasterosteus aculeatus*, *Can J. Zool.* 52, 827, 1974.
194. Özer, A., Ozturk, T., and Ozturk, M.O., Prevalence and intensity *of Gyrodactylus arcuatus* Bychowsky, 1933 (Monogenea) infestations on the three-spined stickleback, *Gasterosteus aculeatus* L., 1758, *Turk. J. Vet. Anim. Sci.* 28, 807, 2004.
195. Harris, P.D., Ecological and genetic evidence for clonal reproduction in *Gyrodactylus gasterostei* Glaser, 1974, *Int. J. Parasitol.* 28, 1595, 1998.
196. Petkeviciute, R., Stunzenas, V., and Staneviciute, G., Cytogenetic and sequence comparison of adult *Phyllodistomum* (Digenea: Gorgoderidae) from the three-spined stickleback with larvae from two bivalves, *Parasitology* 129, 771, 2004.
197. Reimchen, T.E., Incidence and intensity of *Cyathocephalus truncatus* and *Schistocephalus solidus* infection in *Gasterosteus aculeatus*, *Can. J. Zool.* 60, 1091, 1982.

198. Marcogliese, D.J., Metazoan parasites of sticklebacks on Sable Island, northwest Atlantic Ocean: biogeographic considerations, *J. Fish Biol.* 41, 399, 1992.
199. Køie, A., Experimental infections of copepods and sticklebacks *Gasterosteus aculeatus* with small ensheathed and large third-stage larvae of *Anisakis simplex* (Nematoda, Ascaridoidea, Anisakidae), *Parasitol. Res.* 87, 32, 2001.
200. Holland, C.V. and Kennedy, C.R., A checklist of parasitic helminth and crustacean species recorded in freshwater fish from Ireland, *Proc. R. Irish. Acad.* 97B, 225, 1997.
201. Dartnall, H.J.G. and Walkey, M., Parasites of marine sticklebacks, *J. Fish Biol.* 14, 471, 1979.

10 The Use of the Stickleback as a Sentinel and Model Species in Ecotoxicology

Ioanna Katsiadaki

CONTENTS

10.1 INTRODUCTION

Studies of the "bioindicator" aspects of the three-spined stickleback (*Gasterosteus aculeatus*) life cycle have so far been fewer compared to research into its behaviour and evolutionary biology (speciation, sexual selection). Nonetheless, owing to its ubiquity and the ease with which it can be caught and held in a laboratory, a number of studies have examined the value of the stickleback as an environmental sentinel, and the findings are discussed in this chapter. More recently, the stickleback received attention as a model species in the field of endocrine disruption research. This is primarily due to the unique traits it presents for the detection of compounds with androgen-modulating activity. The strong secondary sexual characters (i.e., kidney hypertrophy and spiggin production) together with the typical nest-building behaviour are traits that are under the control of androgens and thus offer scope for use as specific biomarkers. In addition, the ability to determine genetic sex is also regarded as a benefit in endocrine disruption research. Other advantages of the stickleback as a sentinel species are presented together with a discussion on the current status of its use as a model species in chemical testing.

10.2 THE IDEAL MODEL AND SENTINEL SPECIES?

In general, suitable bioindicators of water quality should fulfil at least some of the following basic criteria:

1. They must be resident and specific for a particular aquatic environment at least for some life stages. Fish of unknown migratory history for example complicate the provision and analysis of meaningful data.
2. Their life history should be well documented and representative of a larger number of species.
3. They should be present in adequate numbers and not listed as protected, threatened, or endangered species in the area of interest. Prior knowledge of population status in diverse environments is a substantial advantage.
4. They should be relatively tolerant to pollution, allowing physiological responses to be estimated prior to acutely toxic effects.
5. Field sampling and laboratory holding should be relatively easy and inexpensive.
6. The species should not be targeted by either commercial or leisure fisheries, and neither should it be subject to river stocking, which would complicate any assessment.
7. Laboratory conditions should be able to mimic/reproduce the aquatic environment under controlled conditions that should be easy to perform and reproduce. Preferentially, the species should have a short life cycle, a small body size, known husbandry conditions, and well-documented biology.
8. A battery of validated relevant biomarkers/tools that are either specific to a certain class of contaminants or generic must be available for these species.

The three-spined stickleback largely meets all the preceding criteria:

- Freshwater populations are not known to be migratory.
- Its biology is probably better documented than any other fish (the present volume is a testimony to the wealth of relevant information).

- It is ubiquitous and widespread across the northern hemisphere in many aquatic habitats.
- Its sensitivity to environmental perturbations is generally perceived to be intermediate or low, thus allowing a range of sublethal physiological changes to be measured over a period of time.
- It is easy to catch in the wild with simple hand nets and minnow traps.
- It is easy to keep and can reproduce under laboratory conditions (their eggs and larvae display close to 100% hatching and survival rates).
- It is very tolerant to manipulations such as handling and transport.
- It presents all the advantages of a small teleost (i.e., small aquaria, short life cycle, etc.).

Furthermore, the stickleback is present in environments of all salinities (marine, estuarine, and freshwater) and is one of the few endemic and ubiquitous species in Europe that offers scope for environmental monitoring. In addition, the stickleback has unique traits for the detection of endocrine disrupting chemicals (EDCs), such as a genetic sex marker and a xenoandrogen specific endpoint. The numbers of published, standardised test protocols are increasing, as are the tools for *in situ* biomonitoring. The total species complex provides a further challenge for environmental toxicology related to their freshwater speciation: how different are these populations in terms of pollutant responses, do they adapt, and is there a physiological or genetic basis to this adaptation? The potential value of the stickleback as an environmental sentinel has been recognised for some years now, and a small body of literature has been generated.

Although the species is included in the guidelines for toxicity testing set by the Organisation for Economic Cooperation of Developed countries (OECD), there is a lack of systematic use of the stickleback in global chemical effluent testing. Thus, despite their popular use as environmental sentinels, sticklebacks are still not widely used as model surrogate species for wildlife or human research in toxicology or disease. The fathead minnow (*Pimephales promelas*), the zebra fish (*Danio rerio*), the sheephead minnow (*Cyprinodon variegates*), the medaka (*Oryzias latipes*), and rainbow trout (*Oncorhynchus mykiss*) are examples of more widely used species in chemical screening and testing.

However, we may be entering a time when the use of the stickleback as a universal model in biological studies will sharply increase. The available molecular resources are increasing and the imminent completion of the full genome sequencing will give the species a clear advantage over other models, especially in defining the mechanisms underlining the biological responses to a changing environment. Their relatively small genome, estimated at 675 mega-bases, implies that intron size is small and that a second genome duplication event did not take place in the stickleback, aiding the speed of annotation and the functional characterisation of the genes of interest.

In addition, the presence of a genetic sex marker and the presence of a specific androgen, and antiandrogen biomarker are traits that make the species particularly useful in the detection of EDCs. Research into this relatively new class of contaminants has increased dramatically over the past decade. EDCs are different from classic toxicants as they have the potential to compromise fish reproduction, ultimately affecting populations, and thus are considered as the highest priority in risk assessment. The stickleback has very quickly gained popularity in the field of endocrine disruption, and this aspect will therefore be examined separately in the present chapter.

The wide distribution of the species in temperate and sub-Arctic zones and practically all aquatic environments can bring together laboratory and field data, and allow a better estimation of the risks associated with environmental perturbations, of either chemical, physical, or biological nature. The increasing resources and understanding can dissect field- and laboratory-derived information at the finest molecular level, and provide new revelations in the understanding of interactions between organisms and their environment. It is therefore anticipated that soon the stickleback will develop into a prominent model in the field of ecotoxicology just as it has in evolutionary biology.

One of the criticisms often encountered when presenting a case for the value of the species in ecotoxicology is that the sticklebacks are insensitive to pollution. This statement stems from observations that the stickleback is one the first species to recolonise rivers after a pollution incident. On the other hand, in some areas of the U.S., Canada, Japan, and Europe sticklebacks are extinct or are under threat, and have therefore been proposed for protected species status.[1,2] Unfortunately, information on the state and health of stickleback populations across the globe is scarce.

In the U.K. alone, the Department for Environment, Fisheries, and Rural Affairs (Defra), the National Environment Research Council (NERC), and the Environment Agency (EA) are currently funding numerous programmes that are designed to further develop techniques and approaches which use the stickleback as a universal bioindicator and model species in risk assessment application, and national monitoring programmes.

10.3 ENVIRONMENTAL SENTINEL

The industrial and technological revolution has created a great economic benefit in developed countries but has left a legacy of contamination issues with inorganic and organic pollutants in the environment. The challenges presented by these contaminants and their long-term effects on wildlife and humans will only increase with continuous development in new areas such as biotechnology and nanotechnology. To this end, an informed impact and risk assessment should be based on environmental monitoring and surveillance to understand the extent of the potential problems. The aquatic environment is often the final sink for anthropogenic discharges and, therefore, aquatic organisms play an important role in providing an integrated assessment of the biological effects of pollutants.

Environmental stressors, however, are not only chemical in nature. Habitat destruction, global warming, increased UV radiation, nutrient enhancement or deprivation, and hypoxia are other factors that on their own or in conjunction with chemical contaminants can alter the shape and state of aquatic ecosystems.

To this end, the effects of increased water temperature on survival of adult sticklebacks have been studied,[3] and more recently it has been suggested that species survival and ability to reproduce will be negatively impacted by global warming, even by a mean rise in water temperature of as little as 2°C.[4] The effects of hypoxia in two different stickleback populations, one from a static pond and the other from a fast-flowing river, were investigated by metabolic changes and dominance hierarchies and found to be different.[5] The vast majority of studies using the stickleback as an environmental sentinel involve a chemical challenge.

The description of some toxicological studies and specific tools (biomarkers) presented in the following text is not intended as a comprehensive review of the literature, but to show examples of the use and value of the species in environmental monitoring. The wide distribution of sticklebacks, along with their relative tolerance to pollutants, has led to a significant number of toxicological studies, the vast majority of which were laboratory-based tests. The endpoints of these tests include death (LC50), specific biomarkers to a certain pollutant, general stress responses, growth, histopathology, parasitic load, and alterations in behaviour.

10.3.1 Inorganic Compounds (Metals and Their Salts)

Metals are introduced in aquatic systems as a result of the weathering of soils and rocks, from volcanic eruptions, and from a variety of human activities involving mining, processing, or use of metals and substances that contain metal pollutants. With inorganic chemicals, the extent of long-term bioaccumulation depends on the rate of excretion. Bioaccumulation of cadmium in animals is high compared to most other metals, as it is assimilated rapidly and excreted slowly. Also, the sensitivity of individuals of a particular species to a pollutant may be influenced by factors such as sex, age, or size. In general, the concentrations of metals in invertebrates are inversely related to their body mass. In fish, the embryonic and larval stages are usually the most sensitive to pollutants. The most common heavy metal pollutants are arsenic, cadmium, chromium, copper, nickel, lead, and mercury.

Very early, Jones[6] provided a lucid account of toxicity for over 20 inorganic salts (mainly nitrates and chlorides). Other studies on salt toxicity and speciation include sodium sulphide,[7,8] sodium cyanide,[9] ammonium chloride,[10] sodium fluoride,[11] potassium chromate, and dichromate.[12]

Heavy metal toxicity has been extensively studied in the stickleback. Cadmium is produced as an inevitable by-product of zinc (or occasionally lead) refining, because these metals occur naturally within the raw ore. The most significant use of cadmium is in nickel cadmium batteries. Cadmium coatings provide good corrosion resistance, particularly in high-stress environments such as marine and aerospace applications in which high safety or reliability is required; the coating is preferentially corroded if damaged. Other examples of cadmium use are as pigments, stabilisers for PVC, in alloys, and in compounds used in the electronic industry. Cadmium is also present as an impurity in several products, including phosphate fertilisers, detergents, and refined petroleum products. Cadmium is biopersistent and, once absorbed by an organism, remains resident for many years (over decades for humans), although it is eventually excreted. The cadmium studies conducted in the stickleback so far include toxicity,[13] uptake and distribution,[14] effect on the chloride cells of the gills,[15] histological changes in the gills and kidney,[16] and induction of cadmium-binding protein.[17]

The relative toxicity of lead, zinc, and copper as affected by calcium concentration has been reported in detail.[18] The toxicity of zinc has been well documented, including effects on gills,[19] uptake and loss,[20] and accumulation.[21]

Copper is an essential substance for animal life, but in high doses it can cause anaemia, liver and kidney damage, and stomach and intestinal irritation. Copper normally occurs in drinking water from copper pipes, as well as from additives designed to control algal growth. The acute toxicity of copper in the stickleback has been described in a number of studies.[22,23] More recently, Sanchez et al.[24] described the changes of several hepatic biomarkers after copper exposure and concluded that the species provides a good model for oxidative stress responses. A more comprehensive study on copper toxicity, including comparative data with other fish, was reported recently.[25] The authors concluded that the species was of intermediate sensitivity to copper and more tolerant than other freshwater species. However, the toxicity of copper was not determined for all life stages, and it is therefore important to note that environmental concentrations may still have adverse effects on wild stickleback populations.

Field studies involving metal toxicity are scarcer, although a thorough study on heavy metal bioaccumulation in different tissues was conducted in Belgium,[26] and Falandysz and Kowalewska[27] reported mercury levels in sticklebacks from Poland.

10.3.2 Organic Xenobiotics

Van den Dikkenberg et al.[28] conducted a series of acute toxicity tests for a number of organic pollutants, including 2,4-dinitrotoluene, 2,4-dichloro-aniline, 2,6-dimethylquinoline, 3,4-bis(2-ethylhexyl) phthalate, n-(1-methyethyl)-2-propanamine, tetrapro-pylenebenzenesulfonic acid, and tricresyl phosphate, using the stickleback. The toxicity of organoarsenic-based warfare agents has also been determined.[29]

A comprehensive acute toxicity study for organic insecticides, including aldrin, azinphos-methyl, carbaryl, coumaphos, chlordane, dichloro-diphenyl-trichloroethane (DDT), dieldrin, endrin, heptachlor, lindane, malathion, methoxychlor, and toxaphene, was conducted in the early 1960s.[30] Other insecticides for which acute toxicity and bioconcentration reports are available include chlorpyrifos,[31,32] endosulfan — which was characterised as highly toxic to the stickleback[33] — lindane,[34] the potato-sprouting inhibitor tecnazene,[35] and dichlorobenzenamine, the dominant degradation product of the widely used herbicide dichlobenil.[36] Sturm et al.[37] reported on the natural variability in response to organophosphates by means of cholinesterase inhibition.

The LC50 of two fenitrothion formulations (both characterised as slightly toxic to the stickleback) was reported to be significantly different.[38] We have observed strong inhibition of spiggin induction (the nest glue protein) in androgen-stimulated females by fenitrothion (a widely used organophosphate insecticide) in concentrations much below the toxicity levels.[39] Two further biocides, vinclozolin, a widely used fungicide, and linuron, an herbicide applied to suppress broadleaf and grassy weeds, were also found to be androgen antagonists, though to a lesser extent in comparison to fenitrothion.[39]

The studies on polychlorinated biphenyls (PCBs) using the three-spined stickleback include an attempt to model bioaccumulation using Aroclor[40] and the uptake and elimination of more than 20 PCBs after three different routes of exposure.[41] The levels of polychlorinated naphthalenes (PCNs) in sticklebacks from the Gulf of Gdansk, Poland, have been measured.[42] The effects of persistent organic pollutants such as polybrominated diphenyl ethers (PBDEs, Bromkal 70-5DE), PCNs (Halowax 1014), and PCBs (Clophen A 50) on reproduction, liver morphology, and cytochrome P450 activity under controlled conditions have also been described.[43]

Studies on organic solvents include determination of LC50 for benzene (24.8 μg/l after 96 h of static exposure[44]), the range of lethal body burdens for 1,2,4-trichlorobenzene,[45] and uptake, metabolism, and elimination of diphenyl ether.[46] The toxicity of several oil dispersants such as Berol TL-188, Berol TL-198, BP 1100 x, and Corexit 7664 on the stickleback was generally low[47]; however, the morphological changes[48] and enzyme activity[49] of liver cells after exposure to fuel oils were rather profound. The lack of acute effects of exposure to resuspended harbour sediment was also reported.[50]

The effects of long-term tributyltin oxide (TBTO) exposure (TBT is widely used as an antifouling agent in ships and boats) on reproduction and histopathology have been described in detail,[51] and Falandysz et al.[52] attempted a correlation between organo-tin residues in sticklebacks and sediments from the Gulf of Gdansk, Poland. Bioaccumulation of nonylphenol (a degradation product of nonylphenol ethoxylate, a high-volume surfactant that has been used for more than 40 years as detergent) emulsifier, wetting, and dispersing agent has been described in a number of marine animals including the stickleback.[53] A risk assessment for ditallow dimonium chloride, a cationic fabric softener, has also been conducted[54] as studies on lauryl sulphate.[33,48]

10.3.3 Integrated Indices: Disease, Stress, Growth, and Behaviour

Parasites are often regarded as an interface between environmental stress and fish health and thus have been used as indexes of water quality. Although it is arguable that environmental stressors may ameliorate the effects or intensity of parasitism, more often it is accepted that the interaction of parasites with environmental stressors can lead to increased pathogenicity, due to the impairment of the hosts' immune system.[55] Skin-dwelling ciliate protozoans are a good example of the latter interaction[56] and have been shown to increase in intensity and severity on sticklebacks after exposure to pulp mill effluents.[57,58] These studies also observed an increasing trend in the parasitic load of the highly pathogenic monogenean species, *Gyrodactylus*. Yeomans et al.[59] conducted a preliminary, yet comprehensive, study and confirmed the apparent relationship between protozoan communities and water quality using the stickleback as the model host in the wild. Other parasitic species appear to be less suitable as indicators of pollution. Although parasitic prevalence and pathogenicity present the advantages of an integrated assessment of fish health, the complexity of factors that govern host–parasite interactions does not allow a straightforward link between aquatic pollution and fish disease. A more systematic characterisation of the quality of effluents and water bodies may lead to the incorporation of parasitic indices in field monitoring of fish health.

Histopathological markers are often regarded as the "golden standard" in ecotoxicology. Handy et al.[60] conducted a survey in several rivers of good to fair water quality and proposed the use of histopathology in the stickleback as a sensitive tool for routine monitoring of fish health. They found that gill, liver, and spleen lesions were present in fish caught in rivers that were broadly in compliance with the water quality directives to protect fish health.

Pottinger et al.[61] presented the advantages of the stickleback as an environmental sentinel in a study where among other whole-body indices they examined the value of corticosteroid level as an indicator of environmental stress. Although cortisol excretion increased in the challenged fish, it declined toward the end of the study in the chronically stressed group, suggesting that fish could become acclimated to stress, thus compromising its value as an indicator of environmental stress.

Ali and Wootton[62] evaluated the value of growth predictors with a view to applying them to natural populations. They found a strong correlation of RNA:DNA ratio in white muscle or lipid concentration (percentage dry weight) with short-term growth rates of juvenile sticklebacks fed with different daily rations over 21 d. Although some of these correlates proved biased and inefficient,[62] these assays offered great potential as growth estimators. There are currently concerns over the effects of chemicals on growth and thyroid function because robust endpoints are either absent or difficult to measure and interpret in the wild. Lehtinen[57] reported that exposure of sticklebacks to kraft mill effluent adversely affected their growth. In contrast, cage deployments in a U.K. sewage treatment plant in summer 2004 resulted in significantly increased body weight in the effluent-exposed fish (Figure 10.1).

These results indicate the complexity of effluent chemistry and suggest that different pathways are activated in different effluents by different chemical components.

Despite the fact that behaviour is often regarded as the link between physiological function and ecological processes, the incorporation of behavioural indicators in the study of biological effects of pollutants is a relative new approach in ecotoxicology. Behaviour is a neurological response under the influence of endocrine changes. Subtle alterations in sensory, hormonal, neurological, and metabolic processes may affect behavioural patterns that are essential to fitness and survival. If these patterns are well documented, they have the potential of providing a sensitive and noninvasive assay for the detection of environmental contaminants. Abnormal behaviour is one of the more obvious endpoints produced by contaminants in some species in which the patterns are relatively easy to evaluate, sensitive, and relatively inexpensive.[63,64] Scott and Sloman[65] provide an excellent review of the effects of pollutants on fish reproductive behaviour, highlighting the value of behavioural disruption in the management of aquatic pollution.

A few studies have already used behaviour of the three-spined stickleback as an indicator of chemical or physical disturbances. Sneddon and Yerbury[5] suggest that,

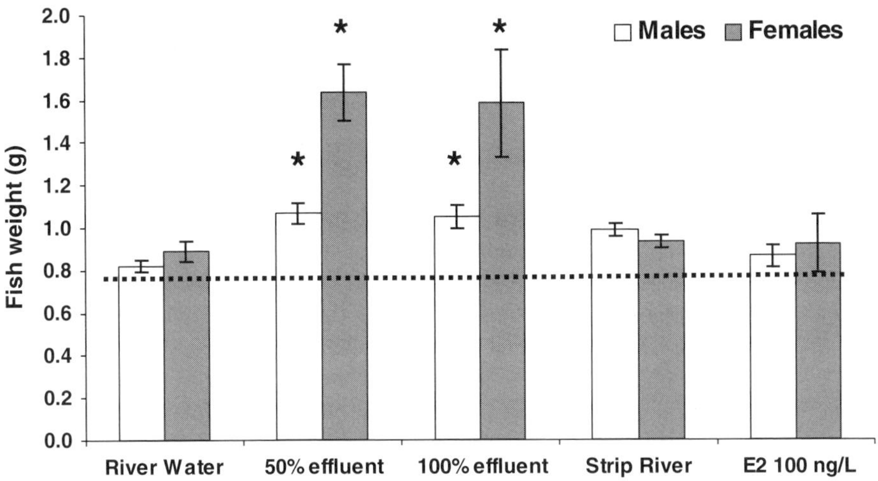

FIGURE 10.1 Stickleback weight gain after 3 months of exposure to sewage effluent. Strip River: River water after having passed through activated charcoal to remove all organic compounds; E_2: Oestradiol.

under hypoxic conditions, social hierarchies are less stable in terms of rank position. Craig and Laming,[66] show that high levels of ammonia result in a decrease of the locomotory activity (swimming) and a small increase in nonlocomotory activity relative to the controls, whereas low levels of ammonia result in a pronounced increase in nonlocomotory activity (ventilatory rate) and a small difference in locomotory activity relative to controls.

Aggregation, bottom-dwelling, and feeding behaviours have been investigated in response to butyl benzyl phtalate (BBP) and p, p-DDE, a DDT metabolite,[67,68] exposures. Fish exposed to 0.1 mg/l of BBP for 26 d aggregated more into a single shoal and spent more time at the bottom of the tank than the control fish, suggesting that aggregation and bottom-dwelling activity may be suitable and sensitive endpoints for the detection of BBP in three-spined sticklebacks.[67] However, 5 weeks after the exposure to DDE (50 μg/l) and BBP (100 μg/l) terminated, the exposed fish displayed significantly higher feeding activity than the control fish.

To date, alterations in the sexual behaviour in the three-spined stickleback have been studied only in response to oestrogenic chemicals. Males exposed to ethinyl-oestradiol over a short-term (10–50 ng/l, range 8–32 d) decreased their aggressive response to a conspecific male over time, whereas control males increased their aggressiveness in breeding conditions.[69] The exogenous oestrogen action might have reduced endogenous androgen action and, thereby, the frequency of androgen-regulated behaviours. Bell[69] suggested that low levels of EDCs in the environment may exert subtle, yet important, effects on animal behaviour. In another study, exposure to 100 ng/l ethinyl-oestradiol accelerated growth rate and increased the levels of behaviour that make individuals more susceptible to predation (activity and foraging under predation risk), particularly in female fish.[70] This approach underlines the importance of chemical impact evaluation on the species survival via a different route to direct toxicity. Finally, in a single exposure (injection in peanut oil to allow slow release) of 17-oestradiol (2 μg/g body mass) over 21 d, exposed and control males presented differences in the onset of the nest-building and impaired paternal care.[71] These differences however, were not evident in courtship patterns.

10.3.4 Population Genetics

Despite the increasing use of the stickleback for effluent and chemical testing, the available information is not systematically documented, thus not allowing direct comparisons with the other test species. Therefore, the relative sensitivities of different stickleback morphs or populations toward chemical, physical, or biological stressors remain largely unknown to date.

The very same phenotypic diversity that sticklebacks display may be a contributing factor to this apparent failure of research to produce systematic toxicological data. Indeed, it is often asked how different stickleback populations are in their responses to pollutants? Can we infer that a species with such an extensive and rapid adaptive radiation will display the same ability to adapt to different environmental pressures, including chemical insults, by increasing their tolerance? If susceptibility is reduced, is there a physiological or genetic basis for such adaptation? What are the implications for the genetic diversity of impacted populations? Evidence from studies in other fish suggests that there is a difference in the expression of xenobiotic detoxification systems in chronically exposed populations and that these may be determined genetically.[72–75]

Although the preceding uncertainties may compromise the value of the stickleback for toxicological studies (where pure genetic strains are preferable for testing), they provide an attractive challenge for ecotoxicology (in the same way as for evolutionary biology). The numerous fragmented and endemic populations of this species offer an excellent model system to study the basis of individual and population susceptibility and adaptation to environmental pollutants. Continuous exposure to pollutants over long periods could lead to the selection of populations of organisms whose individual members have inherited particularly advantageous or unusual detoxification mechanisms and genes. The extent to which such a scenario may occur depends on the initial gene pool, population dynamics of individual species, reproductive strategies, and life history interaction with the contaminated environment. If we further hypothesize that the different adaptations to pollutants between individuals or populations are of genetic nature (i.e., polymorphic variants of key enzymes in detoxification and repair), then genetic diversity in contaminated areas is likely to be lower because of selection.

Sticklebacks are known to adapt to novel environments more rapidly than predicted from conventional methods of biological differentiation.[76–78] However, this divergence appears to be constrained by gene flow.[79] This is an area of research in which the study of allozyme and microsatellite variations, singly and in combination, will play a major role in the discovery of new insights in the future. Allozyme variation continues to be a valuable and economical method for analysis of evolutionary processes of divergence and gene flow.[80] The numerous electrophoretic studies of allozymes and other proteins in *Gasterosteus aculeatus* have been reviewed by Buth and Haglund.[81] However, to my knowledge, no studies to date have used allozyme variation to discriminate between stickleback populations from environments with different pollution histories.

Similarly, a large number of microsatellite markers in the stickleback have been developed and applied in the field and the laboratory.[82–87] Nevertheless, most studies utilise microsatellite polymorphism to confirm provenance and discriminate between anadromous and freshwater populations or even between different sympatric morphs of the species complex. Therefore, these markers may require revision and further development before they find an application in the assessment of population genetic status and restricted gene flow as this might be affected by pollution. A review of microsatellite variation in marine, anadromous, and freshwater systems revealed that freshwater fish displayed lower heterozygosities than anadromous and marine species.[88] Furthermore, mating systems and sexual selection also influence the degree of expected polymorphisms. Fish that provide parental care spawned on average with more mates compared to fish with internal fertilisation and pregnancy,[89] and both nest building and pregnant fish presumably spawn with fewer mates than fish that release their gametes free in the water column during mass spawning. Female sticklebacks are known to optimise their major histocompatibility complex (MHC) allele polymorphism by sexual selection to avoid inbreeding and maximise the parasitic resistance of the offspring.[90]

These factors need consideration before robust sets of allozyme or microsatellite markers can be deployed to address the effects of pollution on fish populations. Nonetheless, this same approach holds the key to doing precisely this, and it appears that the stickleback offers new prospects for environmental risk assessment on this basis.

10.4 ENDOCRINE DISRUPTION IN THE ENVIRONMENT

The issue of endocrine disruption in both wildlife and humans was raised in the early 1990s and quickly attracted the attention of researchers, regulators, and the public, developing rapidly into a young but active discipline. Although reproductive abnormalities in fish were reported as early as the 19th century, the link between these observations and the presence of environmental chemicals with endocrine modulating activity was not made until only a decade ago, most likely because of the great diversity and plasticity displayed by aquatic animals in sexual differentiation and development. Since then, several field and laboratory studies have shown induction of adverse effects in wildlife species and populations upon exposure to EDCs. These effects vary from subtle changes in the physiology and sexual behaviour of species to permanently altered sexual differentiation.[91]

The importance of the issue is highlighted by a large number of studies set to link human adverse effects and exposure to EDCs in the last decade.

10.4.1 EDCs and Their Biological Effects

Evidence for the effect of EDCs in human health includes the increased incidence of idiopathic hypospadias, alterations of the male genitalia, pseudohermaphroditism, decline in sperm counts, and the increasing incidence of breast cancer among females as well as testicular and prostate cancers amongst males.[92–94]

EDCs mainly act by mimicking or antagonising the effect of the endogenous hormones but may also disrupt the synthesis and metabolism of endogenous hormones and their receptors.[93,95,96] Chemicals that interact directly with hormone receptors, to either mimic or block the hormone action, have the potential to produce effects at extremely low concentrations. Although disruption of the endocrine system can occur at many sites (adrenal, thyroid), the most commonly reported system is the reproductive one. Thus, compounds that interfere with oestrogens, androgens, and steroidogenic enzymes involved in sexual differentiation and maturation (i.e., CYP19α and CYP19β, widely known as ovarian and brain aromatase, respectively) have been the main focus of research so far. Because gonadal steroid hormones, generally at very low concentrations, control and regulate embryonic development and sex differentiation, it is during this period that the effects of EDCs can be particularly severe.

Within the group of chemicals that interferes with the reproductive system, the vast majority of research in fish has focused on oestrogenic xenobiotics and has used the presence of the female-specific egg yolk protein precursor, vitellogenin (VTG), in male fish as an unambiguous biomarker for oestrogens in a variety of model species.[97–101] Male fish do not normally synthesise vitellogenin and for this reason this oestrogen-induced protein provides an ideal biomarker for xenoestrogens.

Oestrogenic activity has been identified in a large number of domestic sewage and industrial effluents and the responsible chemicals have been identified through a procedure named *toxicity identification evaluation* (TIE) in which the effluent is fractionated in active and inactive fractions by means of *in vitro* assays (e.g., yeast oestrogen receptor, reported gene assays, etc.), followed by identification by mass spectroscopy. The nature and potency of xenoestrogens exhibit a large diversity and include natural products (oestrone, oestradiol, coumestrol, genistein), pesticides, (dieldrin, toxaphene, endosulfan, phenylphenol, DDT and its metabolites, methoxylchlor, vinclozolin), medical drugs (hydroxyflutamide, nilutamide, tamoxifen, diethylstibestrol, oral contraceptives such as ethinyl-oestradiol), and commercial or industrial chemicals such as bisphenol A, alkylphenols (*p*-nonylphenol), phenolic antioxidants (BHA), polychlorinated biphenyls (PCBs), phthalate plasticisers, dioxins, etc.

The role of environmental androgens and antiandrogens has until recently been overlooked, most likely because xenoestrogens are far more ubiquitous, besides the fact that there is an increased concern for clinical implications of these chemicals in humans.[102] One of the clearest observations of androgenicity in the aquatic environment has been made in female mosquito fish (*Gambusia* sp.) living downstream of kraft mill effluent discharges. These fish develop anal fin appendages (gonopodia) that are normally only found in males.[103–105] The active compounds in the effluent were tentatively identified as bacterial degradation products of the plant sterol, stigmastanol.[106] Moreover, exposure of white sucker (*Catostomus commersoni*) to bleached kraft pulp mill effluent from North American plants affected plasma concentrations of sex-steroid hormones, age at sexual maturity, and gonadal size.[107–109]

In Europe, studies have revealed a decrease in the gonad size of female fish living near coastal waters receiving bleached pulp mill effluent[110,111] and significantly male-biased eelpout (*Zoacres viviparous*) broods near a large pulp mill.[112] The causative agents of masculinisation contained within pulp mill effluents have been shown to be a compound with similar, though not exactly identical, chromatographic properties to androstenedione.[113,114] Furthermore, Parks and coworkers[115] have confirmed that there are compounds in pulp mill effluent that bind agonistically to human androgen receptor (AR) in a mammalian whole-cell assay.

More recently, concern has arisen regarding the presence of trenbolone in the environment. Trenbolone acetate is an anabolic steroid used as a growth promoter in beef cattle in the U.S. and Canada. It is hydrolysed to 17β-trenbolone, which is a potent environmental androgen both *in vitro* and *in vivo*.[116–118]

Detection of androgenic compounds in domestic effluents has also been achieved. However, it appears that secondary treatment (biological) is successful in removing androgens because positive *in vitro* activity was evident only in effluents without any secondary treatment.[119,120]

Although androgenic xenobiotics are restricted to pulp mill, feedlot, and few sewage effluents, compounds with antiandrogenic activity appear to be more widespread.[121,122] Identification of the chemicals that cause high antiandrogenic activity In *in vitro* assays, however, has not yet been achieved. So far, *in vitro* and *in vivo* data (the latter mainly in mammals) have indicated that various chemical classes display such activity, including pesticides such as fenitrothion, vinclozolin, procymidone, linuron, iprodione, chlozolinate, ketoconazole, DDT and its metabolite, p,p,-DDE, as well as several pyrethroids.[123–128]

At present, the only officially accepted *in vivo* test for screening chemicals with suspected androgenic or antiandrogenic activity is the Hershberger castrated male rat assay.[129] The basis of the assay is that castrated sexually mature male rats undergo regression of androgen-sensitive tissues (prostate, epididymis, seminal vesicle and levator-ani muscles). These tissues are restored to their original weight upon treatment with testosterone, and that growth can be blocked by the concomitant administration of an antiandrogen. The OECD validation of this assay has progressed to the point where agreement has been reached on the optimum dose level of testosterone to be used, and work has only recently commenced on the testing of reference agents.

The development of male sex characteristics by female gonochoristic snails, a condition termed *imposex*, has been documented globally and is causally associated with exposure to the ubiquitous environmental contaminant tributyltin (TBT). This phenomenon is a masculinisation process involving the development of a penis and a vas deferens; in some species this development of a vas deferens disrupts oviducal structure and function, preventing normal breeding activity and causing population disappearance locally.[130] In some species, oogenesis is supplanted by spermatogenesis. Field evidence clearly associates these syndromes with marinas where the antifouling agents are used; dose-related effects can be replicated in laboratory exposures to environmentally relevant concentrations of TBT compounds.[131] It has now been established that imposex and intersex are forms of endocrine disruption caused by elevated testosterone titres that masculinise TBT-exposed females. The increase in testosterone titres with TBT exposure has been attributed both to the inhibition of the aromatase enzyme CYP19α,[132] and a decrease in the metabolic elimination of testosterone as sulphate conjugates.[133] TBT-induced masculinisation in gastropods is the clearest example of endocrine disruption in invertebrates to date that is unequivocally linked to a specific environmental pollutant and has affected gastropod populations globally. This discovery led to a ban of the use of TBT in 2004.

10.4.2 The Stickleback Model in EDC Research

The three-spined stickleback has gained rapidly a lot of popularity in this field because of the unique traits it possesses for the characterisation of endocrine-modulating substances. In the past 3 years a number of studies have employed the stickleback as a model for endocrine disruption research.[39,134–136] In addition to the general advantages it offers as a sentinel and test organism outlined earlier, sticklebacks have two further traits that provided the species with clear advantages over other fish models.

One important element is the presence of a rudimentary Y chromosome that can be determined molecularly,[137,138] so that a genetic sex can be assigned. In fish, despite the fact that the vast majority of species are strictly gonochoristic, the presence of morphologically differentiated sex chromosomes is very rare. Even though sex determination seems to have a genetic basis for most teleosts, it is not as strict as in mammals and birds. The genetic mechanisms for sex determination in fish are generally primitive and labile and environmental factors often play an important role in the phenotypic expression of sex. To this end, the stickleback appears to be more "hard-wired" than other teleosts in that it has a Y chromosome similar to mammalian species; thus, a specific genetic marker for sex identification is available.

The second advantage is the specific androgen endpoint it offers, the kidney glue protein spiggin. Spiggin is produced naturally in male fish only during breeding for building a nest and is under the control of androgens. It was first characterised as a 205 kDa glycoprotein by Jakobsson et al.,[139] who named it after the Swedish name for the stickleback, *spigg*. Numerous studies exist on the endocrine control of this particular secondary sexual character of the male stickleback (see Chapter 7). Jones et al.[140] were the first to provide a partial sequence for spiggin. We now know that there is at least one more cDNA encoding for a second spiggin type.[141] The two molecules (spiggin I and II) show marked structural differences but have a high homology in amino acid sequences (80%).

10.4.2.1 Spiggin

Spiggin is to date the only androgen-induced protein in fish, and we were the first to recognise its potential in the stickleback as an androgen biomarker,[142] similar to vitellogenin, a widely used biomarker for oestrogens. Since then, *in vivo* assays were developed to detect androgens[134] and antiandrogens.[39] In brief, female fish are exposed to a series of concentrations of the test chemical (the putative androgen or a model androgen in combination with the putative antiandrogen) through water for a period of 21 d. At the end of this period the fish are sacrificed, the kidneys dissected,

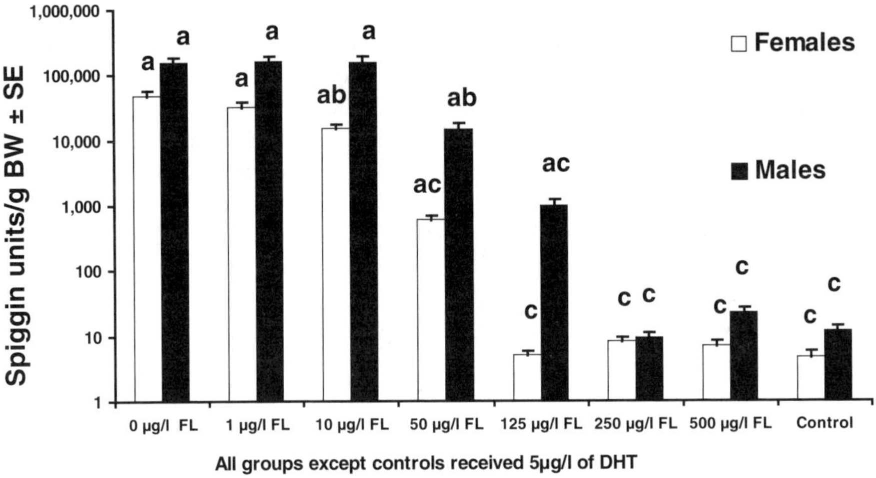

FIGURE 10.2 Inhibition of DHT-induced spiggin production by flutamide in a dose–response manner. Treatments shown with the same letter do not differ significantly from each other ($p < .05$).

and their spiggin content measured by a specific ELISA. The exposure protocols broadly follow the guidelines set by the OECD for screening EDCs.

Although researchers have explored the potential use of the mosquito fish as a biomarker for environmental androgens,[143] there is clearly a lack of sensitive test systems to screen for xenoandrogens. This is especially true in Europe, where the mosquito fish is not an endemic species.

The antiandrogenic activity of flutamide (FL), a clinical drug used in the treatment of prostate cancer, has been detected in male fathead minnows through a reduction in the number of nuptial tubercles.[144] However, an FL concentration of 1000 µg/l was required to elicit this response, four times greater than that used in our study on sticklebacks.[39] In addition, exposure of fathead minnow embryos to vinclozolin (VZ) at concentrations ranging from 90 to 1200 µg/l did not result in any adverse effects on sexual differentiation or reproductive health.[127] These findings call into question the suitability and sensitivity of fathead minnow bioassays for the detection of EDCs with antiandrogenic activity.

Bayley et al.[145] exposed juvenile guppies to FL, VZ, and p,p-DDE (a DDT metabolite) via the food from birth to adulthood and concluded that all three chemicals had a clear demasculinising effect (reduction of orange display colouration, gonopodium development, reduction in sperm count, and suppressed courtship behaviour). However, the length of the study and the route of exposure (not via the water) preclude any comparisons on the sensitivity of the bioassay with the stickleback assay.

In addition, Kinnberg and Toft[146] exposed sexually mature male guppies to a number of oestrogenic and antiandrogenic compounds. Although FL, p,p-DDE, and oestrogens blocked spermatogonial mitosis, VZ did not have any adverse effects.

In the stickleback kidney, previous studies have shown that no specific binding of 11-ketotestosterone (11-KT) or testosterone was detected in either cytosolic or nuclear fractions, although displacement of tritiated 11-KT with unlabelled 11-KT was observed in the kidney membrane fraction.[147] More recently, molecular cloning of a nuclear AR in the stickleback kidney has revealed that it is the classic mammalian type, AR2 or AR beta (NCBI GenBank, Accession numbers AAO83572/3), as we have previously speculated on the basis of androgen potency.[134] The presence of a nuclear AR homologous to the mammalian AR in the stickleback kidney has important implications regarding the use of the species as a model surrogate organism for medical endocrine disruption research.

Using spiggin as a biomarker, we have established that a number of androgens when administered via water are capable of inducing spiggin synthesis in female kidneys. The relative potency of androgens tested in this system on spiggin induction is: 17α-methyltestosterone > dihydrotestosterone > 17β-trenbolone > 11-ketotestosterone > testosterone (own unpublished data). The stability of the test chemical in water and its relative binding affinity for the sex hormone binding globulin are factors (other than receptor affinity) that affect exposure responses. In addition, we have shown that exposure to a 10% dilution of pulp mill effluent also induces spiggin in female fish.[135]

The relative potency of tested antiandrogens on inhibition of spiggin induction by 17α-methyltestosterone (500 ng/l) or dihydrotestosterone (5 μg/l) is flutamide > fenitrothion > vinclozolin > linuron > procymidon > DDE. In addition, we have shown that *in vivo* exposure to high concentrations of oestrogens (i.e., 17β-oestradiol and ethinyl-oestradiol at > 100 ng/l) also inhibit androgen-induced spiggin synthesis. Oestrogens have a relatively high affinity for the AR, so they may act as pure antiandrogens by occupying the receptor but not activating it. It is also possible that another mechanism is operating in the observed inhibition of spiggin synthesis. Indeed Olsson et al.[148] found that AR mRNA levels were suppressed in castrated males that received oestradiol implants.

The results so far suggest that the stickleback is by far the most sensitive and relevant bioassay for the detection of androgenic xenobiotics.

10.4.2.2 Vitellogenin

In fish, the site of VTG synthesis is the liver, from where it is transported to the ovaries via the blood. This makes plasma the natural target for VTG measurement, and this is reflected in the literature, where the vast majority of studies on aquatic organisms express VTG in μg/ml plasma. However, there are several studies involving small fish (< 1 g) that report VTG concentrations in whole-body extracts.[149–151] In all cases, assays of VTG in whole-body extracts appear to yield data of equivalent quality to those studies involving plasma only.

Oogenesis and vitellogenesis have previously been described in the three-spined stickleback.[152–154] Ollevier and Covens[155] have reported the production of an antiserum against stickleback VTG and its cross-reactivity with lipovitellin preparation from the ovary. This antiserum was used for the immunocytochemical detection of yolk proteins[155] and the study of the immunological relationship between VTG and yolk proteins.[156] Covens et al.[157] also described the specific binding of stickleback VTG (blotted on nitrocellulose membrane) to concanavalin A (con A) and suggested that this property could be used for affinity purification of the native VTG molecule.

We developed a VTG assay (ELISA) for the stickleback a few years ago[135] by purifying VTG from a pool of oestrogenised plasma with gel filtration chromatography and raising an antistickleback VTG serum in a rabbit. Since then, the assay has been used for the analysis of more than 5000 samples measuring VTG in plasma, heart, and whole-body homogenates.

In practical terms, collecting plasma from sticklebacks can be a difficult, time-consuming, and low-yield procedure, particularly when a large number of fish or small-sized fish have to be sacrificed on the same day. On the other hand, the procedure leaves the rest of the tissues available for other analyses (i.e., spiggin in kidney, gonad histology, etc.). We have demonstrated that if the fish have been sacrificed by terminal anaesthesia and immediately snap-frozen in liquid nitrogen, the blood trapped in the heart can yield an amount of plasma comparable to that obtained by bleeding (Katsiadaki, unpublished data). We have also shown that VTG can be easily measured in whole-body homogenates of juvenile sticklebacks exposed to oestrogens.[136] Pottinger et al.[61] have also demonstrated that measuring the levels of alkaline-label phosphate (an indirect estimate of VTG) in whole-body extracts of adult fish is a suitable (if insensitive) method.

The results to date suggest that the stickleback VTG assay is a sensitive, easy-to-perform, and reproducible test. Although there is large interspecies variation in terms of magnitude of VTG production after oestrogen exposure, the threshold concentrations at which VTG is traceable do not differ by more than one order of magnitude between different fish species.[158] To this end, the stickleback appears to be a highly sensitive assay, although considerable individual variation is commonly present in response to oestrogen exposures. The latter observation is more true of industrial chemicals than pure steroids. Figure 10.3 displays some of our recent results in laboratory-based exposure, whereas Figure 10.4 relates to VTG titres of fish caught in U.K. rivers of varying oestrogenic load.

10.4.2.3 Other Biomarkers/Tools for Endocrine Disruption Research

Apical endpoints such as survival, growth, development, and reproduction, although integrative in nature and hence useful in predicting adverse effects, do not provide any insight into the mode of action of each chemical. This is of particular importance in the assessment of EDCs where regulation is based on the chemical's mode of action.[158] The specific biomarkers for EDCs commonly used in fish include secondary sexual characteristics, gonadosomatic index, plasma steroids, VTG (and spiggin in the stickleback), and gonadal histology.

Intersex sticklebacks have been reported both before[159] and after[160] endocrine disruption in wildlife became a major issue. In the U.K., despite the fact that no monitoring programmes use this species at the present, random stickleback sampling (i.e., not systematically targeting impacted areas) has revealed the common presence of intersex fish (Figure 10.5). Other diagnostic features include the morphology of

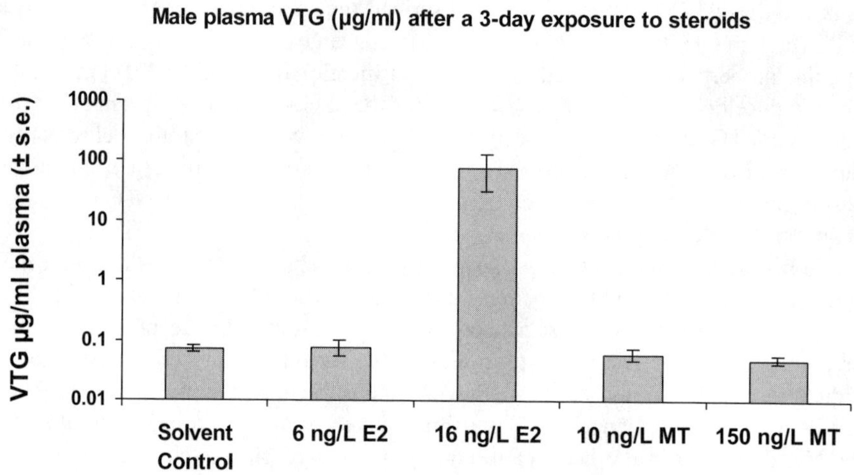

FIGURE 10.3 Plasma VTG in male sticklebacks after exposure to oestradiol (E2) and methyltestosterone (MT).

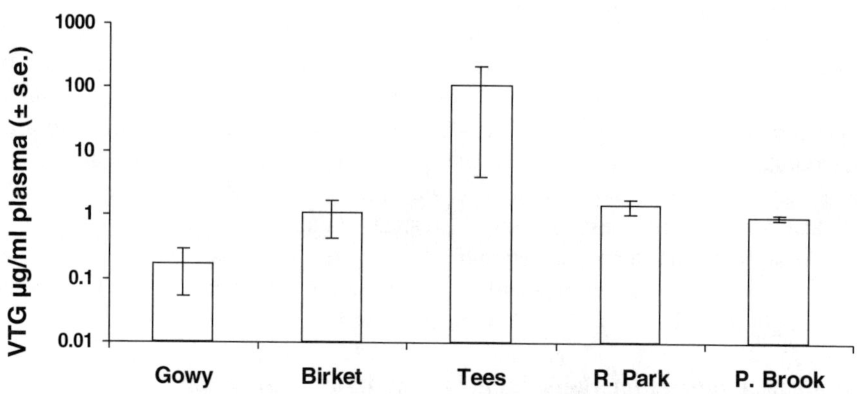

FIGURE 10.4 Plasma VTG in male sticklebacks caught in various U.K. river sites. Gowy was included as a control site of low environmental pollution.

spermatogonia, Sertoli and Leydig cells,[161] or malformations of the gonadal ducts.[162] If developing fish are exposed to an EDC during the critical period of sexual differentiation, then a complete sex reversal may occur, in which case gonadal histology alone may not show any adverse impact. This is where determination of the genetic sex is a major advantage. Intersex and total sex reversal occurred in juvenile sticklebacks exposed to high concentrations of oestradiol during the first 2 weeks of their development; oestradiol was able to elicit the same responses at later stages but not to the same degree as during the first 14 d post hatch.[163] Currently there are at least two published methods for molecular sex determination in the stickleback.[137,138] Another proposed sex marker that might have some value in EDCs research with sticklebacks is the dentition type.[164]

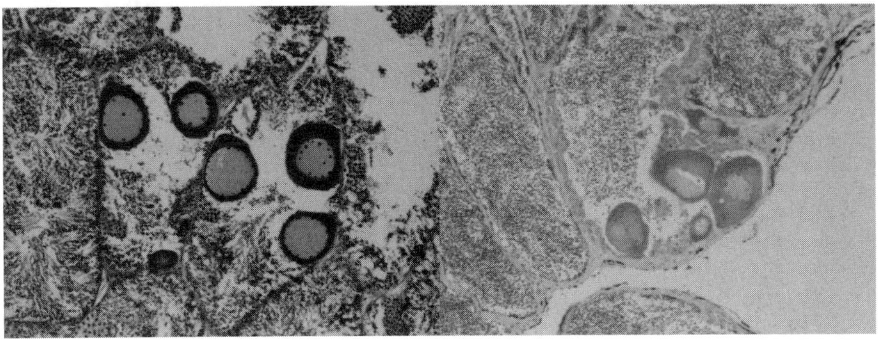

FIGURE 10.5 Oocytes in the testes of sticklebacks (intersex) caught in the wild.

We are currently developing a noninvasive method of measuring steroids in aquaria water (excreted by individual fish) and have found that breeding males excrete large quantities of 11-KT and androstenedione. We intend to apply this assay for the assessment of chemical impact on steroidogenesis. Numerous methods exist on plasma sex steroid determination; however, bleeding sticklebacks is a terminal procedure and thus cannot be used for sequential sampling.

The number of additional molecular biomarkers for EDCs include specific gene transcripts such as the oestrogen and androgen receptors, gonadotropins and steroidogenic enzymes. Of the plethora of enzymes involved in steroid biosynthesis pathway, aromatase, which is responsible for the conversion of C18 androgens to C19 oestrogens, is of primary importance. We have recently cloned in Cefas more than 80% of both CYP19α and β from the stickleback and are currently developing a quantitative PCR assay. The beta subunits of both gonadotropins (LH and FSH) in the stickleback have been sequenced and mRNA expression measured under different photoperiods and temperatures.[165] Two oestrogen receptors (ERα and β) have also been cloned recently in a collaborative effort between U.K. and Japan and real-time PCR assays were validated (Geoghegan, pers. comm.). In addition, one androgen receptor from the stickleback kidney has been sequenced,[148] and a second one was recently cloned in Japan (Nagae, pers. comm.). Higher-throughput approaches than single-gene targets, however, are desirable and to our knowledge the first stickleback 12 k DNA microarray was printed in Birmingham University, U.K., only recently. With this and future tools, the identification of pathways and mechanisms of endocrine disruption and toxicology in general will accelerate. The potential of the "omics" technologies in the unravelling of interactions between fish and environmental chemicals is widely recognised. In Cefas we are developing comprehensive two-dimensional maps of the liver and brain proteomes for both healthy/diseased or control/exposed fish. With this technique, we hope to identify novel markers of exposure and mechanistic information on physiological processes.

Alterations of reproductive behaviour after chemical exposure can be presented as an integrative approach, but also as specific to EDCs. This is especially true for the male stickleback, where the typical phases of reproductive behaviour (nest building, courtship, parental care) have been linked with specific hormonal signals.[166,167] We have only recently exploited the value of behaviour as a specific EDC biomarker and our preliminary results suggest that antiandrogens inhibit not only spiggin synthesis but also nest-building and courtship behaviour. In another study it was found that exposure to oestrogens early in life significantly decreased the number of nests built by adult males (Maunder, pers. comm.).

Finally, it is possible that the adverse effects of environmental contaminants on aquatic wildlife may be mediated via their deleterious effects on gamete quality. The impact of putative EDCs on sperm quality (sperm motility) is presently being investigated in the stickleback using computer-assisted sperm analysis (CASA). Preliminary studies have indicated that exposure to oestrogenic contaminants can impair sperm motility *in vitro* (Mayer, pers. comm.), opening the possibility of using sperm motility as a novel biomarker of endocrine disruption.

10.4.2.4 Screening Tests for EDCs

The OECD endocrine disruptor activity was initiated in November 1996. A task force on endocrine disrupter testing and assessment (EDTA) was established in 1997 to oversee the programme. The initial framework consisted of three tiers — initial assessment, screening, and testing.[168] So far, the most popular species for screening, and testing endocrine disrupters in fish are the fathead minnow (*Pimephales promelas*), the zebra fish (*Danio rerio*), and the medaka (*Oryzias latipes*). In view of the absence of all these species in the European environment, the U.K. in 2004 suggested the inclusion of the stickleback as an additional species for the prevalidation studies.

The OECD is currently in the process of validating a fish screening guideline for endocrine disrupters (phase 1). Several potential screening and testing assays were identified at an expert consultation on fish-testing needs, which needed formal validation. To address this, the OECD secretariat established a Validation Management Group for Ecotoxicological Test Methods for Endocrine Disrupters (VMG-Eco), with the remit to oversee the work on validation of EDC fish tests as well as all other environmental species of concern. The tests that were put forward for consideration as screening tests included fish juvenile growth, gonadal recrudescence assay, sex-reversal assay, and adult terminal reproduction test. The last test was chosen to be trialled first in an intercalibration exercise during the phase 1a validation of the fish-screening assay, using a potent oestrogen (oestradiol) and a potent androgen (trenbolone) over a 21-d exposure period of sexually mature fish. Essentially, the core endpoints used in this screen are VTG induction, gross morphology (including secondary sexual characteristics where relevant, gonadosomatic index [GSI]), and gonadal histopathology. The Department for Environment, Food and Rural Affairs (DEFRA) in its role as the National Coordinator of the OECD Test Guidelines Programme for the U.K. funded a small ring test involving the stickleback in which two U.K. laboratories (Cefas and CEH) along with Bergen University in Norway participated. The results from the interlaboratory exposures of sticklebacks proved highly relevant, sensitive (particularly in respect to trenbolone), and reproducible.

A second intercalibration exercise (phase 1b) using a weak oestrogen, an aromatase inhibitor, and an antiandrogen has just been completed for the three core species; however, none of the core species could detect the antiandrogenic activity of flutamide that was used as the model weak antiandrogen. Although the female fathead minnow develops nuptial tubercles in response to exogenous androgens, the sensitivity of this endpoint and the ability of different laboratories to measure it quantitatively, repeatably, and reproducibly is questionable. The only other androgen endpoint that has potential among the core species is the development of anal fin papillary processes in female Japanese medaka. In the initial phase of screen validation (Phase 1a), the three Japanese laboratories testing the screening protocol with this species found this endpoint to be relevant and reproducible.

Phase 1b of the screening assay for EDCs is just about to start for the stickleback. The participating laboratories this time will be Cefas and CEH in the U.K. and Wildlife International in the U.S. Completion of phase 1b of the screening assay is essential to fully align the stickleback with the other species. Some minor modifications were made to the OECD guideline to allow nesting and spawning of different males within one tank. For this purpose, each 40 l tank is separated into one "female" and one "male" compartment. The male compartment is further divided into five small areas to avoid aggression between male fish (Figure 10.6). The exposure duration is also 21 d long, nesting material is provided on day ten, and mating with females is attempted on day 20 and 21. We have already tested this design and found that fenitrothion inhibited not only spiggin synthesis in the male kidneys but also nest-building and courtship activities.

10.5 SUMMARY

The three-spined stickleback (*Gasterosteus aculeatus*) offers great potential for assessing reproductive disturbances caused by androgenic xenobiotics because of its pronounced androgen-dependent male secondary sexual characters (nuptial colouration, kidney hypertrophy, and territorial and nest-building behaviour). To date,

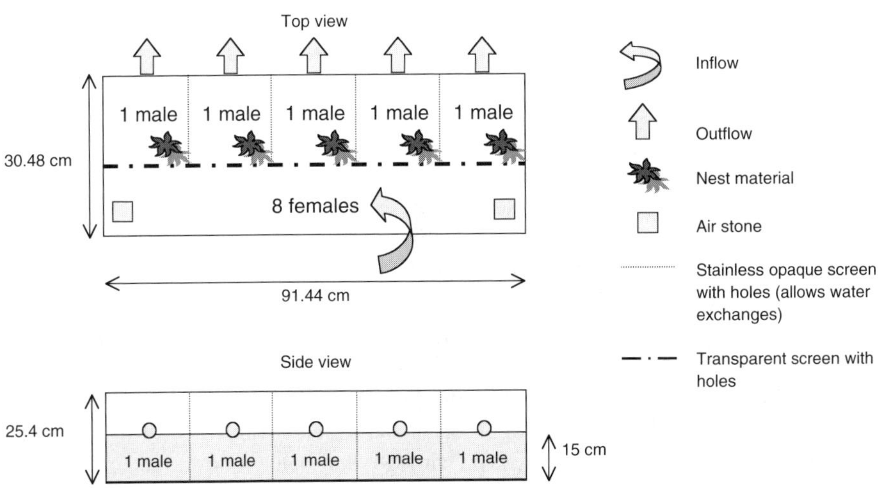

FIGURE 10.6 Schematic representation of the experimental setup for screening endocrine-disrupting chemicals.

spiggin is the only androgen-induced protein that has been isolated from fish. Research at Cefas has firmly established that female stickleback kidneys can produce spiggin in response to androgens added to ambient water.

Although the detection of androgenic compounds in the environment is restricted to relatively few situations (marinas, sewage effluents with primary treatment only, pulp mill effluents), compounds with antiandrogenic activity appear to be widespread in wastewater.[109] The biological impact of antiandrogenic xenobiotics has not been studied in detail, despite the fact that there is an increasing concern for clinical implications of these chemicals in humans.[87]

Spiggin production is inhibited in males (and in androgen-stimulated females) by the model antiandrogen flutamide; linuron, fenitrothion, and vinclozolin have also been tested and showed clear inhibition of spiggin production.[39] In contrast, exposure to flutamide of the three species used in the OECD guidelines for screening EDCs (zebra fish, medaka, fathead minnow) failed to produce reproducible and measurable effects.

A homologue ELISA for stickleback VTG has been developed,[139] validated, and applied in a number of studies including a small ring test involving three laboratories.

In addition to spiggin and VTG, the stickleback has molecular sex markers that can be used to determine the genetic sex of fish in the field and in the laboratory after EDC exposure during the critical periods of sexual differentiation. This tool is not commonly present in other fish species.

Intersex sticklebacks have been reported on numerous occasions despite the fact that no monitoring programmes use this species at the present.

Perhaps the most interesting aspect of stickleback biology is its reproductive behaviour, which has been the subject of extensive research for over 100 years. The typical male behaviour can be divided into three major phases: namely, nest-building, courtship, and parental care. All three phases are under the control of environmental and hormonal factors and can therefore be potentially affected by EDCs. Behavioural endpoints can be easily incorporated during EDC-exposure, providing further information in addition to biomarker data. We are currently designing full life cycle exposure tests that are looking for the most sensitive endpoints of EDC exposure.

In addition to the existing and potential markers for EDCs listed here, the species provides more advantages as a sentinel organism, as reported in a review by Pottinger et al.[61] Most importantly, sticklebacks are present in most aquatic environments (rivers, lakes, estuaries, coastal waters). They are present everywhere in the northern hemisphere and can bring together laboratory and field observations that could answer major questions related to the effects of pollutants on fish populations. Sticklebacks even occur in the Arctic (including Spitzbergen) and potentially can become an important sentinel species for identifying environmental disturbance in the more sensitive Arctic ecosystems. This may be very relevant with respect to the proposed oil and gas exploration and production efforts in Artic regions. Another important factor in this rapid recognition of the species' value is that most aspects of their reproductive endocrinology, physiology, and behaviour are very well documented (see Chapter 7).

The lack of substantial molecular resources has been so far one of the inhibiting factors in the acceptance of the species as a model organism, because molecular tools are of paramount importance in the understanding of mechanisms and pathways of biological responses. However, owing to the current rapid growth of available genomic information, it is anticipated that the species will develop into a prominent model in the field of ecotoxicology. The 675-megabase stickleback genome is quite compact for a vertebrate, less than one quarter the size of a typical mammalian genome; thus intron size is expected to be minimum and annotation fast.

A number of microsatellite markers able to discriminate between stickleback populations are already published. The combination of microsatellite and allozyme markers is a very promising tool to assess the impact of pollution on genetic variability. Therefore, the species can potentially provide important information on genetic adaptation to polluted habitats and the effects of this selection on fitness and biodiversity.

ACKNOWLEDGMENTS

I gratefully acknowledge financial support from the Department for Environment, Food and Rural Affairs (DEFRA) under contract A1149 and, in particular, Paul Leonard from Science Directorate.

I also want to thank Professor A.P. Scott, who first recognised the potential of the stickleback as a biomarker for androgenic xenobiotics, for his constant scientific and moral support.

REFERENCES

1. Foster, S.A., Baker, J.A., and Bell, M.A., The case for conserving threespine stickleback populations: protecting an adaptive radiation, *Fisheries*, 28(5), 10, 2003.
2. Wood, P.M., Will Canadian policies protect British Columbia's endangered pairs of sympatric sticklebacks?, *Fisheries*, 28(5), 19, 2003.
3. Blahm, T.H. and Snyder, G.R., Effect of increased water temperatures on survival of adult threespine stickleback and juvenile yellow perch in the Columbia River, *Northwest Sci.*, 49, 267, 1975.
4. Hopkins, K., The effects of global warming on the freshwater fish, Gasterosteus aculeatus (L.) (the three-spined stickleback), in *4th Symposium for European Freshwater Sciences,* Krakow, Poland, 2005, p. 79.
5. Sneddon, L.U. and Yerbury, J., Differences in response to hypoxia in the three-spined stickleback from lotic and lentic localities: dominance and an anaerobic metabolite, *J. Fish Biol.*, 64, 799, 2004.
6. Jones, J.R.E., The relation between the electrolytic solution pressures of the metals and their toxicity to the stickleback (*Gasterosteus aculeatus* L.), *J. Exp. Biol.*, 16(4), 425, 1939.
7. Jones, J.R.E., The oxygen consumption of *Gasterosteus aculeatus* L. in toxic solutions, *J. Exp. Biol.*, 23, 298, 1947.
8. Jones, J.R.E., A further study of the reactions of fish to toxic solutions, *J. Exp. Biol.*, 25, 22, 1948.
9. Costa, H.H., Responses of freshwater animals to sodium cyanide solutions I. Fish, *Ceylon J. Sci. Biol. Sci.*, 5(2), 41, 1965.

10. Hazel, C.R., Thomsen, W., and Meith, S.J., Sensitivity of striped bass and stickleback to ammonia in relation to temperature and salinity, *Calif. Fish Game*, 57(3), 138, 1971.
11. Smith, L.R., Holsen, T.M., Ibay, N.C., Block, R.M., and de Leon, A.B., Studies on the acute toxicity of fluoride ion to stickleback, fathead minnow, and rainbow trout, *Chemosphere*, 14(9), 1383, 1985.
12. Jop, K.M., Parkerton, T.F., Rodgers, J.H.J., Dickson, K.L., and Dorn, P.B., Comparative toxicity and speciation of two hexavalent chromium salts in acute toxicity tests, *Environ. Toxicol. Chem.*, 6(9), 697, 1987.
13. Pascoe, D. and Mattey, D.L., Studies on the toxicity of cadmium to the three-spined stickleback *Gasterosteus aculeatus* L., *J. Fish Biol.*, 11(2), 207, 1977.
14. Woodworth, J. and Pascoe, D., Cadmium update and distribution in sticklebacks related to the concentration and method of exposure, *Ecotoxicol. Environ. Saf.*, 7(6), 525, 1983.
15. Oronsaye, J.A.O. and Brafield, A.E., The effect of dissolved cadmium on the chloride cells of the gills of the stickleback, *Gasterosteus aculeatus* L., *J. Fish Biol.*, 25(2), 253, 1984.
16. Oronsaye, J.A.O., Histological changes in the kidneys and gills of the stickleback *Gasterosteus aculeatus* L. exposed to dissolved cadmium in hard water, *Ecotoxicol. Environ. Saf.*, 17, 279, 1989.
17. Woodworth, J. and Pascoe, D., Induction of cadmium-binding protein in the three-spined stickleback, *Aquat. Toxicol.*, 3(2), 141, 1983.
18. Jones, J.R.E., The relative toxicity of salts of lead, zinc and copper to the stickleback (*Gasterosteus aculeatus* L.) and the effect of calcium on the toxicity of lead and zinc salts, *J. Exp. Biol.*, 15(3), 394, 1938.
19. Matthiessen, P. and Brafield, A.E., The effects of dissolved zinc on the gills of the stickleback *Gasterosteus aculeatus* (L.), *J. Fish Biol.*, 5(5), 607, 1973.
20. Matthiessen, P. and Brafield, A.E., Uptake and loss of dissolved zinc by the stickleback *Gasterosteus aculeatus* L., *J. Fish Biol.*, 10(4), 399, 1977.
21. Deutch, B., Borg, B., Kloster, L., Meyer, H., and Moller, M.M., The accumulation of 65Zn by various marine organisms, *Ophelia* (Suppl. 1), 235, 1980.
22. Lemma, A. and Yau, P., Studies on the molluscicidal properties of endod (*Phytolacca dodecandra*): II. Comparative toxicity of various molluscicides to fish and snails, *Ethiop. Med. J.*, 12(3), 109, 1974.
23. Svecevicius, G. and Vosyliene, M.Z., Acute toxicity of copper to common freshwater fishes of Lithuania, *Ekologija*, 2, 17, 1996.
24. Sanchez, W., Palluel, O., Meunier, L., Coquery, M., Porcher, J.M., and Aït-Aïssa, S., Copper-induced oxidative stress in three-spined stickleback: relationship with hepatic metal levels, *Environ. Toxicol. Pharmacol.*, 19(1), 177, 2005.
25. Gravenmier, J.J., Johnston, D.W., Santore, R.C., and Arnold, W.R., Acute toxicity of copper to the threespine stickleback, *Gasterosteus aculeatus*, *Environ. Toxicol.*, 20(2), 150, 2005.
26. Bervoets, L., Blust, R., and Verheyen, R., Accumulation of metals in the tissues of three spined stickelback (*Gasterosteus aculeatus*) from natural fresh waters, *Exotoxicol. Environ. Saf.*, 48, 117, 2001.
27. Falandysz, J. and Kowalewska, M., Mercury concentration of stickleback *Gasterosteus aculeatus* from the Gulf of Gdansk, *Bull. Environ. Contam. Toxicol.*, 51, 710, 1993.

28. van den Dikkenberg, R.P., Canton, H.H., Mathijssen-Spiekman, L.A.M., and Roghair, C.J., The Usefulness of *Gasterosteus aculeatus* — The Three-Spined Stickleback — As a Test Organism in Routine Toxicity Testing, Rijksinstitut voor de Volksgezondheid en Milieuhygiene, Bilthoven, The Netherlands, 1989.
29. Henriksson, J., Johanisson, A., Bergqvist, P.A., and Norrgren, L., The toxicity of organoarsenic-based warfare agents: *in vitro* and *in vivo* studies, *Arch. Environ. Contam. Toxicol.*, 30(2), 213, 1996.
30. Katz, M., Acute toxicity of some organic insecticides to three species of salmonids and to the threespine stickleback, *Trans. Am. Fish. Soc.*, 90(3), 264, 1961.
31. van Wijngarden, R., Leeuwangh, P., Lucassen, W.G.H., Romijn, K., Ronday, R., van der Velde, R., and Willigenburg, W., Acute toxicity of chlorpyrifos to fish, a newt, and aquatic invertebrates, *Bull. Environ. Contam. Toxicol.*, 51(5), 716, 1993.
32. Deener, J.W., Bioconcentration of chlorpyrifos by the three-spined stickleback under laboratory and field conditions, *Chemosphere*, 29, 1561, 1994.
33. Ernst, W.R., Jonah, P., Doe, K., Julien, G., and Hennigar, P., Toxicity to aquatic organisms of off-target deposition of endosulfan applied by aircraft, *Environ. Toxicol. Chem.*, 10(1), 103, 1991.
34. Geyer, H.J., Scheuner, R., Bruggemann, R., Matthies, M., Steinberg, C.E.W., Zitko, V., Kettrup, A., and Garrison, W., The relevance of aquatic organisms' lipid content to the toxicity of lipophilic chemicals: toxicity of lindane to different fish species, *Exotoxicol. Environ. Saf.*, 28(1), 53, 1994.
35. Whale, G., Sheahan, D., and Matthiessen, P., The toxicity of tecnazene, a potato sprouting inhibitor, to freshwater fauna, *Chemosphere*, 17(6), 1205, 1988.
36. Girling, A.E., Tattersfield, L.J., Mitchell, G., Pearson, N., Woodbridge, A.P., and Bennett, D., Development of methods to assess the effects of xenobiotics in outdoor artificial streams, *Exotoxicol. Environ. Saf.*, 45(1), 1, 2000.
37. Sturm, A., Wogram, J., Hansen, P.D., and Liess, M., Potential use of cholinesterase in monitoring low levels of organophosphates in small streams: natural variability in three-spined stickleback (*Gasterosteus aculeatus*) and relation to pollution, *Environ. Toxicol. Chem.*, 18(2), 194, 1999.
38. Ernst, W.R. and Doe, K.G., A comparison of the aquatic toxicity of fenitrothion flowable and fenitrothion liquid technical formulations, *Water Pollut. Res. J. Can.*, 24(4), 553, 1989.
39. Katsiadaki, I., Morris, S., Squires, C., Hurst, M.R., James, J.D., and Scott, A.P., A sensitive, *in vivo* test for the detection of environmental antiandrogens, using the three-spined stickleback (*Gasterosteus aculeatus*), *Environ. Health Perspect.*, doi:10.1289/ehp.8063., 2006, in press.
40. van Bavel, B., Andersson, P., Wingfors, H., Ahgren, J., Bergqvist, P.A., Norrgren, L., Rappe, C., and Tysklind, M., Multivariate modelling of PCB bioaccumulation in the three-spined stickleback (*Gasterosteus aculeatus*), *Environ. Toxicol. Chem.*, 15(6), 947, 1996.
41. Andersson, P.L., Berg, A.H., Bjerselius, R., Norrgren, L., Olsén, H., Olsson, P.E., Örn, S., and Tysklind, M., Bioaccumulation of selected PCBs in zebrafish, three-spined stickleback, and arctic char after three different routes of exposure, *Arch. Environ. Contam. Toxicol.*, 40, 519, 2001.
42. Falandysz, J., Strandberg, B., Strandberg, L., and Rappe, C., Polychlorinated naphtalenes in three-spined stickleback *Gasterosteus aculeatus* from the Gulf of Gdansk, *Chemosphere*, 37, 2473, 1998.

43. Holm, G., Norrgren, L., Andersson, T., and Thuren, A., Effects of exposure to food contaminated with PBDE, PCN or PCB on reproduction, liver morphology and cytochrome P450 activity in the three-spined stickleback, *Gasterosteus aculeatus*, *Aquat. Toxicol.*, 27, 33, 1993.
44. Moles, A., Rice, S.D., and Korn, S., Sensitivity of Alaskan freshwater and anadromous fishes to Prudhoe Bay crude oil and benzene, *Trans. Am. Fish. Soc.*, 108(4), 408, 1979.
45. van Wezel, A.P. and Jonker, M.T.O., Use of the lethal body burden in the risk quantification of field sediments: influence of temperature and salinity, *Aquat. Toxicol.*, 42, 287, 1998.
46. Wong, K., Addison, R.F., and Law, F.C.P., Uptake, metabolism, and elimination of diphenyl ether by trout and stickleback, *Bull. Environ. Contam. Toxicol.*, 26(2), 243, 1981.
47. Nagell, B., Notini, M., and Grahn, O., Toxicity of four oil dispersants to some animals from the Baltic Sea, *Mar. Biol.*, 28(4), 237, 1974.
48. Drewa, G., Andruszczak, D., Chesy, M., Kozica-Raszeja, L., Palgen, K., and Zbytniewski, Z., The influence of anionic detergent and diesel fuel oil on arylsulphatase and acid phosphatase activity in *Gasterosteus aculeatus* L. and Dreissena polymorpha Pall, *Pol. Arch. Hydrobiol.*, 39, 161, 1992.
49. Drewa, G., Chesy, M., Brukwicka, A., Grzanka, A., and Sujkowska, R., Morphological changes of liver cells of *Gasterosteus aculeatus* L. exposed to pollutants, *Pol. Arch. Hydrobiol.*, 41(4), 507, 1994.
50. Le Gore, R.S. and Des Voigne, D.M., Absence of acute effects on threespine sticklebacks (*Gasterosteus aculeatus*) and coho salmon (*Oncorhynchus kisutch*) exposed to resuspended harbour sediment contaminants, *J. Fish Res. Board Can.*, 30, 1240, 1973.
51. Holm, G., Norrgren, L., and Linden, O., Reproductive and histopathological effects of long-term experimental exposure to bis(tributyltin)oxide (TBTO) on the three-spined stickleback, *Gasterosteus aculeatus* Linnaeus, *J. Fish Biol.*, 38, 373, 1991.
52. Falandysz, J., Brzostowski, A., Szpunar, J., and Rodriguez-Pereiro, I., Butyltins in sediments and three-spined stickleback (*Gasterosteus aculleatus*) from the marinas of the Gulf of Gdansk, Baltic Sea, *J. Environ. Sci. Health. A*, 37(3), 353, 2002.
53. Ekelund, R., Bergman, A., Granmo, A., and Berggren, M., Bioaccumulation of 4-nonylphenol in marine animals — a re-evaluation, *Environ. Pollut.*, 64, 107, 1990.
54. Roghair, C.J., Buijze, A., and Schoon, H.N.P., Ecotoxilogical risk evaluation of the cationic fabric softener DTDMAC. I. Ecotoxilogical effects, *Chemosphere*, 24(5), 599, 1992.
55. Khan, R.A., Parasitism in marine fish after chronic exposure to petroleum hydrocarbons in the laboratory and to the Exxon Valdez oil spill, *Bull. Environ. Contam. Toxicol.*, 44(5), 759, 1990.
56. Lafferty, K.D. and Kuris, A.M., How environmental stress affects the impacts of parasites, *Limnol. Oceanogr.*, 44 (3, part 2), 925, 1999.
57. Lehtinen, K.J., Survival, growth and disease of three-spined stickleback, *Gasterosteus aculeatus* L., brood exposed to bleached kraft mill effluents (BKME) in mesocosms, *Ann. Zool. Fenn.*, 26, 133, 1989.
58. Axelsson, B. and Norrgren, L., Parasite frequency and liver anomalies in three-spined stickleback, *Gasterosteus aculeatus* (L.), after long-term exposure to pulp mill effluents in marine mesocosms, *Arch. Environ. Contam. Toxicol.*, 21, 505, 1991.
59. Yeomans, W.E., Chubb, J.C., and Sweeting, R.A., Use of protozoan communities for pollution monitoring, *Parassitologia*, 39, 201, 1997.

60. Handy, R.D., Runnalls, T., and Russell, P.M., Histopathologic biomarkers in three spined sticklebacks, *Gasterosteus aculeatus*, from several rivers in Southern England that meet the freshwater fisheries directive, *Ecotoxicology*, 11(6), 467, 2002.
61. Pottinger, T.G., Carrick, T.R., and Yeomans, W.E., The three-spined stickleback as an environmental sentinel: effects of stressors on whole-body physiological indices, *J. Fish Biol.*, 61, 207, 2002.
62. Ali, M. and Wootton, R.J., Correlates of growth in juvenile three-spined sticklebacks: potential predictors of growth rates in natural populations, *Ecol. Freshwater Fish*, 12, 87, 2003.
63. Clotfelter, E.D., Bell, A.M., and Levering, K.R., The role of animal behaviour in the study of endocrine-disrupting chemicals, *Anim. Behav.*, 68(665–676), 2004.
64. Zala, S.M. and Penn, D.J., Abnormal behaviours induced by chemical pollution: a review of the evidence and new challenges, *Anim. Behav.*, 68(4), 649, 2004.
65. Scott, G.R. and Sloman, K.A., The effects of environmental pollutants on complex fish behaviour: integrating behavioural and physiological indicators of toxicity, *Aquat. Toxicol.*, 68, 369, 2004.
66. Craig, S. and Laming, P., Behaviour of the three-spined stickleback, *Gasterosteous aculeatus* (Gasterosteidae, Teleostei) in the multispecies freshwater biomonitor: a validation of automated recordings at three levels of ammonia pollution, *Water Res.*, 38, 2144, 2004.
67. Wibe, Å.E., Billing, A., Rosenqvist, G., and Jenssen, B.M., Butyl benzyl phthalate affects shoaling behavior and bottom-dwelling behavior in threespine stickleback, *Environ. Res. A*, 89(2), 180, 2002.
68. Wibe, Å.E., Fjeld, E., Rosenqvist, G., and Jenssen, B.M., Postexposure effects of DDE and butylbenzylphthalate on feeding behaviour in threespine stickleback, *Exotoxicol. Environ. Saf.*, 57(2), 213, 2004.
69. Bell, A.M., Effects of an endocrine disrupter on courtship and aggressive behaviour of male three-spined stickleback, *Gasteroteus aculeatus*, *Anim. Behav.*, 62, 775, 2001.
70. Bell, A.M., An endocrine disrupter increases growth and risky behavior in threespined stickleback (*Gasterosteus aculeatus*), *Horm. Behav.*, 45(2), 108, 2004.
71. Wibe, Å.E., Rosenqvist, G., and Jenssen, B.M., Disruption of male reproductive behavior in threespine stickleback *Gasterosteus aculeatus* exposed to 17ß-Estradiol, *Environ. Res. A*, 90(2), 136, 2002.
72. Roy, N.K., Kreamer, G.L., Konkle, B., Grunwald, C., and Wirgin, I., Characterization and prevalence of a polymorphism in the 3' untranslated region of cytochrome P4501A1 in cancer-prone Atlantic tomcod, *Arch. Biochem. Biophys.*, 322(1), 204, 1995.
73. Cooper, P.S., Vogelbein, W.K., and van Veld, P.A., Altered expression of the xenobiotic transporter P-glycoprotein in liver and liver tumours of mummichog (*Fundulus heteroclitus*) from a creosote-contaminated environment, *Biomarkers*, 4(1), 48, 1999.
74. Meyer, J. and Di Giulio, R., Patterns of heritability of decreased EROD activity and resistance to PCB 126-induced teratogenesis in laboratory-reared offspring of killifish (*Fundulus heteroclitus*) from a creosote-contaminated site in the Elizabeth River, VA, *Mar. Environ. Res.*, 54, 621, 2002.
75. Meyer, J.N., Nacci, D.E., and Di Giulio, R.T., Cytochrome P4501A (CYP1A) in killifish (*Fundulus heteroclitus*): heritability of altered expression and relationship to survival in contaminated sediments, *Toxicol. Sci.*, 68(1), 69, 2002.
76. Klepaker, T., Morphological changes in a marine population of threespined stickleback, *Gasterosteus aculeatus*, recently isolated in fresh water, *Can. J. Zool.*, 71, 1231, 1993.

77. Ziuganov, V.V., Reproductive isolation among lateral plate phenotypes (low, partial, complete) of the threespine stickleback, *Gasterosteus aculeatus*, from the White Sea basin and the Kamchatka peninsula, Russia, *Behaviour*, 132, 15, 1995.
78. Kristjánsson, B.K., Skúlason, S., and Noakes, D.L.G., Rapid divergence in a recently isolated population of threespine stickleback (*Gasterosteus aculeatus* L.), *Evol. Ecol. Res.*, 4, 659, 2002.
79. Hendry, A.P., Taylor, E.B., and McPhail, J.D., Adaptive divergence and the balance between selection and gene flow: lake and stream stickleback in the misty system, *Evolution*, 56(6), 1199, 2002.
80. Avise, J.C., Systematic value of electrophoretic data, *Syst. Zool.*, 23, 465, 1974.
81. Buth, D.G. and Haglund, T.R., Allozyme variation in the Gasterosteus aculeatus complex, in *The Evolutionary Biology of the Threespine Stickleback*, Bell, M.A. and Foster, S.A., Eds., Oxford University Press, Oxford, 1994, chap. 3.
82. Rico, C., Kuhnlein, U., and FitzGerald, G.J., Male reproduction tactics in the threespine stickleback: an evaluation by DNA fingerprinting, *Mol. Ecol.*, 1(79–87), 1992.
83. Rico, C., Zadworny, D., Kuhnlein, U., and Fitzerald, G.J., Characterization of hypervariable microsatellite loci in the threespine stickleback *Gasterosteus-aculeatus*, *Mol. Ecol.*, 2(4), 271, 1993.
84. Taylor, E.B., Microsatellites isolated from the threespine stickleback *Gasterosteus aculeatus*, *Mol. Ecol.*, 7(7), 930, 1998.
85. Largiader, C.R., Fries, V., Kobler, B., and Bakker, T.C., Isolation and characterization of microsatellite loci from the three-spined stickleback (*Gasterosteus aculeatus* L.)., *Mol. Ecol.*, 8(2), 342, 1999.
86. Peichel, C.L., Nereng, K.S., Ohgi, K.A., Cole, B.L.E., Colosimo, P.F., Buerkle, C.A., Schluter, D., and Kingsley, D.M., The genetic architecture of divergence between threespine stickleback species, *Nature*, 414(6866), 901, 2001.
87. Heckel, G., Zbinden, M., Mazzi, D., Kohler, A., Reckeweg, G., Bakker, T.C.M., and Largiadèr, C.R., Microsatellite markers for the three-spined stickleback (*Gasterosteus aculeatus* L.) and their applicability in a freshwater and an anadromous population, *Conserv. Gen.*, 3, 79, 2002.
88. DeWoody, J.A. and Avise, J.C., Microsatellite variation in marine, freshwater and anadromous fishes compared with other animals, *J. Fish Biol.*, 56, 461, 2000.
89. DeWoody, J.A. and Avise, J.C., Genetic perspectives on the natural history of fish mating systems, *J. Hered.*, 92(2), 167, 2001.
90. Reusch, T.B., Haberli, M.A., Aeschlimann, P.B., and Milinski, M., Female sticklebacks count alleles in a strategy of sexual selection explaining MHC polymorphism, *Nature*, 414(6861), 262, 2001.
91. Vos, J.G., Dybing, E., Greim, H.A., Ladefoged, O., Lambré, C., Tarazona, J.V., Brandt, I., and Vethaak, A.D., Health effects of endocrine-disrupting chemicals on wildlife, with special reference to the European situation, *Clin. Rev. Toxicol.*, 30(1), 71, 2000.
92. Toppari, J., Larsen, J., Christiansen, P., Giwercman, J., Grandjean, P., Guillette, L.J.J., Jegou, B., Jensen, T., Jouannet, P., Keiding, N., Leffers, H., McLachlan, J., Meyer, O., Muller, J., Rajpert-De Meyts, E., Scheike, T., Sharpe, R., Sumpter, J., and Skakkebaek, N., Male reproductive health and environmental xenoestrogens, *Environ. Health Perspect.*, 104(4), 741, 1996.
93. Kelce, W.R. and Wilson, E.M., Environmental antiandrogens: developmental effects, molecular mechanisms, and clinical implications, *J. Mol. Med.*, 75, 198, 1997.

94. Vidaeff, A.C. and Sever, L.E., In utero exposure to environmental estrogens and male reproductive health: a systematic review of biological and epidemiologic evidence, *Reprod. Toxicol.*, 20, 5, 2005.
95. Sonnenschein, C. and Soto, A.M., An updated review of environmental estrogen and androgen mimics and antagonists, *J. Steroid Biochem. Mol. Biol.*, 65(1), 143, 1998.
96. Massaad, C., Entezami, F., Massade, L., Benahmed, M., Olivennes, F., Barouki, R., and Hamamah, S., How can chemical compounds alter human fertility?, *Eur. J. Obstet. Gynecol. Reprod. Biol.*, 100, 127, 2002.
97. Purdom, C.E., Hardiman, P.A., Bye, V.J., Eno, N.C., Tyler, C.R., and Sumpter, J.P., Estrogenic effects of effluents from sewage treatment works, *Chem. Ecol.*, 8, 275, 1994.
98. Sumpter, J.P. and Jobling, S., Vitellogenesis as a biomarker for oestrogenic contamination of the aquatic environment, *Environ. Health Perspect.*, 103 (Suppl. 7), 173, 1995.
99. Folmar, L.C., Denslow, N.D., Rao, V., Chow, M., Crain, D.A., Enblom, J., Marcino, J., and Guillette, L.J.J., Vitellogenin induction and reduced serum testosterone concentrations in feral male carp (*Cyprinus carpio*) captured near a major metropolitan sewage treatment plant, *Environ. Health Perspect.*, 104, 1096, 1996.
100. Allen, Y., Matthiessen, P., Scott, A.P., Haworth, S., Feist, S., and Thain, J.E., The extent of oestrogenic contamination in the U.K. estuarine and marine environments — further surveys of flounder, *Sci. Tot. Environ.*, 233, 5, 1999.
101. Denslow, N.D., Chow, M.C., Kroll, K.J., and Green, L., Vitellogenin as a biomarker of exposure for oestrogen and estrogen mimics, *Ecotoxicology*, 8, 385, 1999.
102. Kelce, W.R., Gray, L.E.J., and Wilson, E.M., Antiandrogens as environmental endocrine disruptors, *Reprod. Fertil. Dev.*, 10, 105, 1998.
103. Howell, W.M., Black, D.A., and Bortone, S.A., Abnormal expression of secondary sex characters in a population of mosquitofish, *Gambusia affinis holbrooki*: evidence for environmentally-induced masculinization, *Copeia*, 1980(4), 676, 1980.
104. Howell, W.M. and Denton, T.E., Gonopodial morphogenesis in female mosquitofish, *Gambusia affinis affinis*, masculinized by exposure to degradation products from plant sterols, *Environ. Biol. Fish*, 24, 43, 1989.
105. Cody, R.P. and Bortone, S.A., Masculinization of mosquitofish as an indicator of exposure to kraft mill effluent, *Bull. Environ. Contam. Toxicol.*, 58, 429, 1997.
106. Denton, T.E., Howell, W.M., McCollum, C.J., Marks, E.B., and Allison, J.J., Masculinization of female mosquitofish by exposure to plant sterols and *Mycobacterium smegmatis*, *Bull. Environ. Contam. Toxicol.*, 35, 627, 1985.
107. McMaster, M.E., van der Kraak, G.J., Portt, C.B., Munkittrick, K.R., Sibley, P.K., Smith, I.R., and Dixon, D.G., Changes in hepatic mixed-function oxidase (MFO) activity, plasma steroid levels and age at maturity of a white sucker (*Catostomus commersoni*) population exposed to bleached kraft pulp mill effluent, *Aquat. Toxicol.*, 21, 199, 1991.
108. van der Kraak, G.J., Munkittrick, K.R., McMaster, M.E., Portt, C.B., and Chang, J.P., Exposure to bleached kraft pulp mill effluent disrupts the pituitary-gonadal axis of White Sucker at multiple sites, *Toxicol. Appl. Pharmacol.*, 115, 224, 1992.
109. Munkittrick, K.R., van Der Kraak, G.J., McMaster, M.E., Portt, C.B., van den Heuvel, M.R., and Servos, M.R., Survey of receiving-water environmental impacts associated with discharges from pulp mills. 2. gonad size, liver size, hepatic EROD activity and plasma sex steroid levels in white sucker, *Environ. Toxicol. Chem.*, 13, 1089, 1994.

110. Sandström, O., Newman, E., and Karås, P., Effects of a bleached pulp mill effluent on growth and gonadal function in Baltic coastal fish, *Water Sci. Technol.*, 20(2), 107, 1988.
111. Andersson, T., Forlin, L., Harding, J., and Larsson, A., Physiological disturbances in fish living in coastal water polluted with bleached kraft mill effluents, *Can. J. Fish Aquat. Sci.*, 45, 1525, 1988.
112. Larsson, D.G., Hällman, H., and Förlin, L., More male fish embryos near a pulpmill, *Environ. Toxicol. Chem.*, 19, 2911, 2000.
113. Jenkins, R., Angus, R.A., McNatt, H., Howell, W.M., Kemppainen, J.A., Kirk, M., and Wilson, E.M., Identification of androstenedione in a river containing paper mill effluent, *Environ. Toxicol. Chem.*, 20(6), 1325, 2001.
114. Durhan, E.J., Lambright, C., Wilson, V., Butterworth, B.C., Kuehl, D.W., Orlando, E.F., Guillette, L.J.J., Gray, L.E., and Ankley, G.T., Evaluation of androstenedione as an androgenic component of river water downstream of a pulp and paper mill effluent, *Environ. Toxicol. Chem.*, 21(9), 1973, 2002.
115. Parks, L.G., Lambright, C.S., Orlando, E.F., Guillette, L.J.J., Ankley, G.T., and Gray, L.E.J., Masculinization of female mosquitofish in Kraft Mill effluent-contaminated Fenholloway River water is associated with androgen receptor agonist activity, *Toxicol. Sci.*, 62, 257, 2001.
116. Jegou, B., Sots, A., Sundlof, S., Stephany, R., Meyer, H., and Leffers, H., Existing guidelines for the use of meat hormones and other food additives in Europe and USA, *APMIS*, 103 (Suppl.), S551, 2001.
117. Wilson, V.S., Lambright, C., Ostby, J., and Gray, L.E.J., *In vitro* and *in vivo* effects of 17ß-Trenbolone: a feedlot effluent contaminant, *Toxicol. Sci.*, 70, 202, 2002.
118. Ankley, G.T., Jensen, K.M., Makynen, E.A., Kahl, M.D., Korte, J.J., Hornung, M.W., Henry, T.R., Denny, J.S., Leino, R.L., Wilson, V.S., Cardon, M.C., Hartig, P.C., and L.E., G., Effects of the androgenic growth promoter 17-beta-trenbolone on fecundity and reproductive endocrinology of the fathead minnow (*Pimephalas promelas*), *Environ. Toxicol. Chem.*, 22, 1350, 2003.
119. Thomas, K.V., Hurst, M.R., Matthiessen, P., McHugh, M., Smith, A., and Waldock, M.J., An assessment of *in vitro* androgenic activity and the identification of environmental androgens in United Kingdom estuaries, *Environ. Toxicol. Chem.*, 20(7), 1456, 2002.
120. Kirk, L.A., Tyler, C.R., Lye, C.M., and Sumpter, J.P., Changes in estrogenic and androgenic activities at different stages of treatment in wastewater treatment works, *Environ. Toxicol. Chem.*, 21(5), 972, 2002.
121. Evans, K. and Hill, E., Occurrence of oestrogenic, anti-oestrogenic, androgenic and anti-androgenic activity in effluents of wastewater treatment works in the U.K., in *SETAC Europe 16th annual meeting,* Lille, France, Society of Environmental Toxicology and Chemistry, 2005, p. 195.
122. Environment Agency, *Assessment of the (Anti-)Oestrogenic and (Anti-)Androgenic Activity of Sewage Treatment Works Final Effluents*, Environment Agency, Bristol, U.K., 2006, in press.
123. Ostby, J., Kelce, W.R., Lambright, C., Wolf, C.J., Mann, P., and Gray, L.E.J., The fungicide procymidone alters sexual differentiation in the male rat by acting as androgen-receptor antagonist *in vivo* and *in vitro*, *Toxicol. Ind. Health*, 15, 80, 1999.

124. Gray, L.E.J., Wolf, C., Lambright, C., Mann, P., Price, M., Cooper, R.L., and Ostby, J., Administration of potentially antiandrogenic pesticides (procymidone, linuron, iprodione, chlozolinate, p,p'-DDE, and ketoconazole) and toxic substances (dibutyl- and diethylexyl phthalate, PCB 169, and ethane dimethane sulphonate) during sexual differentiation produces diverse profiles of reproductive malformations in the male rat, *Toxicol. Ind. Health*, 15, 94, 1999.
125. Makynen, E.A., Kahl, M.D., Jensen, K.M., Tietge, J.E., Wells, K.L., van der Kraak, G.J., and Ankley, G.T., Effects of the mammalian antiandrogen vinclozolin on development and reproduction of the fathead minnow (*Pimephales promelas*), *Aquat. Toxicol.*, 48, 461, 2000.
126. Ashby, J. and Lefevre, P.A., The peripubertial male rat assay as a replacement for the Hershberger castrated male rat assay, *J. Appl. Toxicol.*, 20, 35, 2000.
127. Tamura, H., Maness, S.C., Reischmann, K., Dorman, D.C., Gray, L.E.J., and Gaido, K.W., Androgen receptor antagonism by the organophosphate insecticide fenitrothion, *Toxicol. Sci.*, 60, 56, 2001.
128. Sohoni, P., Lefevre, P.A., Ashby, J., and Sumpter, J.P., Possible androgenic/anti-androgenic activity of the fungicide fenitrothion, *J. Appl. Toxicol.*, 21, 173, 2001.
129. Hershberger, L.G., Shipley, E.G., and Meyer, R.K., Myotrophic activity of 19-nor-testosterone and other steroids determined by modified levator ani muscle method, *Proc. Soc. Exp. Biol. Med.*, 83, 175, 1953.
130. Matthiessen, P. and Gibbs, P.E., Critical appraisal of the evidence for tributyltin-mediated endocrine disruption in molluscs, *Environ. Toxicol. Chem.*, 17(1), 37, 1997.
131. Gibbs, P.E., Pascoe, P.L., and Burt, G.R., Sex change in the female dogwhelk *Nucella lapillus* induced by tributyltin from antifouling paints, *J. Mar. Biol. Assoc. U.K.*, 68, 715, 1988.
132. Spooner, N., Gibbs, P.E., Bryan, G.W., and Goad, L.J., The effect of tributyltin upon steroid titres in the female dogwhelk, *Nucella lapillus*, and the development of imposex, *Mar. Environ. Res.*, 32, 37–49, 1991.
133. Ronis, M.J.J. and Mason, A.Z., The metabolism of testosterone by the periwinkle (*Littorina littorea*) *in vitro* and *in vivo*: effects of tributyltin, *Mar. Environ. Res.*, 42, 161–166, 1996.
134. Katsiadaki, I., Scott, A.P., Hurst, M.R., Matthiessen, P., and Mayer, I., Detection of environmental androgens: a novel method based on enzyme-linked immunosorbent assay of spiggin, the stickleback (*Gasterosteus aculeatus*) glue protein, *Environ. Toxicol. Chem.*, 21(9), 1946, 2002.
135. Katsiadaki, I., Scott, A.P., and Mayer, I., The potential of the three-spined stickleback, *Gasterosteus aculeatus* L., as a combined biomarker for oestrogens and androgens in European waters, *Mar. Environ. Res.*, 54, 725, 2002.
136. Hahlbeck, E., Katsiadaki, I., Mayer, I., Adolfsson-Erici, M., James, J., and Bengtsson, B.E., The juvenile three-spined stickleback (*Gasterosteus aculeatus* L.) as a model organism for endocrine disruption II — kidney hypertrophy, vitellogenin and spiggin induction, *Aquat. Toxicol.*, 70(4), 311, 2004.
137. Griffiths, R., Orr, K.J., Adam, A., and Barber, I., DNA sex identification in the three-spined stickleback, *J. Fish Biol.*, 57, 1331, 2000.
138. Peichel, C.L., Ross, J.A., Matson, C.K., Dickson, M., Grimwood, J., Schmutz, J., Myers, R.M., Mori, S., Schluter, D., and Kingsley, D.M., The master sex-determination locus in threespine sticklebacks is on a nascent Y chromosome, *Curr. Biol.*, 14(16), 1416, 2004.

139. Jakobsson, S., Borg, B., Haux, C., and Hyllner, S.J., An 11-ketotestosterone induced kidney-secreted protein: the nest building glue from male three-spined stickleback, *Gasterosteus aculeatus*, *Fish Physiol. Biochem.*, 20(1), 79, 1999.
140. Jones, I., Lindberg, C., Jakobsson, S., Hellqvist, A., Hellman, U., Borg, B., and Olsson, P.E., Molecular cloning and characterization of spiggin: an androgen-regulated extraorganismal adhesive with structural similarities to Von Willebrand factor-related proteins, *J. Biol. Chem.*, 276, 17857, 2001.
141. Kawasaki, F., Katsiadaki, I., Scott, A.P., Matsubara, T., Osatomi, K., Soyano, K., Hara, A., Arizono, K., and Nagae, M., Molecular cloning of two subtypes of spiggin cDNA in the three-spined stickleback, *Gasterosteus aculeatus*, *Fish Physiol. Biochem.*, 28, 425, 2003.
142. Katsiadaki, I., Scott, A.P., and Matthiessen, P., The use of the three-spined stickleback as a potential biomarker for androgenic xenobiotics, in *Proceedings of the Sixth International Symposium on the Reproductive Physiology of Fish*, Norberg, B., Kjesbu, O.S., Taranger, G.L., Andersson E., and Stefansson, S.O, Eds., Bergen, Norway, 2000, p. 359.
143. Angus, R.A., McNatt, H.B., Howell, W.M., and Peoples, S.D., Gonopodium development in normal male and 11-ketotestosterone-treated female mosquitofish (*Gambusia affinis*): a quantitative study using computer image analysis, *Gen. Comp. Endocrinol.*, 123, 222, 2001.
144. Panter, G.H., Hutchinson, T.H., Hurd, K.S., Sherren, A., Stanley, R.D., and Tyler, C.R., Successful detection of (anti-)androgenic and aromatase inhibitors in pre-spawning adult fathead minnows (*Pimephales promelas*) using easily measured end-points of sexual development, *Aquat. Toxicol.*, 70, 11, 2004.
145. Bayley, M., Junge, M., and Baatrup, E., Exposure of juvenile guppies to three antiandrogens causes demasculinization and a reduced sperm count in adult males, *Aquat. Toxicol.*, 56, 227, 2002.
146. Kinnberg, K. and Toft, G., Effects of estrogenic and antiandrogenic compounds on the testis structure of the adult guppy (*Poecilia reticulata*), *Ecotoxicol. Environ. Saf.*, 54(1), 16, 2003.
147. Jakobsson, S., Mayer, I., Schulz, R.W., Blankenstein, M.A., and Borg, B., Specific binding of 11-ketotestosterone in an androgen target organ, the kidney of the male three-spined stickleback, *Gasterosteus aculeatus*, *Fish Physiol. Biochem.*, 15, 459, 1996.
148. Olsson, P.-E., Berg, A.H., von Hofsten, J., Grahn, B., Hellqvist, A., Larsson, A., Karlsson, J., Modig, C., Borg, B., and Thomas, P., Molecular cloning and characterization of a nuclear androgen receptor activated by 11-ketotestosterone, *Reprod. Biol. Endocrinol.*, 3(37), 2005.
149. Tyler, C.R., van Aerle, R., Hutchinson, T.H., Maddix, S., and Trip, H., An *in vivo* testing system for endocrine disruptors in fish early life stages using induction of vitellogenin, *Environ. Toxicol. Chem.*, 18, 337, 1999.
150. Holbeck, H., Andersen, L., Petersen, G.I., Korsgaard, B., Pedersen, K.L., and Bjerregaard, P., Development of an ELISA for vitellogenin in whole body homogenate of zebrafish (*Danio rerio*), *Comp. Biochem. Physiol.*, 130C, 119, 2001.
151. Panter, G.H., Hutchinson, T.H., Länge, R., Lye, C.M., Sumpter, J.P., Zerulla, M., and Tyler, C.R., Utility of a juvenile fathead minnow screening assay for detecting (anti)-estrogenic substances, *Environ. Toxicol. Chem.*, 21(2), 319, 2002.
152. Wootton, R.J., Evans, G.W., and Mills, L.A., Annual cycle in female three-spined sticklebacks (*Gasterosteus aculeatus* L) from an upland and lowland population, *J. Fish Biol.*, 12, 331, 1978.

153. Wallace, R.A. and Selman, K., Physiological aspects of oogenesis in two species of sticklebacks, *Gasterosteus aculeatus* and *Apeltes quadracus*, *J. Fish Biol.*, 14, 551, 1979.
154. Ollevier, F. and Covens, M., Oestradiol-induced female specific proteins in the plasma and in the oocytes of *Gasterosteus aculeatus F. trachurus*: an immunological approach, in *Proceedings of the International Symposium on Fish,* Richter, C.J.J., Goos, H.J.T., Eds., Wageningen, The Netherlands, 1982, p. 159.
155. Ollevier, F. and Covens, M., Vitellogenins in *Gasterosteus aculeatus*, *Ann. Soc. R. Zool. Bel.*, 113 (Suppl. 1), 327, 1983.
156. Covens, M., Covens, L., Ollevier, F., and de Loof, A., A comparative study of some properties of vitellogenin (Vg) and yolk proteins in a number of freshwater and marine teleost fishes, *Comp. Biochem. Physiol. B: Biochem. Mol. Biol.*, 88(1), 75, 1987.
157. Covens, M., Stynen, D., Ollevier, F., and de Loof, A., Concanavalin A reactivity of vitellogenin and yolk proteins of the three-spined stickleback *Gasterosteus aculeatus* (Teleostei), *Comp. Biochem. Physiol. B: Biochem. Mol. Biol.*, 90(1), 227, 1988.
158. Hutchinson, T.H., Ankley, G.T., Segner, H., and Tyler, C.R., Screening and testing for endocrine disruption in fish — biomarkers as signposts not traffic lights in risk assessment, *Environ. Health Perspect.*, doi:10.1289/ehp.8062, 2006, in press.
159. Borg, B. and van den Hurk, R., Oocytes in the testes of the three-spined stickleback, *Gasterosteus aculeatus*, *Copeia*, 1983(1), 259, 1983.
160. Gercken, J. and Holmer, S., Intersex in feral marine and freshwater fish from north-eastern Germany, *Mar. Environ. Res.*, 54, 651, 2002.
161. van der Ven, L.T.M., Wester, P.W., and Vos, J.G., Histopathology as a tool for the evaluation of endocrine disruption in zebrafish (*Danio rerio*), *Environ. Toxicol. Chem.*, 22(4), 908, 2003.
162. Rodgers-Gray, T.P., Jobling, S., Kelly, C., Morris, S., Brighty, G., Waldock, M.J., Sumpter, J.P., and Tyler, C.R., Exposure of juvenile roach (*Rutilus rutilus*) to treated sewage effluent induces dose-dependent and persistent disruption in gonadal duct development, *Environ. Sci. Technol.*, 35(3), 462, 2001.
163. Hahlbeck, E., Griffiths, R., and Bengtsson, B.E., The juvenile three-spined stickleback (*Gasterosteus aculeatus* L.) as a model organism for endocrine disruption. I. Sexual differentiation, *Aquat. Toxicol.*, 70(4), 287, 2004.
164. Caldecutt, W.J., Bell, M.A., and Buckland-Nicks, J.A., Sexual dimorphism and geographic variation in dentition of threespine stickleback, *Gasterosteus aculeatus*, *Copeia*, 2001(4), 936, 2001.
165. Hellqvist, A., Bornestaf, C., Borg, B., and Schmitz, M., Cloning and sequencing of the FSH-ß and LH ß-subunit in the three-spined stickleback, *Gasterosteus aculeatus*, and effects of photoperiod and temperature on LH-ß and FSH-ß mRNA expression, *Gen. Comp. Endocrinol.*, 135, 167, 2004.
166. Páll, M.K., Mayer, I., and Borg, B., Androgen and behaviour in the male three-spined stickleback, *Gasterosteus aculeatus* I. Changes in 11-ketotestosterone levels during the nesting cycle, *Horm. Behav.*, 41(4), 377, 2002.
167. Páll, M.K., Mayer, I., and Borg, B., Androgen and behaviour in the male three-spined stickleback, *Gasterosteus aculeatus* II. Castration and 11-ketoandrostenedione effects on courtship and parental care during the nesting cycle, *Horm. Behav.*, 42(3), 337, 2002.
168. Huet, M.C., OECD activity on endocrine disrupters test guidelines development, *Ecotoxicology*, 9, 77, 2000.

11 The Biology of Other Sticklebacks

Sara Östlund-Nilsson and Ian Mayer

CONTENTS

11.1 INTRODUCTION

Although the biology of sticklebacks has been extensively studied, resulting in over 2000 research papers, the vast majority of these publications relate solely to the three-spined stickleback, *Gasterosteus aculeatus*. It is surprising how little we know of the biology of the other stickleback species (family Gasterosteidae), considering the wealth of knowledge we now have on the three-spined stickleback, from the traditional fields such as behaviour and evolutionary biology to the more recently emerging fields such as comparative and functional genomics.

In this chapter, we aim to summarise studies performed, mainly within the fields of behavioural ecology and physiology, on some of the less-studied stickleback species. Whereas the phylogeny of the family Gasterosteidae is still a contentious issue, it is generally accepted that it contains five genera with at least six species,

FIGURE 11.1 Three species of sticklebacks. The species shown in the foreground are (from the top) the 15-spined stickleback (*Spinachia spinachia*), the nine-spined stickleback (*Pungitius pungitius*), and two individuals of the three-spined stickleback *Gasterosteus aculeatus*). (From Brehm, A.E., De Kallblodiga Ryggradsdjurens Lif. Norstedt, Stockholm, 1887.)

including species complexes[1–3] (see also Chapter 1). Of the five recognised gasterosteid genera, three (*Spinachia*, *Apeltes*, and *Culaea*) are generally monotypic and geographically restricted, whereas the remaining two, *Gasterosteus* and *Pungitius*, are morphologically strikingly variable and geographically widespread.

To date, there are only two recognised species in the genus *Gasterosteus*, the three-spined stickleback (*G. aculeatus*) and the black-spotted stickleback (*G. wheatlandi*). The number of species belonging to the genus *Pungitius* is still under discussion, with up to eight species or subspecies being recognised,[3] and of those this chapter will cover only the nine-spined stickleback, *P. pungitius*. The other species included in this chapter are the brook stickleback, *Culaea inconstans*, the four-spined stickleback, *Apeltes quadracus*, and what is generally considered to be the most primitive of the sticklebacks, the 15-spined stickleback, *Spinachia spinachia*. Much of this chapter will focus on the biology, in particular the behavioural ecology, of the 15-spined stickleback, given the sparse attention this giant of sticklebacks has gained in the past.

11.2 THE 15-SPINED STICKLEBACK, *SPINACHIA SPINACHIA*

The 15-spined stickleback is a marine or brackish water fish, unique to Europe. Its geographical distribution encompasses the Gulf of Biscay including the southern Galician estuaries,[4] the British Isles, the Norwegian Sea, and the Baltic Sea, where the 0.4% salinity isocline limits its distribution.[5,6] The 15-spined stickleback differs distinctively from the other species of gasterosteids by having a very elongated body, long snout, and small mouth. Moreover, it is the only stickleback that never occurs in freshwater. It is regarded as a primitive member of the Gasterosteidae[2,3,7,8] (see Chapter 1). In this species the sexes are monomorphic. Thus, no extravagant character or sexual colouration is obvious in either males or females. They have 10 to 15 dorsal spines. Females may grow slightly larger than males, their final length usually being around 135 and 119 mm, respectively, in a population on the west coast of Sweden.[9] However, specimens reaching 188 to 191 mm have been found in Scotland and Denmark.[5,10] The life span of the 15-spined stickleback is presumed to be 1 year,[11] but because this conclusion is based on specimens that reached only 117 mm, Kaiser and Croy[10] have suggested that the 15-spined stickleback may live longer. However, they also observed that females die after having spawned. This has not been the case in a well-studied population on the west coast of Sweden, where females lay repeated clutches throughout the spring and early summer, being ready to spawn again only 5 d after the last mating (aquaria observations by Östlund-Nilsson). Also, in a microsatellite study,[12] we found field evidence that females lay clutches repeatedly over the season.

The most important food items in their diet are copepods, isopods, mysids, and amphipods.[13] Prey size correlates with the size of the fish.[14] As fish size increases, mysids are gradually succeeded by amphipods as the most important items in the diet.

During the reproductive period in the spring, 15-spined sticklebacks inhabit shallow eelgrass (*Zostera marina*) meadows and *Fucus* belts. The male builds a nest attached to a macroalgae, mainly *Ascophyllum nodosum*, *Fucus vesiculosus*, and *Halidrys siliquosa*, using epiphytic algae as nest material.[9] He winds secretional threads of glue protein[15–17] around the nest material to tie it together, giving the nest a discrete ball or oblong shape. Once completed, the width of the male's nest is between 2 and 8 cm and the height between 3 and 15 cm (Östlund-Nilsson, unpublished).

The glue protein is called *tangspiggin*, named after the Swedish name of this fish, *tångspigg*, in analogy with the structurally similar glue protein *spiggin* produced by three-spined stickleback males.[18] As in the three-spined stickleback, this protein is produced by transformed kidney cells in the nest-building male. The kidney of 15-spined stickleback males undergoes androgen-dependent hypertrophy during the breeding season, and its function changes from its normal secretory role to one of producing copious amounts of tangspiggin, which is stored in the urinary bladder before being released as conspicuous threads. Whereas plasma levels of androgens (testosterone and 11-ketotestosterone) increase in breeding 15-spined stickleback males, measured levels are much lower than those observed in three-spined stickleback males, and further, in contrast to the three-spined stickleback and, indeed, other paternal teleost species, plasma androgen levels in *Spinachia* do not decline during the parental phase (see Chapter 8).

It has been shown that a female can detect a nesting male through olfactory cues.[19] Because this species occurs at relatively low densities compared to many other sticklebacks, olfaction may be of particular importance for the females in the initial search for nesting males. When a female enters the territory of a male, he swims toward her and bites her in her tail or fins. If she is interested she begins to follow him around in circles for a short while. After circling, the male swims to the nest and puts his snout into the nest. He sometimes fans the nest with his pectoral fins in short bouts. Normally, the male swims between the female and his nest a few times. In this species the males do not show any zig zag dance, but instead the male repeatedly shakes or vibrates his body with a high frequency while pointing his head upwards. The shaking is done both while he bites the female and also while swimming between the female and the nest. If the female decides to spawn, she pushes her body into the nest and the male helps her to enter by holding her tail. After about half a minute the male also enters the nest, and the couple is often seen on top of each other in the nest. The male shivers while fertilising the eggs whereupon he swims out of the nest, usually staying there less than 40 sec in total. The female typically remains in the nest for around 3 min but sometimes stays for a longer time. Thus, the female normally remains in the nest for a while after the male has left (Figure 11.2). When the female leaves the nest the male chases her away.

After one or more females have spawned in the nest, the male alone cares for the eggs until they hatch.[7,20,21] Females release between 200 to 600 eggs per spawning. After spawning, the male is occupied primarily with fanning the eggs with his pectoral fins to oxygenate the developing embryos,[22] although he also guards, cleans, and repairs the nest for about 20 d until the eggs hatch.[7,21] In contrast to the intensively investigated three-spined stickleback, only a handful studies have been carried out on the reproduction of the 15-spined stickleback.[7,9,12,18–25] The earlier of these studies are careful observational studies. All the recent studies (1995–2002)[9,18,19,22,25] have dealt with the reproductive behaviour and behavioural ecology of a population of 15-spined stickleback on the west coast of Sweden. We will here summarise the results from these more recent studies.

11.2.1 Females Can Detect Paternal Skills

It is generally assumed that female mate choice in paternally caring fish species is likely to be for the direct benefits to the offspring. Often, paternal quality is related to male body size, as bigger males may more efficiently defend their offspring. Female choice for larger males has been demonstrated in many fish species with paternal or biparental care[26–31] (see also Chapter 5). However, females may also choose males on paternal competence (i.e., expected hatching success) over

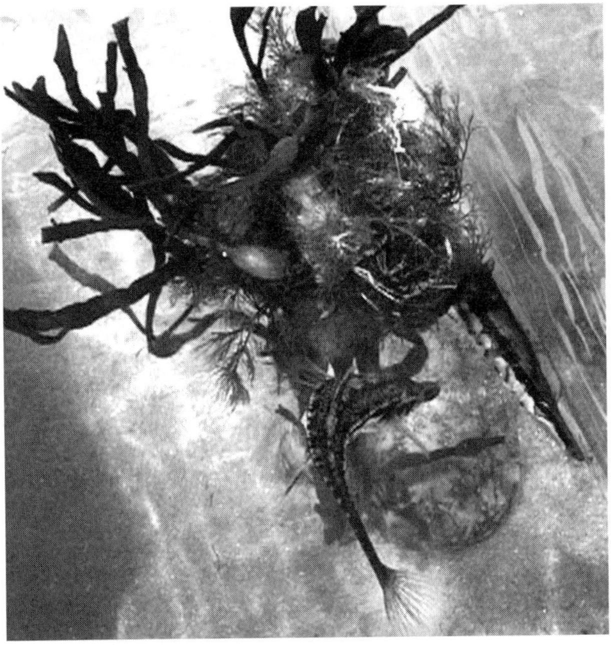

FIGURE 11.2 A video frame showing a female 15-spined stickleback laying her eggs in the nest. The male has just left his nest after fertilising the eggs. (Filmed by Sara Östlund-Nilsson.)

male–male competitive ability.[32] Moreover, a male's courtship may be an honest indicator of his parental quality.[33]

In the 15-spined stickleback, as in other sticklebacks, males provide females with direct benefits in the form of paternal care for offspring. By being choosy as to which male to mate with, females may gain better-skilled and better-nourished fathers, along with safer nests for their eggs. In the population of 15-spined sticklebacks studied on the west coast of Sweden it was found that males differed in their competence as fathers. Ripe females were observed to prefer males that displayed a higher frequency of body shakes during their courtship (shorter but more frequent bouts). Later, these males enjoyed significantly higher hatching success.[22] The connection was that these preferred males were subsequently found to display more frequent fanning bouts during the paternal phase, probably one of the major factors contributing to the greater hatching success of these males. Thus, body shaking seems to be an honest advertisement by males signalling their paternal skills and parental investment in the form of nest fanning to the females. In the same study females did not prefer longer over shorter males, and male size was not correlated to any paternal qualities. Thus, male size seemed irrelevant for parental care and female choice in this population. It should be mentioned that whereas preparental fanning has yet to be correlated to mating success in the three-spined stickleback, an early study found that males showed increased fanning when ripe females entered their territory.[34]

11.2.2 Nests for Safety and Nests as Ornaments Indicating Male Condition

In the 15-spined stickleback, the male's nest is quite conspicuous, being built and exposed high above the surrounding vegetation and constructed with many bright shiny protein (tangspiggin) threads. In the stickleback family (Gasterosteidae), the 15-spined stickleback is the largest species and exhibits by far the largest production of nest-building protein threads.[16] As we shall see, this protein is metabolically expensive to synthesise and therefore its rate of production is related to the level of food intake of the male.

In many fishes, females base their choice on male behavioural cues[33,35] or on morphological traits such as male secondary sexual characters and male size.[26–31] In the three-spined stickleback, it has been observed that receptive females may base their mate choice on nest characters, such as nest concealment,[36–39] water depth,[40] and territory quality.[37] In sticklebacks, findings of female preference for males with eggs in their nests are equivocal.[41–44]

In the 15-spined stickleback, it has been shown that females show preference for male behavioural cues,[22] but do they value male extra-bodily traits as well? This was tested in laboratory experiments in which males were forced to build either in the upper or the lower parts of a plastic plant. All high nests were about 14 cm above the bottom, and all low nests were about 1 cm from the bottom. Twelve pairs of males were matched as closely as possible in standard length, nest volume, and nest age. Female preference was tested in a large outdoor tanks, and it was shown that females clearly preferred males with nests built higher up in the vegetation over males with low nests.[9] One possible reason why females preferred higher nests could be that higher nests offer increased safety for her eggs from predators such as crabs. This hypothesis was confirmed by laboratory studies showing that shore crabs (*Carcinus maenas*) begin to predate the eggs in low nests before they attack those in higher nests.[9]

It appears that female choice and egg safety from predation are two factors contributing to why *Spinachia* males build their conspicuous nests high up in the seaweed bed. However, does male–male competition affect the positioning of the nests? This question was examined in a field study in which the height of nests over the bottom and over the surrounding vegetation, and the distance to the closest neighbouring nest were measured at seven different localities not more than 1.3 km apart. It was found that male–male competition may further explain why males build nests high up, because the results showed that the closer the distance to another nest, the higher the males built their nests.[9] Furthermore, larger males had more distantly positioned neighbours; that is, such males occupied larger territories than smaller males.[9]

11.2.3 Tangspiggin, an Extra-Bodily Ornament?

In species in which the male offers parental care, an ornament (morphological or behavioural) may signal to the female his ability to do so.[45,46] Could the amount of tangspiggin that a 15-spined stickleback male produces, which is primarily used in constructing the nest, also function as an extra-bodily ornament, signalling the condition of the male to the female? It has been noticed, both in the field and under laboratory conditions, that nest-building 15-spined stickleback males often use extravagant and conspicuous amounts of tangspiggin threads.

In a study aimed at answering this question,[19] male pairs were matched for standard length and weight and then placed in 50 litre aquaria. Randomly, one of the males within each pair was given a high-food ration and the other male a low-food ration. The males in the low-food group were given food twice a day every third day, and the males in the high-food group were supplied with food twice every day. Seventeen male pairs were tested for 12 d. Each male was supplied with a plastic plant to attach filamentous red algae for nest building, and the same amount of a mixture of two species of algae (*Polysiphonia fibrillosa* and *Ceramium rubrum*) to exclude a possible effect of algae availability on tangspiggin production. On the 12th day, all tangspiggin threads and nest-building material, including tangspiggin threads found on the bottom of the aquarium, were collected.

Well-fed males produced significantly more tangspiggin than starved males during the 12-d period.[19] To estimate the proportion of a male's resting metabolic rate that is devoted to producing tangspiggin, the resting metabolic rates of individual males were measured in a closed respirometer and the tangspiggin production cost was calculated. It was found that the high-food group used a significantly larger fraction of their resting metabolic rate (about 7%) for tangspiggin production than the low-food group (about 3%).[19]

It is likely that females would gain several direct benefits from males producing more tangspiggin, because it would indicate that the male is in good physical condition and could therefore provide better care for the eggs. Moreover, a nest constructed with more tangspiggin may also be a sturdier, and therefore a safer, nest.

Consequently, whether or not female choice was influenced by the quantity of tangspiggin contained in a nest was tested. Females were allowed to choose between a pair of males that were similar in standard length and weight. When the males had begun building a nest, they were given another nest to adopt that had a different tangspiggin content than their own nest. One of the males in each pair had three times more tangspiggin in his adopted nest than the other. The males were leashed and an opaque half-wall was placed between them to block physical and visual contact between the males. Females were observed until mating occurred. All six females chose to spawn with the male with more tangspiggin in his nest,[19] indicating that this protein functions as an extra-bodily ornament for the males.

11.2.4 Male–Male Interactions: Sneaked Fertilisations and Egg Stealing

Male–male interactions may spur the evolution of armaments or ornaments,[47] but may also reduce male fitness through cuckoldry or egg theft from neighbouring nests.[48] The advantage of sneaking to a sneaker is easily understood, but it is more difficult to explain the evolutionary benefit of stealing eggs from other males (see also Chapter 5). One explanation for egg stealing is that in some species females prefer males with eggs already in their nests.[49]

In the population of 15-spined stickleback studied on the Swedish west coast, frequent interactions between males are seen both in the field and in the laboratory. This species often raids nests, and when raiding they may sneak fertilisations, cannibalise other males' offspring, and steal nest material and nests. In this population, males also conduct filial cannibalism (Östlund-Nilsson, unpublished). In a study utilising microsatellite markers for relatedness,[12] the frequency of egg theft in this population was documented, the identity of the sneaker males were revealed, and these findings were placed in a spatial context by analysing nest positions. Here, six microsatellite markers were developed and employed to assay a total of 1307 embryos from 28 nests. Males, with or without nests, were collected, marked, and their position was recorded on a map. All nest-holding males in the study were genotyped for two to six loci. Analysis of male and embryo genotypes revealed that 5 of the 28 nests contained progeny from sneaked fertilisations, and that 4 of the 24 nests (17%) with nest-holding males contained stolen egg clutches. However, only 66 (5%) of the total number of eggs assayed came from sneaked fertilisations. Every sneaker male, however, fertilised on average 40% of the eggs in a clutch, suggesting that sneaking is beneficial to males adopting this strategy. Comparisons of the composite DNA genotypes of nest-holding males against the inferred sneakers indicated that one nest-holder had sneaked a nest 7 m from his own. This suggests that sneaking in this species can be a conditional, facultative strategy that nest-holding males engage in opportunistically.

11.2.5 The Effects of Paternity on Paternal Care

In the same Swedish population, a laboratory study[25] was conducted to investigate how sneaking and egg stealing may affect paternal efforts in the 15-spined stickleback. Male parental effort was estimated by measuring fanning frequency and predator defence. Two species of nest predators were used: goldsinny wrasse (*Ctenolabrus rupestris*) and shore crabs (*Carcinus maenas*). Male attack rate toward the predators was measured with predators 40, 30, 20, or 10 cm from the nest.

Paternity among the males was manipulated in the following way:

1. Full-paternity males: Males mated alone with a female.
2. Shared paternity males: The male shared paternity with another male. This was accomplished by introducing another male who sneak-fertilised the eggs when the nest owner was about to mate with the female.
3. No paternity male: Males had no paternity at all. To achieve this, the entire clutches were switched between two males that had just mated.

When comparing males with different degrees of paternity, no significant differences in fanning bout duration or hatching success was found. However, whereas the intensity of predator defence against goldsinny wrasses did not vary with the degree of paternity, the defence intensity increased significantly in all males over time as eggs developed. The same pattern was found with shore crabs; whereas defence did not vary with paternity, it increased as eggs grew older. It was concluded that there was no relationship between the degree of paternity and fanning activity, hatching success, or nest defence. However, males adjusted their level of defence to the amount of energy and time already invested in the clutch.

11.2.6 Can Female Preference for More Eggs Explain Egg Stealing?

To answer this question, female preference for males that have more eggs in their nests were examined, because this might explain egg stealing among males.[25] Here, females were given a choice between three size-matched males with different clutch sizes in their nests. In each replicate, the three males had their newly built nests removed, whereupon they were each given a nest from another male to adopt. The size and age of all three adopted nests were the same. After each male had mated with a female in the adopted nest, the eggs were removed completely from one male. Eggs were mixed and distributed from the other two males so that one of them received half a normal clutch and the other a clutch 1.5 times the size of a normal egg clutch (which weighs about 2 g). In a subsequent trial, it was shown that females preferred males with reduced clutches to males with enlarged clutches.[25] Thus, female choice is unlikely to be a driving force behind egg stealing in this species. Another reason for why males steal eggs could be that they use eggs as a food supply. Although the present data suggest that males cannot recognise their own eggs, it is still possible that they can recognise stolen eggs by how they position them in the nest.

11.2.7 Conclusions

The 15-spined stickleback males provide females with direct benefits in the form of paternal care for offspring. By being choosy as to which male to mate with, females gain better-skilled and better-nourished fathers, along with safer nests for their eggs. Fanning of eggs is important in this species as it correlates with hatching success. Moreover, a male signals his fanning skills to the female not by his body size but by shaking his body intensively during courtship. Also, male nest-building skills affect female choice, and males use their nest as an extra-bodily ornament, signalling the condition of the male and the safety of the eggs. Females also choose nests built high, rather than low, above the surrounding vegetation possibly because this reduces the risk of egg predation by other animals.

Male–male interactions in this species not only influence the position of the nests, but may also reduce paternity for males through stolen fertilisations and eggs. Microsatellite markers revealed that sneak matings are common among males, and they also steal eggs from each other. Moreover, there is a substantial fitness cost to males victimised by sneakers, as males do not recognise their own eggs. Indeed, clutch age and not the degree of paternity determines the willingness of the male to defend his eggs. Females do not show any preference for more eggs in the nests — rather the opposite — so this cannot explain why the nesting males steal eggs from each other. Males possibly use the stolen eggs as food, and they may recognise the stolen eggs by their position in the nest.

Thus, female choice is based on multiple cues and results in better paternal care. Males invest in courtship, male–male competition, nest construction, and paternal care, the outcome determining their success in mate attraction.

11.3 THE BLACK-SPOTTED STICKLEBACK, *GASTEROSTEUS WHEATLANDI*

The black-spotted stickleback, *Gasterosteus wheatlandi*, is one of two recognised species in the genus *Gasterosteus*, the other being the three-spined stickleback, *G. aculeatus*. The black-spotted stickleback is found in marine coastal habitats off eastern North America, from Newfoundland to New York.[7] It is easily distinguished from the three-spined stickleback by its nuptial colouration, breeding males being gold in colour with distinctive black spots and red pelvic spines. The black-spotted stickleback may be up to 75 mm long. It is characterised by having a small number of anterior lateral plates (5–11), and it lacks a caudal peduncle keel and posttemporal and supracliethra bones.

In a comparison between sympatric populations of *G. aculeatus* and *G. wheatlandi* in Quebec, it was found that besides a number of morphological differences, *G. wheatlandi* was smaller and produced fewer, smaller eggs than *G. aculeatus*.[50] Although sympatric populations of *G. wheatlandi* and *G. aculeatus* have been widely reported, it appears that these populations are to a large extent reproductively isolated. Cleveland[51] showed that there is competition between the two species in nest site habitat and nest material. Whereas *G. wheatlandi* is a generalist that puts low demands on vegetation, *G. aculeatus* was found to be dependent on particular vegetation for nesting. It was found that the coexistence of these species was possible because of the generalist behaviour of *G. wheatlandi*, which changed their preferred habitat and nest material and also built significantly fewer nests than *G. aculeatus*, which was not affected by the presence of *G. wheatlandi*. In an early study, it was reported that there was little overlap in the breeding season in sympatric populations of *G. wheatlandi* and *G. aculeatus* occurring in coastal waters off Long Island, because *G. wheatlandi* was breeding later than *G. aculeatus*.[52] Reisman[53] proposed that these two *Gasterosteus* species are potentially reproductively isolated owing to, for example, clear species differences in male courtship and nuptial colouration. However, whereas both *G. wheatlandi* and *G. aculeatus* females showed a strong preference for conspecific males during courtship, males did not discriminate between con- and heterospecific females.[54] Whoriskey and FitzGerald[55] found that, despite the similar ecology of *G. wheatlandi* and *G. aculeatus*, they showed several differences in their patterns of egg cannibalism. Thus, *G. wheatlandi* showed no density-dependent pattern in cannibalism, and males consumed more eggs than females, whereas the opposite was seen in *G. aculeatus*.

11.4 THE NINE-SPINED STICKLEBACK, *PUNGITIUS PUNGITIUS*

The nine-spined stickleback, or as it is often called, the ten-spined stickleback, belongs to the genus *Pungitius*, for which as many as eight species have been described.[3] It is long and slender, has between 7 and 10 dorsal spines, and never exceeds 100 mm in total length, normally being half that size.[7] The geographical distribution of *Pungitius* spp. is similar to that of *Gasterosteus aculeatus*, being widely distributed in freshwater and brackish-water habitats throughout the temperate areas of the northern hemisphere, including Japan and much of the Western Pacific. One species, *P. hellenicus*, is restricted to a small number of freshwater habitats in central Greece.

Despite its wide distribution[7] and abundance, only few studies have been carried out on the life histories in the nine-spined stickleback and on its reproduction from an evolutionary perspective.[56–59]

However, its reproductive behaviour has been relatively well described,[7,60] and we will therefore only give an overview here. During the breeding season when males become reproductively active, they turn black, which beautifully contrasts with the white pelvic spines.[60] However, other colour forms do occur.[61] The nine-spined stickleback prefers vegetated areas for nesting and like the other sticklebacks, it is reproductively active during spring and summer. Leiner[62] observed autumn spawning in his laboratory, but this has not been verified in nature. The nine-spined stickleback is ready to reproduce the year following hatching.[11]

The construction of a nest is similar to that of the 15-spined stickleback. They build their nest out of soft filamentous algae that they shape into a ball or a cylinder. Although the nest may sometimes be built on the bottom, they are normally built above the bottom and attached to water plants of different kinds. Ransom (1865),[63] Landois (1871),[64] Leiner (1931 and 1934),[62,65] Sevenster (1949),[66] and Morris (1952 and 1958)[60,67] provided some of the earliest descriptions and observations of nest building in the nine-spined stickleback. Interestingly, like *G. aculeatus* (see Chapter 5), the nine-spined stickleback shows preference for particular colours of the nest-building material. Morris[60] gave a few individuals of nine-spined stickleback and three-spined sticklebacks cotton threads of different colours. In one series of experiments, they were given white, two shades of grey, and black cotton threads, and the fish showed preference for white as nest-building material. In another series, they were allowed to choose between coloured cotton threads: yellow, blue, red, and green, and yellow was preferred. *Gasterosteus* also preferred yellow to blue, red, and green. Morris's own interpretation of these results was that sticklebacks choose nest materials by their lightness of shade and special hue.

When a gravid female enters the territory and nest of a male, he dances to her, a dance very similar to that of the three-spined stickleback but with the head pointing downwards. He also jumps and his pelvic spines become erect. His courtship includes nest fanning, biting the female, and swimming in the direction of the nest but sometimes in circles.[7]

Females spawn repeatedly during a season (for reproductive cycle, see Sokolowska and Skora[56]) and like the rest of the sticklebacks, she leaves the male alone to care for the eggs.

Lately, comparative studies between the three-spined stickleback, the black-dotted stickleback, and the nine-spined stickleback have been performed. FitzGerald[68] conducted the first study in field, comparing the reproductive ecology of all three sympatric species living in a salt marsh at Isle of Verte, Quebec. He found that the parental cycle was similar for all three species, but that fanning and aggression varied between them. He noted that the nine-spined stickleback was the only one of the three that was nesting in dense vegetation. Their cannibalistic tendencies varied as well. The nine-spined stickleback preyed on one another's fry, whereas the others preyed primarily on the eggs. Recent field studies[69,70] compared three-spined stickleback and the nine-spined stickleback living sympatrically in streams in eastern England. Habitat choice differed between the two species. The three-spined stickleback was most widely distributed, and in line with FizGerald's result,[68] they found that the nine-spined stickleback preferred the more vegetated areas. The fishes had different reproductive strategies and to compensate for its smaller size, the nine-spined stickleback matured earlier and had a lower annual investment in reproduction, producing smaller clutches more frequently compared to the three-spined stickleback.

11.5 THE BROOK STICKLEBACK, *CULAEA INCONSTANS*

The genus *Culaea,* represented by the brook stickleback (*C. inconstans*), is the only stickleback genus confined entirely to freshwater. The brook stickleback occurs only in North America, where its geographical range extends from the Atlantic coast westwards to the Rocky Mountains. However, it appears to be spreading westwards of the continental divide, and a recent report shows that the brook stickleback has established itself in Rock Creek, Spokane County, Washington.[71] It may grow up to 85 mm but normally does not exceed 65 mm. It lives for about 2 years. Its body is long and slender. The majority have four to six dorsal spines, although the number can vary between two and ten dorsal spines.[7] Like the nine-spined stickleback, this species seems to prefer vegetated areas. The reproductive season stretches from May to July, like most other species of sticklebacks. The colouration is cryptic when outside the breeding season, but reproductively mature males get darker bodies, their eyes turn yellow, and they develop a black vertical bar across each eye (for a review, see Wootton[7]). McLennan[72] showed that the male brook stickleback signals his reproductive stage through four distinct body patterns. Also, female brook sticklebacks develop nuptial colouration following ovulation, and males were shown to spend more time with those compared to females in the interspawning stage.[73] Like in the 15-spined, the nine-spined, and the four-spined sticklebacks, the males of the brook stickleback build their nest in the vegetation, but commonly close to the bottom. When a gravid female enters the territory of a male, he performs a courtship dance, and when she has spawned in his nest he is left to care for the offspring.[7]

The male fans the nest with his pectoral fins throughout the paternal phase. Interestingly, this species shows sexual dimorphism in the pectoral fins, with males having larger fins, which may relate to the importance of the male fins during courtship and nest fanning during the parental phase.[74] However, a female's fecundity in the brook stickleback has been shown to correlate to her fluctuating asymmetry in her pectoral fin ray number.[74]

McLennan[75–77] showed that olfaction is important in the reproduction of the brook stickleback (see also Chapter 6). Females of the brook stickleback use olfactory cues, and their responses to scent from conspecific males were strongest when they were ready to spawn. McLennan hypothesized that scent might work as a long-distance signal to help the females find males, whereupon vision may guide them.[75] Also, males may use their olfaction to differentiate between conspecific males and conspecific females that are ovulating.[76]

Olfactory cues may not only be used in finding a partner. They may also help the fish escape predation. Trials conducted both in laboratory and field have found that the brook stickleback reacts to alarm substance(s), i.e., odours released from conspecifics or other species while being attacked and injured by a predator.[78–81]

High predation risk has been found to influence the choice of whom to shoal with, and the strongly armoured brook stickleback prefers to shoal with minnows (*Pimephales promelas*), which are not armoured and would therefore be preferred by predators.[82] Predation may drive the evolution of spines and other defence structures, which may explain the unusually long spines of the brook stickleback in the Mad River drainage in Ohio.[83] However, not all environments under intense predator pressure may drive the spines and other defensive structures to become bigger and more robust. Excessive armaments may negatively affect swimming ability (escape response), and there might be a trade-off situation between armament and agility.[84,85]

11.6 THE FOUR-SPINED STICKLEBACK, *APELTES QUADRACUS*

Apeltes is one of only two stickleback genera, the other being *Culaea*, that are restricted to North America. The four-spined stickleback, *A. quadracus*, occurs widely in coastal waters off the coast of Eastern North America, from Newfoundland to Virginia in the south.[86] Whereas originally thought to be confined to brackish and marine waters, *A. quadracus* has now been found to occur widely in many lakes along the northern seaboard of its geographical range. Interestingly, this stickleback functions as a cleaner fish by feeding on ectoparasites on the skin of the rainwater fish, *Lucania parva*.[87] The four-spined stickleback may become 3 years old. It grows up to 60 mm (but is often found to be smaller); it has between two and five dorsal spines and totally lacks lateral plates.[7] They retain a cryptic colouration throughout the year, but when males become reproductively mature, their pelvic spines turn red. In some populations, they may develop a black band over the eye in addition to the red pelvic spines.[7,88]

Like the brook, the 15-spined, and the nine-spined sticklebacks, the four-spined stickleback male builds his nest up in the vegetation, but the nest is much smaller than those of the other species.[7] In contrast to other sticklebacks, this species does not fan the nest during nest building, courtship, or parental care. Instead, the male sucks water through the nest to ventilate the eggs.[88] During courtship the four-spined stickleback uses a unique spiral dance in which he zigzags very rapidly.[89] During the reproductive phase the males raid one another's nests and steal fertilisations. Thus, presence of rival males during courtship and mating may call for substantial postspawning nest repairs in this species. This may be one reason (perhaps, together with uncertain paternity) for the reduced hatching success seen when rival males are present.[90] The four-spined stickleback has not gained much attention in reproductive ecology compared to the three-spined stickleback (but see for example, Rowland[89,91] and Willmott and Foster[90]), although its reproductive behaviour has been thoroughly reviewed.[7]

11.7 EVOLUTION OF STICKLEBACKS: FURTHER EVIDENCE OF A MARINE ANCESTOR

From phylogenetic evidence, it is generally accepted that the sticklebacks were originally marine in origin,[92] and that over time a number of species invaded brackish water habitats, or, as in the case of the three-spined stickleback, became habitat generalists, occupying habitats from fully marine to freshwater. Both geographical and genetic evidence indicate that the present freshwater populations of the three-spined stickleback were derived repeatedly from a common marine ancestor.[92] One intriguing question is why some sticklebacks, notably those of the genera *Culaea* and *Gasterosteus*, managed to invade freshwater habitats, whereas others such as *Spinachia* remained restricted to their original marine habitats.

One of the unique characteristics of the three-spined stickleback is that it can spawn in a wide range of aquatic habitats, from freshwater to fully marine. In contrast, the 15-spined stickleback, *Spinachia spinachia*, is strictly confined to marine waters. Owing to the restrictions imposed by the physiological demands of ion and osmotic regulation, many teleosts spawn in an aquatic habit of the same salinity from which that species originated. It is quite unusual for a species to possess gametes (eggs and sperm) that can withstand the physiological demands springing from a wide range of salinities. A number of recent studies that investigated the sperm characteristics of both the three-spined and 15-spined sticklebacks have provided further evidence of their having a marine ancestor, and this fact also goes some way in explaining how *Gasterosteus* but not *Spinachia* has been able to successfully invade freshwater habitats.[93,94]

A number of sperm quality characteristics, including sperm longevity, curvilinear velocity (VSL), and percentage of sperm active were measured in mature three-spined sticklebacks sampled from three different habitats (freshwater, brackish, and marine), following dilution of the sperm in water of different salinities[93] (Table 11.1). Whereas the sperm from brackish water and marine sticklebacks showed prolonged longevity (123 and 76 min, respectively) when diluted in brackish and full seawater, respectively, sperm of freshwater sticklebacks remained active for only 30 sec when diluted with freshwater. Further, the other measured parameters of sperm quality

TABLE 11.1
Sperm Quality Variables (sperm longevity, curvilinear velocity, and percentage active) Shown by Three-Spined Sticklebacks Sampled in Different Aquatic Habitats, and Tested with Different Water Salinities with or without Ovarian Fluid

Population Dilution Media	n	Longevity of Sperm (min; Mean ± SEM)	Curvilinear Sperm Velocity ($\mu m\ sec^{-1}$)	Percentage of Sperm Active
Freshwater sticklebacks				
Freshwater	20	0.5	10.3 ± 4.1	19.8 ± 5.6
Brackish water	10	116 ± 12.8	43.9 ± 3.1	69.1 ± 4.6
Seawater	10	0	0	0
Freshwater + ovarian fluid	10	245 ± 43.0	49.2 ± 6.9	62.4 ± 4.9
Brackish water sticklebacks				
Freshwater	10	0.5	17.7 ± 1.2	48.1 ± 3.6
Brackish water	30	123 ± 22.5	42.4 ± 4.0	64.0 ± 3.5
Seawater	10	0	0	0
Brackish water + ovarian fluid	10	> 483	42.8 ± 4.2	71.0
Seawater sticklebacks				
Freshwater	10	0	0	0
Brackish water	10	173 ± 19.8	61.7 ± 5.4	59.7 ± 3.2
Seawater	20	76 ± 6.5	41.5 ± 2.4	34.9 ± 7.2
Seawater + ovarian fluid	10	92 ± 4.8	40.3 ± 2.6	42.2 ± 3.7

Source: Modified from Elofsson, H. et al., *J. Fish Biol.,* 63, 240, 2003.

(VSL and percentage of sperm active) were also impaired in freshwater compared to the other tested salinities. The observation that hypo-osmotic conditions are physiologically very demanding on stickleback sperm supports the view that sticklebacks are of marine origin. In the same study, it was observed that the addition of ovarian fluid to water at all tested salinities had the effect of increasing sperm quality. This was most noticeable for sperm of freshwater sticklebacks. When diluted with freshwater alone, the sperm from freshwater sticklebacks remained active for just 30 sec. However, the addition of ovarian fluid (25% by volume) to the freshwater increased sperm longevity to 245 min. In addition, the addition of ovarian fluid also significantly increased both VSL and the percentage of sperm active (Table 11.1). In contrast to these results with the three-spined stickleback, *Spinachia* sperm was completely inactive in freshwater, with or without the addition of ovarian fluid.[94] Taken together, these results provide a possible answer to how the three-spined stickleback, but not *Spinachia*, was able to colonise freshwater habitats. These results show that the ovarian fluid of the three-spined stickleback formed a favourable microenvironment for the sperm, protecting it from the physiologically demanding hypo-osmotic surroundings. The fact that the ovarian fluid of *Spinachia* does not provide similar advantages for its sperm could explain why this primitive stickleback species remained restricted to marine habitats.

REFERENCES

1. Brooks, D.R. and McLennan, D.A., *Phylogeny, Ecology, and Behaviour*, The University of Chicago Press, Chicago and London, 1991.
2. McLennan D.A., Phylogenetic relationships in the Gasterosteidae: an updated tree based on behavioral characters with a discussion of homoplasy, *Copeia,* 1993, 318, 1993.
3. Keivany, Y. and Nelson, J.S., Phylogenetic relationships of sticklebacks (Gasterosteidae), with emphasis on ninespine sticklebacks (*Pungitius spp.*), *Behaviour,* 141, 1485, 2004.
4. Fernandez, C.A., Fifteen-spined stickleback (*Spinachia spinachia* Linnaeus, 1758) in waters off the NW Iberian peninsula, southern limit of its range, *Misc. Zool.,* 20(1), 65, 1997.
5. Johnsen, S., Studies on variation in fish in North-European waters. I. Variation in size, *Bergens Mus. årbok,* 4, 14, 1945.
6. Gross, H.P., Observations on the geographic variation of the marine coastal fish *Spinachia spinachia*, *Mar. Biol.,* 47, 297, 1978.
7. Wootton, R.J., *The Biology of the Sticklebacks*, Academic Press, London, 1976.
8. McLennan D.A., Brooks D.R., and McPhail, J.D., The benefits of communication between comparative ethology and phylogenetic systematics: a case study using gasterosteid fishes, *Can. J. Zool.,* 66, 2177, 1988.
9. Östlund-Nilsson, S., Are nest characters of importance when choosing a male in the fifteen-spined stickleback (*Spinachia spinachia*)? *Behav. Ecol. Sociobiol.,* 48, 229, 2000.
10. Kaiser, M.J. and Croy, M.I., Population structure of the fifteen-spined stickleback, *Spinachia spinachia* (L.), *J. Fish Biol.,* 39, 129, 1991.
11. Jones, J.W. and Hynes, H.B.N., The age and growth of *Gasterosteus aculeatus*, *Pygosteus pungitius* and *Spinachia vulgaris,* as shown by their otholiths, *J. Anim. Ecol.,* 19, 59, 1950.
12. Jones, A., Östlund-Nilsson S., and Avise J., A microsatellite assessment of sneaked fertilizations and egg thievery in the fifteen-spined stickleback, *Evolution,* 52, 848, 1998.
13. Kaiser, M.J., Westhead, A.P., Hughes, R.N., and Gibson, R.N., Are digestive characteristics important contributors to the profitability of prey? A study of diet selection in the fifteen-spined stickleback *Spinachia spinachia* L., *Oecologia*, 90, 61, 1992.
14. Kislalioglu, M. and Gibson, R.N., Field and laboratory observations on prey-size selection in *Spinachia spinachia* (L.), in *Proceedings of the 9th European Marine Biology Symposium*, Barnes, H., Ed., Aberdeen University Press, Aberdeen, Scotland, 1975.
15. Hentschel, H., The kidney of *Spinachia spinachia* (L.) Flem. (Gasterosteidae, Pisces) 1. Investigations of juvenile sticklebacks: anatomy, circulation and fine structure, *Z. mikrosk. Anat. Forsch. Leipzig.,* 91, 4, 1977.
16. Hentschel, H., The kidney of a teleost, *Spinachia spinachia* II. Histochemical identification of sialic acid-containing glycoprotein and fine structure of mucus secreting cells, *Tissue Cell,* 3, 517, 1979.
17. De Ruiter, A.J.H. and Mein, C.G., Testosterone-dependent transformation of nephronic tubule cells into serous and mucus gland cells in stickleback kidney *in vivo* and *in vitro*, *Gen. Comp. Endocrinol.*, 47, 70, 1982.

18. Östlund-Nilsson, S., Fifteen-spined stickleback females prefer males with more secretional threads in their nests: an honest condition display by males, *Behav. Ecol. Sociobiol.*, 50(3), 263, 2001.
19. Östlund, S., Female 15-spined sticklebacks detect males with empty nests by non-visual cues, *J. Fish Biol.*, 47, 1106, 1995.
20. Prince, E.E., On the nest and development of *Gastrosteus spinachia* at the St. Andrews Marine laboratory, *Ann. Mag. Nat. Hist.*, 16, 487, 1885.
21. Sevenster, P., The mating of the sea stickleback, *Discovery* (London), 12, 52, 1951.
22. Östlund, S. and Ahnesjö, I., Female fifteen-spined sticklebacks prefer better fathers, *Anim. Behav.*, 56, 1177, 1998.
23. Potts, G.W., Parental behaviour in temperate marine teleosts with special reference to the development of nest structures, in *Fish Reproduction: Strategies and Tactics*, Potts G.W. and Wootton R.J. Eds., Academic Press, London, 1984.
24. Potts, G.W., Keenleyside M.H.A., and Edwards, J.M., The effect of silt on the parental behaviour of the sea stickleback, *Spinachia spinachia*, *J. Mar. Biol. Assoc. U.K.*, 68, 277, 1988.
25. Östlund Nilsson, S., Does paternity or paternal investment determine the level of paternal care and does female choice explain egg stealing in the fifteen-spined stickleback? *Behav. Ecol.*, 13(2), 188, 2002.
26. Downhower, J.F. and Brown, L., Mate preferences of female mottled sculpins, *Cottus bairdi*, *Anim. Behav.*, 28, 728, 1980.
27. Noonan, K., Female mate choice in a cichlid fish *Cichlasoma nigrofasciatum*, *Anim. Behav.*, 31, 1005, 1983.
28. Keenleyside, M.H.A., Rangeley, R.W., and Kuppers, B.U., Female mate choice and male parental defense behaviour in the cichlid fish *Cichlasoma nigrofasciatum, Can. J. Zool.*, 63, 2489, 1985.
29. Bisazza, A. and Marconato, A., Female mate choice, male-male competition and parental care in the river bullhead, *Cottus gobio* L. (Pisces, Cottidae), *Anim. Behav.*, 36, 1352, 1988.
30. Côte, I.M. and Hunte, W., Male and female mate choice in the redlip blenny: why bigger is better, *Anim. Behav.*, 38, 78, 1989.
31. Rogers, W. and Barlow, G.W., Sex differences in mate choice in a monogamous biparental fish, the Midas cichlid (*Cichlasoma citrinellum*), *Ethology*, 87, 249, 1991.
32. Forsgren, E., Female sand gobies prefer good fathers over dominant males, *Proc. R. Soc. Lond. B*, 264, 1283, 1997.
33. Knapp, R.A. and Kovach, J.T., Courtship as an honest indicator of male parental quality in the bicolor damselfish, *Stegastes partitus*, *Behav. Ecol.*, 2, 295, 1991.
34. van Iersel, J.J.A., An analysis of the parental behaviour of the male threespined stickleback (*Gasterosteus aculeatus* L.), *Behaviour*, Suppl. 3, 1, 1953.
35. Knapp, R.A. and Warner, R.R., Male parental care and female choice in the bicolor damselfish, *Stegastes partitus*: bigger is not always better, *Anim. Behav.*, 41, 747, 1991.
36. Sargent, R.C. and Gebler, J.B., Effects of nest site concealment on hatching success, reproductive success, and paternal behaviour of the threespine stickleback, *Gasterosteus aculeatus*, *Behav. Ecol. Sociobiol.*, 7, 137, 1980.
37. Sargent, R.C., Territory quality, male quality, courtship intrusions, and female nest-choice in the threespine stickleback, *Gasterosteus aculeatus*, *Anim. Behav.*, 30, 364, 1982.
38. Candolin, U. and Voigt, H.R., Predator-induced nest site preference: safe nests allow courtship in sticklebacks, *Anim. Behav.*, 56, 1205, 1998.

39. Kraak, S.B.M., Bakker, T.C.M., and Mundwiler, B., Sexual selection in sticklebacks in the field: correlates of reproductive, mating, and paternal success, *Behav. Ecol.*, 10, 696, 1999.
40. Mori, S., Nest site choice by the three-spined stickleback, *Gasterosteus aculeatus* (form *leiurus*), in spring fed waters, *J. Fish. Biol.*, 45, 279, 1994.
41. Ridley, M. and Rechten, C., Female sticklebacks prefer to spawn with males whose nests contain eggs, *Behaviour,* 76, 152, 1981.
42. Jamieson, I.G. and Colgan, P.W., Eggs in the nest of males and their effect on mate choice in the three-spined stickleback, *Anim. Behav.*, 38, 859, 1989.
43. Jamieson, I.G. and Colgan, P.W., Sneak spawning and egg stealing by male threespine sticklebacks, *Can. J. Zool.*, 70, 963, 1992.
44. Goldschmidt, T. and Bakker, T.C.M., Determinants of reproductive success of male sticklebacks in the field and in the laboratory, *Neth. J. Zool.*, 40(4), 664–687, 1990.
45. Hoelzer, G.A., The good parent process of sexual selection, *Anim. Behav.*, 38, 1067, 1989.
46. Wolf, J.B., Moore, A.J., and Brodie, E.D., The evolution of indicator traits for parental quality, the role of maternal and paternal effects, *Am. Nat.,* 150(5), 639, 1997.
47. Berglund, A., Bisazza, A., and Pilastro, A., Armaments and ornaments: an evolutionary explanation of traits of dual utility, *Biol. J. Linn. Soc.,* 58, 385, 1996.
48. Mori, S., Factors associated with and fitness effects of nest-raiding in the three-spined stickleback, *Gasterosteus aculeatus*, in a natural situation, *Behaviour,* 132, 1011, 1995.
49. Rohwer, S., Parental cannibalism of offspring and egg raiding as a courtship strategy, *Am. Nat.*, 112, 429, 1978.
50. Coad, B.W. and Power, G., Observations on ecology and meristic variation of ninespine stickleback, *Pungitius pungitius* (L. 1758) of Matamek-River system, Quebec, *Am. Midl. Nat.*, 90, 498, 1973.
51. Cleveland, A., Nest site habitat preference and competition in *Gasterosteus aculeatus* and *G. wheatlandi*, *Copeia*, 1994(3), 698, 1994.
52. Perlmutter, A., Observations of fishes of the genus *Gasterosteus* in the waters of Long Island, New York, *Copeia*, 1963, 168, 1963.
53. Reisman, H.M., Reproductive isolating mechanisms of blackspotted stickleback *Gasterosteus wheatlandi*, *J. Fish. Res. Board Can.*, 25, 2703, 1968.
54. McInerney, J.E., Reproductive behaviour of blackspotted btickleback, *Gasterosteus wheatlandi*, *J. Fish. Res. Board Can.*, 26, 2061, 1969.
55. Whoriskey, F.G. and FitzGerald, G.J., Sex, cannibalism and sticklebacks, *Behav. Ecol. Sociobiol.,* 18, 15, 1985.
56. Sokolowska, E. and Skora, K.E., Reproductive cycle and the related spatial and temporal distribution of the nine-spined stickleback (*Pungitius pungitius* L.) in Puck Bay, *Oceanologia,* 44(4), 475, 2002.
57. Heins, D.C., Johnson, J.M., and Baker, J.A., Reproductive ecology of the nine-spined stickleback from south-central Alaska, *J. Fish Biol.*, 63, 1131, 2003.
58. Heins, D.C., Baker, J.A., DeSilva, G., and Birden, E.L., Clutch characteristics of two populations of nine-spined stickleback from south-central Alaska, *J. Fish Biol.,* 67, 873, 2005.
59. Takaomi, A. and Akira, G., Flexible life history strategies of ninespine sticklebacks, genus *Pungitius*, *Environ. Biol. Fish.*, 74, 43, 2005.
60. Morris, D., The reproductive behaviour of the ten-spined stickleback (*Pygosteus pungitius* L.), *Behaviour,* Suppl. 6, 1, 1958.

61. McKenzie, J.A. and Keenlyside, M.H.A., Reproductive behaviour of ninespine sticklebacks (*Pungitius pungitius* (L)) in South Bay, Manitoulin, Ontario, *Can. J. Zool.*, 48, 55, 1970.
62. Leiner, M., Die drei europäischen Stichlinge (*Gasterosteus aculeatus* L., *Gasterosteus pungitius* L. und *Gasterosteus Spinachia* L.) und ihre Kreuzungsprodukte, *Z. Morphol. Ökol. Tiere,* 28, 107, 1934.
63. Ransom, W.H., On the nest of the ten-spined stickleback, *Ann. Mag. Nat. Hist.,* 16, 449, 1865.
64. Landois, H., Der kleine Stichling (*Gasterosteus pungitius*) und sein Nestbau, *Zool. Garten.*, 12, 1, 1871.
65. Leiner, M., Ökologisches von *Gasterosteus aculeatus* L., *Zool. Anz.*, 93, 317, 1931.
66. Sevenster, P., Modderbaarsjes, *De Levende Nat.*, 52, 161, 184, 1949.
67. Morris, D., Homosexuality in the ten-spined stickleback, *Behaviour*, 4, 233, 1952.
68. FitzGerald, G.J., The reproductive ecology and behaviour of three sympatric stickleback (*Gasterosteidae*) in a saltmarsh, *Biol. Behav.*, 8, 67, 1982.
69. Copp, G.H., Kovac, V., and Blacker, F., Differential reproductive allocation in sympatric stream-dwelling sticklebacks *Gasterosteus aculeatus* and *Pungitius pungitius*, *Folia Zool.,* 51(4), 337, 2002.
70. Copp, G.H. and Kovac, V., Sympatry between threespine *Gasterosteus aculeatus* and ninespine *Pungitius pungitius* sticklebacks in English lowland streams, *Ann. Zool. Fenn.,* 40(4), 341, 2003.
71. Scholz, A.T., Lang, B.Z., Black, A.R., McLellan, H.J., and Peck, R.L., Brook stickleback established in eastern Washington, *Northwest Sci.*, 77(2), 110, 2003.
72. McLennan, D.A., Temporal changes in the structure of the male nuptial signal in the brook stickleback *Culaea inconstans* (Kirtland), *Can. J. Zool.*, 71(6), 1111, 1993.
73. McLennan, D.A., Male mate choice based upon female nuptial coloration in the brook stickleback, *Culaea inconstans*, *Anim. Behav.*, 50(1), 213, 1995.
74. Hechter, R.P. and Moodie, P.F., Pectoral fin asymmetry, dimorphism and fecundity in the Brook stickleback, *Culaea inconstans*, *Behaviour*, 137(7–8), 999, 2000.
75. McLennan, D.A., The importance of olfactory signals in the gasterosteid mating system: sticklebacks go multimodal, *Biol. J. Linn. Soc.*, 80(4), 555, 2003.
76. McLennan, D.A., Male brook sticklebacks (*Culaea inconstans*) response to olfactory cues, *Behaviour*, 141(11–12), 1411, 2004.
77. McLennan, D.A., Changes in response to olfactory cues across the ovulatory cycle in brook sticklebacks, *Culaea inconstans*, *Anim. Behav.*, 69(1), 181, 2005.
78. Gelowitz, C.M., Mathis, A., and Smith, R.J.F., Chemosensory recognition of northern pike (*Esox lucius*) by brook stickleback (*Culaea inconstans*): population differences and the influence of predator diet, *Behaviour*, 127(1–2), 105, 1993.
79. Mathis, A. and Smith, R.J.F., Intraspecific and cross-superorder responses to chemical alarm signals by brook stickleback, *Ecology*, 74(8), 2395, 1993.
80. Wisenden, B.D., Chivers, D.P., and Smith, R.J., Risk sensitive habitat use by brook stickleback (*Culaea inconstans*) in areas associated with minnow alarm pheromone, *J. Chem. Ecol.*, 20(11), 2959, 1994.
81. Wisenden, B.D., Chivers, D.P., Brown, G.E., and Smith, R.J.F., The role of experience in risk assessment: avoidance of areas chemically labelled with fathead minnow alarm pheromone by conspecifics and heterospecifics, *Ecoscience*, 2(2), 116, 1995.
82. Mathis, A. and Chivers, D.P., Overriding the oddity effect in mixed-species aggregations: group choice by armored and nonarmored prey, *Behav. Ecol.*, 14(3), 334, 2003.
83. Andraso, G.M. and Barron, J.N., Unusually long spines in brook stickleback (*Culaea inconstans*), *Am. Midl. Nat.*, 147(1), 162, 2002.

84. Andraso, G.M. and Barron, J.N., Evidence for a trade-off between defensive morphology and startle response performance in the brook stickleback (*Culaea inconstans*), *Can. J. Zool.*, 73(6), 1147, 1995.
85. Andraso, G.M., A comparison of startle response in two morphs of the brook stickleback (*Culea inconstans*): further evidence for a trade-off between defensive morphology and swimming ability, *Evol. Ecol.,* 11(1), 83, 1997.
86. Scott, W.B. and Crossman, E.J., Freshwater fishes of Canada, *Bull. Fish. Res. Board Can.,* 184, 1, 1973.
87. Tyler, A.V., A cleaning symbiosis between rainwater fish, *Lucania parva*, and the stickleback, *Apeltes quadracus*, *Chesapeake. Sci.,* 4, 105, 1963.
88. Hall, M.A., Comparative Study of the Reproductive Behaviour of the Sticklebacks (*Gasterosteidae*), D. Phil. thesis, Department of Zoology, University of Oxford, 1956.
89. Rowland, W.J., Reproductive behaviour of the four-spine stickleback *Apeltes quadracus*, *Copeia*, 1974, 788, 1974.
90. Willmott, H.E. and Foster, S.A., The effects of rival male interaction on courtship and parental care in the fourspine stickleback, *Apeltes quadracus*, *Behaviour,* 132(13–14), 997, 1995.
91. Rowland, W.J., Ground nest construction in the four-spine stickleback *Apeltes quadracus*, *Copeia,* 1974, 788, 1974.
92. Bell, M.A. and Foster, S., *The Evolutionary Biology of the Threespine Stickleback,* Oxford University Press, Oxford, 1994.
93. Elofsson, H., McAllister, B., Kime, D., Mayer, I., and Borg, B., Long lasting sperm in sticklebacks; ovarian fluid a key to success in freshwater? *J. Fish Biol.,* 63, 240, 2003.
94. Elofsson, H., Van Look, K., Borg, B., and Mayer, I., Influence of salinity and ovarian fluid on sperm motility in the 15-spined stickleback, *J. Fish Biol.,* 63, 1429, 2004.

Index

A

B

C

H

I

M

N

O

P

S

T

U

V

W

X

Y

Z